Advanced D... and Networks

Advanced Data Communications and Networks

W. Buchanan
Senior Lecturer
Department of EECE
Napier University
Edinburgh
UK

Published by Chapman & Hall, 2–6 Boundary Row, London SE1 8HN, UK

Chapman & Hall, 2–6 Boundary Row, London SE1 8HN, UK

Chapman & Hall GmbH, Pappelallee 3, 69469 Weinheim, Germany

Chapman & Hall USA, 115 Fifth Avenue, New York, NY 10003, USA

Chapman & Hall Japan, ITP-Japan, Kyowa Building, 3F, 2-2-1 Hirakawacho, Chiyoda-ku, Tokyo 102, Japan

Chapman & Hall Australia, 102 Dodds Street, South Melbourne, Victoria 3205, Australia

Chapman & Hall India, R. Seshadri, 32 Second Main Road, CIT East, Madras 600 035, India

First edition 1997

Printed in Great Britain by the Alden Press, Osney Mead, Oxford

ISBN 0 412 80630 4

A catalogue record for this book is available from the British Library

∞ Printed on permanent acid-free text paper, manufactured in accordance with ANSI/NISO Z39.48-1992 and ANSI/NISO Z39.48-1984 (Permanence of Paper).

This book is dedicated to Andy and Liz for happy holidays, gallons of wine and beer, and buckets of snacks.

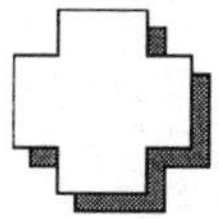

Table of Contents

Preface

Data communication and networks are fast becoming the largest industry in the world. It is one area of work with which every virtual person has some involvement, whether it be using the telephone or connecting to the Internet.

An IT specialist must know existing technologies, but must also know where the industry will go. This book is intended to cover the existing technologies and also the emerging technologies. I believe that the growth areas in the coming years will be in:

- Windows NT networking and associated networking programs.
- TCP/IP communications.
- NetWare 4.
- Data encryption.
- Data compression.
- Digitization of video, audio and images.
- Fast Ethernet.
- ATM.
- FDDI.
- HTML/Java/JavaScript.
- Virtually anything to do with the Internet and especially Intranets.

This book discusses each of these technologies in great detail and covers areas which are currently relevant, such as Ethernet, UNIX, RS-232 and electronic mail.

The book splits into seven main sections and some appendices, such as:

- General data compression (Chapters 2 and 3).
- Video, images and sound (Chapters 4–10).
- Error coding and encryption (Chapters 11–15).
- TCP/IP and the Internet (Chapters 16–23).
- Network operating systems (Chapters 24–26).
- LANs/WANs (Chapters 27–33).
- Cables and connectors (Appendix B).

Further information on related subjects, such as emerging technologies and the diagrams from the text are available on the WWW page:

```
http://www.eece.napier.ac.uk/~bill_b/adcbook.html
```

Help from myself can be sought using the email address:

```
w.buchanan@napier.ac.uk
```

or, if unavailable, send an email to:

```
bill_b@pollux.eece.napier.ac.uk
```

or, if that isn't available, try:

```
bill_b@www.eece.napier.ac.uk
```

else, try snail mail:

Dr William J Buchanan
Senior Lecturer
Dept. of E&EE
Napier University
219 Colinton Road
Edinburgh EH14 1DJ

1 Introduction

1.1 Introduction

The usage of data communications and computer networks is ever increasing. It is one of the few technological areas which brings benefits to most of the countries and the peoples of the world. Without it many industries could not exist. It is the objective of this book to discuss data communications in a readable form that both students and professionals all over the world can understand.

The book splits into seven main sections and some appendices, such as:

- General data compression.
- Video, images and sound.
- Error coding and encryption.
- TCP/IP and the Internet.
- Network operating systems.
- LANs/WANs.
- Cables and connectors.

In the past, most electronic communication systems transmitted analogue signals. On an analogue telephone system the voltage level from the phone varies with the voice signal. Unwanted signals from external sources easily corrupt these signals. In a digital communication system a series of digital codes represents the analogue signal. These are then transmitted as 1's and 0's, which are less likely to be affected by noise and thus become the predominant form of communications.

Digital communication also offers a greater number of services, greater traffic and allows for high speed communications between digital equipment. The usage of digital communications includes cable television, computer networks, facsimile, mobile digital radio, digital FM radio and so on.

1.2 A little bit of history

First, a little bit of history, which will lay the ground for the more technical coverage of the preceding chapters. Students and professionals in many other areas of work know of the important people of the past, so why not in data communications.

Communications, whether from smoke signals or in the form of pictures or the written word is as old as mankind. Fire, smoke and light have all been used to transmit messages over long distances. For example, Claude Chappe's semaphore system was developed in 1792 and has since been used to transmit messages with flags and light.

The time taken to transmit messages has changed from days and weeks (with carrier pigeons) to fractions of a second.

The history of communication can be traced to three main stages:

- The foundation of electrical engineering and radio wave transmission which owes a lot to the founding fathers of electrical engineering who were Coulomb, Ampère, Ohm, Gauss, Faraday, Henry and Maxwell. They laid down the basic principles of electrical engineering.
- The electronics revolution which brought increased reliability, improved operations, improved sensitization and increased miniaturization.
- The desktop computer revolution has accelerated the usage of digital communication and has finally integrated all form electronic communications: text, speech, images and video.
- The usage of modern communications techniques, such as satellite communications, local area networks and digital networks.

1.2.1 History of electrical engineering

The Greek philosopher Thales appears to have been the first to document observations of electrical force. He noted that when he rubbed a piece of amber with fur it attracted feathers. The Greek name for amber was *elektron* and the name has since been used in electrical engineering.

The one thing that must be realized with electrics is that electrical energy is undoubtedly tied to magnetic energy. When there is an electric force, there is a magnetic force. The growth in understanding of electrics and magnetics began during the 1600s when the court physician of Queen Elizabeth I, William Gilbert, investigated magnets and found that the Earth had a magnet field. He found that when a magnetic was freely suspended its tends to align itself with the magnetic field lines of the Earth. From then on, travellers around the world could easily plot their course because they knew which way was North.

Much of the early research in magnetics and electrics was conducted in the old world, mainly in England, France and Germany. But, in 1752, Benjamin Franklin put American science on the map when he flew a kite in an electrical storm and discovered the flow of electrical current.

In 1785, the French scientist Charles Coulomb showed the force of attraction and repulsion of electrical charges varies inversely with the square of the distance between them. He also showed that two similar charges repel each other, while two dissimilar charges attract.

In 1820, the French scientist André Ampère studied electrical current in wires and the forces between them. Then in 1827, the German scientist Georg Ohm studied the resistance to electrical flow. From this he determined that resistance in a conductor was equal to the voltage across applied across the material divided by current through.

In 1830, English scientist Michael Faraday produced an electric generator when he found that electricity could be generated by motion of a wire through an electric field. Thus he was the first to mathematically express that magnetism and electricity were related.

The root of modern communication can be traced back to the work of Henry, Maxwell, Hertz, Bell, Marconi and Watt. American Joseph Henry produced the first electromagnet with a coil of insulated electrical wire wound around a metal inner. Henry, unfortunately, like many other great scientists, did not patent his discovery and its first application was the beginning of the communications industry: telegraphy. It was the first real usage of electrical engineering to communications. Henry built the first telegraph system which sent coded electrical pulses over telegraph wires to an electromagnet at the other end. The system worked well but Henry got no credit for his invention and it was left to the artist Samuel Morse (the American Leonardo, according to one of his biographers) to develop the code of dots and dashes, and to take most of the credit with his Morse code.

Morse installed the first commercial telegraph system from Washington to Baltimore. The first message transmitted was "What hath God wrought." It received good publicity and after eight years there were over 23 000 miles (37 000 km) of telegraph wires over the USA. Several of the first companies to develop telegraph systems went on to become very large corporations. These include the Mississippi Valley Printing Telegraph Company which later became the Western Union. One of the first non-commercial uses of telegraph was the Crimean War and the American Civil War. A line from New York to San Francisco was an important mechanism for transmitting information to and from troops.

Other important developers of telegraph systems around the world were P.L. Shilling in Russia, Gauss and Weber in Germany and Cooke and Wheatstone in Britain.

One of the all time greats was the Scot James Clerk Maxwell. He was born in Edinburgh in 1831 and rates amongst the greatest of all the scientists. His importance to science puts him on par with Isaac Newton, Albert Einstein, James Watt and Michael Faraday. Maxwell's most famous formulation was a set of four equations that define the basic laws of electricity and magnetism. These are commonly known as Maxwell's equations.

Before Maxwell's work, many scientists had observed the relationship between electricity and magnetism. But, it was Maxwell, who finally derived the mathematical link between these forces. His four short equations described exactly the behavior and interaction of electric and magnetic fields. From this work he also proved that all electromagnetic waves, in a vacuum, travel at 300 000 km per second (or 186 000 miles per second). This, Maxwell recognized, was equal to the speed of light and from this he deduced that light was also an electromagnetic wave.

He then reasoned that the electromagnetic wave spectrum contained many invisible waves, each with its own wavelength and characteristic. Other practical scientists, such as Hertz and Marconi soon discovered these 'unseen' waves. They electromagnetic spectrum was soon filled with infrared waves, ultraviolet, gamma ray, X-rays and radio waves.

Another Scot Alexander Graham Bell had a great interest in the study of speech and elocution. In the USA he opened the Boston School for the Deaf. His other interest was in multiple telegraphy and he worked on a device he called a harmonic telegraph. This was used to aid the teaching of speech to deaf people. In 1876, out of this research he produced the first telephone with an electromagnet for the mouthpiece and the receiver. "It talks" was the headline (it hasn't stopped since). The great Maxwell was even amazed that anything so simple could reproduce the human voice. In 1877, Queen Victoria acquired a telephone. In 1877 Edison enhanced it using carbon powder in the diaphragm that produced more current. Bell along with several others formed the Bell Telephone Company which then developed the telephone so that, by 1915, long-distance telephone calls were possible. Bell's patent number 174465 is the most lucrative ever issued. At the time a reporter wrote, about the telephone, "It is an interesting toy ... but it can never be of any practical value."

The great inventor Edison, along with D.E. Hughes, produced a carbon transmitter which was a basic microphone. Edison also first patented the phonograph which consisted on a tinfoil wrapped round rotating cylinder with grooves in it.

Around 1888, German Heinrich Hertz detect radio waves (as predicted by Maxwell). He found that current was induced into a coil produced by a spark on the other side of a room. Then, Marconi, in 1896, succeeded in transmitting a radio wave over a distance of two miles. He soon managed to transmit a radio wave across the Atlantic.

Around 1851, the brothers Jacob and John Watkins Brett laid a cable across the English Channel between Dover and Cape Griz Nez. It was the first use of electrical communications between England and France (unfortunately it was soon trawled up by a French fisherman who mistook it for a sea monster). The British maintained a monopoly on submarine cables and laid cables across the Thames, Scotland to Ireland, England to Holland, as well as cables under the Black Sea, the Mississippi River and the Gulf of St Lawrence.

Submarine cables have since been placed under most of the major seas and oceans around the world.

Scot Robert Watson-Watt made RADAR (radio detection and ranging) practicable in 1935. It is typically used to detect planes, but is also used for weather forecasts and scientific measurement. The next great revolution occurred with electronics.

1.2.2 History of electronics

The science of electronics began in 1895 when H. Lorentz postulated the existence of charges called electrons. By 1897 Braun had built the first electron valve, which was a simple cathode-ray tube. Then at the beginning of the century, Fleming invented a diode which he called a valve. This device used a heated cylindrical plate in a vacuum. A positively charged heater plate caused a current to flow, but when negatively charged no current flowed.

In 1907, Lee De Forest made a triode by adding a grid so that a small control voltage could control a large current. Its main advantage was that it could amplify electrical signals. By the 1940s several scientists at the Bell Laboratories were investigating materials called semiconductors. These substances, such as silicon and germanium, conducted electricity only moderately well, but, when doped with impurities their resistance changed. From this work they made a crystal called a diode which worked like a valve. The diode had many advantages over the valve, including the fact that it did not require a vacuum and was much smaller. It also worked well at room temperature, required little current and had no warm-up time. This was the start of microelectronics.

In 1948, William Shockley at the Bell Labs produced a transistor that could act as an amplifier. This was a germanium crystal with a thin p-type section sandwiched between two n-type materials and it could operate like a triode. Shockley received a Nobel prize in 1956 for this work.

By 1953, transistors were small enough to fit into hearing aids that fitted into the ear. Transistors soon operated over higher frequencies and within larger temperature ranges. Eventually they became so small that many were placed on a single piece of silicon. Typically known as microchips, they started the microelectronics industry.

In 1947 the invention of the transistor caused a storm. In fact, scientists at the Bell Laboratories kept its invention secret for over seven months so that they could fully understand its operation. On 30 June 1948 the transistor was finally revealed to the world. Unfortunately, as with many other great inventions, it received little public attention and even less press coverage (the *New York Times* gave it 4½ inches on page 46).

Transistors had initially been made from germanium which is not a robust material and cannot withstand high temperatures. The first company to propose a method of using silicon transistors was a geological research company named Texas Instruments (which had diversified into transistors). Soon many companies were producing silicon transistors, and by 1955 the electronic valve market had peaked, while the market for transistors was rocketing. The larger electronic valve manufacturers, such as Western Electric, CBS, Raytheon and Westinghouse failed to adapt to the changing market and quickly lost their market share to the new transistor manufacturing companies, such as Texas Instruments, Motorola, Hughes and RCA.

1.2.3 History of computing

In 1959, IBM built the first commercial transistorized computer named the IBM 7090/7094 series. It was so successful that it dominated the computer market for many years. In 1965, they produced the famous IBM system 360 which was built with integrated circuits. Then in 1970 IBM introduced the 370 system, which included semiconductor memories. Unfortunately, these computers were extremely expensive to purchase and maintain.

Around the same time the electronics industry was producing cheap pocket calculators. The development of affordable computers happened when Busicon commissioned what was then a small company named Intel to produce a set of between eight and twelve ICs for a calculator. Instead of designing a complete set of ICs, Intel produced a set of ICs which could be programmed to perform different tasks. These were the first ever microprocessors. Soon Intel (short for *Int*egrated *El*ectronics) produced a general-purpose 4-bit microprocessor, named the 4004 and a more powerful 8-bit version, named the 8080. Other companies, such as Motorola, MOS Technologies and Zilog were soon also making microprocessors.

IBM's virtual monopoly on computer systems soon started to slip as many companies developed computers based around the newly available 8-bit microprocessors, namely MOS Technologies 6502 and Zilog's Z-80. IBM's main contenders were Apple and Commodore who introduced a new type of computer – the personal computer (PC). The leading systems where the Apple I and the Commodore PET. These spawned many others, including the Sinclair ZX80/ZX81, the BBC microcomputer, the Sinclair Spectrum, the Commodore Vic-20 and the classic Apple II (all of which where based on the 6502 or Z-80).

IBM realized the potential of the microprocessor and used Intel's 16-bit 8086 microprocessor in their version of the PC. It was named the IBM PC and has since become the parent of all the PCs ever produced. IBM's main aim was to make a computer which could run business applications, such as word processors, spreadsheets and databases. To increase the production of this software they made information on the hardware freely available. This resulted in many software packages being developed and helped clone manufacturers to copy the original design. So the term 'IBM compatible' was born and it quickly became an industry standard by sheer market dominance.

On previous computers IBM had written most of their programs for their systems. For the PC they had a strict time limit, so they went to a small computer company called Microsoft to develop the operating system program. This program was named the Disk Operating System (DOS) because of its original purpose of controlling the disk drives. It accepted commands from the keyboard and displayed them to the monitor. The language of DOS consisted of a set of commands which were entered directly by the user and interpreted to perform file management tasks, program execution and system configuration. The main functions of DOS were to run programs, copy and remove files, create directories, move within a directory structure and to list files.

Microsoft has since gone on develop industry-standard software such as Microsoft Windows Version 3, Microsoft Office and Microsoft Windows 95. Intel has also benefited greatly from the development of the PC and has developed a large market share for their industry-standard microprocessors, such as the 80286, 80386, 80486 and Pentium processors. Microsoft's operating systems Windows NT has networking built into its core.

The power of modern desktop computers has allowed great advances in electronic communications. For many computers networking is now a standard part of the computer, whether it be through a modem or directly onto a digital network.

1.2.4 History of modern communications

After the telephone's initial development, call switching was achieved by using operators. But in 1889, Strowger, a Kansas City undertaker, patented an automatic switching system. It was advertised as "a girl-less, cuss-less, out-of-orderless, wait-less telephone system," and it used a pawl-and-ratchet system to move a wiper over a set of electrical contacts. This led to the development of the Strowger exchange which was used extensively until the 1970s.

Another important improvement came with the crossbar, which allowed many inputs to be connected to many outputs simply by addressing the required connection. The first inventor is claimed to be J.N Reynolds of Bell Systems, but it is normally given to G.A. Betulander.

Radio transmission was developed over World War I. The by-product of radio work in World War I was frequency modulation (FM) and amplitude modulation (AM). These allowed signals to be carried on (modulated) a high frequency carrier wave which traveled through the air better than the unmodulated wave. Another by-product was frequency division multiplexing (FDM) which allowed many signals to be transmitted over the same channel, but with a different carrier frequency.

In 1956, the first telephone cable across the Atlantic was laid from Oban, in Scotland to Clarenville in Newfoundland. The first Pacific cable was laid in 1902. A cable laid in 1963 and stretched from Australia to Canada.

A great revolution was started with the launch of the ATT-owned Telstar satellite. This satellite allowed communications over large distances using microwave signals which could propagate through rain and clouds and bounce off the satellite.

Pulse code modulation (PCM) had been invented by A.H. Reeves in the 1930s, but was finally used from the 1960s onwards to transmit analogue signals in a digital form.

Baudot developed a 5-unit standard code for telegraph systems. Unfortunately it had a limited alphabet of upper-case letters and a few punctuation symbols. In 1966 a standard code, known as ASCII, was defined. This has since become the standard coding system for text-based communications.

The amount of information that can be transmitted varies with the bandwidth of the system. This is normally limited by the transmission system. The usage of microwaves and radio communication increased the amount of data that could be transmitted. A great leap forward has happened with the usage of fiber-optic communications which has the potential for almost limitless amounts of data capacity.

1.3 Information

Information is available in an analogue form or in a digital form. These forms are shown in Figure 1.1. Computer-generated data can be easily stored in a digital format, but analogue signals, such as speech and video, require to be sampled at fixed intervals and then converted to a digital form. This process is known as digitization. The advantages of converting analogue information into digital are that:

- Digital data is less affected by noise, as illustrated in Figure 1.2.
- Extra data can be added so that errors can either be detected or corrected.
- The data does not degrade over time.
- Processing is relatively easy, either in real-time or non real-time.
- A single type of media can be used to store many different types of information (such as a hard disk or a CD-ROM).
- A digital system has a dependable response, whereas an analogue system's accuracy depends on parameters such as component tolerance, temperature, power supply variations, and so on. Analogue systems thus produce a variable response and no two analogue systems are identical.
- Digital systems are more adaptable and can be reprogrammed with software. Analogue systems normally require a change of hardware for any functional changes.

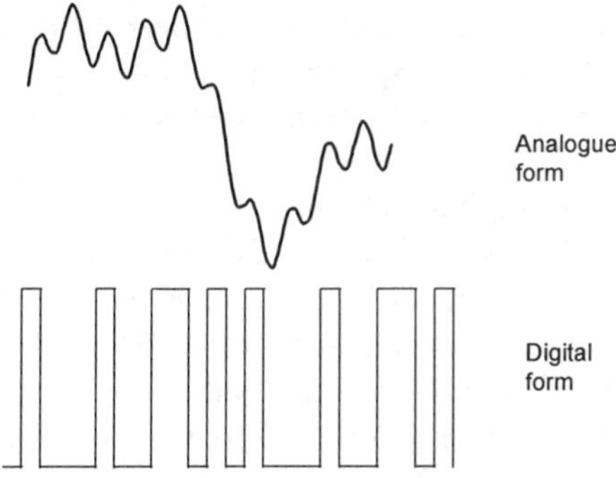

Figure 1.1 Analogue and digital format

The main disadvantage with digital conversion is:

- Data samples must be quantized to given levels; this adds an error called quantization error. The larger the number of bits used to represent each sample, the smaller the quantization error.

1.4 Digital versus analogue

The main advantage of digital technology over analogue is that digital signals are less affected by noise. Any unwanted distortion added to a signal is described as noise. This could be generated by external equipment producing airborne static, from other signals coupling into the signal's path (crosstalk), from within electrical components, from recording and playback media, and so on. Figure 1.2 shows an example of a digital signal with noise added to it. The comparator outputs a HIGH (a '1') if the signal voltage is greater than the threshold voltage, else it outputs a LOW. If the noise voltage is less than the threshold voltage then the noise will not affect the recovered signal. Even if the noise is greater than this threshold there are techniques which can reduce its effect. For example, extra bits can be added to the data either to detect errors or to correct the bits in error.

Large amounts of storage are required for digital data. For example, 70 minutes of hi-fi quality music requires over 600 MB of data storage. The data once stored tends to be reliable and will not degrade over time (extra data bits can also be added to correct or detect any errors). Typically, the data is stored either as magnetic fields on a magnetic disk or as pits on an optical disk.

The accuracy of digital systems depends on the number of bits used for each sample, whereas an analogue system's accuracy depends on component tolerance. Analogue systems also produce a differing response for different systems whereas a digital system has a dependable response.

It is very difficult (if not impossible) to recover the original analogue signal after it is affected by noise (especially if the noise is random). Most methods of reducing noise involve some form of filtering or smoothing of the signal.

A great advantage of digital technology is that once the analogue data has been converted to digital then it is relatively easy to store it with other purely digital data. Once stored in digital form it is relatively easy to process the data before it is converted back into analogue form.

An advantage of analogue technology is that it is relatively easy to store. For example, video and audio signals are stored as magnetic fields on tape and a picture is stored on photographic paper. These media tend to add noise (such as tape hiss) during storage and recovery. Unfortunately, it is not possible to detect whether an analogue signal has an error in it.

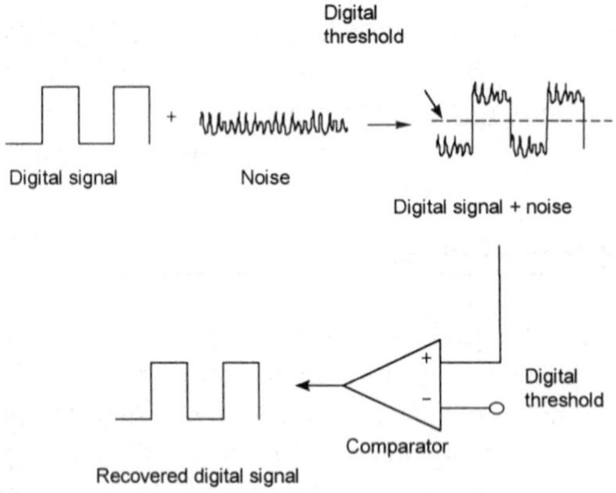

Figure 1.2 Recovery of a digital signal with noise added to it.

1.5 Conversion to digital

Figure 1.3 outlines the conversion process for digital data (the upper diagram) and for analogue data (the lower diagram). When data is already in a digital form (such as text or animation) it is converted into a given data format (such as BMP, GIF, JPG, and so on). It can be further compressed before it is either stored, transmitted or processed. The lower diagram shows how an analogue signal (such as speech or video) is first sampled at regular intervals of time. These samples are then converted into a digital form with an ADC (analogue-to-digital converter). It can then be compressed and/or stored in a defined digital format (such as WAV, JPG, and so on). This digital form is then converted back into an analogue form with a DAC (digital-to-analogue converter).

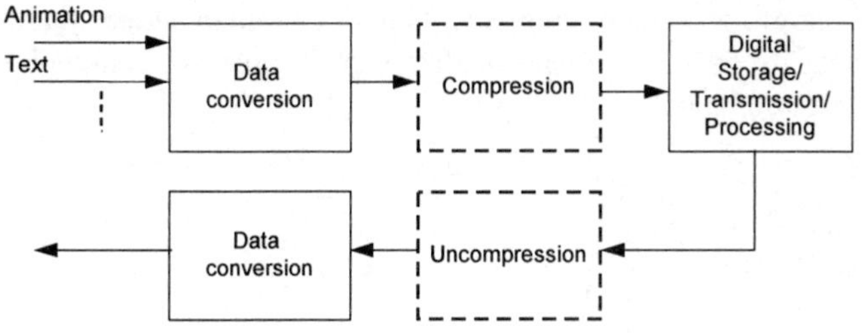

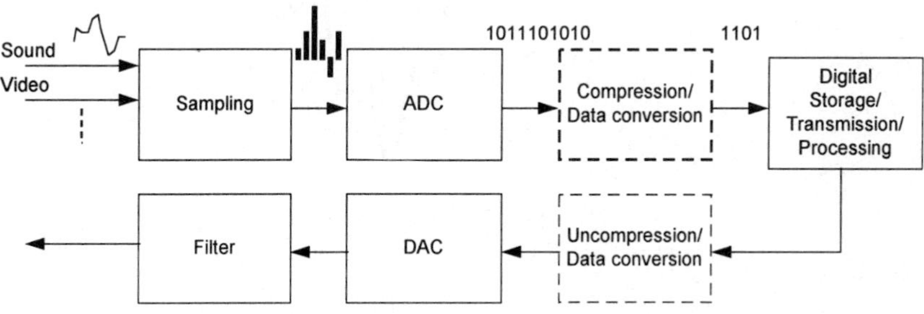

Figure 1.3 Information conversion into a digital form

1.6 Sampling theory

As a signal may be continually changing, a sample of it must be taken at given time intervals. The rate of sampling depends on its rate of change. For example, the temperature of the sea will not vary much over a short time but a video image of a sports match will. To encode a signal digitally it is normally sampled at fixed time intervals. Sufficient information is then extracted to allow the signal to be processed or reconstructed. Figure 1.4 shows a signal sampled every T_S seconds.

If a signal is to be reconstructed as the original signal it must be sampled at a rate defined by the Nyquist criterion. This states:

the sampling rate must be twice the highest frequency of the signal

For telephone speech channels, the maximum signal frequency is limited to 4 kHz and must thus be sampled at least 8 000 times per second (8 kHz). This gives one sample every 125 μs. Hi-fi quality audio has a maximum signal frequency of 20 kHz and must be sampled at least 40 000 times per second (many professional hi-fi sampling systems sample at 44.1 kHz). Video signals have a maximum frequency of 6 MHz, thus a video signal must be sampled at 12 MHz (or once every 83.3 ns).

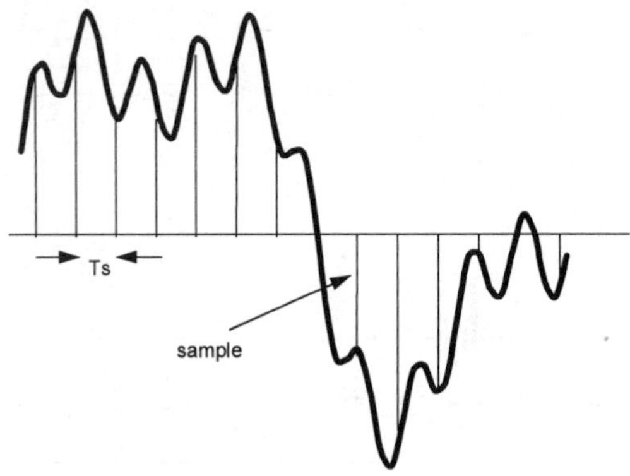

Figure 1.4 The sampling process

1.7 Quantization

Quantization involves converting an analogue level into a discrete quantized level. Figure 1.5 shows the conversion of an example waveform into a 4-bit digital code. In this case there are 16 discrete levels which are represented with the binary values 0000 to 1111. Value 1111 represents the maximum voltage level and value 0000 the minimum. It can be seen, in this case, that the digital codes for the four samples are 1011, 1011, 1001, 0111.

The quantization process approximates the level of the analogue level to the nearest quantized level. This approximation leads to an error known as quantization error. The greater the number of levels the smaller the quantization error. Table 1.1 gives the number of levels for a given number of bits.

The maximum error between the original level and the quantized level occurs when the original level falls exactly halfway between two quantized levels. The maximum error will be a half of the smallest increment or

$$\text{Max error} = \pm\frac{1}{2}\frac{\text{Maximum range}}{2^N}$$

Table 1.1 states the quantization error (as a percentage) of a given number of bits. For example the maximum error with 8 bits is 0.2%, while for 16 bits it is only 0.000 76%.

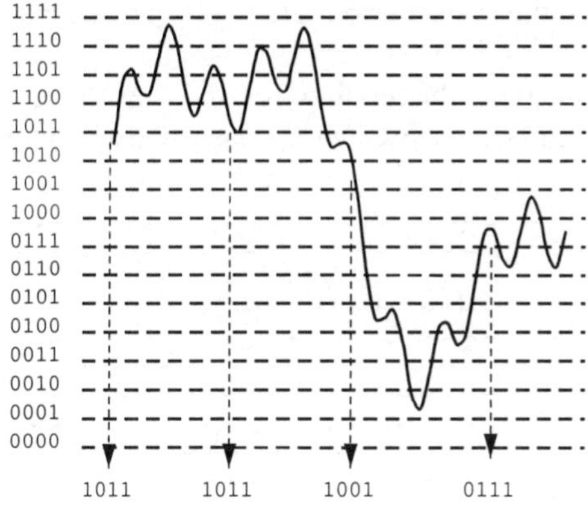

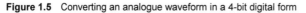

Figure 1.5 Converting an analogue waveform in a 4-bit digital form

Table 1.1 Number of quantization levels as a function of bits

Bits (N)	*Quantization levels*	*Accuracy (%)*
1	2	50
2	4	25
3	8	12.5
4	16	6.25
8	256	0.2
12	4 096	0.012
14	16 384	0.003
16	65 536	0.000 76

1.8 Exercises

1.1 Outline the main advantages of converting analogue information into a digital format.

1.2 For an analogue-to-digital conversion, determine the maximum percentage error for the following number of coding bits:

(i) 10 bits.
(ii) 18 bits.
(iii) 24 bits.

1.3 If speech is sampled at 8 kHz and each sample has eight bits then determine the transmission rate in bits per second (bps).

1.4 Determine the 4-bit coding for the waveform in Figure 1.6 (the sample points are shown with a dashed arrow pointing downwards).

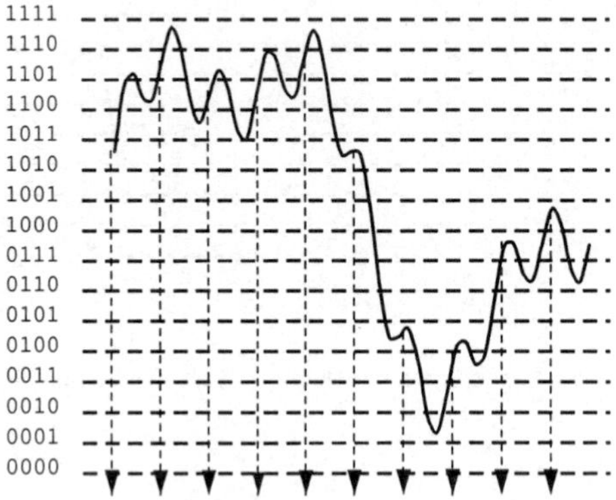

Figure 1.6 Analogue waveform

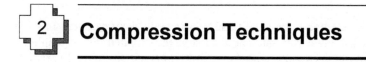

2 Compression Techniques

2.1 Introduction

Most sources of information contain redundant information or information that adds little to the stored data. An example of this might be the storage of a rectangular image. As a single color bitmapped image it could be stored as:

```
000000000000000000000000000000000000000000000
000000000000000000000000000000000000000000000
000000000000000000000000000000000000000000000
000000000011111111111111111111111111000000000
000000000010000000000000000000000001000000000
000000000010000000000000000000000001000000000
000000000010000000000000000000000001000000000
000000000010000000000000000000000001000000000
000000000010000000000000000000000001000000000
000000000011111111111111111111111111000000000
000000000000000000000000000000000000000000000
000000000000000000000000000000000000000000000
```

An improved method might be to store the image as a graphics metafile, such as:

```
Rectangle 11, 4, 26, 7
;     rectangle at start co-ordinate (11,4) of width 26 and length 7
;     pixels
```

or to compress long sequences of identical bit states, such as:

```
0,45       ; 45 bits of a '0'
0,45       ; 45 bits of a '0'
0,45       ; 45 bits of a '0'
0,10 1,26 0,9  ; 10 bits of a '0', 26 1's and 9 0's
0,10 1,1 0,24 1,1 0,9
0,10 1,1 0,24 1,1 0,9
0,10 1,1 0,24 1,1 0,9
0,10 1,1 0,24 1,1 0,9
0,10 1,1 0,24 1,1 0,9
0,10 1,26 0,9
0,45
0,45
```

Data compression is becoming an important subject as more and more digital information is required to be stored and transmitted. Many data transmission systems have a limited data rate and data storage is limited. For example an uncompressed digitized video stored on CD-ROM could give only 30 seconds of video (approximately 648 MB of data).

Data compression normally involves analyzing the source of the information and determining if there is redundant data in it.

2.2 Compression methods

When compressing data it is important to take into account the type of data and how it is interpreted. For example pixels in an image may be distorted but it would still contain the required information. Whereas a single erroneous bit in a computer data file could cause severe problems.

Video and sound images are normally compressed with a lossy compression whereas computer-type data has a lossless compression. The basic definitions are:

- Lossless compression – where the information, once uncompressed, will be identical to the original uncompressed data. This will obviously be the case with computer-type data, such as data files, computer programs, and so on. Any loss of data will cause the file to be corrupted.
- Lossy compression – where the information, once uncompressed, cannot be fully recovered. Lossy compression normally involves analyzing the data and determining which information has little effect on the resulting compressed data. For example, there is little difference, to the eye, between an image with 16.7 million colors (24-bit color information) and an image stored with 1024 colors (10-bit color information), but the storage will be reduced to 41.67% (many computer systems cannot even display 24-color information in certain resolutions). Compression of an image might also be used to reduce the resolution of the image. Again, the human eye might compensate for the loss of resolution.

2.3 Letter probabilities

The English language has a great deal of redundancy in it, thus common occurrences in text can be coded with short bit sequences. The probability of each letter also varies. For example a letter '*e*' occurs many more times than the letter '*z*'. Program 2.1 determines the number of occurrences of the characters from 'a' to '*z*' in an example text input. Sample run 2.1 shows a sample run with an example piece of text. It can be seen that the most common character in the text is an '*e*' (11 occurrences), followed by a '*t*' (7 occurrences), followed by '*n*' (5 occurrences).

Program 2.1

```
#include    <stdio.h>
#include    <string.h>

int   get_occurances(char c, char txt[]);

int   main(void)
{
char  ch, text[BUFSIZ];
int   occ;

   printf("Enter text >>");
   gets(text);

   for (ch='a';ch<='z';ch++)
   {
      occ=get_occurances(ch,text);
      printf("%c %d\n",ch,occ);
   }

   return(0);
}

int   get_occurances(char c, char txt[])
{
int   occ=0,i;

   for (i=0;i<strlen(txt);i++)
      if (c==txt[i]) occ++;

   return(occ);
}
```

Sample run 2.1

```
Enter text >>  this is an example text input to determine the
number of occurrences of a character within a piece of text. it
shows how the characters used in the english language vary in
their occurrences. the e character is the most common character in
the english language.

a 17
b 1
c 16
d 2
e 30
f 3
g 6
h 17
i 15
j 0
k 0
l 5
m 5
n 15
```

```
o 11
p 3
q 0
r 16
s 12
t 22
u 7
v 1
w 3
x 3
y 1
z 0
```

Program 2.6 in Section 2.10 gives a simple C program which determines the probability of letters within a text file. This program can be used to determine typical letter probabilities. Sample run 2.2 shows a sample run using the text from Chapter 1 as the input to the program. It can be seen that the highest probability is with the letter '*e*', which occurs, on average, 94.3 times every 1000 letters. Table 2.1 lists the letters in order of their probability. Notice that the letters which are worth the least in the popular board game Scrabble (such as, '*e*', '*t*', '*a*', and so on) are the most probable and the letters with the highest scores (such as '*x*', '*z*' and '*q*') are the least probable.

Sample run 2.2

```
Enter text file>> adcom_c01.txt
Char. Occur. Prob.
  a    1963   0.0672
  b     284   0.0097
  c     914   0.0313
  d     920   0.0315
  e    2752   0.0943
  f     471   0.0161
  g     473   0.0162
  h     934   0.0320
  i    1680   0.0576
  j      13   0.0004
  k      96   0.0033
  l     968   0.0332
  m     724   0.0248
  n    1541   0.0528
  o    1599   0.0548
  p     443   0.0152
  q      49   0.0017
  r    1410   0.0483
  s    1521   0.0521
  t    2079   0.0712
  u     552   0.0189
  v     264   0.0090
  w     383   0.0131
  x      57   0.0020
  y     278   0.0095
  z      44   0.0015
  .     292   0.0100
  SP   4474   0.1533
  ,     189   0.0065
```

Table 2.1 Letters and their occurrence in a sample text file.

Character	*Occurrences*	*Probability*	*Character*	*Occurrences*	*Probability*
SPACE	4474	0.1533	g	473	0.0162
e	2752	0.0943	f	471	0.0161
t	4079	0.0712	p	443	0.0152
a	1963	0.0672	w	383	0.0131
i	1680	0.0576	.	292	0.0100
o	1599	0.0548	b	284	0.0097
n	1541	0.0528	y	278	0.0095
s	1521	0.0521	v	264	0.0090
r	1410	0.0483	,	189	0.0065
l	968	0.0332	k	96	0.0033
h	934	0.0320	x	57	0.0020
d	920	0.0315	q	49	0.0017
c	914	0.0313	z	44	0.0015
m	724	0.0248	j	13	0.0004
u	552	0.0189			

2.4 Coding methods

Apart from lossy and lossless compression, data encoding is normally classified into two main areas: entropy encoding and source encoding.

2.4.1 Entropy coding

Entropy coding does not take into account the characteristics of the data and treats all the bits in the same way; it produces lossless coding. Typically it uses:

- Statistical encoding – where the coding analyses the statistical pattern of the data. For example if a source of text contains many more 'e' characters than 'z' characters then the character 'e' could be coded with very few bits and the character 'z' with many bits.
- Suppressing repetitive sequences – many sources of information contain large amount of receptive data. For example this page contains large amounts of 'white space'. If the image of this page were to be stored, a special character sequence could represent long runs of 'white space'.

2.4.2 Source encoding

Source encoding normally takes into account characteristics of the information. For example images normally contain many repetitive sequences, such as common pixel colors in neighboring pixels. This can be encoded as a special coding sequence. In video pictures, also, there are very few changes between one frame and the next. Thus typically the data encoded only stores the changes from one frame to the next.

2.5 Statistical encoding

Statistical encoding is an entropy technique which identifies certain sequences within the data. These 'patterns' are then coded so that they are coded in fewer bits. Frequently used patterns are coded with fewer bits than less common patterns. For example, text files normally contain many more '*e*' characters than '*z*' characters. Thus the '*e*' character could be encoded with a few bits and the '*z*' with many bits. Statistical encoding is also known as arithmetic compression.

A typical statistical coding scheme is Huffman encoding. Initially the encoder scans through the file and generates a table of occurrences of each character. The codes are assigned to minimize the number of encoded bits, then stored in a codebook which must be transmitted with the data.

Table 2.2 shows a typical coding scheme for the characters '*a*' to '*z*'. It uses the same number of bits for each character. Morse code is an example of statistical encoding. It uses dots (a zero) and dashes (a one) to code characters, where a short space in time delimits each character. It uses short codes for the most probable letters and longer codes for less probable letters. In the form of zeros and ones it is stated in Table 2.3.

Thus the message:

```
this an
```

would be encoded as:

```
Message:      t      h      i      s             a      n
Simple code:  10011  00111  01000  10010  11010  01000  10010
Morse code:   1      0000   00     000    0011   01     10
```

This has reduced the number of bits used to represent the message from 35 (7×5) to 18.

Table 2.2 Simple coding scheme

a	00000	b	00001	c	00010	d	00011	e	00100
f	00101	g	00110	h	00111	i	01000	j	01001
k	01010	l	01011	m	01100	n	01101	o	01110
p	01111	q	10000	r	10001	s	10010	t	10011
u	10100	v	10101	w	10110	x	10111	y	11000
z	11001	SP	11010						

Table 2.3 Morse coding scheme

a	01	b	1000	c	1010	d	100	e	0
f	0010	g	110	h	0000	i	00	j	0111
k	101	l	0100	m	11	n	10	o	111
p	0110	q	1101	r	010	s	000	t	1
u	001	v	0001	w	011	x	1001	y	1011
z	1100	SP	0011						

2.6 Repetitive sequence suppression

Repetitive sequence suppression involves representing long runs of a certain bit sequence with a special character. A special bit sequence is then used to represent that character, followed by the number of times it appears in sequence. Typically 0s (zero) and ' ' (spaces) occur repetitively in text files. For example the data:

```
8.3200000000000
```

could be coded as:

```
8.32F11
```

where F represents the flag. In this case the number of stored characters has been reduced from 16 to 7. Many text sources have other characters which occur repetitively. Run-length encoding (RLE) uses this to encode any character sequence with a special flag followed by the number of characters and finally the character which is repeated. For example

```
Fred        has     when.........
```

could be coded as:

```
FredF7 hasF7 whenF9.
```

where F represents the flag. In this case the number of stored characters has been reduced from 32 to 20. The 'F7 ' character code represents seven ' ' (spaces) and 'F9.' represents nine '.' characters.

Program 2.2 is a very simple program which scans a file IN.DAT and, using RLE, stores to a file OUT.DAT. The special character sequence is:

ZZ*cxx*

where ZZ is the flag sequence, *c* is the repeating character and *xx* the number of times the character occurs. The ZZ flag sequence is chosen because, in a text file, it is unlikely to occur within the file. File listing 2.1 shows a sample IN.DAT and File listing 2.2 shows the RLE encoded file (OUT.DAT).

Program 2.2

```
/* ENCODE.C    */
#include <stdio.h>

int   main(void)
{
FILE  *in,*out;
char  previous,current;
int      count;

   if ((in=fopen("in.dat","r"))==NULL)
   {
      printf("Cannot open <in.dat>");
      return(1);
   }
   if ((out=fopen("out.dat","w"))==NULL)
   {
      printf("Cannot open <out.dat>");
      return(1);
   }

   do
   {
      count=1;
      previous=current;
      current=fgetc(in);
      do
      {
         previous=current;
         current=fgetc(in);
         if (previous!=current) ungetc(current,in);
         else count++;
      } while (previous==current);

      if (count>1) fprintf(out,"ZZ%c%02d",previous,count);
      else fprintf(out,"%c",previous);
   }  while (!feof(in));

   fclose(in);

   fclose(out);

   return(0);
}
```

File list 2.1

```
The     bbbbbbboy stood onnnnn the burning
deck         and still did.
1.000000000
3.000000010
5.000000000
```

File list 2.2

```
TheZZ 05ZZb07oy stZZo02d oZZn05 the burning
deckZZ 09and stiZZl02 did.
1.ZZ009
3.ZZ00710
5.ZZ009
```

Program 2.3 is a simple C program which decodes the RLE file produced by the previous program.

Program 2.3

```
/* UNENCODE.C     */
#include <stdio.h>

int   main(void)
{
FILE  *in,*out;
char  ch;
int   count,i;
   if ((in=fopen("out.dat","r"))==NULL)
   {
      printf("Cannot open <out.dat>");
      return(1);
   }
   if ((out=fopen("in1.dat","w"))==NULL)
   {
      printf("Cannot open <in1.dat>");
      return(1);
   }

   do
   {
      ch=fgetc(in);

      if (ch=='Z')
      {
         ch=fgetc(in);
         if (ch=='Z')
         {
            fscanf(in,"%c%02d",&ch,&count);
            for (i=0;i<count;i++)
               fprintf(out,"%c",ch);
         }
         else ungetc(ch,in);
      }
      else fprintf(out,"%c",ch);

   } while (!feof(in));
   fclose(in);
   fclose(out);
   return(0);
}
```

The ZZ flag sequence is inefficient as it uses two characters to store the flag, a better flag could be an 8-bit character which cannot occur, such as 11111111b, or ffh. Program 2.4 is an example of this and Program 2.5 shows the decoding program.

Program 2.4

```
#include <stdio.h>

#define  FLAG  0xff  /* 1111 1111b  */

int   main(void)
{
FILE  *in,*out;
char  previous,current;
int   count;

  ;;; ;;;;;

      if (count>1) fprintf(out,"%c%c%02d",FLAG,previous,count);
      else fprintf(out,"%c",previous);
   }  while (!feof(in));

   fclose(in);
   fclose(out);
   return(0);
}
```

Program 2.5

```
/* UNENCODE.C     */
#include <stdio.h>

#define  FLAG  0xff  /* 1111 1111b  */

int   main(void)
{
FILE  *in,*out;
char  ch;
int   count,i;

   ;;; ;;;;
   ;;; ;;;;
   do
   {
      ch=fgetc(in);

      if (ch==FLAG)
      {
         ch=fgetc(in);
         fscanf(in,"%c%02d",&ch,&count);
         for (i=0;i<count;i++)
               fprintf(out,"%c",ch);
      }
      else fprintf(out,"%c",ch);

   }  while (!feof(in));
   fclose(in);
   fclose(out);
   return(0);
}
```

In a binary file any bit sequence can occur. To overcome this a flag sequence, such as 10101010 can be used to identify the flag. If this sequence occurs within the data then it will be coded with two flags, two consecutive flags in the data are coded with three flags, and so on. For example:

```
00000000 10101010 10101010 00011100 01001100
```

would be encoded as:

```
00000000 10101010 10101010 10101010 00011100 01001100
```

thus when the three flags are detected then one of them is deleted.

Repetitive sequence suppression is an excellent general-purpose compression technique for images, as most images tend to have long sequences of the same pixel intensity or color.

2.7 Differential encoding

Differential coding is a source coding method which is used when there is a limited change from one value to the next. It is well suited to video and audio signals, especially audio, where the sampled values can only change within a given range. It is typically used in PCM (pulse code modulation) schemes to encode audio and video signals. PCM converts analogue samples into a digital code. Examples are:

- Delta modulation PCM – delta PCM uses a single-bit code to represent an analogue signal. With delta modulation a '1' is transmitted (or stored) if the analogue input is higher than the previous sample or a '0' if it is lower. It must obviously work at a higher rate than the Nyquist frequency but because it only uses 1 bit it normally results in a lower output bit rate (because the factor of increasing the sampling rate is normally less than the factor of reducing the number of encoded bits).

- Adaptive delta modulation PCM – unfortunately delta modulation cannot react to very rapidly changing signals and will thus take a relatively long time to catch up (known as slope overload). It also suffers when the signal does not change much as this produces a square wave signal (known as granular noise). One method of reducing granular noise and slope overload is to use adaptive delta PCM. With this method the step size is varied by the slope of the input signal. The larger the slope, the larger the step size.

Algorithms usually depend on the system and the characteristics of the signal. A typical algorithm is to start with a small step and increase it by a multiple until the required level is reached. The number of slopes will depend on the number of coded bits, such as 4 step sizes for 2 bits, 8 for 3 bits, and so on.

- Differential PCM (DPCM) – speech signals tend not to change much between two samples. Thus similar codes are sent, which leads to a degree of redundancy. Certain signals, such as speech, have a limited range for a sample amplitude for a given sample time. DPCM reduces the redundancy by transmitting the difference in the amplitude of two consecutive samples. Since the range of sample differences is typically less than the range of individual samples, fewer bits are required for DPCM than for conventional PCM.

2.8 Transform encoding

Transform encoding is a source-encoding scheme where the data is transformed by a mathematical transform in order to reduce the transmitted (or stored) data. A typical method is to conduct a Fourier transform to determine the frequency information from the data. The coefficients of the transform can then be stored. The strongest coefficients are accurately coded and the less significant coefficients can be coded less accurately (thus fewer bits are used to code these coefficients).

Transform encoding is suitable for compressing images and a typical transform is the discrete cosine transform (DCT).

2.9 Exercises

2.1 Explain the difference between metafile format and bitmapped format. State which format is likely to be used from the following:

Picture of a sea front
CAD drawing.
3D wireframe image.

2.2 The two basic types of compression are lossless compression and lossy compression. Discuss the difference between the two types and give examples of where they would be used.

2.3 The letters on the left-hand side of a QWERTY keyboard are:

```
QWERTASDFGZXCVB
```

and on the right-hand side (excluding non-letters) are:

```
POIUYHJKLNM,.
```

Using the probabilities given in Table 2.1 determine the probability of using the left-hand side of the keyboard, and compare this with the probability of using the right-hand side. Thus determine which hand is used more often by a typist.

2.4 Explain the differences between source coding and entropy coding. Give examples of where they could possibly be used.

2.5 Code the following message with pure binary (Table 2.2) and with Morse code (Table 2.3):

```
TAKE ME TO YOUR LEADER SAID THE ALIEN INVADER
```

Thus determine the saving in the number of bits used to encode the message.

2.6 How could the following text file be coded with RLE coding.

```
1.000000000000000000000
10.00000000000000000000
6.000000000000000000000
5.600000000000000000000
88.80000000000000000000
```

Explain why an image which has many consecutive pixels of the same color can be highly compressed with RLE coding.

2.10 Letter probablity program

Program 2.6

```
#include <stdio.h>
#include <string.h>
#include <ctype.h>

#define  NUM_LETTERS 29

int   get_occurances(char c, char txt[]);
```

```
int   main(void)
{
char  ch, fname[BUFSIZ];
int   occ[NUM_LETTERS]={0,0,0,0,0,0,0,0,0,0,0,0,0,0,0,
                        0,0,0,0,0,0,0,0,0,0,0,0,0,0};
unsigned int   total,i;
FILE           *in;

   printf("Enter text file>>");
   gets(fname);

   if ((in=fopen(fname,"r"))==NULL)
   {
      printf("Can't find file %s\n",fname);
      return(1);
   }

   do
   {
      ch=tolower(getc(in));

      if (isalpha(ch))
      {
         (occ[ch-'a'])++;
         total++;
      }
      else if (ch=='.') { occ[NUM_LETTERS-3]++; total++; }
      else if (ch==' ') { occ[NUM_LETTERS-2]++; total++; }
      else if (ch==',') { occ[NUM_LETTERS-1]++; total++; }
   } while (!feof(in));

   fclose(in);

   puts("Char. Occur. Prob.");

   for (i=0;i<NUM_LETTERS;i++)
   {
      printf("  %c   %5d %5.4f\n",'a'+i,occ[i],(float)occ[i]/(float)total);
   }

   return(0);
}

int   get_occurances(char c, char txt[])
{
int   occ=0,i;

   for (i=0;i<strlen(txt);i++)
      if (c==txt[i]) occ++;

   return(occ);
}
```

3 Huffman/Lempel-Ziv Compression Methods

3.1 Introduction

Normally, general data compression does not take into account the type of data which is being compressed and is lossless. It can be applied to computer data files, documents, images, and so on. The two main techniques are statistical coding and repetitive sequence suppression. This chapter discusses two of the most widely used methods for general data compression: Huffman coding and Lempel-Ziv coding.

3.2 Huffman coding

Huffman coding uses a variable length code for each of the elements within the information. This normally involves analyzing the information to determine the probability of elements within the information. The most probable elements are coded with a few bits and the least probable coded with a greater number of bits.

The following example relates to characters. First, the textual information is scanned to determine the number of occurrences of a given letter. For example:

'b'	'c'	'e'	'i'	'o'	'p'
12	3	57	51	33	20

Next the characters are arranged in order of their number of occurrences, such as:

'e'	'i'	'o'	'p'	'b'	'c'
57	51	33	20	12	3

Next the two least probable characters are assigned either a 0 or a 1. Figure 3.1 shows that the least probable ('c') has been assigned a 0 and the next least probable ('b') has been assigned a 1. The summation of the two occurrences is then taken to the next column and the occurrence values are again arranged in descending order (that is, 57, 51, 33, 20 and 15). As with the first column, the least probable occurrence is assigned a 0 and the next least probable occurrence is assigned a 1. This continues until the last column. The Huffman-coded values are then read from left to right and the bits are listed from right to left.

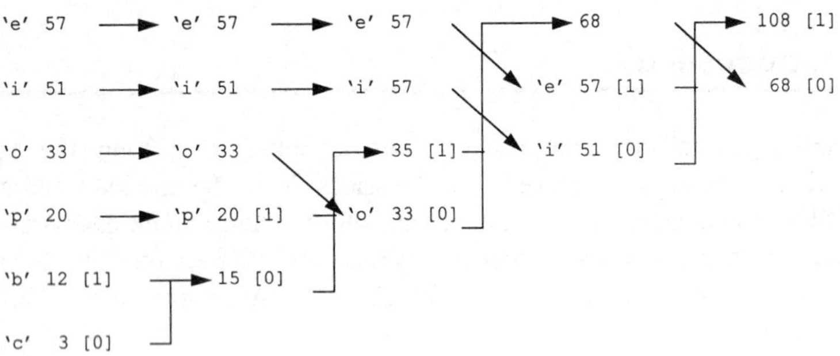

Figure 3.1 Huffman coding example

The final coding will be:

```
'e'   11
'i'   10
'o'   00
'p'   011
'b'   0101
'c'   0100
```

The great advantage of Huffman coding is that, although each character is coded with a different number of bits, the receiver will automatically determine the character whatever their order. For example if a 1 is followed by a 1 then the received character is an 'e'. If it is followed by a 0 then it is an 'o'. Here is an example:

```
11000110100100110100
```

will be decoded as:

'e' 'o' 'p' 'c' 'i' 'p' 'c'

When transmitting or storing Huffman-coded data, the coding table needs to be stored with the data (if the table is generated dynamically). It is generally a good compression technique but it does not take into account higher order associations between characters. For example, the character 'q' is normally followed by the character 'u' (apart from words such as Iraq). An efficient coding scheme for text would be to encode a single character 'q' with a longer bit sequence than a 'qu' sequence.

3.3 Adaptive Huffman coding

Adaptive Huffman coding was first conceived by Faller and Gallager and then further refined by Knuth (so it is often called the FGK algorithm). It uses defined word schemes which determine the mapping from source messages to code words based upon a running estimate of the source message probabilities. The code is adaptive and changes so as to remain optimal for the current estimates. In this way, the adaptive Huffman codes respond to locality and the encoder thus learns the characteristics of the source data. The decoder must then learn along with the encoder by continually updating the Huffman tree so as to stay in synchronization with the encoder.

A second advantage of adaptive Huffman coding is that it only requires a single pass over the data. In many cases the adaptive Huffman method actually gives a better performance, in terms of number of bits transmitted, than static Huffman coding. This does not contradict the optimality of the static method as the static method is optimal only over all methods, which assumes a time-invariant mapping. The performance of the adaptive methods can also be worse than that of the static method. Upper bounds on the redundancy of these methods are presented in this section. As discussed in the introduction, the FGK method is the basis of the UNIX compact program.

3.3.1 FGK algorithm

The basis for the FGK algorithm is the Sibling Property: a binary code tree has the sibling property if each node (except the root) has a sibling and if the nodes can be listed in order of non-increasing weight with each node adjacent to its sibling.

3.4 Lempel-Ziv coding

Around 1977, Abraham Lempel and Jacob Ziv developed the Lempel-Ziv class of adaptive dictionary data compression techniques. Also known as LZ-77 coding, they are now some of the most popular compression techniques.

The LZ coding scheme takes into account repetition in phases, words or parts of words. These repeated parts can either be text or binary. A flag is normally used to identify coded and unencoded parts. An example piece of text could be:

'The receiver requires a receipt which is automatically sent when it is received.'

This has the repetitive sequence 'recei'. The encoded sequence could be modified with the flag sequence #*m*#*n* where *m* represents the number of characters to trace back to find the character sequence and *n* the number of replaced characters. Thus the encoded message could become:

'The receiver requires a #20#5pt which is automatically sent wh#6#2 it #30#2 #47#5ved.'

Normally a long sequence of text has many repeated words and phases, such as 'and', 'there', and so on. Note that in some cases this could lead to longer files if short sequences were replaced with codes that were longer than the actual sequence itself.

3.5 Lempel-Ziv-Welsh coding

The Lempel-Ziv-Welsh (LZW) algorithm (also known LZ-78) builds a dictionary of frequently used groups of characters (or 8-bit binary values). Before the file is decoded, the compression dictionary must be sent (if transmitting data) or stored (if data is being stored). This method is good at compressing text files because text files contain ASCII characters (which are stored as 8-bit binary values) but not so good for graphics files, which may have repeating patterns of binary digits that might not be multiples of 8 bits.

A simple example is to use a 6 character alphabet and a 16 entry dictionary, thus the resulting code word will have 4 bits. If the transmitted message is:

```
ababacdcdaaaaaaef
```

Then the transmitter and receiver would initially add the following to its dictionary:

```
0000        'a'
0001        'b'
0010        'c'
```

```
0011        'd'
0100        'e'
0101        'f'
0110–1111   empty
```

First the 'a' character is sent with 0000, next the 'b' character is sent and the transmitter checks to see that the 'ab' sequence has been stored in the dictionary. As it has not, it adds 'ab' to the dictionary, to give:

```
0000        'a'
0001        'b'
0010        'c'
0011        'd'
0100        'e'
0101        'f'
0110        'ab'
0111–1111   empty
```

The receiver will also add this to its table (thus the transmitter and receiver will always have the same tables). Next the transmitter reads the 'a' character and checks to see if the 'ba' sequence is in the code table. As it is not, it transmits the 'a' character as 0000, adds the 'ba' sequence to the dictionary, which will now contain:

```
0000        'a'
0001        'b'
0010        'c'
0011        'd'
0100        'e'
0101        'f'
0110        'ab'
0111        'ba'
1000–1111   empty
```

Next the transmitter reads the 'b' character and checks to see if the 'ab' sequence is in the table. As it is, it will transmit the code table address which identifies it, i.e. 0110. When this is received, the receiver detects that it is in its dictionary and it knows that the addressed sequence is 'ab'.

Next the transmitter reads an 'a' and checks for the sequence 'ba' in its dictionary. As it is included, it transmits its address, i.e. 0111. When this is received, the receiver checks its dictionary and locates the sequence 'ba'. This then continues with the transmitter and receiver maintaining identical copies of their dictionaries. A great deal of compression occurs when sending a sequence of one character, such as a long sequence of 'a'.

Typically, in a practical implementation of LZW, the dictionary size for LZW starts at 4K (4096). The directory then stores bytes from 0 to 255 and the addresses 256 to 4095 are used for strings (which can contain two or more characters). As there are 4096 entries then it is a 12-bit coding scheme (0 to 4096 gives 0 to $2^{12}-1$ different addresses).

3.6 Variable-length-code LZW compression

The Variable-length-code LZW (VLC-LZW) uses a variation of the LZW algorithm where variable-length codes are used to replace patterns detected in the original data. It uses a dictionary constructed from the patterns encountered in the original data. Each new pattern is entered into it and its indexed address is used to replace it in the compressed stream. The transmitter and receiver maintain the same dictionary.

The VLC part of the algorithm is based on an initial code size (the LZW initial code size), which specifies the initial number of bits used for the compression codes. When the number of patterns detected by the compressor in the input stream exceeds the number of patterns encodable with the current number of bits then the number of bits per LZW code is increased by one. The code size is initially transmitted (or stored) so that the receiver (or uncompressor) knows the size of the dictionary and the length of the codewords.

In 1985 the LZW algorithm was patented by the Sperry Corp. It is used by the GIF file format and is similar to the technique used to compress data in V.42bis modems.

3.7 Disadvantages with LZ compression

LZ compression substitutes the detected repeated patterns with references to a dictionary. Unfortunately the larger the dictionary, the greater the number of bits that are necessary for the references. The optimal size of the dictionary also varies for different types of data; the more variable the data, the smaller the optimal size of the directory.

3.8 Practical Lempel-Ziv/Huffman coding

This section contains practical examples of programs which use Lempel-Ziv and/or Huffman coding. Most compression programs use either one or both of these techniques. As previously mentioned, both techniques are lossless. In general, Huffman is the most efficient but requires two passes over the data, while Lempel-Ziv uses just one pass. This feature of a single pass is obviously important when saving to a hard disk drive or when encoding and decoding data in real-time communications. One of the most widely used variants is LZS, owned by Stac Electronics (who were the first commercial company to produce a compressed drive, named Stacker). Microsoft have included a variation of this program, called DoubleSpace in DOS Version 6 and DriveSpace in Windows 95.

The LZS technique is typically used in mass backup devices, such as tape drives, where the compression can either be implemented in hardware or in software. This typically allows the tape to store at least twice the quoted physical capacity of the tape.

The amount of compression, of course, depends on the type of file being compressed. Random data, such as executable programs or object code files, typically has low compression (resulting in a file which is 50 to 95% of the original file size). Still images and animation files tend to have high compression and typically result in a file which is between only 2 and 20% of the original file size. It should be noted that once a file has been compressed there is virtually no gain in compressing it again (unless a differential method is used). Thus storing or transmitting compressed files over a system which has further compression will not increase the compression ratio (unless another algorithm is used).

Typical files produced from LZ77/LZ78 compression methods are ZIP, ARJ, LZH, Z, and so on. Huffman is used in ARC and PKARC utilities, and in the UNIX compact command.

3.8.1 Lempel-Ziv/Huffman practical compression

DriveSpace and DoubleSpace are programs used in PC systems to compress files on hard disk drives. They use a mixture of Huffman and Lempel-Ziv coding, where Huffman codes are used to differentiate between data (literal values) and back references and LZ coding is used for back references.

A DriveSpace disk starts with a 4 byte magic number (52 B2 00 00 08h). This identifies the that the disk is using DriveSpace. Following this there are either literal values or back references which are preceded by control bits of either `0`, `11`, `100`, `1010` or `1011` (these are Huffman values coded to differentiate them). If the control bit is 0 then the following 7 bits of `abcdefg` correspond to data of `0gfedcba`, else if it is 11 then the following 7 bits of `abcdefg` correspond to data `1gfedcba`. Thus the data:

```
10110101 01111110 11100000 11111111
```

would be encoded as:

`11 1010110 0 0111111 11 0000011 11 1111111`

The back-reference values are preceded by `100`, `1010` or `1011`. If preceded by `100` then followed by `abcdef`*X*, which is a 6-bit back reference of a length given by *X*. A `1010` followed by `abcdefgh`*X* is an 8-bit back reference of 64+`hgfedcba` with length given by *X*. A `1011` followed by `abcdefghijkl`*X* is a 12-bit back reference of 64+256+`lkjihgfedcba` with length given by *X*.

The back reference consists of a code indicating the number of bits back to find the start of the referenced data, followed by the length of the data itself. This code consist of *N* zeros followed by a 1. The number of zeros, *N*, indicates the number of bits of length data and the length of the back reference is $M+2^N+2$, where *M* is the *N*-bit unsigned number comprising the data length. Thus the minimum length of a back reference will be when $M=0$ and $N=0$ giving a value of 3. An example format of a back pointer is:

`100  abcdef 000001 ghijk`

where *N* will be 5 since there are five zeros after the 5-bit back reference and `fedcba` corresponds to the back reference `fedcba`. The length of the reference values will be $M+2^5+2$, where *M* is the 5-digit unsigned binary number `kjihg`. For example if the stored bit field were:

```
010100101011001000000000000000000000001001000010011001000001O
0001001100101000010
```

It would be decoded as:

0101001010110010000000000000000000001001
MAGIC NUMBER

`0 0001001`	**`100`** `100000 1`
'H' (100 1000)	As the control bit field is **`100`** then it has 6-bit back reference of `000001` (one place back) followed by `1` which shows that the back reference length of bits is 0. Thus, using the formula $M+2^N+2$ gives $0+2^0+2=3$. The back reference has a length of 3 bytes, giving the output 'HHH'

`0 0001001`	**`100`** `1010000 10`
'E' (`01010001`)	As the control bit field is **`100`**, it has a 6-bit back reference of `000101` (five places back) followed by `01` which shows that the back-reference length of bits is 1. Thus, using the formula $M+2^N+2$ gives $0+2^1+2=4$. The character five places back is an 'H', thus 'H' is repeated four times.

This then gives the sequence 'HHHHEHHHH'.

In DriveSpace, each of the fields after the magic number is a group. A group consist of a control part (the Huffman code) and an item. An item may be either a literal item or a copy item (i.e. a 6, 8 or 12 bit back reference). The end of a file in DriveSpace is identified with a special 12-bit back-reference value of 1111 1111 1111 1111 (FFFFh).

3.8.2 GIF files

The graphic interface format (GIF) uses a compression algorithm based on the Lempel-Ziv-Welsh (LZW) compression scheme. When compressing an image the compression program maintains a list of substrings that have been found previously. When a repeated string is found, the referred item is replaced with a pointer to the original. Since images tends to contain many repeated values, the GIF format is a good compression technique. The format of the data file will be discussed in Chapter 5.

3.8.3 UNIX compress/uncompress

The UNIX programs `compress` and `uncompress` use adaptive Lempel-Ziv coding. They are generally better than `pack` and `unpack` which are based on Huffman coding. Where possible, the `compress` program adds a `.Z` onto a file when compressed. Compressed files can be restored using the `uncompress` or `zcat` programs.

3.8.4 UNIX archive/zoo

The UNIX-based `zoo` freeware file compression utility employs the Lempel-Ziv algorithm. It can store and selectively extract multiple generations of the same file. Data can thus be recovered from damaged archives by skipping the damaged portion and locating undamaged data (using the `fiz` program).

3.9 Exercises

3.1 Determine the Huffman coding for the following characters:

'a'	[60]	'e'	[120]
'f'	[30]	'g'	[25]
'.'	[10]	'p'	[55]

Note that the number of occurrences of the character is given within brackets.

3.2 The top line of letters of a QWERTY keyboard is `QWERTYUIOP`, the second line is `ASDFGHJKL` and the bottom line is `ZXCVBNM,..` Using Table 3.1 determine the most used line and the least used line on the keyboard.

3.3 Using the occurrences given in Table 3.1 determine the Huffman code for each of the characters.

3.4 Using the Huffman code developed in the previous question determine the number of bits used for the following message:

```
In modern electronic systems, hardware, software and
firmware interconnect in the design process. Formal rules on
the design of software have been developed over the years,
these are now being applied to the development of hardware
systems.
System development is normally a multistaged process. It is
iterative and a not sequential process and feedback from any
part updates earlier stages. At the first stage a need for
the system is identified. If there is, a design team
produces a design specification to define the functionality
of the complete system.
```

Table 3.1 recaps the table of probabilities from Chapter 2. This is used in some of the questions in this section.

3.5 Explain how LZH coding operates with a 4K dictionary size.

Table 3.1 Letters and their occurrence in a sample text file.

Character	*Occurrences*	*Probability*	*Character*	*Occurrences*	*Probability*
SPACE	4474	0.1533	g	473	0.0162
e	2752	0.0943	f	471	0.0161
t	2079	0.0712	p	443	0.0152
a	1963	0.0672	w	383	0.0131
i	1680	0.0576	.	292	0.0100
o	1599	0.0548	b	284	0.0097
n	1541	0.0528	y	278	0.0095
s	1521	0.0521	v	264	0.0090
r	1410	0.0483	,	189	0.0065
l	968	0.0332	k	96	0.0033
h	934	0.0320	x	57	0.0020
d	920	0.0315	q	49	0.0017
c	914	0.0313	z	44	0.0015
m	724	0.0248	j	13	0.0004
u	552	0.0189			

4 Image Compression (GIF/TIFF/ PCX)

4.1 Introduction

Data communication increasingly involves the transmission of still and moving images. Great savings in the number of bits can be made by compressing the images into a standard form. Some of these forms are outlined in Table 4.1. The main parameters in a graphics file are:

- The picture resolution. This is defined by the number of pixels in the x- and y-directions.
- The number of colors per pixel. If N bits are used for the bit color then the total number of displayable colors will be 2^N. For example an 8-bit color field defines 256 colors, a 24-bit color field gives 2^{24} or 16.7M colors.
- Palette size. Some systems reduce the number of bits used to display a color by reducing the number of displayable colors for a given palette size.

Table 4.1 Typical standard compressed graphics formats

File	*Compression type*	*Max. resolution or colors*	
TIFF	Huffman RLE and/or LZW	48-bit color	TIFF (tagged image file format) is typically used to transfer graphics from one computer system to another. It allows high resolutions and colors of up to 48 bits (16 bits for red, green and blue).
PCX	RLE	$65\,536 \times 65\,536$ (24-bit color)	Graphics file format which uses RLE to compress the image. Unfortunately it make no provision for storing gray scale or color-correcting tables.
GIF	LZW	$65\,536 \times 65\,536$ (24-bit color, but only 256 displayable colors)	Standardized graphics file format which can be read by most graphics packages. It has similar graphics characteristics to PCX files and allows multiple images in a single file and interlaced graphics.

JPG	JPEG compression (DCT, Quantization and Huffman)	Depends on the compression	Excellent compression technique which produces lossy compression. It normally results in much greater compression than the methods outlined above.

4.2 Comparison of the different methods

This section uses example bitmapped images and shows how much the different techniques manage to compress them. Figure 4.1 shows an image and Table 4.2 shows the resultant file size when it is saved in different formats. It can be seen that the BMP file format has the largest storage. The two main forms of BMP files are RGB (red, green, blue) encoded and RLE encoded. RGB coding saves the bit-map in an uncompressed form, whereas the RLE coding will reduce the total storage by compressing repetitive sequences. Next is the PCX file which has limited compression abilities (the format used in this case is version 5). The GIF format manages to compress the file to around 40% of its original size and the TIF file achieves similar compression (mainly because both techniques use LZH compression). It can be seen that by far the best compression is achieved with JPEG which in both forms has compressed the file to under 10% of its original size.

The reason that the compression ratios for GIF, TIF and BMP RLE are relatively high is that the image in Figure 4.1 contains a lot of changing data. Most images will compress to less than 10% because they have large areas which do not change much.

Figure 4.1 Sample graphics image

Table 4.2 Compression on a graphics file

Type	*Size(B)*	*Compression(%)*	
BMP	308 278	100.0	BMP, RBG encoded (640 × 480, 256 colors)
BMP	301 584	97.8	BMP, RLE encoded
PCX	274 050	88.9	PCX, Version 5
GIF	124 304	40.3	GIF, Version 89a, non-interlaced
GIF	127 849	41.5	GIF, Version 89a, interlaced
TIF	136 276	44.2	TIF, LZW compressed
TIF	81 106	26.3	TIF, CCITT Group 3, MONOCHROME
JPG	28 271	9.2	JPEG - JFIF Complaint (Standard coding)
JPG	26 511	8.6	JPEG - JFIF Complaint (Progressive coding)

Figure 4.2 shows a simple graphic of 500 × 500, 24-bit, which has large areas with identical colors. Table 4.3 shows that, in this case, the compression ratio is low. The RLE encoded BMP file is only 1% of the original as the graphic contains long runs of the same color. The GIF file has compressed to less than 1%. Note that the PCX, GIF and BMP RLE files have saved the image with only 256 colors. The JPG formats have the advantage that they have saved the image with the full 16.7M colors and give compression rates of around 2%.

Table 4.3 Compression on a graphics file with highly redundant data

Type	*Size (B)*	*Compression (%)*	
BMP	750 054	100.0	BMP, RBG encoded (500 × 500, 16.7M colors)
BMP	7 832	1.0	BMP, RLE encoded (256 colors)
PCX	31 983	4.3	PCX, Version 5 (256 colors)
GIF	4 585	0.6	GIF, Version 89a, non-interlaced (256 colors)
TIF	26 072	3.5	TIF, LZW compressed (16.7M colors)
JPG	15 800	2.1	JPEG (Standard coding, 16.7M colors)
JPG	12 600	1.7	JPEG (Progressive coding 16.7M colors)

4.3 GIF coding

The graphics interchange format (GIF) is the copyright of CompuServe Incorporated. Its popularity has increased mainly because of its wide usage on the Internet. CompuServe Incorporated, luckily, has granted a limited, non-exclusive, royalty-free license for the use of GIF (but any software using the GIF format must acknowledge the ownership of the GIF format).

Figure 4.2 Sample graphics image

Most graphics software supports the Version 87a or 89a format (the 89a format is an update the 87a format). Both have basic specification:

- A header with GIF identification.
- A logical screen descriptor block which defines the size, aspect ratio and color depth of the image place.
- A global color table.
- Data blocks with bitmapped images the possibility of text overlay.
- Multiple images, with image sequencing or interlacing. This process is defined in a graphic-rendering block.
- LZW compressed bitmapped images.

4.3.1 Color tables

Color tables store the color information of part of an image (a local color table) or they can be global (a global table).

4.3.2 Blocks, extensions and scope

Blocks can be specified into three groups: control, graphic-rendering and special purpose. Control blocks contain information used to control the process of the data stream or information used in setting hardware parameters. They include:

- GIF Header – which contains basic information on the GIF file, such as the version number and the GIF file signature.
- Logical screen descriptor – which contains information about the active screen display, such as screen width and height, and the aspect ratio.

- Global color table – which contains up to 256 colors from a palette of 16.7M colors (i.e. 256 colors with 24-bit color information).
- Data subblocks – which contain the compressed image data.
- Image description – which contains, possibly, a local color table and defines the image width and height, and its top left coordinate.
- Local color table – an optional block which contains local color information for an image as with the global color table, it has a maximum of 256 colors from a palette of 16.7M.
- Table-based image data – which contains compressed image data.
- Graphic control extension – an optional block which has extra graphic-rendering information, such as timing information and transparency.
- Comment extension – an optional block which contains comments ignored by the decoder.
- Plain text extension – an optional block which contains textual data.
- Application extension – which contains application-specific data. This block can be used by a software package to add extra information to the file.
- Trailer – which defines the end of a block of data.

4.3.3 GIF header

The header file is 6 bytes long and identifies the GIF signature and the version number of the chosen GIF specification. Its format is:

- 3 bytes with the characters 'G', 'I' and 'F'.
- 3 bytes with the version number (such as 87a or 89a). Version numbers are ordered with two digits for the year, followed by a letter ('a', 'b', and so on).

Program 4.1 is a C program for reading the 6-byte header. Sample run 4.1 shows a sample run with a GIF file. It can be seen that the file in the test run has the required signature and has been stored with Version 89a.

Program 4.1

```
#include    <stdio.h>

int   main(void)
{
FILE  *in;
char  fname[BUFSIZ], str[BUFSIZ];

   printf("Enter GIF file>>");
   gets(fname);

   if ((in=fopen(fname,"r"))==NULL)
   {
      printf("Can't find file %s\n",fname);
```

```
        return(1);
    }

    fread(str,3,1,in);
    str[3]=NULL; /* terminate string */
    printf("Signature: %s\n",str);
    fread(str,3,1,in);
    str[3]=NULL; /* terminate string */
    printf("Version: %s\n",str);

    fclose(in);
    return(0);
}
```

Sample run 4.1

```
Enter GIF file>> clouds.gif
Signature: GIF
Version: 89a
```

4.3.4 Logical screen descriptor

The logical screen descriptor appears after the header. Its format is:

- 2 bytes with the logical screen width (unsigned integer).
- 2 bytes with the logical screen height (unsigned integer).
- 1 byte of a packed bit field, with 1 bit for global color table flag, 3 bits for color resolution, 1 bit for sort flag and 3 bits to give an indication of the number of colors in the global color table
- 1 byte for the background color index.
- 1 byte for the pixel aspect ratio.

Program 4.2 is a C program which reads the header and the logical descriptor field, and Sample run 4.2 shows a sample run. It can be seen, in this case, that the logic screen size is 640×480. The packed field, in this case, has a hexadecimal value of F7h, which is 1111 0111b in binary. Thus all the bits of the packed bit field are set, apart from the sort flag. If this is set then the global color table is sorted in order of decreasing importance (the most frequent color appearing first and the least frequent color last). The total number of colors in the global color table is found by raising 2 to the power of 1+the color value in the packed bit field:

$$\text{Number of colors} = 2^{\text{Color value in packed bit field}+1}$$

In this case, there is a bit field of seven colors, thus the total number of colors is 2^8, or 256.

It can be seen that the aspect ratio in Sample run 4.2 is zero. If it is zero then no aspect ratio is given. If it is not equal to zero then the aspect ratio of the image is computed by:

$$\text{Aspect ratio} = \frac{\text{Pixel aspect ratio} + 15}{64}$$

where the pixel ratio is the pixel's width divided by its height.

Program 4.2

```
#include    <stdio.h>

int   main(void)
{
FILE  *in;
char  fname[BUFSIZ], str[BUFSIZ];
int   x,y;
char  color_index, aspect, packed;

   printf("Enter GIF file>>");
   gets(fname);

   if ((in=fopen(fname,"r"))==NULL)
   {
      printf("Can't find file %s\n",fname);
      return(1);
   }

   fread(str,3,1,in);    str[3]=NULL; /* terminate string */
   printf("Signature: %s\n",str);
   fread(str,3,1,in);    str[3]=NULL; /* terminate string */
   printf("Version: %s\n",str);

   fread(&x,2,1,in); str[3]=NULL; /* terminate string */
   printf("Screen width: %d\n",x);
   fread(&y,2,1,in); str[3]=NULL; /* terminate string */
   printf("Screen height: %d\n",y);

   fread(&packed,1,1,in);
   printf("Packed: %x\n",packed & 0xff); /* mask-off the bottom 8 bits */
   fread(&color_index,1,1,in);
   printf("Color index: %d\n",color_index);
   fread(&aspect,1,1,in);
   printf("Aspect ratio: %d\n",aspect);

   fclose(in);
   return(0);
}
```

Sample run 4.2

```
Enter GIF file>> clouds.gif
Signature: GIF
Version: 89a
Screen width: 640
Screen height: 480
Packed: f7
Color index: 0
Aspect ratio: 0
```

4.3.5 Global color table

After the header and the logical display descriptor comes the global color table. It contains up to 256 colors from a palette of 16.7M colors. Each of the colors is defined as a 24-bit color of red (8 bits), green (8 bits) and blue (8 bits). The format in memory is:

```
RRRRRRRR
GGGGGGGG
BBBBBBBB
RRRRRRRR
GGGGGGGG
BBBBBBBB
 :     :
RRRRRRRR
GGGGGGGG
BBBBBBBB
```

Thus the number of bytes that the table will contain will be:

$$\text{Number of bytes} = 3 \times 2^{\text{Size of global color table}+1}$$

The 24-bit color scheme allows a total of 16777216 (2^{24}) different colors to be displayed. Table 4.4 defines some colors in the RGB (red/green/blue) strength. The format is `rrggbbh`, where `rr` is the hexadecimal equivalent for the red component, `gg` the hexadecimal equivalent for the green component and `bb` the hexadecimal equivalent for the blue component. For example, in binary:

```
000000000000000000000000  represents black (000000h)
111111111111111111111111  represents white (FFFFFFh)
011101110111011101110111  represents gray (777777h)
111111001110010100000011  represents yellow (FCE503h)
001110100000101101011001  represents purple (3A0B59h)
```

Table 4.4 Hexadecimal colors for 24-bit color representation

Color	*Code*	*Color*	*Code*
White	FFFFFFh	Dark red	C91F16h
Light red	DC640Dh	Orange	F1A60Ah
Yellow	FCE503h	Light green	BED20Fh
Dark green	088343h	Light blue	009DBEh
Dark blue	0D3981h	Purple	3A0B59h
Pink	F3D7E3h	Nearly black	434343h
Dark gray	777777h	Gray	A7A7A7h
Light gray	D4D4D4h	Black	000000h

Program 4.3 is a C program which reads the header, the image descriptor and the color table. Sample run 4.3 shows a truncated color table. The first three are:

0111 1011 1010 1101 1101 0110 (7BADD6h)
1000 0100 1011 0101 1101 1110 (84B5DEh)
0111 0011 1010 1101 1101 0110 (73ADD6h)

These colors have a strong blue component (D6h and DEh) and reduced strength red and green components. The image itself is a picture of clouds on a blue sky, thus the image is likely to have strong blue colors.

Program 4.3

```
#include    <stdio.h>

int   main(void)
{
FILE  *in;
char  fname[BUFSIZ], str[BUFSIZ];
int   x,y,i;
char  color_index, aspect, packed,red,blue,green;

   printf("Enter GIF file>>");
   gets(fname);

   if ((in=fopen(fname,"r"))==NULL)
   {
      printf("Can't find file %s\n",fname);
      return(1);
   }

   fread(str,3,1,in);   str[3]=NULL; /* terminate string */
   printf("Signature: %s\n",str);
   fread(str,3,1,in);   str[3]=NULL; /* terminate string */
   printf("Version: %s\n",str);

   fread(&x,2,1,in); str[3]=NULL; /* terminate string */
   printf("Screen width: %d\n",x);
   fread(&y,2,1,in); str[3]=NULL; /* terminate string */
   printf("Screen height: %d\n",y);

   fread(&packed,1,1,in);
   printf("Packed: %x\n",packed & 0xff); /* mask-off the bottom 8 bits */
   fread(&color_index,1,1,in);
   printf("Color index: %d\n",color_index);
   fread(&aspect,1,1,in);
   printf("Aspect ratio: %d\n",aspect);

   for (i=0;i<64;i++)
   {
      fread(&red,1,1,in);
      printf("Red: %x ",red & 0xff);      /* display 8 bits */
      fread(&green,1,1,in);
      printf("Green: %x ",green & 0xff); /* display 8 bits */
      fread(&blue,1,1,in);
      printf("Blue: %x\n",blue & 0xff);   /* display 8 bits */
   }

   fclose(in);

   return(0);
}
```

Sample run 4.3

```
Enter GIF file>> clouds.gif
Signature: GIF
Version: 89a
Screen width: 640
Screen height: 480
Packed: f7
Color index: 0
Aspect ratio: 0
Red: 7b Green: ad Blue: d6
Red: 84 Green: b5 Blue: de
Red: 73 Green: ad Blue: d6
Red: 7b Green: ad Blue: de
Red: 94 Green: bd Blue: de
Red: 7b Green: b5 Blue: de
Red: 8c Green: b5 Blue: de
Red: 8c Green: bd Blue: de
Red: 9c Green: c6 Blue: de
Red: ce Green: de Blue: ef
Red: de Green: e7 Blue: ef
Red: a5 Green: c6 Blue: e7
   ::::::
Red: 8c Green: bd Blue: e7
Red: ff Green: ff Blue: f7
Red: ad Green: d6 Blue: ef
Red: 8c Green: b5 Blue: e7
Red: 84 Green: b5 Blue: e7
```

4.3.6 Image descriptor

After the global color table is the image descriptor. Its format is:

- 1 byte for the image separator (always 2Ch).
- 2 bytes for the image left position (unsigned integer).
- 2 bytes for the image top position (unsigned integer).
- 2 bytes for the image width (unsigned integer).
- 2 bytes for the image height (unsigned integer).
- 1 byte of a packed bit field, with 1 bit for local color table flag, 1 bit for interlace flag, 1 bit for sort flag, 2 bits are reserved and 3 bits for the size of the local color table.

Program 4.4 is a C program which searches for the image separator (2Ch) and displays the image descriptor data that follows. Sample run 4.4 shows a sample run. It can be seen from this sample run that the image is to be displayed at (0,0), its width is 640 pixels and its height is 480 pixels. The packed bit field contains all zeros, thus there is no local color table (and the global color table should be used).

Program 4.4

```
#include    <stdio.h>

int   main(void)
{
FILE  *in;
```

```
char   fname[BUFSIZ];
int    i,left,top,width,height;
char   ch,packed;

   printf("Enter GIF file>>");
   gets(fname);

   if ((in=fopen(fname,"r"))==NULL)
   {
      printf("Can't find file %s\n",fname);
      return(1);
   }

   do
   {
      fread(&ch,1,1,in);
   } while (ch!=0x2C); /* find image seperator */

   fread(&left,2,1,in);
   printf("Image left position: %d\n",left);
   fread(&top,2,1,in);
   printf("Image top position: %d\n",top);
   fread(&width,2,1,in);
   printf("Image width: %d\n",width);
   fread(&height,2,1,in);
   printf("Image height: %d\n",height);
   fread(&packed,1,1,in);
   printf("Packed: %x\n",packed & 0xff);
   fclose(in);
   return(0);
}
```

Sample run 4.4

```
Enter GIF file>> clouds.gif
Image left position: 0
Image top position: 0
Image width: 640
Image height: 480
Packed: 0
```

4.3.7 Local color table

The local color table is an optional block which defines the color map for the image that precedes it. The format is identical to the global color map, i.e. 3 bytes for each of the colors.

4.3.8 Table-based image data

The table-based image data follows the local color table. This table contains compressed image data. It consists of a series of subblocks of up to 255 bytes. The data consists of an index to the color table (either global or local) for each pixel in the image. As the global (or local) color table has 256 entries, the data value (in its uncompressed form) will range from 0 to 255 (8 bits). The tables format is:

- 1 byte for the LZW minimum code size, which is the initial number of bits used in the LZW coding.

- *N* bytes for the LZW compressed image data. The first block is preceded by the data size.

To recap from Chapter 3, GIF coding uses the variable-length-code LZW technique where a variable-length code replaces image data (pixel color references). These variable-length codes are specified in a Huffman code table. The encoder replaces the data from the input and builds a dictionary with the patterns in the data. Every new pattern is entered into the dictionary and the index value of the table is added to coded data. When a previously stored pattern is encountered, its dictionary index value is added to the coded data. The decoder takes the compressed data and builds the dictionary which is identical to the encoder. It then replaces indexed terms from the dictionary.

The VLC algorithm uses an initial code size to specify the initial number of bits used for the compression codes. When the number of patterns detected by the encoder exceeds the number of patterns encodable with the current number of bits then the number of bits per LZW is increased by 1.

Program 4.5 reads the LZW code size byte. The byte after this is the block size, followed by the number of bytes of data as defined in the block size byte. Sample run 4.5 gives a sample run. It can be seen that the initial LZW code size is 8 and that the block size of the first block is 254 bytes. The dictionary entries will thus start at entry 256 (2^{8+1}).

Program 4.5

```
#include    <stdio.h>

int    main(void)
{
FILE   *in;
char   fname[BUFSIZ];
int    i,left,top,width,height;
char   ch,packed,code,block;

   printf("Enter GIF file>>");
   gets(fname);

   if ((in=fopen(fname,"r"))==NULL)
   {
      printf("Can't find file %s\n",fname);
      return(1);
   }

   do
   {
      fread(&ch,1,1,in);
   } while (ch!=0x2C);
   fread(&left,2,1,in);
   printf("Image left position: %d\n",left);
   fread(&top,2,1,in);
   printf("Image top position: %d\n",top);
   fread(&width,2,1,in);
   printf("Image width: %d\n",width);
   fread(&height,2,1,in);
   printf("Image height: %d\n",height);
```

```
    fread(&packed,1,1,in);
    printf("Packed: %x\n",packed & 0xff);
    fread(&code,1,1,in);
    printf("LZW code size: %d\n",code & 0xff);
    fread(&block,1,1,in);
    printf("Block size: %d\n",block & 0xff);
    fclose(in);
    return(0);
}
```

Sample run 4.5

```
Enter GIF file>> clouds.gif
Image left position: 0
Image top position: 0
Image width: 640
Image height: 480
Packed: 0
LZW code size: 8
Block size: 254
```

4.3.9 Graphic control extension

The graphic control extension is optional and contains information on the rendering of the image that follows. Its format is:

- 1 byte with the extension identifier (21h).
- 1 byte with the graphic control label (F9h).
- 1 byte with the block size following this field and up to but not including, the end terminator. It always has a fixed value of 4.
- 1 byte with a packed array of which the first 3 bits are reserved, 3 bits define the disposal method, 1 bit defines the user input flag and 1 bit defines the transparent color flag.
- 2 bytes with the delay time for the encode wait, in hundreds of a seconds, before encoding the image data.
- 1 byte with the transparent color index.
- 1 byte for the block terminator (00h).

4.3.10 Comment extension

The comment extension is optional and contains information which is ignored by the encoder. Its format is:

- 1 byte with the extension identifier (21h).
- 1 byte with the comment extension label (FEh).
- *N* bytes, with comment data.
- 1 byte for the block terminator (00h).

4.3.11 Plain text extension

The plain text extension is optional and contains text information. Its format is:

- 1 byte with the extension identifier (21h).
- 1 byte with the plain text label (01h).
- 1 byte with the block size. This is the number of bytes after the block size field up to but not including the beginning of the plain text data block. It always contains the value 12.
- 2 bytes for the text grid left position.
- 2 bytes for the text grid top position.
- 2 bytes for the text width.
- 2 bytes for the text height.
- 1 byte for the character cell width.
- 1 byte for the character cell height.
- 1 byte for the text foreground color.
- 1 byte for the text background color.
- *N* bytes for the plain text data.
- 1 byte for the block terminator (00h).

4.3.12 Application extension

The application extension is optional and contains information for application programs. Its format is:

- 1 byte with the extension identifier (21h).
- 1 byte with the application extension label (FFh).
- 1 byte for the block size. This is the number of bytes after the block size field up to but not including the beginning of the application data. It always contains the value 11.
- 8 bytes for the application identifier.
- 3 bytes for the application authentication code.
- *N* bytes, for the application data.
- 1 byte for the block terminator (00h).

4.3.13 Trailer

The trailer indicates the end of the GIF file. Its format is:

- 1 byte identifying the trailer (3Bh).

4.4 TIFF coding

Tag image file format (TIFF) is an excellent method of transporting images between file systems and software packages. It is supported by most graphics import packages. It has a high resolution and thus is typically used when scanning images. There are two main types of TIFF coding, baseline TIFF and extended TIFF. It can also use different compression methods and different file formats, depending on the type of data stored.

In TIFF 6.0, defined in June 1992, the pixel data can be stored in several different compression formats, such as:

- Code number 1, no compression.
- Code number 2, CCITT Group 3 modified Huffman RLE encoding (see Section 33.9.1).
- Code number 3, Fax-compatible CCITT Group 3 (see Section 33.9).
- Code number 4, Fax-compatible CCITT Group 4 (see Section 33.9).
- Code number 5, LZW compression (see Chapter 3).

Codes 1 and 2 are baseline TIFF files whereas the others are extended.

4.4.1 File structure

TIFF files have a three-level hierarchy:

- A file header.
- One or more IFDs (image file directories). These contain codes and their data (or pointers to the data).
- Data.

The file header contains 8 bytes: a byte order field (2 bytes), the version number field (2 bytes) and the pointer to the first IFD (4 bytes). Figure 4.3 shows the file header format. The byte order field defines whether Motorola architecture is used (the character sequence is 'MM', or 4D4Dh) or Intel architecture (the character sequence is 'II', or 4949h). The Motorola format defines that the bytes are ordered from the most significant to the least significant, the Intel format defines that the bytes are organized from least significant to the most significant.

The version number field always contains the decimal number 42 (maybe related to Douglas Adam's *Hitchhikers Guide to the Galaxy*, where 42 is described as the answer to the life, the universe and everything). It is used to identify that the file is TIFF format.

The first IFD offset pointer is a 4-byte pointer to the first IFD. If the format is Intel then the bytes are arranged from least significant to most significant else they are arranged from most significant to least significant.

Program 4.6 is a C program which reads the header of a TIFF file and Sample run 4.6 shows that, in this case, it uses the Intel format and the second byte field contains 2Ah (or 42 decimal).

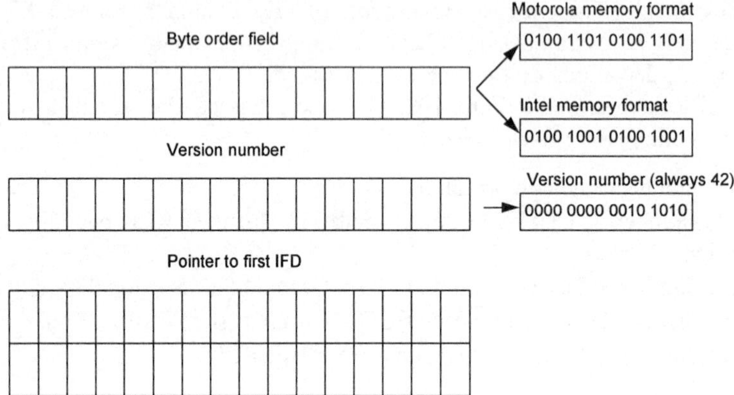

Figure 4.3 TIFF file header

Program 4.6

```
#include    <stdio.h>

int   main(void)
{
FILE  *in;
char  ch1,ch2,fname[BUFSIZ];

   printf("Enter TIFF file>>");
   gets(fname);

   if ((in=fopen(fname,"r"))==NULL)
   {
      printf("Can't find file %s\n",fname);
      return(1);
   }

   ch1=fgetc(in); ch2=getc(in);
   printf("Memory model %c%c\n",ch1,ch2);
   ch1=fgetc(in); ch2=getc(in);
   printf("Version %x%x\n",ch2,ch1);

   fclose(in);
   return(0);
}
```

Sample run 4.6

```
Enter TIFF file>> image1.tif
Memory model II
Version 02a
```

4.4.2 IFD

Typically the first IFD will be the only IFD, which is pointed to by the first IFD in the header field.

4.4.3 Compression code 2: Huffman RLE coding

TIFF compression code 2 uses the CCITT Group 3 type compression which is a modified Huffman coding and is used in many fax transmissions. It specifies a 1-bit monochrome code with alternate black and white sequences of pixels. Tables 4.5 and 4.6 give the predefined coding table for white and black sequence runs. These tables contain codes in which the most frequent run lengths are coded with a short code. The compressed code always starts on white code. Codes themselves range from 0 to 63. Values from 64 to 2560 use two codes. The first gives the multiple of 64 followed by the normally coded remainder. There is no special end-of-line identifier because the size of the image is known by the defined `ImageWidth` tag field. There are thus `ImageWidth` pixels on a line.

Table 4.5 White run-length coding

Run length	*Coding*	*Run length*	*Coding*	*Run length*	*Coding*	*Run length*	*Coding*
0	00110101	1	000111	2	0111	3	1000
4	1011	5	1100	6	1110	7	1111
8	10011	9	10100	10	00111	11	01000
12	001000	13	000011	14	110100	15	110101
16	101010	17	101011	18	0100111	19	0001100
61	00110010	62	00110011	63	00110100	64	110011

Table 4.6 Black run-length coding

Run length	*Coding*	*Run length*	*Coding*	*Run length*	*Coding*	*Run length*	*Coding*
0	0000110111	1	010	2	11	3	10
4	011	5	0011	6	0010	7	00011
8	000101	9	000100	10	0000100	11	0000101
12	0000111	13	00000100	14	00000111	15	000011000
16	0000010111	17	0000011000	18	0000001000	19	00001100111
61	000001011010	62	0000001100110	63	000001100111	64	0000001111

For example, if the data were:

16 white 4 black 16 white 2 black 63 white 10 black 63 white

it would be coded as:

```
101010 011  101010 11 00110100 000010000110100
```

This would take 40 bits to code, whereas it would take 304 bits if coded with pixel colors (i.e., 16+4+16+2+128+10+128). This results in a compression ratio of 7.6:1.

4.4.4 Compression code 5: LZW compression

The compression technique used by TIFF code 5 is the same as is used in GIF files, but has a fixed code size of 8 (refer to Chapter 2 for more information on LZW code sizes). The dictionary starts with the values 0 to 255 stored in the entries 0 to 255. There are two codes for Clear (at 256) and EndOfInformation (at 257) and the dictionary is then built up from 258 to 4095. The Clear code is a special code which resets the dictionary entries to the original entries from 0 to 255.

A basic encoding algorithm could be:

```
Byte: byte;
Buffer, Test, String: string;
Table: array[1..4096] of string;

begin
   clear Table;  clear Buffer; clear Test; clear String;

   write ClearCode code;

   while (valid data)
   begin
      read Byte;
      Test=String+Byte;
      if (Test in Table) then String=String+Byte;
      else
      begin
         write String code;
         add Test to Table;
         String=Byte;
      end;
   end;

   write String code;
   write EndOfInformation code.
end.
```

4.5 GIF interlaced images

GIF images can be stored in an interlaced manner. This facility is useful when receiving information over a relatively slow transmission line, as it allows an outline of an image to be displayed before all of the image has been encoded (or received). The images stored are:

Group 1: Starting at row 0, every 8th row.
Group 2: Starting at row 4, every 8th row.
Group 3: Starting at row 2, every 4th row.
Group 4: Starting at row 1, every 2nd row.

For example if the image has 16 rows (0–15) then the following would be stored:

	Scanned line displayed			
	1	2	3	4
Row 0	X			
Row 1				X
Row 2			X	
Row 3				X
Row 4		X		
Row 5				X
Row 6			X	
Row 7				X
Row 8	X			
Row 9				X
Row 10			X	
Row 11				X
Row 12		X		
Row 13				X
Row 14			X	
Row 15				X

It can be seen that the first 1/8 of the data will display an outline of the image. The next 1/8 will improve the quality. After this the next 1/4 will improve the quality further and then the final 1/2 will give the completed image.

4.6 PCX coding

PCX is a well-supported graphics file format. It uses RLE coding to compress the data. As it uses RLE coding the amount of compression normally depends on the amount of data which is repeated. The less the image changes, the more compression can be achieved. This can be seen from Tables 4.2 and 4.3. The image in Figure 4.1 has a lot of changes in color, whereas Figure 4.2 has very few changes. Table 4.2 shows that the compression for the image in figure is only 88.9% of the original file size, while Table 4.3 shows that the compression for the image in Figure 4.2 is 4.3%.

PCX Version 0 only allows for a basic 2-color or 4-color image, Version 2 supports 16-color images and Version 5 supports 256 colors from 24-bit palettes.

4.6.1 File structure

A PCX file consists of three main parts:

- A file header.
- Bitmapped data.
- A color palette of 256 colors.

4.6.2 Header file

The header always contains 256 bytes and is defined in Figure 4.4. Program 4.7 outlines a C program which can be used to read the header. It can be seen from Sample run 4.2 shows that a sample file IMAGE1.PCX has the flag set to 0Ah (it is thus a PCX file), the version number is set to 5, the encoding field is set to 1, the image size ranges from (0,0) to (639,479), the bits per pixel is 8 and the printable dots per inch in the x- and y-direction is 150. In this case the number of displayable colors will be 256 as there are 8 bits per pixel. Also the screen dimensions are set for 640×480 (to fit a standard VGA screen on a PC).

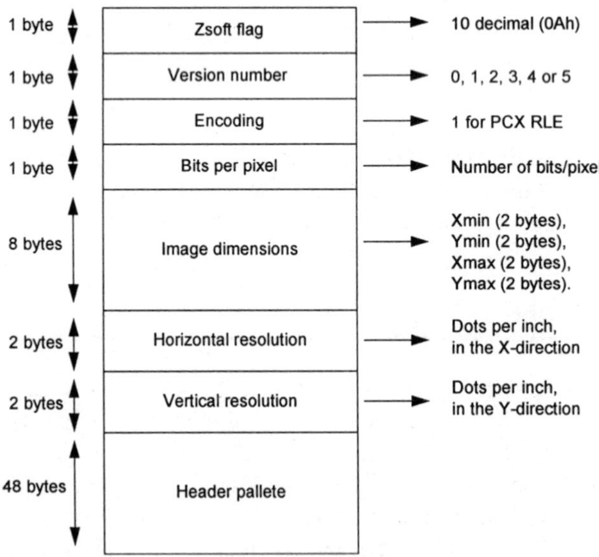

Figure 4.4 PCX file header

Program 4.7

```
#include    <stdio.h>

int   main(void)
{
FILE  *in;
char  flag,version,coding,bits,fname[BUFSIZ];
int   xmin,ymin,xmax,ymax, xres, yres;


   printf("Enter PCX file>>");
   gets(fname);

   if ((in=fopen(fname,"r"))==NULL)
   {
      printf("Can't find file %s\n",fname);
      return(1);
   }

   fread(&flag,1,1,in);    /* read 1 byte from file and put into flag */
   fread(&version,1,1,in); /* read 1 byte from file and put in version */

   fread(&coding,1,1,in);  /* read 1 byte from file and put into coding */
   fread(&bits,1,1,in); /* read 1 byte from file and put into bits */

   fread(&xmin,2,1,in); /* read 2 bytes from file and put into xmin */
   fread(&ymin,2,1,in); /* read 2 bytes from file and put into ymin */

   fread(&xmax,2,1,in); /* read 2 bytes from file and put into xmax */
   fread(&ymax,2,1,in); /* read 2 bytes from file and put into ymax */

   printf("Flag %X\n",flag);
   printf("Version %X\n",version);

   printf("Coding %X\n",coding);

   printf("Bits per pixel %X (%d decimal)\n",bits,bits);
   printf("Min (%d,%d) Max (%d,%d)\n",xmin,ymin,xmax,ymax);

   fread(&xres,2,1,in); /* read 2 bytes from file and put into xres */
   fread(&yres,2,1,in); /* read 2 bytes from file and put into yres */
   printf("Resolution (%d,%d)\n",xres,yres);

   fclose(in);
   return(0);
}
```

Sample run 4.7

```
Enter PCX file>> image1.pcx
Flag A
Version 5
Coding 1
Bits per pixel 8 (8 decimal)
Min (0,0) Max (639,479)
Resolution (150,150)
```

4.6.3 Bitmapped data

If the file does not use the a palette then the data contains pixel colors (if a palette is used then the data relates to pointers for the color palette).

4.7 Exercises

4.1 Explain why the BMP RLE encoding in Figure 4.2 (Table 4.3) gave much greater compression than Figure 4.1 (Table 4.2).

4.2 The image in Figure 4.1 is 640×480 with 256. Show that the bit-mapped image data is approximately equal to the actual file size (i.e. the file size is 308 278 B). From this show that the extra overhead for addition information (such as the BMP header and the color table) is 1078 B.

4.3 Repeat Question 4.2 for the image in Figure 4.2.

4.4 Investigate paint packages and determine the types of files that can be exported. If possible, generate an image and determine the different file sizes for different file formats

4.5 Explain why interlaced GIF images are useful when downloading images over the Internet.

4.6 Find a GIF image and verify the following:

(a) the first few characters contain the GIF file signature (such as GIF79a)
(b) its logical screen descriptor
(c) its global color table

4.7 Explain how 24-bit color information can generate colors. Estimate the color of the following 24-bit colors:

00000000 011111111 011111111
10000000 011111111 011111111
11111111 111111111 000000000
01000000 010000000 010000000

4.8 State the main advantages of TIFF files over the GIF format.

4.9 Explain how TIFF compression code 2 compresses black and white images.

4.10 Explain how TIFF compression code 5 compresses images.

5 Image Compression (JPEG)

5.1 Introduction

JPEG is an excellent compression technique which produces lossy compression (although in one mode it is lossless). As seen from the previous chapter it has excellent compression ratio when applied to a color image. This chapter introduces the JPEG standard and the method used to compress an image. It also discusses the JFIF file standard which defines the file format for JPEG encoded images. Along with GIF files, JPEG is now one of the most widely used standards for image compression.

5.2 JPEG coding

A typical standard for image compression has been devised by the Joint Photographic Expert Group (JPEG), a subcommittee of the ISO/IEC, and the standards produced can be summarized as follows:

It is a compression technique for gray-scale or color images and uses a combination of discrete cosine transform, quantization, run-length and Huffman coding.

It has resulted from research into compression ratios and the resultant image quality. The main steps are:

- Data blocks — Generation of data blocks
- Source-encoding — Discrete cosine transform and quantization
- Entropy-encoding — Run-length encoding and Huffman encoding

Unfortunately, compared with GIF, TIFF and PCX, the compression process is relatively slow. It is also lossy in that some information is lost in the compression process. This information is perceived to have little effect on the decoded image.

GIF files typically take 24-bit color information (8 bits for red, 8 bits for green and 8 bits for blue) and convert it into an 8-bit color palette (thus reducing the number of bits stored to approximately one-third of the original). It then uses LZW compression to further reduce the storage. JPEG operates differently in that it stores changes in color. As the eye is very sensitive to brightness changes it is particularly sensitive to changes in brightness. If these changes are similar to the original then the eye will perceive the recovered image as very similar to the original.

5.2.1 Color conversion and subsampling

In the first part of the JPEG compression, each color component (red, green and blue) is separated in luminance (brightness) and chrominance (color information) separately. JPEG allows more losses on the chrominance and less on the luminance. This is because the human eye is less sensitive to color changes than to brightness changes. In an RGB image, all three channels carry some brightness information but the green component has a stronger effect on brightness than the blue component.

A typical scheme for converting RGB into luminance and color is known as CCIR 601, which coverts the components in Y (can be equated to brightness), C_b (blueness) and C_r (redness). The Y component can be used as a black and white version of the image. The components are computed from the RGB components:

$$Y=0.299R+0.587G+0.114B$$
$$C_b=0.1687\mathrm{R}-0.3313G+0.5B$$
$$C_r=0.5R-0.4187G+0.0813B$$

For the brightness it can be seen that green has the most effect and blue has the least effect. For the redness, the red color (of course) has the most effect and green the least. For the blueness, the blue color has the most effect and green the least. Note that the YC_bC_r components are often known as YUV.

A subsampling process is then conducted which samples the C_b and C_r components at a lower rate than the Y component. A typical sampling rate is 4 samples of the Y component to a single sample on the C_b and C_r component. This sampling rate is normally set with the compression parameters the lower the sampling, the smaller the compressed data and the shorter the compression time. The JPEG header contains all the information necessary to properly decode the JPEG data.

5.2.2 DCT coding

The DCT (discrete cosine transform) converts intensity data into frequency data, which can be used to tell how fast the intensities vary. In JPEG coding the image is segmented into 8×8 pixel rectangles, as illustrated in Figure 5.1. If the image contains several components (such as Y,C_b,C_r or R,G,B), then each of the components in the pixel blocks is operated on separately. If an image is subsampled, there will be more blocks of some components than of others. For example, for 2×2 sampling there will be four blocks of Y data for each block of C_b or C_r data.

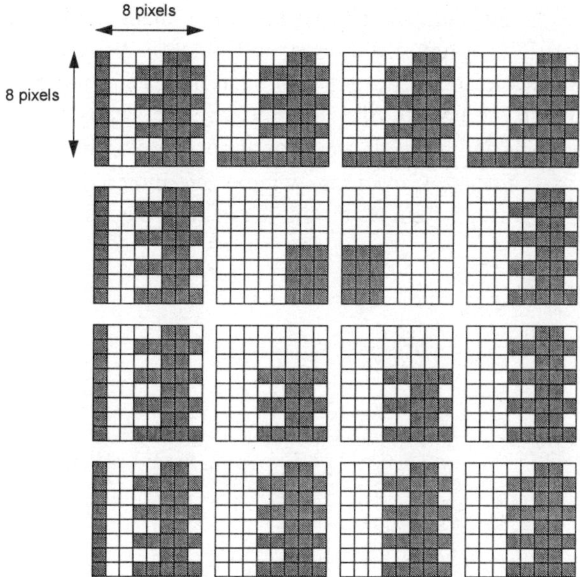

Figure 5.1 Segment of an image in 8x8 pixel blocks

The data points in the 8×8 pixel array starts at the upper right at $(0,0)$ and finish at the lower right at $(7,7)$. At the point (x,y) the data value is $f(x,y)$. The DCT produces a new 8×8 block $(u\times v)$ of transformed data using the formula:

$$F(u,v)=\frac{1}{4}C(u)C(v)\left[\sum_{x=0}^{7}\sum_{y=0}^{7}f(x,y)\cos\frac{(2x+1)u\pi}{16}\cos\frac{(2y+1)v\pi}{16}\right]$$

$$\text{where}\quad C(z)=\frac{1}{\sqrt{2}}\quad \text{if } z=0$$

$$\text{or}\qquad =1\quad \text{if } z\neq 0$$

This results in an array of space frequency $F(u,v)$ which gives the rate of change at a given point. These are normally 12-bit values which give a range of 0 to 1024. Each component specifies the degree to which the image changes over the sampled block. For example:

- F(0,0) gives the average value of the 8×8 array.

- $F(1,0)$ gives the degree to which the values change slowly (low frequency).
- $F(7,7)$ gives indicates the degree to which the values change most quickly in both directions (high frequency).

The coefficients are equivalent to representing changes of frequency within the data block. The value in the upper left block (0,0) is the DC or average value. The values to the right of a row have increasing horizontal frequency and the values to the bottom of a column have increasing vertical frequency. Many of the bands end up having zero or almost zero terms.

Program 5.1 gives a C program which determines the DCT of an 8×8 block and Sample run 5.1 shows a sample run with the resultant coefficients.

Program 5.1

```
#include <stdio.h>
#include <math.h>
#define  PI 3.1415926535897

int   main(void)
{
int   x,y,u,v;
float in[8][8]=   {{144,139,149,155,153,155,155,155},
            {151,151,151,159,156,156,156,158},
            {151,156,160,162,159,151,151,151},
            {158,163,161,160,160,160,160,161},
            {158,160,161,162,160,155,155,156},
            {161,161,161,161,160,157,157,157},
            {162,162,161,160,161,157,157,157},
            {162,162,161,160,163,157,158,154}};
float out[8][8],sum,Cu,Cv;

   for (u=0;u<8;u++)
   {
      for (v=0;v<8;v++)
      {
         sum=0;
         for (x=0;x<8;x++)
            for (y=0;y<8;y++)
            {
               sum=sum+in[x][y]*cos(((2.0*x+1)*u*PI)/16.0)*
                  cos(((2.0*y+1)*v*PI)/16.0);
            }
         if (u==0) Cu=1/sqrt(2); else Cu=1;
         if (v==0) Cv=1/sqrt(2); else Cv=1;

         out[u][v]=1/4.0*Cu*Cv*sum;
         printf("%8.1f ",out[u][v]);
      }
      printf("\n");
   }
   printf("\n");
   return(0);
}
```

The program uses a fixed 8×8 block of:

```
144  139  149  155  153  155  155  155
151  151  151  159  156  156  156  158
151  156  160  162  159  151  151  151
158  163  161  160  160  160  160  161
158  160  161  162  160  155  155  156
161  161  161  161  160  157  157  157
162  162  161  160  161  157  157  157
162  162  161  160  163  157  158  154
```

Sample run 5.1

```
1257.9     2.3    -9.7    -4.1     3.9     0.6    -2.1     0.7
 -21.0   -15.3    -4.3    -2.7     2.3     3.5     2.1    -3.1
 -11.2    -7.6    -0.9     4.1     2.0     3.4     1.4     0.9
  -4.9    -5.8     1.8     1.1     1.6     2.7     2.8    -0.7
   0.1    -3.8     0.5     1.3    -1.4     0.7     1.0     0.9
   0.9    -1.6     0.9    -0.3    -1.8    -0.3     1.4     0.8
  -4.4     2.7    -4.4    -1.5    -0.1     1.1     0.4     1.9
  -6.4     3.8    -5.0    -2.6     1.6     0.6     0.1     1.5
```

Notice that the values of the most significant values are in the top left-hand corner and that many terms are near to zero. It is this property which allows many values to become zeros when quantized. These zeros can then be compressed using run-length coding and Huffman codes.

5.2.3 Quantization

The next stage of the JPEG compression is quantization where bias is given to lower-frequency components. JPEG divides each of the DCT values by a quantization factor, which is then rounded to the nearest integer. As the DCT factors are 8×8 then a table of 8×8 of quantization factors are used, corresponding to each term of the DCT output. The JPEG file then stores this table so that the decoding process may use this table or a standard quantization table. Note that files with multiple components must have multiple tables, such as one each for the Y, C_b and C_r components.

For example the values of the quantized high-frequency term (such as $F(7,7)$) could have a term of around 100, while the low-frequency term could have a factor of 16. These values define the accuracy of the final value. When decoding, the original values are (approximately) recovered through multiplying by the quantization factor.

Figure 5.2 shows that, for a factor of 100, the values between 50 and 150 would be quantized as a 1, thus the maximum error would be ±50. The maximum error for the factor of 16 is ±8. Thus the maximum error of the final unquantized value for a scale factor of 100 is 1.22% (5000/4096), while a factor of 16 gives a maximum error of 0.20% (800/4096). So, using the factors of 100 for $F(7,7)$ and 16 for $F(0,0)$, and a 12-bit DCT, the $F(0,0)$ term would range from 0 to 256 and the $F(7,7)$ term would range from 0 to 41. The $F(0,0)$ term could be coded with 8 bits (0 to 255) and the $F(7,7)$ term with 6 bits (0 to 63).

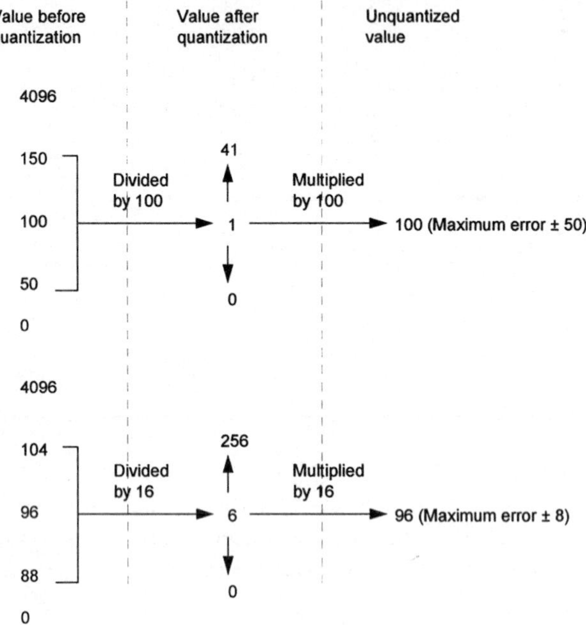

Figure 5.2 Example of quantization

Program 5.2 normalizes and quantizes (to the nearest integer) the example given previously. To recap, the input 8×8 block is:

```
144  139  149  155  153  155  155  155
151  151  151  159  156  156  156  158
151  156  160  162  159  151  151  151
158  163  161  160  160  160  160  161
158  160  161  162  160  155  155  156
161  161  161  161  160  157  157  157
162  162  161  160  161  157  157  157
162  162  161  160  163  157  158  154
```

The applied normalization matrix is:

```
5    3    4    4    4    3    5    4
4    4    5    5    5    6    7    12
8    7    7    7    7    15   11   11
9    12   13   15   18   18   17   15
20   20   20   20   20   20   20   20
20   20   20   20   20   20   20   20
20   20   20   20   20   20   20   20
20   20   20   20   20   20   20   20
```

Program 5.2

```
#include <stdio.h>
#include <math.h>
#define  PI 3.1415926535897

int   main(void)
{
int   x,y,u,v;
float in[8][8]=   {
      {144,139,149,155,153,155,155,155},
      {151,151,151,159,156,156,156,158},
      {151,156,160,162,159,151,151,151},
      {158,163,161,160,160,160,160,161},
      {158,160,161,162,160,155,155,156},
      {161,161,161,161,160,157,157,157},
      {162,162,161,160,161,157,157,157},
      {162,162,161,160,163,157,158,154}};

float norm[8][8]= {
   {5,3,4,4,4,3,5,4},
   {4,4,5,5,5,6,7,12},
   {8,7,7,7,7,15,11,11},
   {9,12,13,15,18,18,17,15},
   {20,20,20,20,20,20,20,20},
   {20,20,20,20,20,20,20,20},
   {20,20,20,20,20,20,20,20},
   {20,20,20,20,20,20,20,20}};

int   out[8][8];
float sum,Cu,Cv;

   for (u=0;u<8;u++)
   {
      for (v=0;v<8;v++)
      {
         sum=0;
         for (x=0;x<8;x++)
            for (y=0;y<8;y++)
            {
               sum=sum+in[x][y]*cos(((2.0*x+1)*u*PI)/16.0)*
                  cos(((2.0*y+1)*v*PI)/16.0);
            }
         if (u==0) Cu=1/sqrt(2); else Cu=1;
         if (v==0) Cv=1/sqrt(2); else Cv=1;

         out[u][v]=(int)1/4.0*Cu*Cv*sum/norm[u][v];
         printf("%8d ",out[u][v]);
      }
      printf("\n");
   }

   printf("\n");
   return(0);
}
```

It can be seen from Sample run 5.2 that most of the normalized and quantized components are zero. This helps in the next stages of the compression which involve either LZW or RLE. Thus a scheme which stores similar values will result in a larger compression than non-arranged values. It can also be seen from Sample run 5.2 that most of the non-zero values are in the top left-hand corner.

To achieve the compression, the DC components are stored as the difference in the DC value from one block to the next. This is because DC components, from block to block tend to be similar. The AC components are then stored logically in a zigzag order, as illustrated in Figure 5.3. This puts the lowest-frequency components first. Typically the quantized high-frequency components will be zero, and the final compression stage will compress these to a high degree.

Sample run 5.2 would be stored as:

```
251, 0, -5, -1, -3, -2, 0, -1, 0, 0, 0, 0, -1, 0, 0, 0, 0,..., 0
```

which has a run of 51 zero (as well as an earlier run of 4 zeros).

Sample run 5.2

251	0	-2	-1	0	0	0	0
-5	-3	0	0	0	0	0	0
-1	-1	0	0	0	0	0	0
0	0	0	0	0	0	0	0
0	0	0	0	0	0	0	0
0	0	0	0	0	0	0	0
0	0	0	0	0	0	0	0
0	0	0	0	0	0	0	0

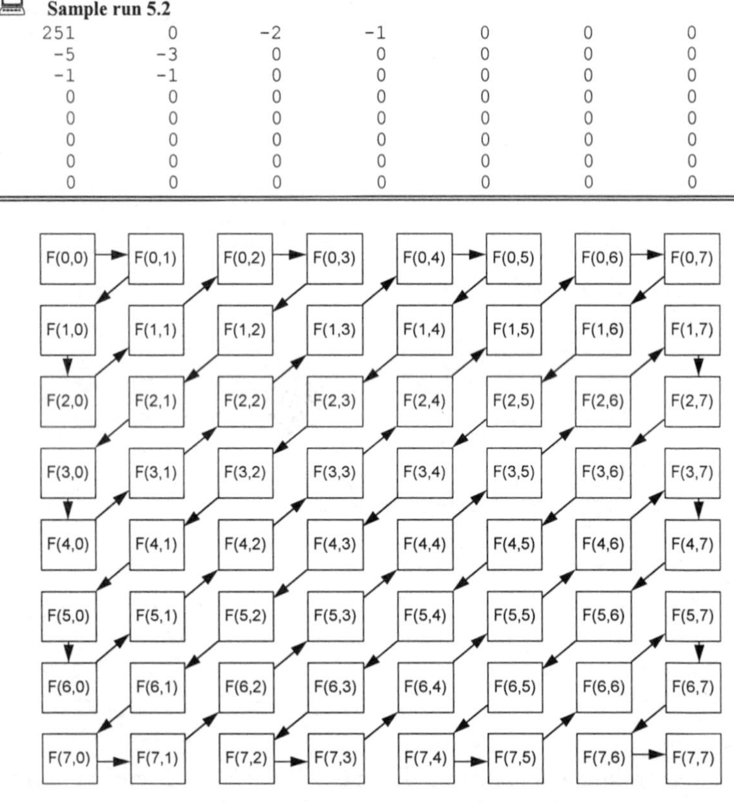

Figure 5.3 Zigzag storage

5.2.4 Final compression

The final part of JPEG compression uses either a modified Huffman coding or arithmetic coding. Huffman coding is by far the most popular technique, but tends to lead to a larger compressed file.

Data values are coded with a modified Huffman code called the variable-length code (VLC). This encodes values as the difference between consecutive values. A positive value is stored with its binary equivalent and a negative value as the one's complement (all the bits inverted) equivalent, such as:

5	101	–5	010
10	1010	–10	0101
1	1	–1	0
23	10111	–23	01000

This difference value is preceded by a 4-bit binary value which defines the number of bits in the data values. Figure 5.4 gives an example. In this case the data is 12, 10, 11, 11 and 11. The initial value for the difference encoding is taken as zero, thus the difference values will be 12, –2, 1, 0, 0. The first four bits will be 0100 (4) as the value 12 requires four bits. Next the value of 12 is stored (1010). The next difference is –2, which is 01 in 1's complement. This requires two bits, thus the next four bits will be 0010 (to define two bits), followed by the –2 value (01). This then continues. Note that a zero value is stored as a single 4-bit value of 0000, so no other bits follow it.

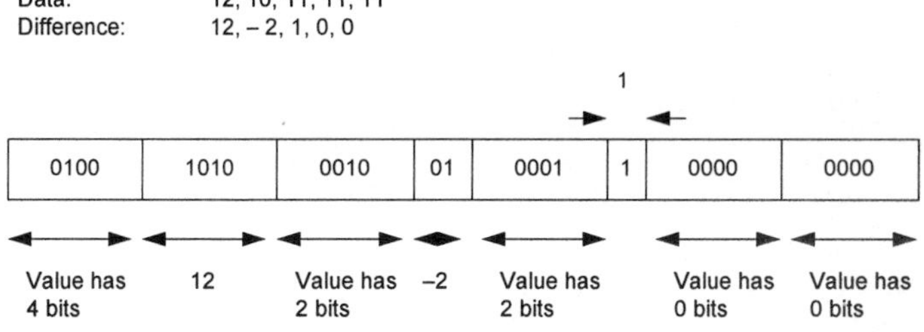

Figure 5.4 VLC storage

AC components (such as $F(0,1)$, $F(1,0)$, and so on) are stored as an 8-bit Huffman code followed by a variable-length integer. The Huffman code is made up of high 4 bits which give the number of zero values preceding this value, and the low 4 bits which give the length of the variable-length integer. Figure 5.5 shows an example of AC coding with the data 0, 0, 0, 0, 2, 0, 0, 6. In this case, the first four bits of the 8-bit Huffman code (before it is converted to a Huffman code) is 0101, because there are four consecutive zeros. The value after these zeros is 2, which requires only two bits. Thus the second part of the Huffman code (before coding) will be 0010 to specify two bits. Next the data contains two zeros so the Huffman code (before coding) will be 0010.

The data after the two zeros is a 6 which requires four bits, thus the second part of the Huffman code (before coding) is 0100. After this the data value for 6 is represented in binary (1010). The AC components contain many runs of zeros so the code produced will tend to be extremely compressed.

The binary value of 0000 0000 (00h) can never occur in the AC coding scheme. This code is used as a special code to identify that all of the values until the end of a block are zero. This is a common occurrence and thus saves coding bits.

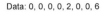

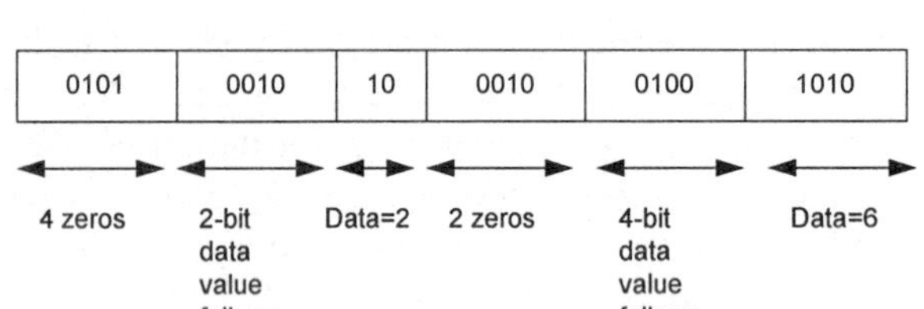

Figure 5.5 Storage of AC components

5.3 JPEG decoding

JPEG decoding involves reversing the process:

- Uncompression.
- Unquantizing (using the stored table or a standard table of factors).
- Reverse DCT.
- Block regeneration.

The DCT is reversed with the transform:

$$f(x,y) = \frac{1}{4}\left[\sum_{u=0}^{7}\sum_{v=0}^{7} C(u)C(v)F(u,v)\cos\frac{(2x+1)u\pi}{16}\cos\frac{(2y+1)v\pi}{16}\right]$$

Program 5.3 uses this inverse transform and contains the normalized and quantized coefficients from the previous example. To recap, the input data was:

```
144  139  149  155  153  155  155  155
151  151  151  159  156  156  156  158
151  156  160  162  159  151  151  151
158  163  161  160  160  160  160  161
158  160  161  162  160  155  155  156
161  161  161  161  160  157  157  157
162  162  161  160  161  157  157  157
162  162  161  160  163  157  158  154
```

The applied normalization matrix was:

```
5    3    4    4    4    3    5    4
4    4    5    5    5    6    7    12
8    7    7    7    7    15   11   11
9    12   13   15   18   18   17   15
20   20   20   20   20   20   20   20
20   20   20   20   20   20   20   20
20   20   20   20   20   20   20   20
20   20   20   20   20   20   20   20
```

Program 5.3

```
#include <stdio.h>
#include <math.h>
#define  PI 3.1415926535897

int   main(void)
{
int   x,y,u,v;
int   in[8][8]=   {{251,0,-2,-1,0,0,0,0},
            {-5,-3,0,0,0,0,0,0},
            {-1,-1,0,0,0,0,0,0},
            {0,0,0,0,0,0,0,0},
            {0,0,0,0,0,0,0,0},
            {0,0,0,0,0,0,0,0},
            {0,0,0,0,0,0,0,0}};

float norm[8][8]= {{5,3,4,4,4,3,5,4},
   {4,4,5,5,5,6,7,12},
   {8,7,7,7,7,15,11,11},
   {9,12,13,15,18,18,17,15},
   {20,20,20,20,20,20,20,20},
   {20,20,20,20,20,20,20,20},
   {20,20,20,20,20,20,20,20},
   {20,20,20,20,20,20,20,20}};

float out[8][8];
float sum,Cu,Cv;

   for (x=0;x<8;x++)
   {
      for (y=0;y<8;y++)
      {
         sum=0;
         for (u=0;u<8;u++)
            for (v=0;v<8;v++)
            {
               if (u==0) Cu=1/sqrt(2); else Cu=1;
               if (v==0) Cv=1/sqrt(2); else Cv=1;
```

```
      sum=sum+Cu*Cv*norm[u][v]*in[u][v]*cos(((2.0*x+1)*u*PI)/16.0)*
                             cos(((2.0*y+1)*v*PI)/16.0);
            }

         out[x][y]=1/4.0*sum;
         printf("%8.0f ",out[x][y]);
      }
      printf("\n");
   }
   printf("\n");
   return(0);
}
```

Sample run 5.3 shows that the values are similar to the input data values.

Sample run 5.3

```
146     148     151     153     154     154     155     156
148     150     153     154     155     155     155     156
153     154     156     157     157     156     156     156
157     158     159     159     158     157     156     156
159     160     161     161     159     157     156     156
160     161     162     162     160     158     157     156
159     160     162     161     160     158     157     157
158     160     161     161     160     158     157     157
```

The errors in the decoding are thus:

```
-2   -9   -2   2    -1   1    0    -1
3    1    -2   5    1    1    1    2
-2   2    4    5    2    -5   -5   -5
1    5    2    1    2    3    4    5
-1   0    0    1    1    -2   -1   0
1    0    -1   -1   0    -1   0    1
3    2    -1   -1   1    -1   0    0
4    2    0    -1   3    -1   1    -3
```

It can be seen that these errors are all less than 10, thus the decoding as not produced any significant errors.

5.4 JPEG file format

JPEG is a standard compression technique. A JPEG file normally complies with JFIF (JPEG file interchange format) which is a defined standard file format for storing a gray scale or YC_bC_r color image. Data within the JFIF contains segments separated by a 2-byte marker. This marker has a binary value of 1111 1111 (FFh) followed by a defined marker field. If a 1111 1111 (FFh) bit field occurs anywhere within the file (and it isn't a marker), the stuffed 0000 0000 (00h) byte is inserted after it so that it cannot be read as a false marker. The uncompression program must then discard the stuffed 00h byte.

Table 5.1 outlines some of the markers. For example the code FFC0h the file is a baseline DCT frame with Huffman coding. Program 5.4 uses these codes to display the markers in a sample JPG file, and Sample run 5.1 show a sample run with a JPG file.

It can be seen that the markers in the test run are:

- Start of image (FFD8h). The segments can be organized in any order but the start-of-image marker is normally the first 2 bytes of the file.
- Application-specific type 0 (FFE0h). The JFIF header is placed after this marker.
- Define quantization table (FFDBh). Lists the quantization table(s).
- Baseline DCT, Huffman coding (FFC0h). Defines the type of coding used.
- Define Huffman table (FFC4h). Defines Huffman table(s).

5.4.1 JFIF header information

The header information of the JFIF file is contained after the application-specific type 0 marker (FFE0h). Figure 5.6 shows its format and Program 5.5 reads some of the header information from a JFIF file. It can be seen that in Sample run 5.5 the length of the segment is 4096 bytes and the file is a JFIF file, as it has the JFIF string at the correct location. The version in this case is 1.01 and the units are given in pixels per inch. Finally it shows that the horizontal and vertical pixel density are both 11265. For comparison another sample run for a different file is shown in Sample run 5.6.

Table 5.1 Typical standard compressed graphics formats

Marker	*Description*	*Marker*	*Description*
C0h	Baseline DCT frame, Huffman coded	C1h	Baseline sequential DCT frame, Huffman coded
C2h	Extended sequential DCT frame, Huffman coded	C3h	Progressive DCT frame, Huffman coded
C4h	Define Huffman table	C5h	Differential sequential DCT frame, Huffman coded
C6h	Differential progressive DCT frame, arithmetic coded	C7h	Differential lossless frame, Huffman coded
C8h	Reserved	C9h	Extended sequential DCT frame, arithmetic coded
CAh	Progressive DCT frame, arithmetic coded	CBh	Lossless frame, arithmetic coded

CDh	Differential extended sequential DCT frame, arithmetic coded	CEh	Differential progressive DCT frame, arithmetic coded
CFh	Differential lossless frame, arithmetic	D8h	Start of image
D9h	End of image	E0h	Application-specific type 0

Program 5.4

```
#include    <stdio.h>

#define     NO_MARKS 19

int   main(void)
{
FILE  *in;
int   i,ch;
int   markers[NO_MARKS]={0xC0,0xC1,0xC2,0xC3,0xC4,
      0xC5,0xC6,0xC7,0xC8,0xC9,0xCA,0xCB,
      0xCD,0xCE,0xCF,0xD8,0xD9,0xDB,0xE0};
char  fname[BUFSIZ];
char  *msgs[NO_MARKS]={"Baseline DCT, Huff","Extended DCT, Huff",
      "Progress DCT, Huff","Lossless frame, Huff",
      "Define Huffman table", "Diff encoded DCT frame, Huff coded",
      "Diff progressive DCT frame, Huff", "Diff lossless frame, Huff",
      "Reserved", "Extended sequential DCT frame, arith coded",
      "Progressive DCT frame, arith coded",
      "Lossless frame, arith coded",
      "Diff extended sequential DCT frame, arith coding",
      "Diff progressive DCT frame, arith coding",
      "Diff lossless frame, arith coding",
      "Start of image", "End of image",
      "Define Quantization Tables", "Application specific type 0"};

   printf("Enter JPG file>>");
   gets(fname);

   if ((in=fopen(fname,"r"))==NULL)
   {
      printf("Can't find file %s\n",fname);
      return(1);
   }

   do
   {
      ch=getc(in);

      if (ch==0xff)
      {
         ch=getc(in);
         printf("%x",ch);
         for (i=0;i<NO_MARKS;i++)
            if (ch==markers[i])  printf("Found:%s\n",msgs[i]);
      }

   } while (!feof(in));

   fclose(in);
   return(0);
}
```

Sample run 5.4

```
Enter JPG file>> marble.jpg
d8:Found:Start of image
e0:Found:Application specific type 0
db:Found:Define Quantization Tables
db:Found:Define Quantization Tables
c0:Found:Baseline DCT, Huff
c4:Found:Define Huffman table
c4:Found:Define Huffman table
```

The reason that the segment is 4096 bytes is that a thumbnail version of the image is stored within the segment, after the basic header information defined in Figure 5.6.

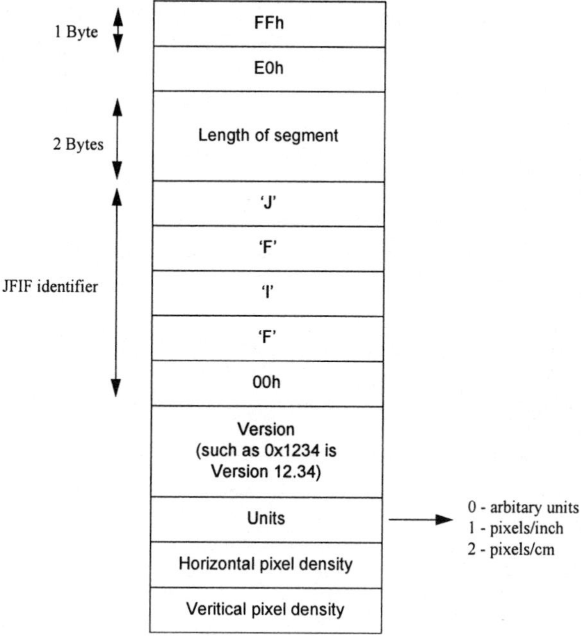

Figure 5.6 JFIF header information

Program 5.5

```
#include    <stdio.h>

int   main(void)
{
FILE  *in;
int   i,ch,version,length,units,pixelden_X,pixelden_Y;
char  fname[BUFSIZ],str[BUFSIZ];
char  *Units[3]={"Artibrary","Pixels per inch","Pixels per cm"};

   printf("Enter JPG file>>");
   gets(fname);
```

```
    if ((in=fopen(fname,"r"))==NULL)
    {
       printf("Can't find file %s\n",fname);
       return(1);
    }

    do
    {
       ch=getc(in);

       if (ch==0xff)
       {
          ch=getc(in);
          if (ch==0xe0)
          {
             fread(&length,2,1,in); printf("Length: %d\n",length);
             fread(str,5,1,in); printf("Marker: %s\n",str);
             fread(&version,2,1,in); printf("Version: %0x\n",version);
             fread(&units,1,1,in); printf("Units: %s\n",Units[units]);
             fread(&pixelden_X,2,1,in); printf("X den: %d\n",pixelden_X);
             fread(&pixelden_Y,2,1,in); printf("Y den: %d\n",pixelden_Y);
          }
       }

    } while (!feof(in));

    fclose(in);
    return(0);
}
```

Sample run 5.5

```
Enter JPG file>> marble.jpg
Length: 4096
Marker: JFIF
Version: 101
Units: Pixels per inch
X den: 11265
Y den: 11265
```

Sample run 5.6

```
Enter JPG file>> test.jpg
Length: 4096
Marker: JFIF
Version: 101
Units: Artibrary
X den: 256
Y den: 256
```

5.4.2 Quantization table

The quantization table is defined after the quantization table marker (FFDBh). Its format, after the marker, is:

- 2 bytes for the length of the segment.
- 1 byte, of which the high 4 bits define the precision (a 0 defines a table with 8-bit entries, a 1 defines 16-bit entries), and the low 4 bits give the table's ID (such as 0 for the first, 1 for the second, and so on).

- 64 entries for the table (either 8-bit or 16-bit entries). These entries are stored in a zigzag manner (see Figure 5.3).

Program 5.6 can be used to list the quantization table and Sample run 5.7 gives a sample run from a JPG file. It can be seen that in this case the factor varies from 3 to 28.

Program 5.6

```
#include    <stdio.h>
int   main(void)
{
FILE  *in;
int      i,ch,length;
char  fname[BUFSIZ], table, entry;

   printf("Enter JPG file>>");
   gets(fname);

   if ((in=fopen(fname,"r"))==NULL)
   {
      printf("Can't find file %s\n",fname);
      return(1);
   }

   do
   {
      ch=getc(in);

      if (ch==0xff)
      {
         ch=getc(in);
         if (ch==0xdb)
         {
            fread(&length,2,1,in); printf("Length: %d\n",length);
            fread(&table,1,1,in); printf("Marker: %x\n",table);

            for (i=0;i<64;i++)
            {
               fread(&entry,1,1,in);
               printf("%d ",entry);
            }
         }
      }

   } while (!feof(in));

   fclose(in);

   return(0);
}
```

Sample run 5.7

```
Enter JPG file>> test.jpg
Length: 17152
Marker: 0
5 3 4 4 4 3 5 4 4 4 5 5 5 6 7 12 8 7 7 7 7 15 11 11 9 12 17 15 18 18 17
15 17 17 19 22 28 23 19 20 20 20 20 20 20 20 20 20 20 20 20 20 20 20 20
20 20 20 20 20 20 20 20 20
```

5.4.3 Huffman tables

Huffman tables are defined after the define Huffman table marker (FFC4h). One table defines the DC components, $F(0,0)$, and the other defines the AC components. Its format, after the marker, is:

- 2 bytes for the length of the segment.
- 1 byte, of which the high 4 bits defines the table class (0 for DC codes and 1 for AC codes), and the low 4 bits gives the table's ID (such as 0 for the first, 1 for the second, and so on).
- 16 bytes for the code lengths.
- A variable number of bytes which contain the Huffman codes (the code length defines the number of bits used for each code).

For example if the code lengths are:

3, 3, 3, 4, 4, 4

the packed Huffman code of:

000001010000100101000

would be separated as:

000 001 010 0001 0010 1000

5.5 JPEG modes

JPEG has three main modes of compression:

- Progressive mode – this mode allows a rough outline of an image to be viewed while decoding the rest of the file. This is useful when an image is being received over a relatively slow transfer channel (such as over a modem or from the Internet). There are two main methods used: spectral-selection mode and successive-approximation mode. The successive-approximation mode first sends the high-order bits of each of the encoded values and then the lower-order lower bits. Spectral-selection mode first sends the low-frequency components of each of the 8×8 blocks then sends the high-frequency terms.
- Hierarchical mode – in this mode the image is stored in increasing resolution. For example a 1280×960 image might be stored as 160×120,

320×240, 640×480 and 1280×960. The viewing program can then show the image in increasing resolution as it reads (or receives) the file. Most systems do not implement this facility.

- Lossless mode – this mode allows data to be stored and recovered in exactly in its original state. It does not use DCT conversion or subsampling.

5.6 Exercises

5.1 Explain how RGB information is converted to YC_bC_r (*YUV*).

5.2 Using CCIR 601 color conversion complete Table 5.2.

Table 5.2 RGB to YC_bC_r. and YC_bC_r to RGB

Red	*Green*	*Blue*	*Y*	C_b	C_r
1	1	1			
0.5	1	0			
1	0	0			
0	1	0			
0.5	0.5	0.5			
			0.6924	0.3	0.04878
			0.2	0.06748	0.03252
			0.369	0.05122	−0.12683
			0.3979	0.13557	0.37439

5.3 Explain how subsampling reduces the amount of data stored.

5.4 Determine the DCT conversion matrix for an 8×8 input block which is filled with the value 128. Explain why some of the factors are much greater than others.

5.5 Determine the DCT conversion matrix for an 8×8 input block which is filled with the following random numbers:

```
 43   91    6   32  115  210  111    3
 91   32   12   54  200  211  100   43
 31   72  100    1    4  140   23   92
 31  103   22   98  222   21    2  109
250  252  111   68   85    1   33  229
133   25  123  128   11   75   23   44
 78   77  10   139   32   11   29   92
```

Explain why some of the factors are much greater than others.

5.6 Using the DCT conversion determine the $F[u,v]$ matrix for the following 8×8 block:

132	120	115	115	116	110	109	109
131	121	110	111	120	122	110	110
131	122	121	110	115	125	115	155
123	125	120	108	110	115	113	114
130	121	115	111	124	122	120	110
123	125	120	108	110	115	123	114
132	123	114	117	116	110	129	109

5.7 Using the results from Question 5.6, and using the quantization factors at the bottom of page 66, determine the resulting quantized values.

5.8 Using the results from Question 5.7, conduct an inverse DCT and thus determine the error between the input and the unencoded values.

5.9 Explain the difference between progressive mode JPEG and hierarchical JPEG.

5.10 With a JPEG file and using Program 5.5 determine the JFIF header information.

5.11 With a JPEG file and using Program 5.6 determine the quantization table.

5.12 With a JPEG file and using Program 5.4 determine the JPEG markers with the file.

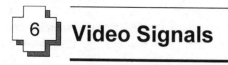

6 Video Signals

6.1 Introduction

This chapter discusses the main technologies using in TV signals and outlines the usage of digital TV. The five main classes of motion video are:

- High-definition television (HDTV).
- Studio-quality digital television (SQDV).
- Current broadcast-quality television.
- VCR-quality television.
- Low-speed video conferencing quality.

There are three main types of TV-type video signal:

- NTSC (American National Television Standards Committee).
- PAL (phase alternation line).
- SECAM (séquential couleur à mémoire).

These signals are based on composite color video which uses gaps in the black and white signal to transmit bursts of color information. In the UK this video signal is known as PAL (phase alternation line), whereas in the USA it is known as NTSC (named after the American National Television Standard Committee, who defined the original standard). Most of the countries of the world now use either PAL or NTSC. The only other standard format is SECAM, popular in French-speaking parts of the world.

6.2 Color-difference signals

A motion video is basically single images scanned at a regular rate. If these images are displayed fast enough, the human eye sees the repetitive images in a smooth way. Video signals use three primary colors: red, green and blue.

An image is normally scanned only a row, one pixel at a time. When a signal row has been scanned, the scanner goes to the next row, and so on until it has scanned the last row, when it goes back to the start. The samples are then split into the three primary colors: red, green and blue, as illustrated in Figure 6.1.

The three main characteristics of a color signals are:

- Its brightness (luminance).
- Its colors (hues).
- Its color saturation (the amount of color).

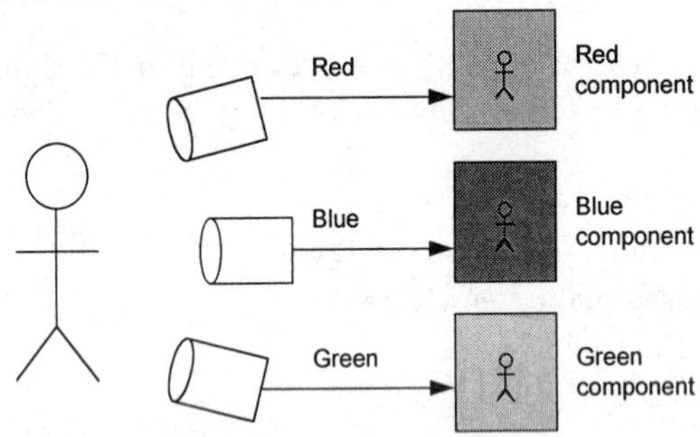

Figure 6.1 Conversion of an image into RGB components

Any hue can vary from very pale to very deep; the amount of the color is its saturation. When transmitting a TV signal it is important to isolate the luminance signal (the black and white element) so it can be sent to monochrome receivers. This is done by carefully adding a mixture of each of the RGB signals. PAL, NSTC and SECAM use the mixture:

$$\text{Y}=0.3R+0.59G+0.11B$$

where the red, green and blue signals vary between 0 and 1. Thus the maximum level for luminance will be 1 (fully saturated, pure white color) and the minimum level will be 0 (black). It can be seen from this formula that the green signal has a much greater effect than the red, which in turn has a much higher influence than the blue. In terms of luminance, the green signal has almost six times the effect of the blue signal and twice the effect of the red.

Color information is then added to the luminance signal so that color receivers can obtain the additional color information for hue and saturation (often known as chroma signal). These signals are transmitted on a subcarrier which is suppressed at the transmitter and recreated at the receiver. The subcarrier is modulated with the three color-difference signals (but only two color-difference signals need to be transmitted).

The three color-difference signals are:

Red–luminance or $(R-Y)$
Green–luminance $(G-Y)$
Blue–luminance $(B-Y)$

There is no need to send all of the signals because if the $R-Y$ and $B-Y$ signals are sent then the $G-Y$ difference signal can be easily recovered. This is because the Y signal is sent in the luminance signal. Thus the red and blue signal are recovered by adding the luminance signal to red – luminance and blue – luminance signals, then using the correct RGB weighting given to give:

$$G = \frac{Y - 0.3R - 0.11B}{0.59}$$

The reason why the $R-Y$ and $B-Y$ signals are sent is that as the Y signal contains a great deal of green information, thus the $G-Y$ signal is likely to be smaller than the $R-Y$ and the $B-Y$ signals. The small $G-Y$ signal is then more likely to be affected by noise in the transmission system.

The receiver must be told when the signal is at its start, thus the transmitter sends a strong sync pulse to identify the start of the image. Figure 6.2 illustrates the transmitter with the baseband video signals, often known as composite video signals.

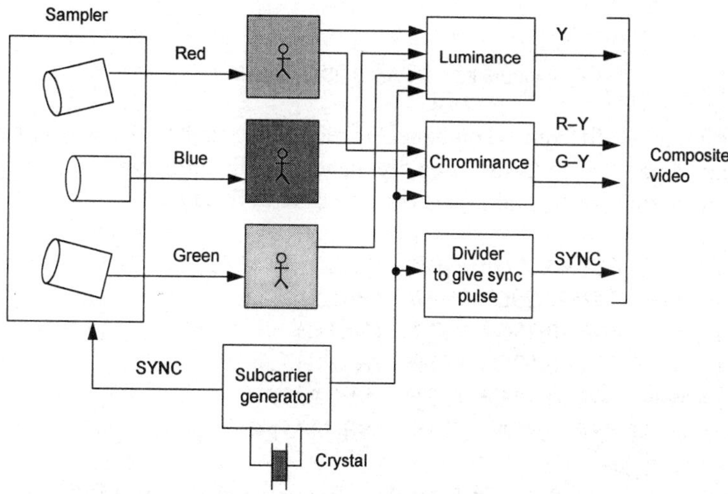

Figure 6.2 Conversion of an image into composite video components

6.3 Quadrature modulation

PAL and NSTC use an amplitude-modulated subcarrier wave. This subcarrier wave (4.43361 87 MHz for PAL and 3.579 545 MHz for NSTC) is amplitude modulated using weighted $R-Y$ and by a weighted $B-Y$ signals shifted by 90° (quadrate modulation), as illustrated in Figure 6.3.

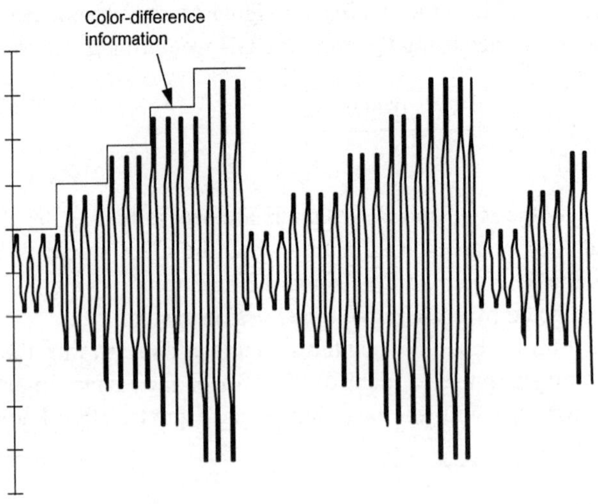

Figure 6.3 A PAL or NSTC waveform

In PAL these difference components are called U and V, whereas in NSTC they are I and Q. SECAM uses frequency modulation, where the color-difference terms are D_R and D_B. These terms are defined by:

$$
\begin{aligned}
U &= 0.62R-0.52G-0.10B \quad \text{(PAL)}\\
V &= -0.15R-0.29G+0.44B \quad \text{(PAL)}\\
I &= 0.60R-0.28G-0.32B \quad \text{(NSTC)}\\
Q &= 0.21R-0.52G+0.31B \quad \text{(NSTC)}\\
D_R &= -1.33R+1.11G+0.22B \quad \text{(SECAM)}\\
D_B &= -0.45R-0.88G+1.33B \quad \text{(SECAM)}
\end{aligned}
$$

It can be seen that if R, G and B are the same value, the color-difference terms will be zero (no color difference). PAL, NSTC and SECAM use:

$$Y=0.30R+0.59G+0.11B$$

Then, in PAL, these weighting values are:

$$V = 0.877(R - Y)$$
$$U = 0.493(B - Y)$$

In NSTC these are:

$$I = 0.74(R - Y) - 0.27(B - Y)$$
$$Q = 0.48(R - Y) + 0.41(B - Y)$$

The fundamental feature of this mode of signal addition is that by special detection at the receiver end it becomes possible to isolate the V/U or I/Q signals again and thus extract the original $R-Y$ and $B-Y$ modulation signals.

Figure 6.4 shows examples of the color-difference signals in relation to various colors. In PAL, it can be seen that yellow has a strong Y component, a positive V component $(R-Y)$ and a negative U component $(B-Y)$. Whereas blue has a low Y component, a strong negative U component and a positive V component. Table 6.1 outlines the components for each color.

Figure 6.5 shows a phasor diagram with the V component at a phase difference of 90° from the U component. When the red signal is greater than the luminance, the V signal will be positive; and if the blue signal is greater than the luminance, the U signal will be positive. The signal vector will thus be in the first quadrant. The colors in this quadrant are likely to be lacking in green, but strong in reds and blues. Thus the first quadrant contains purple-type colors. The example in Figure 6.5 shows how a U component of +0.5 and a V component of +0.5 give a color of magenta.

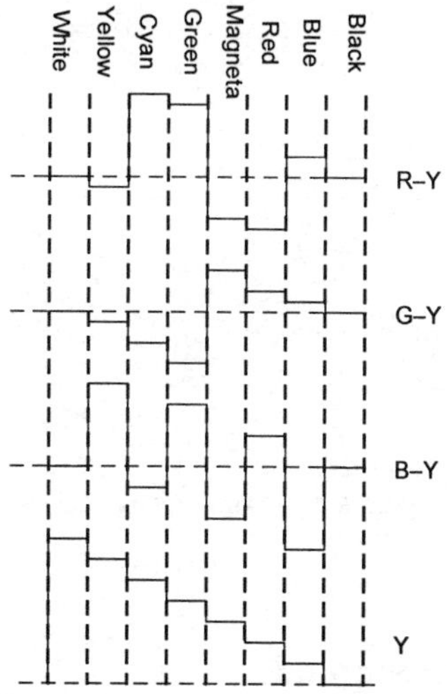

Figure 6.4 Color-difference signals for various colors

Table 6.1 Color examples with luminance and chrominance values

Color	*Y*	*B – Y*	*R – Y*	*U*	*V*	*Amp.*	*Angle*
Black	0	0	0	0	0	0	–
Blue	0.11	+0.89	–0.11	+0.439	–0.096	0.44	347
Red	0.3	-0.3	+0.7	–0.148	–0.614	0.63	103
Magenta	0.41	+0.59	+0.59	+0.291	+0.517	0.59	61
Green	0.59	–0.59	–0.59	–0.291	–0.517	0.59	241
Cyan	0.7	+0.3	–0.7	+0.148	–0.614	0.63	283
Yellow	0.89	–0.89	+0.11	–0.449	+0.096	0.44	167
White	1.0	0	0	0	0	0	–

The colors in the second quadrant have a positive *V* component and a negative *U* component. For this to happen the red component must be greater than the luminance (for *V* to be positive) and the luminance must be greater than the blue component. This gives bright colors with a strong red component. An example of the color yellow is given in Figure 6.5.

The third quadrant has a negative value for the *U* and *V* component. This means that the luminance is greater than both the blue and red components. Thus the colors have a strong green factor.

In the fourth quadrant the U component is positive (the blue component is greater than the luminance) and the V component is negative (the luminance is greater than the red component). The colors in this quadrant will be strong in blues with a low luminance. An example of cyan is shown in Figure 6.5.

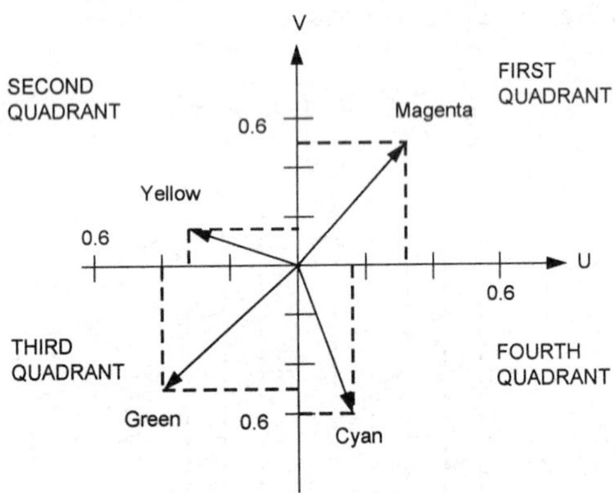

Figure 6.5 Phasors for four colors

6.4 Baseband video signals

Baseband video signals, or composite video, are signals in a form which is used in TV systems where the video is traced in interlacing lines. Initially, the top left-hand corner pixel is transmitted, followed by each pixel in turn on a single line. After the last pixel on the first line, the video traces back to the start of the next displayable line. This continues until it reaches the bottom of the screen. After this the video trace returns to the top left-hand pixel and starts again, as illustrated in Figure 6.6.

With PAL the screen refresh rate is based on the 50 Hz mains frequency while in NTSC it is based on the 60 Hz mains frequency. A 50 Hz refresh rate causes the screen to be updated 50 times every second. Each frame (or picture) is sent by sending the odd lines of the frame on the first update and then the even lines on the next screen update, and so on. This technique is known as interlaced scanning and is illustrated in Figure 6.7. For a 50 Hz system, the frame rate (or picture rate) is thus 25 Hz which means that one picture is drawn every 1/25 of a second (20 ms).

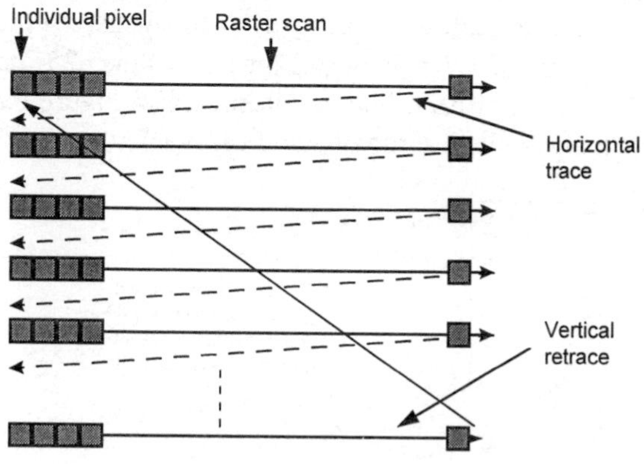

Figure 6.6 Video screen showing raster lines and retrace

A raster line is the smallest subdivision of this horizontal line. PAL systems have 625 video raster lines at a rate of 25 frames per second and NSTC uses 525 lines at a rate of 30 frames per second.

Thus, if the screen refresh rate is 50 Hz, the screen is updated once every 20 ms. Each update contains half a picture so one complete picture is sent every 40 ms. Thus 625 lines are sent every 40 ms, the time to transmit one raster line will thus be:

$$t_{raster} = \frac{40}{625} ms = 64\ \mu s$$

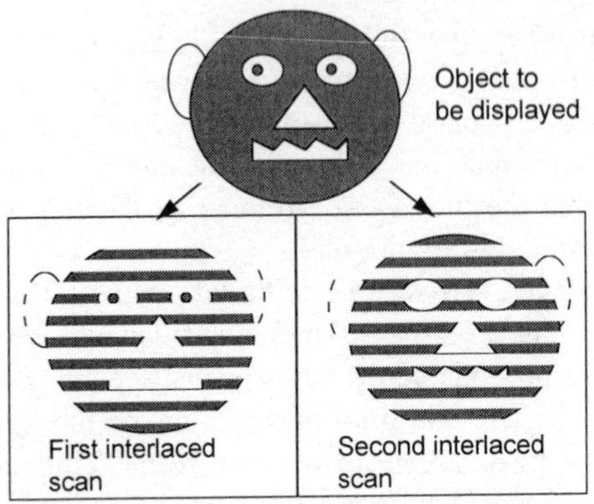

Figure 6.7 Interlaced scanning

With a black and white signal the voltage amplitude of the waveform at any instant gives the brightness of each part of the displayed picture. A negative voltage sync pulse of 4.7 μs indicates the start of the 625 lines. A blanking level indicates the start of each line, as illustrated in Figure 6.8. The largest voltage defines white while a zero voltage gives black.

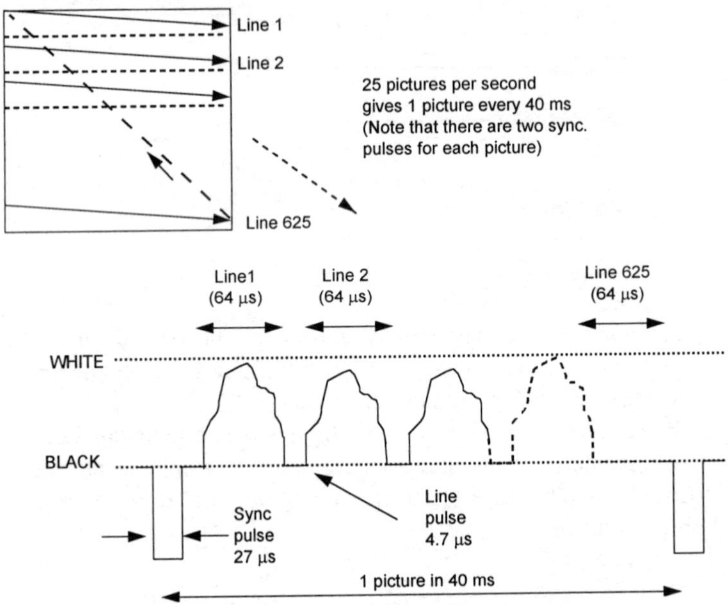

Figure 6.8 TV line waveform

Composite color video signals use gaps in the black and white signal to transmit bursts of color information. A composite video signal consists of a luminance (brightness) component and two chromatic (color) components. These chromatic components are transmitted simultaneously as the amplitude modulation sidebands of a pair of suppressed subcarriers which are identical in frequency but are in phase quadrature.

Figure 6.9 shows the frequency band for a composite video signal for PAL. The black and white (luminance) signal takes up the band from DC to 4.2 MHz, and a subcarrier is added in the higher-frequency portion of the band, at 4.43361875 MHz (3.58 MHz for NTSC); the modulated subcarrier can be thought of as superimposing itself on the luminance signal, as illustrated in Figure 6.10. The bandwidth for a PAL system is approximately 5.5 MHz.

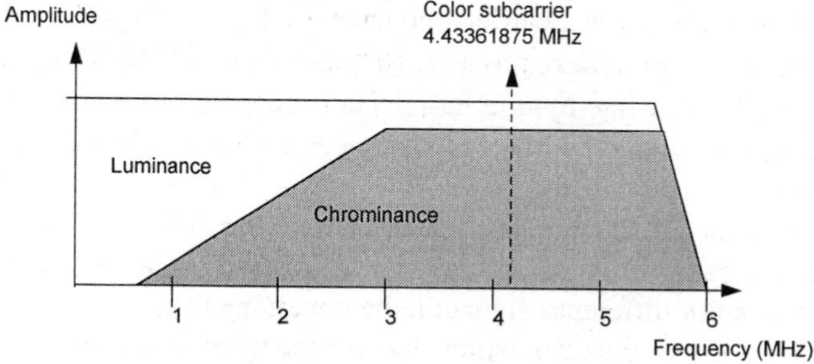

Figure 6.9 Composite video frequency spectrum

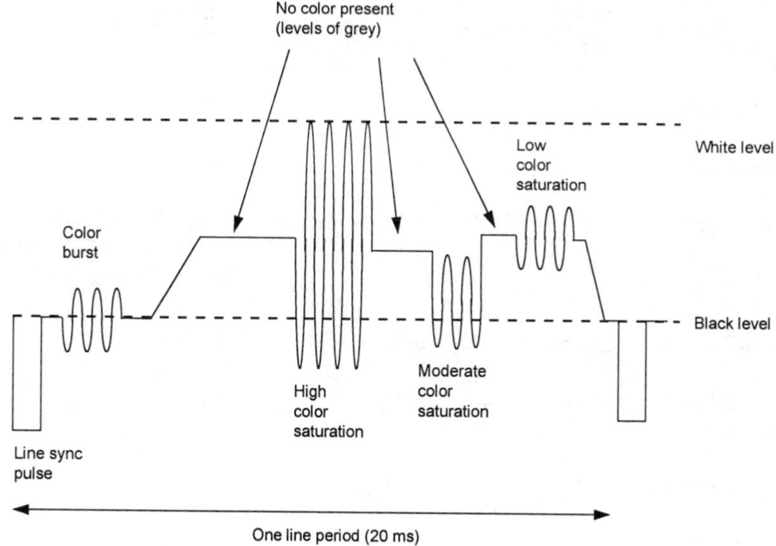

Figure 6.10 Composite video signal

6.4.1 Differences in color modulation between NSTC, PAL and SECAM

The start of the horizontal line is preceded by a sync pulse and then a color burst. This provides a reference for the phase information in the color signal. Figure 6.11 shows the horizontal blanking interval.

In NSTC, the chrominance subcarrier is suppressed-carrier amplitude modulated on a 3.579 545 MHz. It is modulated by the *I* (for In-phase) and *Q* (for Quadrature) components, with the *I* component modulating the subcarrier at 0° and the *Q* component modulating at 90°. The reference burst lasts 2.67 μs and is at an angle of 57° with respect to the *I* carrier. In NSTC, the *Y* bandwidth is 4.5 MHz while the *I* bandwidth is only 1.5 MHz and the *Q* bandwidth is 0.5 MHz (although many TV receivers allow a bandwidth of 0.5 MHz for *I* and *Q* and give perfectly acceptable results).

PAL is similar but has a carrier frequency of 4.433 618 7 MHz. The modulating components are referred to as *U* (0°) and *V* (90°). The V component is alternated 180° on a line-by-line basis. The reference burst last 2.25 μs and also alternates on a line-by-line basis between an angle of +135° and –135° relative to the *U* component.

SECAM uses a frequency-modulated (FM) color subcarrier for the chromatic signals, transmitting one of the color difference signals every other line, and the other color difference signal on the remaining lines.

Table 6.2 outlines the main parameters for the three technologies.

Table 6.2 The main differences between NSTC, PAL and SECAM

	NSTC	PAL	SECAM
Lines/frame	525	625	625
Frame/second	30	25	25
Interlace ratio	2:1	2:1	2:1
Color subcarrier	3.579 545 MHz	4.433 619 MHz	Two FM-modulated carriers at 4.40625 MHz and 4.25 MHz
Line period	63.55 μs	64.0 μs	64.0 μs
Horizontal sync pulse width	4.7 μs	4.7 μs	4.7 μs
Color burst width	2.67 μs	2.25 μs	
Vertical sync pulse width	27.1 μs	27.3 μs	27.3 μs

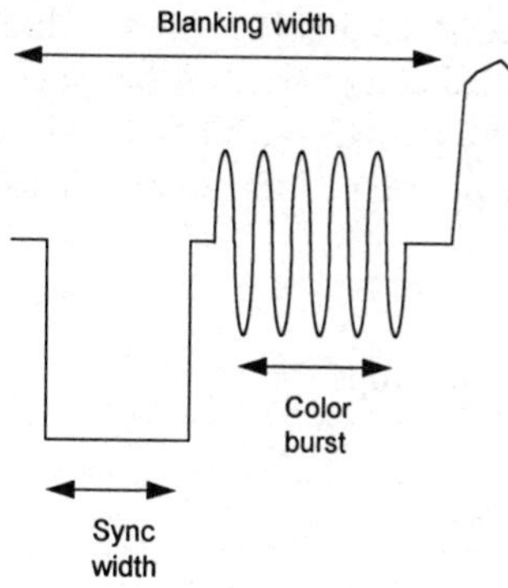

Figure 6.11 Color burst

6.5 Digitizing TV signals

The composite video signal must be sampled at twice the highest frequency of the signal. To standardize this sampling, the ITU CCIR-601 (often known as ITU-R) has been devised. It defines three signal components: Y (for luminance), C_r (for $R-Y$) and C_b (for $B-Y$).

The biggest problem with sampling is that SECAM and PAL use 625 lines and NTSC use 525 lines. The ITU-R 601 standard defines that SECAM and PAL are sampled 858 times for each line and NTSC is sampled 864 times for each line. This produces the same scanning frequency in both cases:

$$\text{Scanning frequency (PAL, SECAM)} = 864 \times 625 \times 25 = 13.5\,\text{MHz}$$

$$\text{Scanning frequency (NTSC)} = 858 \times 525 \times 30 = 13.5\,\text{MHz}$$

With this technique, the luminance (the black and white level) is digitized at a rate for 13.5 MHz and color is sampled at an equivalent rate of 6.75 MHz. Each luminance and color sample is coded as 8 bits, thus the digitized rate is:

$$\text{Digitized video signal rate} = 13.5 \times 8 + 6.75 \times 8 = 162\,\text{Mbps}$$

Thus the sample rate for all three systems will be:

$$\text{Sample rate} = \frac{1}{13.5 \times 10^6} = 74\,\text{ns}$$

Normally the Y is sampled at 13.5 MHz and the chromatic components (such as U/V or C_b/C_r) are sampled at a quarter of this rate (that is, 3.375 MHz). The bits are then interleaved to give 12-bit YUV bundles, as illustrated in Figure 6.12. This shows that there are 12 bits transmitted every 74 ns, thus the bit rate is:

$$\text{Bit rate} = \frac{12}{74 \times 10^{-9}} = 162\ \text{Mbps}$$

74 ns

Y_0	Y_1	Y_2	Y_3	Y_4	Y_5	Y_6	Y_7	U_0	U_1	Y_0	Y_1
Y_0	Y_1	Y_2	Y_3	Y_4	Y_5	Y_6	Y_7	U_2	U_3	Y_2	Y_3
Y_0	Y_1	Y_2	Y_3	Y_4	Y_5	Y_6	Y_7	U_4	U_5	Y_4	Y_5
Y_0	Y_1	Y_2	Y_3	Y_4	Y_5	Y_6	Y_7	U_6	U_7	Y_6	Y_7
Y_0	Y_1	Y_2	Y_3	Y_4	Y_5	Y_6	Y_7	U_0	U_1	Y_0	Y_1
Y_0	Y_1	Y_2	Y_3	Y_4	Y_5	Y_6	Y_7	U_2	U_3	Y_2	Y_3

One U and one V sample every 296 ns

Figure 6.12 Interleaved luminance and chrominance data

6.5.1 Subsampling

Subsampling reduces the digitized bit rate by sampling the luminance and chrominance at different rates. The standard format uses integer values as a ratio, such as:

$$Y \text{ sampling frequency} : C_{d1} \text{ sampling frequency} : C_{d2} \text{ sampling frequency}$$

where C_d is the sampling frequency for the chominance. For example, standard studio-quality TV uses a 4:2:2 sampling ratio for the Y:C_r:C_b ratio. This means that the number of samples for the color difference is half of the luminance. For example, if the image is:

R (720×486), G (720×486) and B (720×486) (NSTC)
R (720×576), G (720×576) and B (720×576) (PAL)

then using 4:2:2 gives the resultant YC_rC_b form:

Y (720×486), C_b (360×486) and C_r (360×486) (NSTC)
Y (720×576), C_b (360×576) and C_r (360×576) (PAL)

Thus, in NSTC, the number of terms has been reduced from 1049760 to 699840 (a saving of 33%).

The H.261 standard, as used in video conferencing, uses a 4:1:1 ratio. Thus the color difference components are sampled at one-quarter of the luminance rate. For example, if the image is:

$$\underline{R}\,(352\times288),\ G\,(352\times288) \text{ and } B\,(352\times288)$$

then using 4:1:1 gives the resultant YC_rC_b form:

$$Y\,(352\times288),\ C_b\,(176\times144) \text{ and } C_r\,(176\times144)$$

Thus, in NSTC, the number of terms has been reduced from 304128 to 152064 (a saving a 50%).

6.5.2 CCIR-601 active lines

CCIR defines a constant sampling rate of 13.5 MHz. Unfortunately, a delay is required to go from the end of one line to the start of the next (called horizontal retrace). A delay is also required when the scanning reaches the end of a frame and returns to the top of the frame (called vertical retrace). The time for active pixels has been defined as 720 per line, or 53 ns (in CCIR-601 the time of scan for a line is 64 μs). This is the same for NSTC (which samples at 858 per line) and PAL (which samples at 864 per line).

6.5.3 CCIR-601 quantization

Each of the samples for luminance and chrominance has 8 bits, which gives 256 levels. Only 220 values are used, the black level is coded as 16 and the peak white levels are coded as 235. Values from 0 to 16 and 235 to 255 are reserved for special code words. The color-difference signal can take on 225 different values; the zero corresponds to coded value 128 and the peak saturation to values 16 and 240. The values from 0 to 16 and 240 to 255 are reserved for special code words.

6.6 100 Hz pictures

Digital transmission of video signals not only improves the transmission of the video signals, it can also be used to increase picture quality. Two typical problems with PAL, NSTC and SECAM systems are:

- Interline flicker – the flickering of sharp horizontal edges around the edges of objects.

- Large-area flicker – most noticeable on large screens.

Research indicates these problems disappear when the frame rate is 90 Hz eliminates all flickering. Thus as the video data is digitized it can be stored in a memory and recalled at any rate. A possible technique, for PAL/625, is to store the incoming video data in memory and then to read it out at a rate of 100 Hz. This will then be displayed to the speed at double-speed lines and field scan rates.

6.7 Compressed TV

TV signals and motion video have a massive amount of redundant information. This is mainly because each image has redundant information and because there are very few changes from one image to the next. A typical standard is the MPEG standard which allows compression of about 130:1.

With MPEG-2 this transmission rate can be reduced to 4 Mbps for PAL and SECAM and 3 Mbps for NTSC, thus giving a compression ratio of 40:1 to high quality TV. MPEG-1 typically compresses TV signals to 1.2 Mbps, giving a compression ratio of 130:1. Unfortunately the quality is reduced to near VCR-type quality. Table 6.3 outlines these parameters.

The base bit rate for a standard Ethernet network is 10 Mbps. This allows compressed video to be transmitted over the network when there is no other traffic on the network. The 4 Mbps rate will load the network by approximately 50%. Standardized and compression techniques will be discussed in the next chapter.

Table 6.3 Motion video compression

Type	*Bit rate*	*Compression*	*Comment*
Uncompressed TV	162 Mbps	1:1	
MPEG-1	4 Mbps	40:1	VCR quality
MPEG-2	1.2 Mbps	130:1	PAL, SECAM TV quality

6.8 HDTV quality

HDTV (high-definition TV) has been supported by many companies for many years. Standards such as the European High-Definition MAC, a mainly analogue-based system, have been promoted then abandoned.

The main parameters in a TV system are the frame rate and the picture resolution; the main improvements are:

- HDTV-quality gives a higher picture resolution with a higher frame rate (1920×1080 at 60 frames per second). This gives excellent images of 1920 pixels per lines, 1080 lines per frame and 60 frames per second.
- HDTV-quality with a high resolution and a conventional frame rate (1920×1080 at 24 frames per second). The advantage with this system is that is gives much high resolution with a frame rate which is similar to the frame rate of current system (25 frames per second for PAL and 30 frames per second for NTSC).
- Improved resolution/conventional frame rate (1280×730 at 30 frames per second). The advantage of this system is that it gives an intermediate response between conventional systems and the alternatives given above. This technology may allow the best intermediate migration between current TV and high-resolution technology. Its screen resolution is also similar to SVGA monitors.

Another important parameter in TV systems is the aspect ratio. Conventional TVs use an aspect ratio of 4:3 (which is defined as the width of the screen divided by its height). This aspect ratio does not really suit showing movies or sports events, and HDTV improves this to a 16:9 aspect ratio. HDTV will be covered in the next chapter.

6.9 Exercise

6.1 For PAL with U, V components, estimate the color of the following U, V components:

(i) relatively large positive U value with a relatively small positive V value.
(ii) relatively large negative U value with a relatively small positive V value.
(iii) relatively small positive U value with a relatively large positive V value.
(iv) relatively large negative U value with a relatively small negative V value.

6.2 For PAL show that the equations:

$$U=0.62R-0.52G-0.10B$$
$$V=-0.15R-0.29G+0.44B$$

$Y=0.30R+0.59G+0.11B$

can be approximated to:

$V=0.877(R-Y)$, and
$U=0.493(B-Y)$.

6.3 For NSTC show that the equations:

$I=0.60R-0.28G-0.32B$
$Q=0.21R-0.52G+0.31B$
$Y=0.30R+0.59G+0.11B$

$I=0.74(R-Y)+0.27(B-Y)$, and
$Q=0.48(R-Y)+0.41(B-Y)$.

6.4 For PAL fill in the gaps of Table 6.4. The first two rows are already complete.

Table 6.4 Color examples with RGB, luminance and chrominance values

R	*G*	*B*	*Y*	*U*	*V*	*Amp.*	*Angle*
0.8	0.1	0.6	0.365	0.384	0.115	0.4	73.33
0.6	0.5	0.5	0.53	0.062	–0.015	0.06	103.6
0.1	0.8	0.6					
0.5	0.5	0.1					
			0.605	–0.446	–0.1		
			0.321	0.424	–0.061		
			0.532			0.25	–57.15
			0.475			0.26	154.52

6.5 Explain the technique of interlacing raster lines and thus show that for a 50 Hz refresh rate and 625 lines that the time for one raster line is 64 μs. Determine also the time for one raster line for a 60 Hz refresh rate and 525 lines.

6.6 Explain, with reference to quadrature modulation, the content of a PAL (or NSTC) composite video signal. The explanation should contain references to the color burst, the line sync, the quadrature color modulation and the luminance level.

6.7 Derive the scanning digital scanning rate for the ITU-R 601 standard. Use it to show that the data rate is 162 Mbps for both PAL and NSTC.

7 Motion Video Compression

7.1 Motion video

Motion video contains massive amounts of redundant information. This is because each image has redundant information and also because there are very few changes from one image to the next.

Motion video image compression relies on two facts:

- Images have a great deal of redundancy (repeated images, repetitive, superfluous, duplicated, exceeding what is necessary, and so on).
- The human eye and brain have limitations on what they can perceive.

Chapter 5 discussed how JPEG encodes still images. Motion video is basically a series of still images. The basis of motion JEPG (MPEG) is to treat the video information as a series of compressed images and to allow for compression of around 130:1.

7.2 MPEG-1 overview

The Motion Picture Experts Group (MPEG) developed an international open standard for the compression of high-quality audio and video information. At the time, CD-ROM single-speed technology allowed a maximum bit rate of 1.2 Mbps and it is this rate that the standard was built around. These days, ×8 and ×10 CD-ROM bit rates are common.

MPEG's main aim was to provide good quality video and audio using hardware processors (and in some cases, on workstations with sufficient computing power, to perform the tasks using software). Figure 7.1 shows the main processing steps of encoding:

- Image conversion – normally involves converting images from RGB into YUV (or YC_rC_b) terms with optional color sub-sampling.
- Conversion into slices and macroblocks – a key part of MPEG-1's compression is the detection of movement within a frame. To detect motion a

frame is subdivided into slices then each slice is divided into a number of macroblocks. Only the luminance component is then used for the motion calculations. In the subblock, luminance (*Y*) values use a 16×16 pixel macroblock, whereas the two chrominance components have 8×8 pixel macroblocks.

- Motion estimation – MPEG-1 uses a motion estimation algorithm to search for multiple blocks of pixels within a given search area and tries to track objects which move across the image.
- DCT conversion – as with JPEG, MPEG-1 uses the DCT method. This transform is used because it exploits the physiology of the human eye. It converts a block of pixels from the spatial domain into the frequency domain. This allows the higher-frequency terms to be reduced as the human eye is less sensitive to high-frequency changes.
- Encoding – the final stages are run-length encoding and fixed Huffman coding to produce a variable-length code.

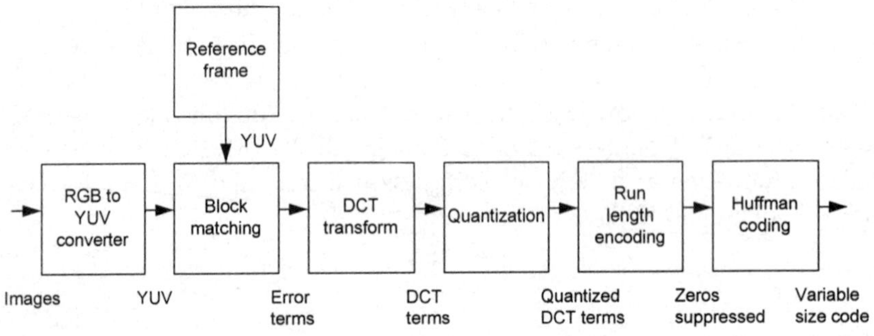

Figure 7.1 MPEG encoding with block matching

7.3 MPEG-1 video compression

MPEG-1 typically uses the CIF format for its input, which has the following parameters:

- For NTSC, 352×240 pixels for luminance and 176×120 pixels for *U* and *V* color-difference components (that is, 4:1:1 subsampling).
- For PAL/SECAM, 352×288 pixels for luminance and 176×144 pixels for *U* and *V* color difference components (i.e. 4:1:1 subsampling).

This gives a picture quality which is similar to VCR technology. MPEG-1 differs from conventional TV in that it is non-interlaced (known as progressive scanning), but the frame rate is the same as conventional TV, i.e. 25 fps (for PAL and SECAM) and 30 fps (for NSTC). Note that MPEG-1 can also use larger pixel frames, such as CCIR-601 740 × 480, but the CIF format is the most frequency used.

Taking into account the interlacing effect, the CIF format is actually derived from the CCIR-601 format. The CCIR-601 digital television standard defines a picture size of 720 × 243 (or 240) by 60 fields per second. Note that a frame actually comprises two fields, where the odd and even information is interlaced to create the full picture. When the interlaced luminance information occupies the full 720 × 480 frame, the chrominance components are reduced by 4:2:2 subsampling to give 360 × 243 (or 240) by 60 fields per second.

MPEG-1 also reduces the chrominance components by reducing the pixel data by half in the vertical, horizontal and time directions. It also reduces the image size so that the number of pixels for it is divisible by 8 or 16. This is because the motion analysis and DCT conversion operate on 16 × 16 or 8 × 8 pixel blocks. As a result, the number of lines changes for an MPEG-1 encoded move between the NSTC standard and PAL and SECAM standards. The final figure for PAL and SECAM is 288 at 50 fps; for NSTC it is 240 at 60 fps. These require the same number of bits to encode the streams.

The MPEG encoded bitstream comprises three components: compressed video, compressed audio and system-level information. To provide easier synchronization and lip synching the audio and video streams are time stamped using a 90 kHz reference clock.

7.4 MPEG-1 compression process

7.4.1 Color space conversion

The first stage of MPEG encoding is to convert a video image into the correct color space format. In most cases, the incoming data is in 24-bit RGB color format and is converted in 4:2:0 YC_rC_b (or YUV) form. Some information will obviously be lost but it results in some compression.

7.4.2 Slices and macroblocks

MPEG-1 compression tries to detect movement within a frame. This is done by subdividing a frame into slices and then subdividing each slice into a number of macroblocks. For example, a PAL format which has:

352 × 288 pixel frame (101 376 pixels)

can, when divided into 16×16 blocks, give a whole number of 396 macroblocks. Dividing 288 by 16 gives a whole number of 18 slices. Dividing 352 gives 22. Thus the image is split into 22 macroblocks in the x-direction and 18 in the y-direction, as illustrated in Figure 7.2.

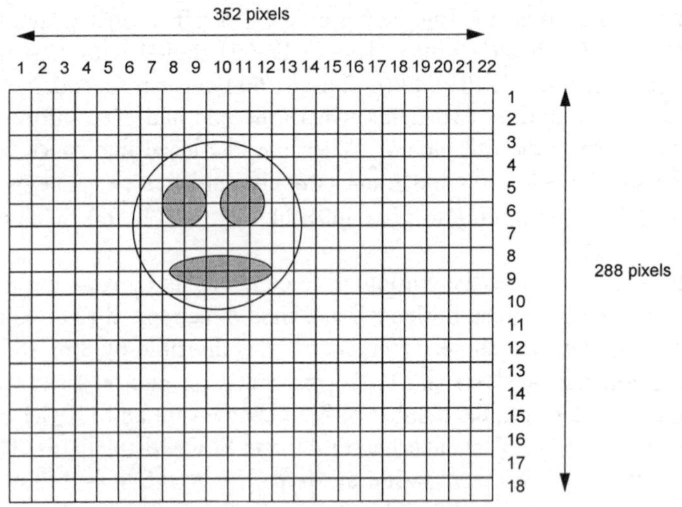

Figure 7.2 Segmentation of an image into subblocks

Luminance (Y) values use a 16×16 pixel macroblock, whereas the two chrominance components have 8×8 pixel macroblocks. Note that only the luminance component is used for the motion calculations.

7.4.3 Motion estimation

MPEG-1 uses a motion estimation algorithm to search for multiple blocks of pixels within a given search area and tries to track objects which move across the image. Each luminance (Y) 16×16 macroblock is compared with other macroblocks within either a previous or future frame to find a close match. When a close match is found, a vector is used to describe where this block should be located as well as any difference information from the compared block. As there tend to be very few changes from one frame to the next, it is far more efficient than using the original data.

Figure 7.3 shows two consecutive images of 2D luminance made up into 16×5 megablocks. Each of these blocks has 16×16 pixels. It can be seen that, in this example, there are very few differences between the two blocks. If the previous image is transmitted in its entirety then the current image can be transmitted with reference to the previous image. For example, the megablocks for $(0,1)$, $(0,2)$ and $(0,3)$ in the current block are the same as in the previous blocks. Thus they can be coded simply with a reference to the previous image. The (0,4) megablock is different to the previous image, but the (0,4) block is identical to the (0,3) block of the previous image, thus a reference to this block is made. This can continue with most of the block in the image being identical to the previous image. The only other differences in the current image are at (4,0) and (4,1); these blocks can be stored in their entirety or specified with their differences to a previous similar block.

Each macroblock is compared mathematically with another block in a previous or future frame. The offset to find another block could be over a macroblock boundary or even a pixel boundary. The comparison then repeats until a match is found or the specified search area within the frame has been exhausted. If no match is available, the search process can be repeated using a different frame or the macroblock can be stored as a complete set of data. As previously stated, if a match is found, the vector information specifying where the matching macroblock is located is specified along with any difference information.

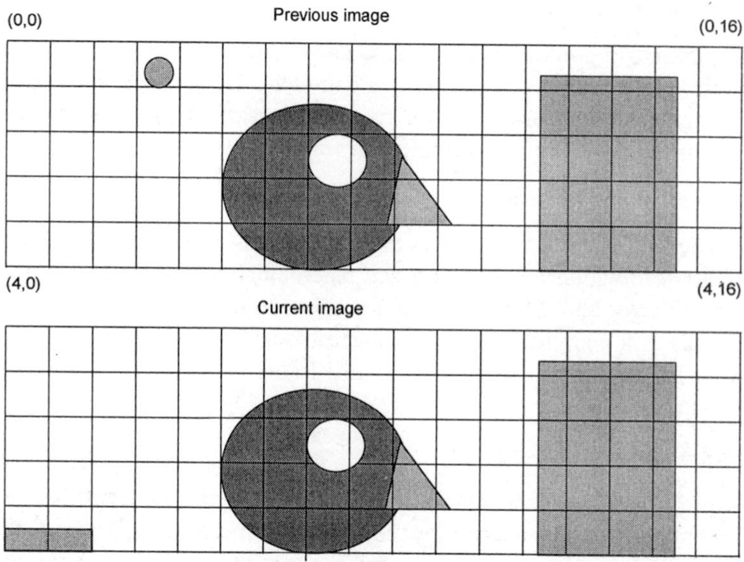

Figure 7.3 Two consecutive images

As the technique involves very many searches over a wide search area and there are many frames to be encoded, the encoder must normally be a high-powered workstation. This has several implications:

- An asymmetrical compression process is adopted, where a relatively large amount of computing power is required for the encoder and much less for the decoder. Normally the encoding is also done in non-real time whereas the decoder reads the data in real-time. As processing power and memory capacity increase, more computers will be able to compress video information in real-time.

- Encoders influence the quality of the decoded image dramatically. Encoding shortcuts, such as limited search areas and macroblock matching, can generate poor picture quality, irrespective of the quality of the encoder.
- The decoder normally requires a large amount of electronic memory to store past and future frames, which may be needed for motion estimation.

With the motion estimation completed, the raw data describing the frame can now be converted using the DCT algorithm ready for Huffman coding.

7.4.4 I, P and B frames

MPEG video compression uses three main types of frames: I-frame, P-frame and B-frame.

Intra frame (I-frame)

An intra frame, or I-frame, is a complete image and does not require any extra information to be added to it to make it complete. Thus no motion estimation processing has been performed on the I-frame. Mainly used to provide a known starting point, it is usually the first frame to be sent.

Predictive frame (P-frame)

The predictive frame, P-frame, uses the preceding I-frame as its reference and has motion estimation processing. Each macroblock in this frame is supplied either as a vector and difference with reference to the I-frame, or if no match was found, as a completely encoded macroblock (called an intracoded macroblock). The decoder must thus retain all I-frame information to allow the P-frame to be decoded.

Bidirectional frame (B-frame)

The bidirectional frame, B-frame, is similar to the P-frame except that its reference frames are to the nearest preceding I- or P-frame and the next future I- or P-frame. When compressing the data, the motion estimation works on the future frame first, followed by the past frame. If this does not give a good match, an average of the two frames is used. If all else fails, the macroblock can be intracoded.

Needless to say, decoding B-frames requires that many I- and P-frames are retained in memory.

MPEG-1 frame sequence

MPEG-1 allows frames to be ordered in any sequence. Unfortunately a large amount of reordering requires many frame buffers that must be stored until all dependencies are cleared.

The MPEG-1 format allows random access to a video sequence, thus the file must contain regular I-frames. It also allows enhanced modes such as fast forward, which means that an I-frame is required every 0.4 seconds, or 12 frames between each I-frame (at 30 fps).

At 30 fps, a typical sequence is a starting I-frame, followed by two B-frames, a P-frame, followed by two B-frames, and so on. This is known as a group of picture (GOP).

I⇒B⇒B⇒P⇒B⇒B⇒I⇒B⇒B⇒P⇒B⇒B⇒I⇒B⇒B⇒P⇒...

When decoding, the decoder must store the I-frame, the next two B-frames are also stored until the B-frame arrives. The next two B-frames have to be stored locally until the P-frame arrives. The P-frame can be decoded using the stored I-frame and the two B-frames can be decoded using the I- and P-frames. One solution of this is to reorder the frames so that the I- and P-frames are sent together followed by the two intermediate B-frames. Another more radical solution is not to send B-frames at all, simply to use I- and P-frames.

On computers with limited memory and limited processing power, the B-frames are diffucult because:

- They increase the encoding computational load and memory storage. The inclusion of the previous and future I- and P-frames as well as the arithmetic average greatly increases the processing needed. The increased frame buffers to store frames allow the encode and decode processes to proceed. this argument is again less valid with the advent of large and high-density memories.
- They do not provide a direct reference in the same way that an I- or P-frame does.

The advantage of B-frames is that they lead to an improved signal-to-noise because of the averaging out of macroblocks between I- and P-frames. This averaging effectively reduces high-frequency random noise. It is particularly useful in lower bit rate applications, but is of less benefit with higher rates, which normally have improved signal-to-noise ratios.

7.4.5 DCT conversion

As with JPEG, MPEG-1 uses the DCT. It transforms macroblocks of luminance (16×16) and chrominance (8×8) into the frequency domain. This allows the higher-frequency terms to be reduced as the human eye is less sensitive to high-frequency changes. This type of coding is the same as used in JPEG still image conversion, that was described in the previous chapter.

Frames are broken up into slices 16 pixels high, and each slice is broken up into a vector of macroblocks having 16×16 pixels. Each macroblock contains luminance and chrominance components for each of four 8×8 pixel blocks. Color decimation can be applied to a macroblock, which yields four 8×8 blocks for luminance and two 8×8 blocks (C_b and C_r) of chrominance, using one chrominance value for each of the four luminance values. This is called the 4:2:0 format; two other formats are available (4:2:2 and 4:4:4, respectively known as two luminance per chrominance and one to one), which require data rates.

For each macroblock, a spacial offset difference between a macroblock in the predicted frame and the reference frame(s) is given if one exists (a motion vector), along with a luminance value and/or chrominance difference values (an error term) if needed. Macroblocks with no differences can be skipped except in intra frames. Blocks with differences are internally compressed, using a combination of a discrete cosine transform (DCT) algorithm on pixel blocks (or error blocks) and variable quantization on the resulting frequency coefficient (rounding off values to one of a limited set of values).

The DCT algorithm accepts signed, 9-bit pixel values and produces signed 12-bit coefficient. The DCT is applied to one block at a time, and works much as it does for JPEG, converting each 8×8 block into an 8×8 matrix of frequency coefficients. The variable quantization process divides each coefficient by a corresponding factor in a matching 8×8 matrix and rounds to an integer.

7.4.6 Quantization

As with JPEG the converted data is divided, or quantized, to remove higher-frequency components and to make more of the values zero. This results in numerous zero coefficients, particularly for high-frequency terms at the high end of the matrix. Accordingly, amplitudes are recorded in run-length form following a diagonal scan pattern from low frequency to high frequency.

7.4.7 Encoding

After the DCT and quantization state, the resultant data is then compressed using Huffman coding with a set of fixed tables. The Huffman code not only specifies the number of zeros, but also the value that ended the run of zeros. This is extremely efficient in compressing the zigzag DCT encoding method.

7.5 MPEG-1 decoder

The resultant encoded bitstream contains both video and audio data. These two elements are identified using system-level coding, which specifies a multiplex data format that allows multiplexing of multiple simultaneous audio and video streams as well as privately defined data streams. This coding includes the following:

- Synchronization data for decoded audio and video frames. Each frame contains a time stamp of frames so that a decoder can synchronize the decoding and playback of audio with the correct video sequence to achieve lip synchronization. The time-stamping gives the decoder a great flexibility in the playback. It even allows variable data rates, where frames can be dropped when they cannot be processed in time, and there is no loss of synchronization. The synchronization is achieved with a 90 kHz reference clock.
- Random frame access within the stream with absolute time identification. This is important when decoding in that the time reference can be independent of the environment.
- Data buffer management to prevent overflow and underflow errors. Frames are not necessarily stored in the consequentive time sequence. Buffers must be set-up to hold data temporarily for future decoding.

7.6 MPEG-1 audio compression

This will be covered in the Chapter 10.

7.7 MPEG-2

The orginal MPEG-1 specification proved so successful that, as soon as it was published, the MPEG committee started work on three derivatives called MPEG-2, MPEG-3 and MPEG-4. MPEG-2 has since been published. MPEG-3 was incorporated into MPEG-2 and work continues on the MPEG-4 standard.

The main drawback with the MPEG-1 standard are:

- It did not directly support broadcast television pictures as in the CCIR-601 specification. In particular, it did not support the interlaced mode of operation, although it could support the larger picture size of 720 × 480 at 30 fps.
- It was designed for a 1.5 Mbps bitstream.

Interlacing dramatically affects the motion estimation process because components could move from one field to another, and vice versa. As a result, the

MPEG-1 was poor at handling interlaced images.

The main objective of the MPEG-2 standard was to make it flexible so that it supported a number of modes, called profiles, with a wide range of options. These different profiles define algorithms that may be used. Each profile has a number of associated levels which define the parameters used.

7.7.1 MPEG-2 profiles and levels

MPEG-2 defines several profiles to provide a set of known configurations for different applications. It can be used from low-level video conferencing to high-definition television. If it were a unitary standard then each encoder and decoder would have to process the signals for the entire range of applications. This would, for example, burden a video conferencing system with the capability to handle very high definition images. The cost of doing this would make MPEG-2 unworkable for video conferencing.

Table 7.1 outlines the valid profiles and modes. The four main profiles are:

- Main – supports the main area of current development.
- Simple – same as the main profile, but the B-frames are not supported (so it is mainly used in software-based applications).
- SNR – enhanced signal-to-noise ratio.
- Spacial – enhanced main profile.
- High.

There are four main levels, these are:

- Low – the low level is similar to MPEG-1 standards and supports the CIF standard of 352×240 at 30 fps (or 352×288 at 25 fps for PAL). This equates to 3.05 Mpixels per second and a bit rate of up to 4 Mbps. The low-profile applications are aimed at the consumer market and offer quality similar to a domestic VCR.
- Main – the main level is able to support a maximum frame size of 720×480 at 30 fps (as defined in the CCIR-601 specification). This equates to 10.4 Mpixels per second and a bit rate of up to 15 Mbps. This level is aimed at the higher-quality consumer market.
- High 1440 – the high 1440 supports a maximum frame size of 1440×1152 at 30 fps. This is frame size is four times the CCIR-601 specification and equates to 47 Mpixels per second, giving a bit rate of up to 60 Mbps. This level is aimed at the high-definition TV (HDT) consumer market.
- High – the high level is able to support a maximum frame size of 1920×1080 at 30 fps. As with high 1440, the frame size is four times the CCIR-601 specification and gives a bit rate of up to 80 Mbps. This level is also aimed at the HDTV consumer market.

Table 7.1 MPEG-2 profiles and levels

	Simple (SP)	*Main (MP)*	*SNR*	*Spatial*	*High*
HIGH (HL)	Illegal	1920 × 1152, 60 fps	Illegal	Illegal	1920 × 1152, 60 fps 960 × 576, 30 fps
HIGH-1440 (H-14)	Illegal	1920 × 1152, 60 fps	Illegal	1440×1152, 60 fps 720 × 576, 30 fps	1440 × 1152, 60 fps 720 × 576, 30 fps
Main (ML)	720 × 576, 30 fps	720 × 576, 30 fps	720 × 576, 30 fps	Illegal	720 × 576, 30 fps 352×288, 30 fps
Low (LL)	Illegal	352×288, 30 fps	352×288, 30 fps	Illegal	Illegal

7.8 MPEG-2 system layer

The MPEG data stream consists of two layers:

- A compression layer.
- A system layer.

The system decoder splits the data stream into video and audio, each to be processed by separate decoders. Every 700 ms (or faster) a 33-bit system clock reference (SCR) is inserted into the data. For synchronization, the video and audio clocks are periodically set to the same value, every 700 ms (or faster), using 33-bit presentation time stamps (PSTs). These serve to invoke a particular picture or audio sequence.

The topmost layer of MPEG-1, the video sequence layer, can be expressed as:

```
video sequence is
{
   next start code
```

```
    repeat
    {
        sequence header
        repeat
        {
            group of pictures
        } while (next word in stream is group start code)
    } while (next word in stream is sequence header code)
    sequence end code
}
```

The video stream comprises a header, a series of frames, an end-of-sequence code. The stream contain periodic I-frames. These provide full images to be used as periodic references, and so allow reasonably random access to the data stream. Other frames are "predicted" using either preceding I-pictures (which create P-pictures) or a combination of preceding and following I-pictures (which creates B-pictures). The encoder decides how I, B and P pictures are interspersed and ordered. A typical sequence would be I-B-B-P. The order of pictures in the data stream is not the order of display; for example, the previous sequence would be sent as I-P-B-B.

7.9 Other MPEG-2 enhancements

MPEG-2 adds, among other features, an alternate scan order which further improves compression. All control data, vectors and DCT coefficients are further compressed using Huffman-like variable-length encoding.

7.10 MPEG-2 bit rate

With MPEG-2 this transmission rate can be reduced to 4 Mbps for PAL and SECAM and 3 Mbps for NTSC, thus giving a compression ratio of 40:1 to high-quality TV. MPEG-1 typically compresses TV signals to 1.2 Mbps, giving a compression ratio of 130:1. Unfortunately the quality is reduced to near VCR-type quality. Table 7.2 outlines these parameters. The base bit rate for a standard Ethernet network is 10 Mbps. This allows compressed video to be transmitted over the network when there is no other traffic on the network. The 4 Mbps rate will load the network by approximately 50%. Standardized and compression techniques will be discussed in the next chapter.

Table 7.2 Motion video compression.

Type	*Bit rate*	*Compression*	*Comment*
Uncompressed TV	162 Mbps	1:1	
MPEG-1	4 Mbps	40:1	VCR quality
MPEG-2	1.2 Mbps	130:1	PAL, SECAM TV quality

7.11 Exercises

7.1 Explain the main steps in MPEG coding.

7.2 State how standard TV differs in its interlacing from MPEG-1.

7.3 Explain why a frame must be divisible by 16 in both the x- and y-directions. Also give an example of a frame split into a number of macroblocks.

7.4 Explain how I, P and B frames might be used.

8 Speech and Audio Signals

8.1 Introduction

Speech and audio signals are normally converted into PCM, which can be stored or transmitted as a PCM code, or compressed to reduce the number of bits used to code the samples. Speech generally has a much smaller bandwidth than audio.

8.2 PCM parameters

Digital systems tend to be less affected by noise than analogue. The main source of noise is quantization noise which is caused by the finite number of quanitization levels converting to a digital code.

The main parameters in determining the quality of a PCM system are the dynamic range (DR) and the signal-to-noise ratio (SNR).

8.2.1 Quantization error

The maximum error between the original level and the quantized level occurs when the original level falls exactly halfway between two quantized levels. This error will be half the smallest increment or

$$\text{Max error} = \pm\frac{1}{2}\frac{\text{Full scale}}{2^N}$$

8.2.2 Dynamic range (DR)

The dynamic range is the ratio of the largest possible signal magnitude to the smallest possible signal magnitude. If the input signal uses the full range of the ADC then the maximum signal will be the full-scale voltage. The smallest signal which can be reproduced is one which toggles between one quantization level and the level above, or below. This signal amplitude, for an n-bit ADC, is the full-scale voltage divided by the number of quantization levels (that is, 2^n). Thus, for a linearly quantized signal:

$$\text{Dynamic range} = \frac{V_{max}}{V_{min}}$$

$$\text{Number of levels} = 2^n - 1$$

$$\text{Dynamic range} = 20\log\frac{V_{max}}{V_{max}/2^n - 1} \quad \text{dB}$$

$$= 20\log(2^n - 1) \quad \text{dB}$$

if 2^n is much greater that 1, then

$$\text{Dynamic range} \approx 20\log 2^n \ \text{dB}$$

$$= 20n\log 2 \ \text{dB}$$

$$\approx 6.02n \quad \text{dB}$$

Table 8.1 outlines the DR for a given number of bits. Normally the maximum number of bits is less than 20. The voltage ratio of a given number of bits is also given in square brackets [*ratio*]. For example an 8-bit system has a DR of 48.18 dB and the largest voltage amplitude is 256 times the smallest voltage amplitude. A 16-bit system has a DR of 96.33 dB and the largest voltage amplitude is 65 536 times the smallest voltage amplitude.

Table 8.1 Dynamic range of a digital system

Number of bits	*DR (dB) [ratio]*	*Number of bits*	*DR (dB) [ratio]*
1	6.02 [2]	11	66.23 [2 048]
2	12.04 [4]	12	72.25 [4 096]
3	18.06 [8]	13	78.27 [8 192]
4	24.08 [16]	14	84.29 [16 384]
5	30.10 [32]	15	90.31 [32 768]
6	36.12 [64]	16	96.33 [65 536]
7	42.14 [128]	17	102.35 [131 072]
8	48.16 [256]	18	108.37 [262 144]
9	54.19 [512]	19	114.39 [524 288]
10	60.21 [1 024]	20	120.41 [1 048 576]

8.2.3 Signal-to-noise ratio (SNR)

It can be shown that the SNR for a linearly quantized digital system is:

$$\text{SNR} = 1.76 + 6.02\,n\,\text{dB}$$

This proof is given in Appendix D. Table 8.2 outlines the SNR for a given number of bits. Normally the maximum number of bits is less than 20. The voltage ratio of a given number of bits is also given in square brackets [*ratio*]. For example an 8-bit system has an SNR of 49.92 dB and the largest rms voltage is 313.33 times the smallest rms voltage. A 16-bit system has an SNR of 96.33 dB and the largest rms voltage is 80 167.81 times the smallest rms voltage.

Table 8.2 Signal-to-noise ratio of a digital system

Number of bits	*SNR (dB) [ratio]*	*Number of bits*	*SNR (dB) [ratio]*
7	43.90 [156.68]	14	86.04 [20 044.72]
8	49.92 [313.33]	15	92.06 [40 086.67]
9	55.94 [626.61]	16	98.08 [80 167.81]
10	61.96 [1 253.14]	17	104.10 [160 324.5]
11	67.98 [2 506.11]	18	110.12 [320 626.9]
12	74.00 [5 011.87]	19	116.14 [641 209.6]
13	80.02 [10 023.05]	20	122.16 [1 282 331]

8.3 Differential encoding

Differential coding is a source-coding method which is used when there is a limited change from one value to the next. It is well suited to video and audio signals, especially audio, where the sampled values can only change within a given range. It is typically used in PCM (pulse code modulation) schemes to encode audio and video signals.

8.3.1 Delta modulation PCM

PCM coverts analogue samples into a digital code. Delta PCM uses a single-bit code to represent an analogue signal. With delta modulation a '1' is transmitted (or stored) if the analogue input is higher than the previous sample or a '0' if it is lower. It must obviously work at a higher rate than the Nyquist frequency, but because it uses only 1 bit, it normally uses a lower output bit rate. Figure 8.1 shows a delta modulation transmitter.

Initially the counter is set to zero. A sample is taken and if it is greater than the analogue value on the DAC output, the counter is incremented by 1, or it is decremented. This continues at a time interval given by the clock. Each time the present sample is greater than the previous sample, a '1' is transmitted; otherwise a '0' is transmitted. Figure 8.2 shows an example signal. The sampling frequency is chosen so that the tracking DAC can follow the input signal. This results in a higher sampling frequency, but because it only transmits one bit at a time, the output bit rate is normally reduced.

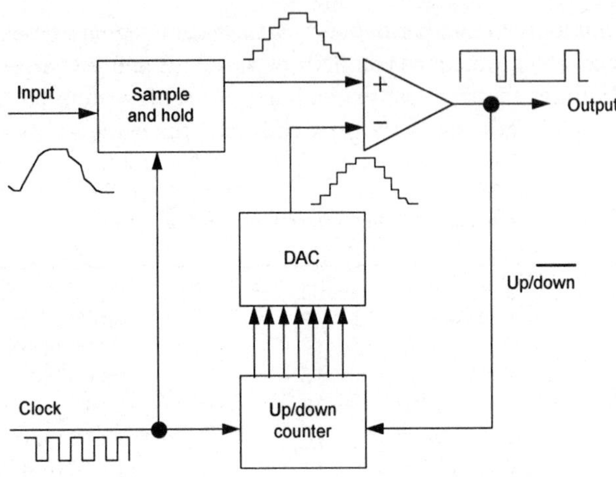

Figure 8.1 Delta modulation

Figure 8.3 shows that the receiver is almost identical to the transmitter except that it has no comparators.

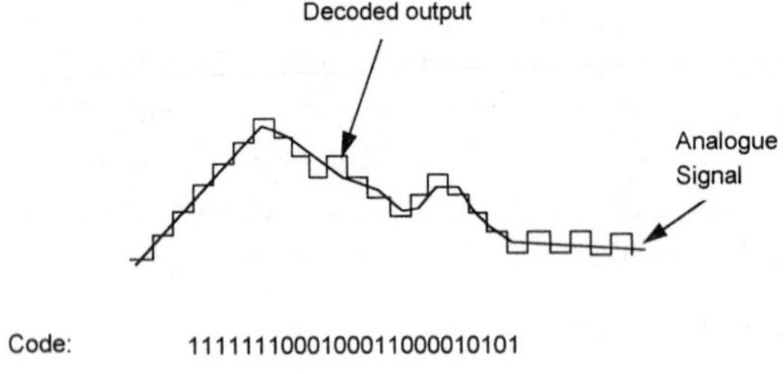

Figure 8.2 Delta modulator signal

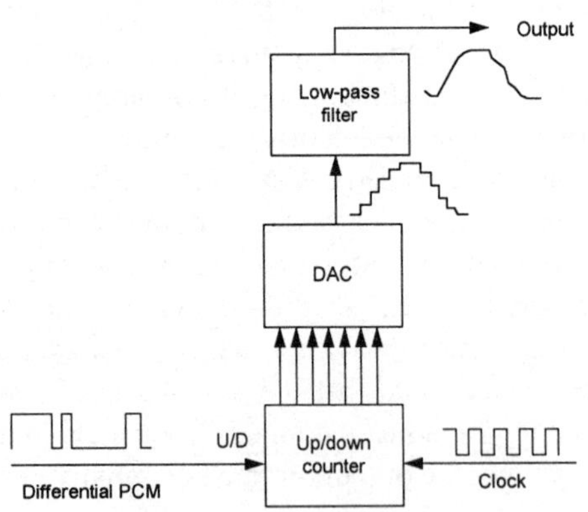

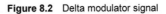

Figure 8.3 Delta modulator receiver

Two problems with delta modulation are granular noise and slope overload. Slope overload occurs when the signal changes too fast for the modulator to keep up; see Figure 8.4. It is possible to overcome this problem by increasing the clock frequency or increasing the step size.

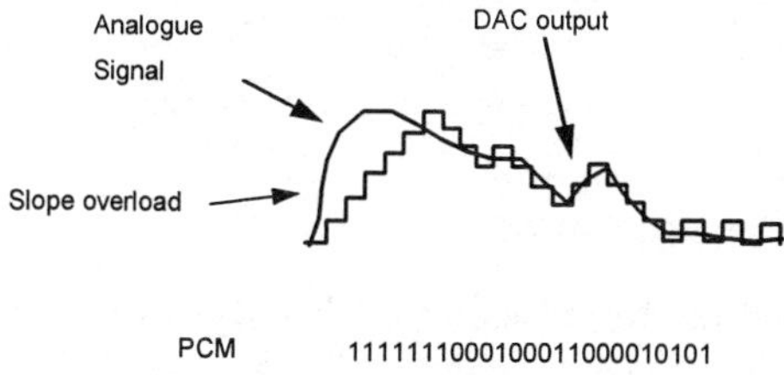

Figure 8.4 Slope overload

Granular noise occurs when the signal changes slowly in amplitude, as illustrated in Figure 8.5. The reconstructed signal contains a noise which is not present at the input. Granular noise is equivalent to quantization noise in a PCM system. It can be reduced by decreasing the step size, though there is a compromise between smaller step size and slope overload.

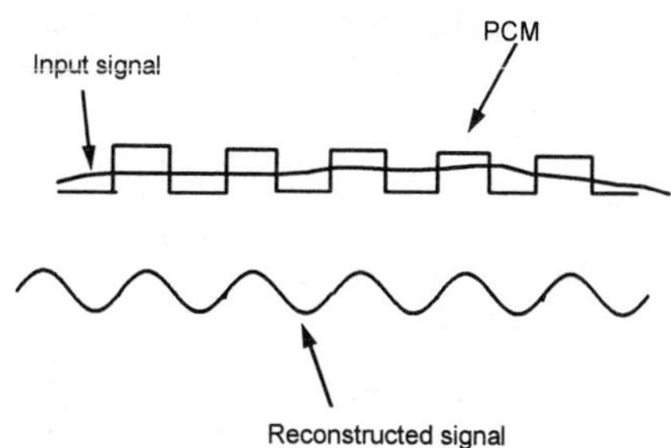

Figure 8.5 Granular noise

8.3.2 Adaptive delta modulation PCM

Unfortunately delta modulation cannot react to very rapidly changing signals and will thus take a relatively long time to catch them up (known as slope overload). It also suffers when the signal does not change much as this ends up in a square wave signal (known as granular noise). One method of reducing granular noise and slope overload is to use adaptive delta PCM. With this method the step size is varied by the slope of the input signal. The larger the slope, the larger the step size; see Figure 8.6. The algorithms usually depend on the system and the characteristics of the signal. A typical algorithm is to start with a small step and increase it by a multiple until the required level is reached. The number of slopes will depend on the number of coded bits, such as 4 step sizes for 2 bits, 8 for 3 bits, and so on.

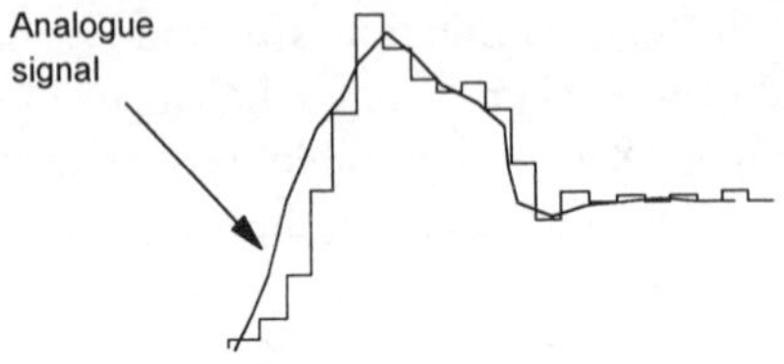

Figure 8.6 Variation of step size

8.3.3 Differential PCM (DPCM)

Speech signals tend not to change much between two samples. Thus similar codes are sent, which leads to a degree of redundancy. For example, in a certain sample it is likely the signal will only change within a range of voltages, as illustrated in Figure 8.7.

DPCM reduces the redundancy by transmitting the difference in the amplitude of two consecutive samples. Since the range of sample differences is typically less than the range of individual samples, fewer bits are required for DPCM than for conventional PCM.

Figure 8.8 shows a simplified transmitter and receiver. The input signal is filtered to half the sampling rate. This filter signal is then compared with the previous DPCM signal. The difference between them is then coded with the ADC.

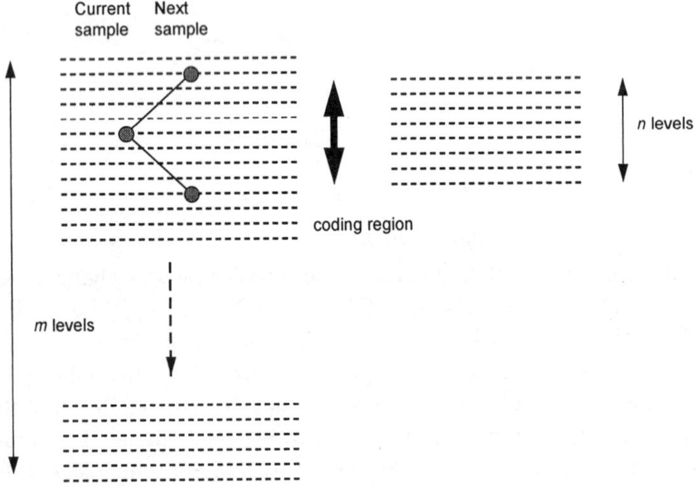

Figure 8.7 Normal and differential quantization

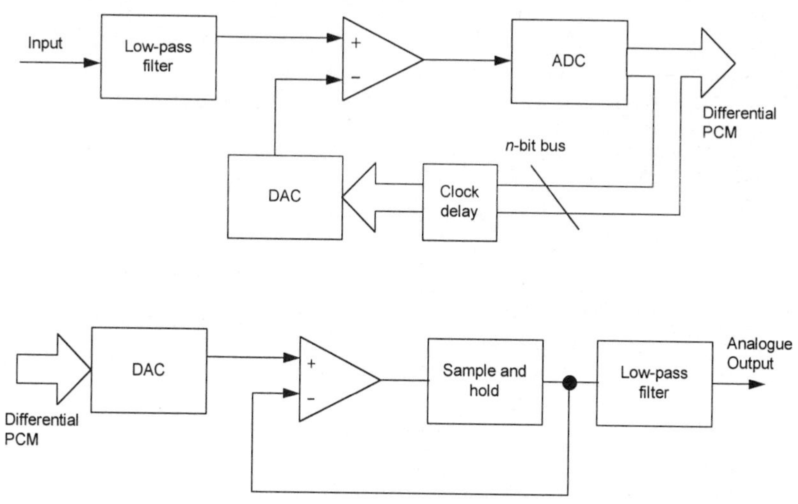

Figure 8.8 DPCM transmitter/receiver

8.3.4 Adaptive differential PCM (ADPCM)

ADPCM allows speech to be transmitted at 32 kbps with little noticeable loss of quality. As with differential PCM the quantizer operates on the difference between the current and previous samples. The adaptive quantizer uses a uniform quantization step M, but when the signal moves towards the limits of the quantization range, the step size M is increased. If it is around the center of the ranges, the step size is decreased. Within any other regions the step size hardly changes. Figure 8.9 illustrates this operation with a signal quanitized to 16 levels. This results in 4-bit code.

The change of the quantization step is done by multiplying the quanitization level, M, by a number slightly greater, or less, than 1 depending on the previously quantized level.

8.4 Speech compression

Subjective and system tests have found that 12-bit coding is required to code speech signals, which gives 4096 quantization levels. If linear quantization is applied then the quantization step is the same for quiet levels as for loud levels. Any quantization noise in the signal will be more noticeable at quiet levels than at loud levels. When the signal is loud, the signal itself swamps the quantization noise, as illustrated in Figure 8.10. Thus an improved coding mechanism is to use small quantization steps at low input levels and a higher one at high levels. This is achieved using non-linear compression.

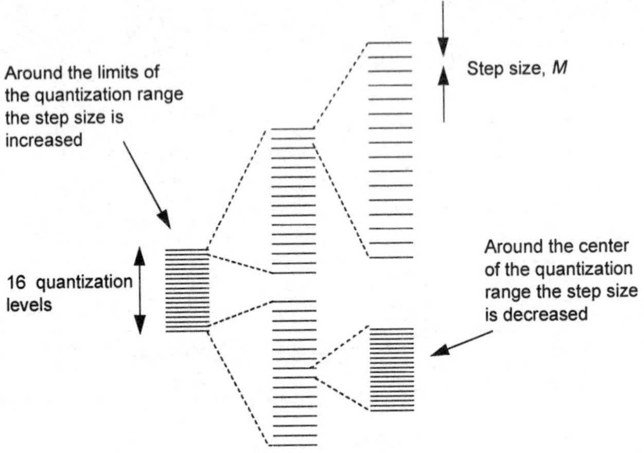

Figure 8.9 ADPCM quantization

The two most popular types of compression are A-Law (in European systems) and μ-Law (in the USA). These laws are similar and compress the 12-bit quantized speech code into an 8-bit compressed code. An example compression curve is shown in Figure 8.11.

As an approximation the two laws are split into 16 line segments. Starting from the origin and moving outwards, left and right, each segment has half the slope of the previous.

Using an 8-bit compressed code at a sample rate of 8000 samples per second gives a bit rate of 64 kbps. ISDN uses this bit rate to transmit digitized speech. Figure 8.12 shows a basic transmission system.

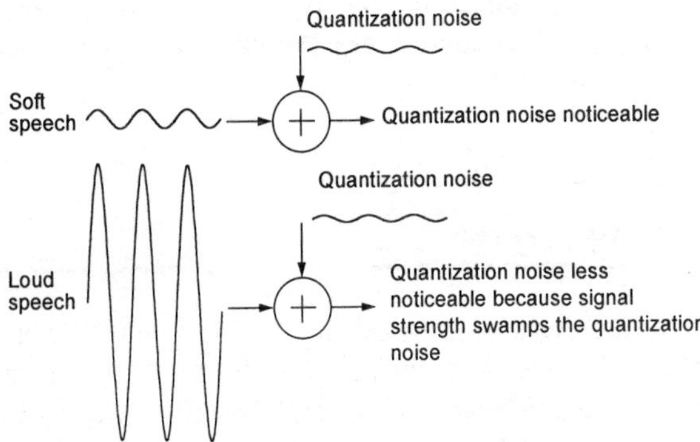

Figure 8.10 Quantization noise is more noticeable with low signal levels

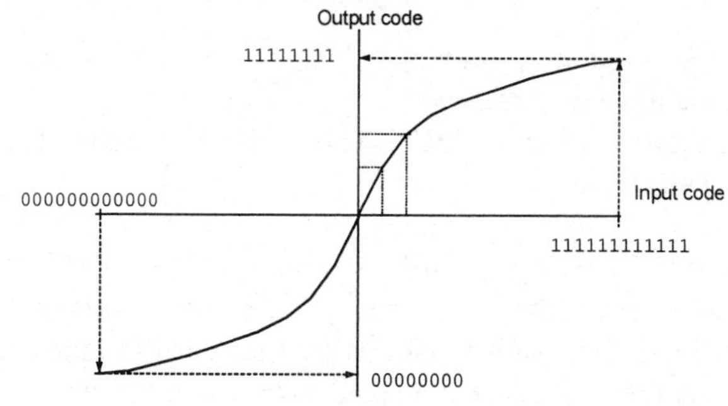

Figure 8.11 12-bit to 8-bit non-linear compression

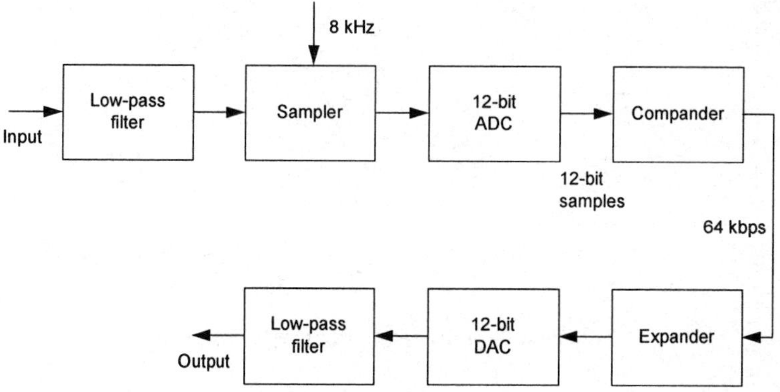

Figure 8.12 Typical PCM speech system

8.5 A-Law and μ-Law companding

The companding and expansion encoding is normally implemented using either μ-Law or A-Law. A-Law is used in Europe and in many other countries, whereas μ-Law is used in North America and Japan. Both were defined by the CCITT in the G.711 recommendation and both use non-uniform quantization step sizes which increase logarithmically with signal level. μ-Law uses the compression characteristic of:

$$y = \frac{\log(1+\mu x)}{\log(1+x)} \quad \text{for } x \geq 0$$

where y is the output magnitude
x is the input magnitude
μ is a positive factor which is chosen for the required compression characteristics

Figure 8.13 shows an example of μ-Law using $\mu=1$, $\mu=50$ and $\mu=255$. Using $\mu=0$ gives uniform conversion (linear quantization). Normally speech systems use $\mu=255$ as this characteristic is well matched to human hearing. An 8-bit implementation can achieve a small SNR a and dynamic range equivalent to that of a 12-bit uniform system.

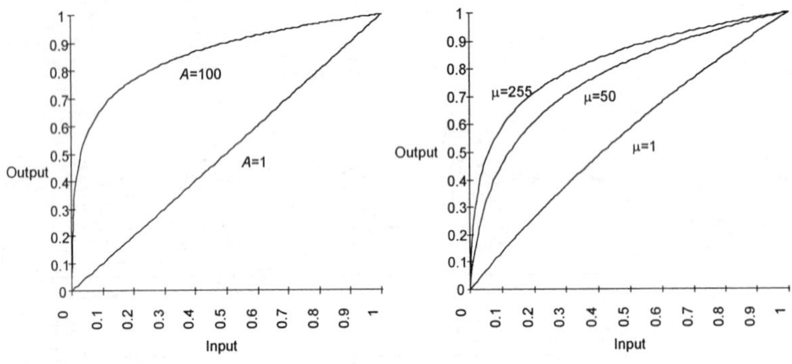

Figure 8.13 A-Law and μ-Law characteristics

The A-law also uses quantization characteristics that vary logarithmatically. Figure 8.13 shows an example of A-Law using $A=1$ and $A=100$. Most A-Law speech systems use $A=87.56$. The compression characteristic is:

$$y=\begin{cases}\dfrac{Ax}{1+\log A} & \text{for } 0\le|x|\le\dfrac{1}{A}\\[2ex] \dfrac{1+\log(Ax)}{1+\log A} & \text{for } \dfrac{1}{A}\le|x|\le 1\end{cases}$$

where A is a positive integer.

Figure 8.13 shows two input waveforms, 1 V peak to peak and 0.1 V peak to peak. It can be seen that the companding processes amplifies the lower amplitudes more than the large amplitudes. This causes low-amplitude speech signals to be boosted compared with loud speech. Also notice that the waveform has been distorted because the low amplitudes are amplified more than the large amplitudes.

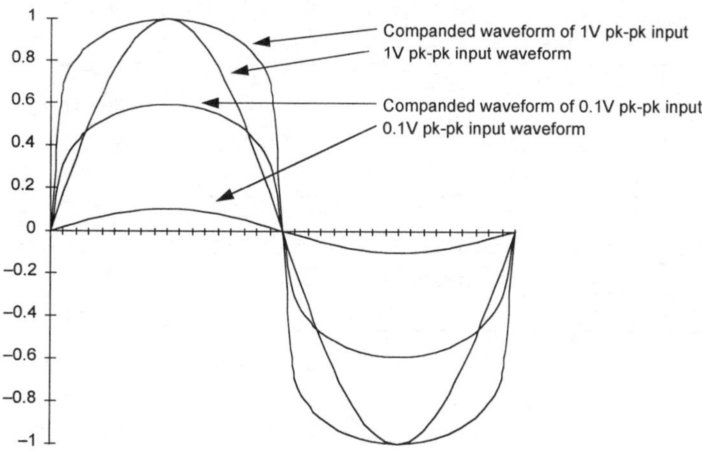

Figure 8.14 Effects of waveforms with μ-255 encoding

8.5.1 Digitally linearizable log-companding

The mathematical formulas for A-Law and μ-Law are normally approximated to a series of linear segments. This permits more precise control of the quantization characteristics. The chosen approximation used is to make the step sizes in consecutive segments change by a factor of 2. Figure 8.15 shows the characteristic of the piecewise linear conversion. It can be seen that the slope of each segment is twice the slope of the previous segment (although in A-Law 98.56, segment 0 and segment 1 have the same slope). Each segment has 16 quantization levels and there are 16 segments (8 for positive inputs and 8 for negative inputs). Thus 1 bit identifies the sign bit, 3 bits identify the segment (in the positive or negative part) and 4 bits identifies the quantization level. The 8-bit companded values thus take the form:

```
SLLLQQQQ
```

where `S` is the sign bit, `LLL` is the segment number and `QQQQ` is the quantization level within the segment.

Table 8.3 shows the conversion for A-Law 87.56. For example, if the input value is between 16 and 17, the companded value will be 001 0000. If this value is positive then the most significant bit will be a 1, thus the companded value will be 1001 0000.

Table 8.3 shows that the step sizes for the first two segments are the same (unity step size). Table 8.4 shows the μ-Law encoding table.

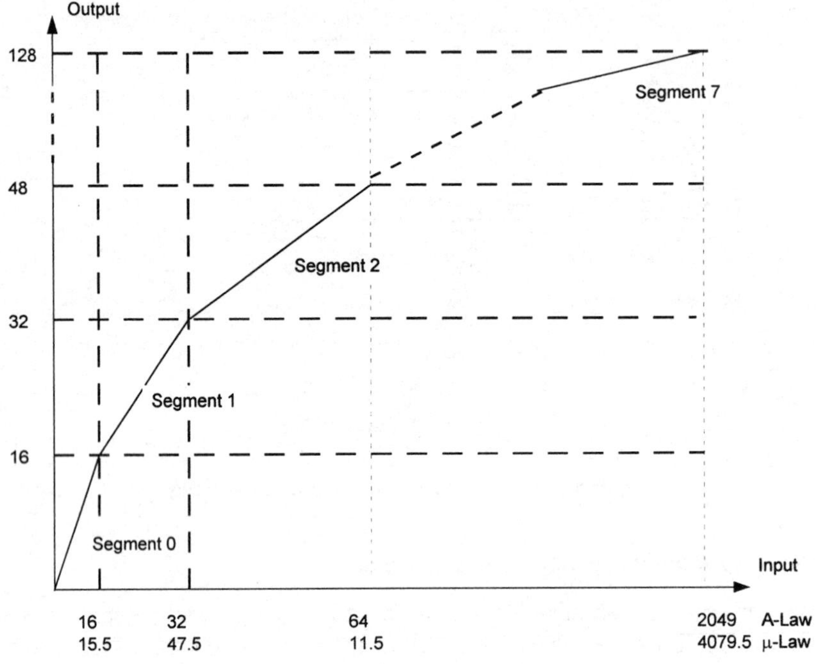

Figure 8.15 Piecewise linear compression for A-Law and μ-Law

Consider A-Law with the input range between +5 V and –5 V. An input voltage of +1 V will correspond to the input level of:

$$\text{Input} = \frac{1}{5} \times 2048 = 409.6$$

Referring to Table 8.3, this is within the segment from 256 to 512. The code will thus be S101XXXX. The level within the segment will be:

$$\text{Level} = \frac{409.6 - 256}{16} = 9.6$$

which corresponds to quantization level 9. Thus the companded value is:

```
01011001
```

Table 8.3 A-Law 87.56 encoding/decoding

Input	*Companded*	*Decoder level*	*Decoded level number*	*Step size*
0–1	000 0000	0	0.5	1
...	...	...	...	
15–16	000 1111	15	15.5	
16–17	001 0000	16	16.5	1
...	...	...	...	
31–32	001 1111	31	31.5	
32–34	010 0000	32	33	2
...	...	...	...	
62–64	010 1111	47	63	
64–68	011 0000	48	66	4
...	...	...	...	
124–128	011 1111	63	126	
128–136	100 0000	64	132	8
...	...	...	...	
248–256	100 1111	79	252	
256–272	101 0000	80	264	16
...	...	...	...	
496–512	101 1111	95	504	
512–544	110 0000	96	528	32
...	...	...	...	
992–1024	110 1111	111	1008	
1024–1088	111 0000	112	1056	64
...	...	...	...	
1984–2048	111 1111	127	2016	

Table 8.4 A-Law 87.56 encoding/decoding

Input	*Companded*	*Decoder level*	*Decoded level number*	*Step size*
0–0.5	000 0000	0	0	1
...	...	...	...	
14.5–15.5	000 1111	15	15	
15.5–17.5	001 0000	16	16.5	2
...	...	...	...	
45.5–47.5	001 1111	31	46.5	
47.5–51.5	010 0000	32	49.5	4
...	...	...	...	
107.5–111.5	010 1111	47	109.5	
111.5–119.5	011 0000	48	115.5	8
...	...	...	...	
231.5–239.5	011 1111	63	235.5	
239.5–255.5	100 0000	64	247.5	16
...	...	...	...	
479.5–495.5	100 1111	79	487.5	
497.5–527.5	101 0000	80	511.5	32
...	...	...	...	
975.5–1007.5	101 1111	95	991.5	

1007.5–1071.5	110 0000	96	1039.5	64
…	…	…	…	
1967.5–2031.5	110 1111	111	1999.5	
2031.5–2159.5	111 0000	112	2095.5	128
…	…	…	…	
3951.5–4079.5	111 1111	127	4015.5	

8.6 Speech sampling

With telephone-quality speech the signal bandwidth is normally limited to 4 kHz, thus it is sampled at 8 kHz. If each sample is coded with 8 bits then the basic bit rate will be:

$$\text{Digitized speech signal rate} = 8 \times 8 \text{ kbps} = 64 \text{ kbps}$$

Table 8.3 outlines the main compression techniques for speech. The G.722 standard allows the best-quality signal. The maximum speech frequency is 7 kHz rather than 4 kHz in normal coding systems; this is equivalent of 14 coding bits. The G.728 allows extremely low bit rates (16 kbps).

Table 8.5 Speech compression standards

ITU standard	*Technology*	*Bit rate*	*Description*
G.711	PCM	64 kbps	Standard PCM
G.721	ADPCM	32 kbps	Adaptive delta PCM where each value is coded with 4 bits
G.722	SB-ADPCM	48, 56 and 64 kbps	Subband ADPCM allows for higher-quality audio signals with a sampling rate of 16 kHz
G.728	LD-CELP	16 kbps	Low-delay code excited linear prediction for low bit rates

8.7 PCM-TDM systems

Multiple channels of speech can be sent over a single line using time division multiplexing (TDM). In the UK a 30-channel PCM system is used, whereas the USA uses 24.

With a PCM-TDM system, several voice band channels are sampled, converted to PCM codes, these are then time division multiplexed onto a single transmission media.

Each sampled channel is given a time slot and all the times slots are built up into a frame. The complete frame usually has extra data added to it such as synchronization data, and so on. Speech channels have a maximum frequency content of 4 kHz and are sampled at 8 kHz. This gives a sample time of 125 μs. In the UK a frame is built up with 32 time slots from TS0 to TS31. TS0 and TS16 provide extra frame and synchronization data. Each of the time slots has 8 bits, therefore the overall bit rate is:

Bits per time slot = 8
Number of time slots = 32
Time for frame = 125 μs

$$\text{Bit rate} = \frac{\text{No of bits}}{\text{Time}} = \frac{32 \times 8}{125 \times 10^{-6}} = 2048 \text{ kbps}$$

In the USA and Japan this bit rate is 1.544 Mbps. These bit rates are known as the primary rate multipliers. Further interleaving of several primary rate multipliers increases the rate to 6.312, 44.736 and 139.264 Mbps (for the USA) and 8.448, 34.368 and 139.264 Mbps (for the UK).

The UK multiframe format is given in Figure 8.16. In the UK format the multiframe has 16 frames. Each frame time slot 0 is used for synchronization and time slot 16 is used for signaling information. This information is submultiplexed over the 16 frames. During frame 0 a multiframe-alignment signal is transmitted in TS16 to identify the start of the multiframe structure. In the following frames, the eight binary digits available are shared by channels 1–15 and 16–30 for signaling purposes. TS16 is used as follows:

Frame 0 `0000XXXX`
Frames 1–15 `1234 5678`

where `1234` are the four signaling bits for channels 1,2,3, …, 15 in consecutive frames, and `5678` are the four signaling bits for channels 16,17, 18,… 31 in consecutive frames.

Thus in the first frame the `0000XXXX` code word is sent, in the next frame the first channel and the 16th channel appear in TS16, the next will contain the second and the 17th, and so on. Typical 4-bit signal information is:

`1111` – circuit idle/busy
`1101` – disconnection

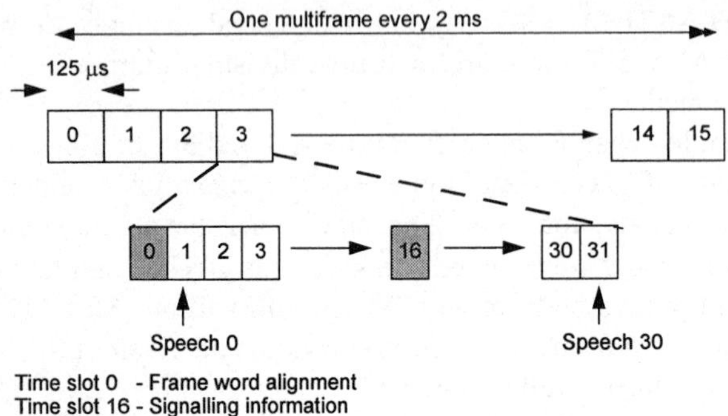

Figure 8.16 PCM-TDM multiframe format with 30 speech channels

TS0 contains a frame-alignment signal which enables the receiver to synchronize with the transmitter. The frame-alignment signal (X0011011) is transmitted in alternative frames. In the intermediate frames a signal known as a not-word is transmitted (X10XXXXX). The second binary digit is the complement of the corresponding binary digit in the frame-alignment signal. This reduces the possibility of demultiplexed misalignment to imitative frame-alignment signals.

Alternative frames:

TS0: X0011011
TS0: X10XXXXX

where X stands for don't care conditions.

8.8 Exercises

8.1 Show how the standard rate for PCM transmission is 64 kbps.

8.2 For μ-Law 255 plot the graph of the output voltage against input voltage and confirm Figure 8.15.

8.3 For A-Law with A=87.5 plot the graph is output voltage against output voltage, and confirm Figure 8.15.

8.4 Using A-Law 87.5 complete Table 8.6 for a voltage range between +5 V and –5 V.

Table 8.6 Conversion

Input voltage (V)	*Companded value*
5	
2.4	
–0.2	
–3.15	
	01010110
	10001100
	00111111

8.5 Using μ-Law 255 complete Table 8.7 for a voltage range between +5 V and –5 V.

Table 8.7 Conversion

Input voltage	*Companded value*
5V	
2.4V	
–0.2V	
–3.15V	
	01010110
	10001100
	00111111

8.6 Using a spreadsheet confirm the graph given in Figure 8.14.

9 Audio Signals

9.1 Introduction

The benefits of converting from analogue audio to digital audio are:

- The quality of the digital audio system only depends on the conversion process, whereas the quality of an analogue audio system depends on the component parts of the system.
- Digital components tend to be easier and cheaper to produce than high-specification analogue components.
- Copying digital information is relatively easy and does not lead to a degradation of the signal.
- Digital storage tends to use less physical space than equivalent analogue forms.
- It is easier to transmit digital data.
- Information can be added to digital data so that errors can be corrected.
- Improved signal-to-noise ratios and dynamic ranges are possible with a digital audio system.

Audio signals normally use PCM codes which can be compressed to reduce the number of bits used to code the samples.

For high-quality monochannel audio, the signal bandwidth is normally limited to 20 kHz, thus it is sampled at 44.1 kHz. If each sample is coded with 16 bits then the basic bit rate will be:

$$\text{Digitized audio signal rate} = 44.1 \times 16 \text{ kbps} = 705.6 \text{ kbps}$$

For stereo signals the bit rate would be 1.4112 Mbps. Many digital audio systems add extra bits from error control and framing. This increases the bit rate.

9.2 Principles

Digital audio normally involves the processes of:

- Filtering the input signal.
- Sampling the signal at twice the highest frequency.
- Converting it into a digital form with an ADC (analogue-to-digital converter).
- Converting the parallel data into a serial form.
- Storing or transmitting the serial information.
- When reading (or receiving) the data the clock information is filtered out using a PLL (phase-locked loop).
- The recovered clock is then used with a SIPO (serial-in parallel-out) converter to convert the data back into a parallel form.
- Converting the digital data back into an analogue voltage.
- Filtering the analogue voltage.

These steps are illustrated in Figure 9.1. The clock recovery part is important; there is no need to save or transmit separate clock information because it can be embedded into the data. It also has the advantage that a clock becomes jittery when it is affected by noise, thus if the clock information is transmitted over relatively long distances it will be jittery.

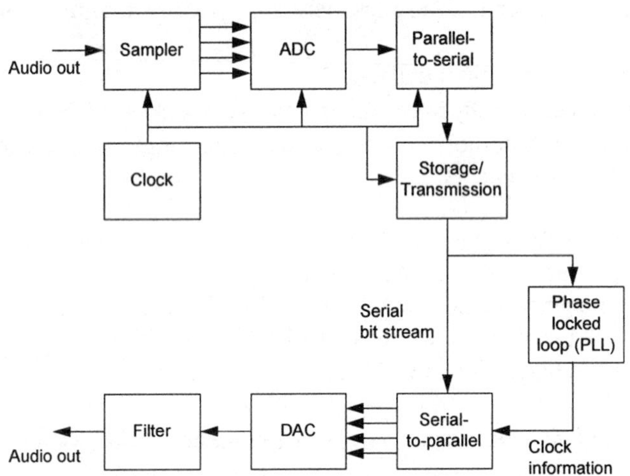

Figure 9.1 A digital audio system

9.3 Digital audio standards

Many techniques are available for compressed audio, such as:

- Broadcast audio distribution with a sample rate of 32 kHz (to provide a 15 kHz bandwidth) and with 14-bit linear coding.
- Professional recording with a sample rate of 48 kHz with 16- or 20-bit coding.
- Compact disk recording with a sample rate of 44.1 kHz (CD/EIAJ/RDAT) with 14- or 16-bit coding.

Most methods of coding for hi-fi applications use 16 bits per sample, but professional applications use 20 bits to give an extremely wide dynamic range of 120.41 dB.

Hi-fi audio is either sampled at 32 kHz, 44.1 kHz or 48 kHz. The 48 kHz sampling rate gives the highest bandwidth. It is possible to change from one sampling rate to another, but this normally degrades the signal quality.

9.3.1 SDIF-2 interconnection

The SDIF-2 (Sony Digital Interface) protocol was developed by the Sony Corporation and allows for a standard interface between professional audio equipment. In its standard form it uses a single-ended signal with three 75 Ω coaxial cables, one for each audio channel and another for data word clock synchronization. This clock is a symmetrical square wave at the sampling frequency and is common to both channels. Thus for a sampling frequency of 44.1 kHz the word clock period will be 22.67574 μs. This is the time to transmit a single 32-bit word.

The bits are transmitted over a serial communications line using the NRZ (non-return to zero) line code, which represents a 1 as a high level and a 0 as a low level.

Figure 9.2 shows the format of the 32-bit word. The bit fields are:

- Bits 1–20 are a 20-bit PCM data sample with the most significant bit sent first. If any bits are unused, they are padded with 0's.
- Bits 21–25 are used for control purposes and are reserved for future development.
- Bits 26 and 27 are emphasis bits which are determined when the sample is converted to a digital value (00 specifies that emphasis is not used and 11 specifies that emphasis is used).
- Bit 28 is the dubbing prohibition bit (0 specifies that dubbing is possible, whereas 1 specifies that dubbing is prohibited).
- Bit 29 is the block flag bit. This is used to identify the beginning of an SDIF-2 block. If it is a 1 then it specifies the start of a 256-word block, else it is a 0.
- Bits 30–32 are used for the synchronization pattern. The time period of the field is $3T$ and it consists of one transition from low to high, if bit 29 is high, or from high to low, if bit 29 is low. This transition always occurs

halfway through the bit field (that is at 1.5*T*). When the receive detects this pattern it knows to expect a following data block (which contains 256 data samples.

Data is transmitted in blocks of 256 PCM samples. The start of a block is identified with a 1 in bit 29 and a low-to-high transition in bit positions 30 to 32. Each of the following samples in the block then have a 0 in bit 29 and a high-to-low transition.

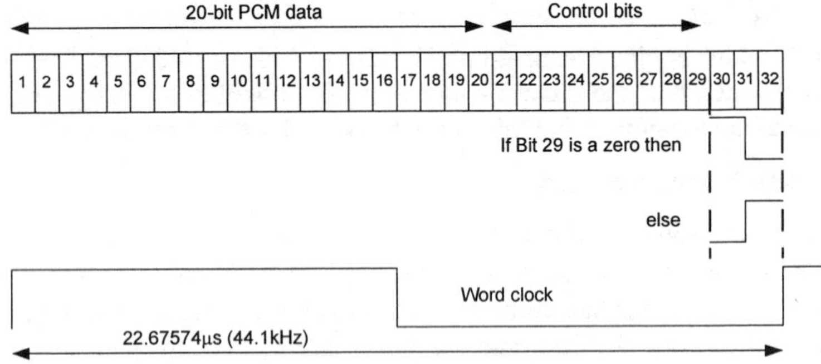

Figure 9.2 SDIF-2 word format

9.3.2 AES/EBU/IEC professional interface format

In 1985 the Audio Engineering Society (AES) defined a standard interface known as the AES3. In 1992 it was further enhanced to create the AES3-1995 standard. It uses serial transmission formation with linear PCM. It has the advantage over SDIF-2 in that a single channel carries the two audio channels and also allows for non-audio data. It is also self-clocking and self-synchronizing.

AES3 is similar to the EBU Tech. 3250-E standard developed by the European Broadcast Union (EBU) and the interface is commonly known as the AES/EBU digital interface. Other similar standards include:

- ANSI S4.40-1985 which was developed by the American National Standards Institute (ANSI).
- CCIR-Rec. 647 which was developed by the International Radio Consultative Committee (CCIR) and provides Rec. 647.
- EIAJ CP-340-type I which was developed by the Electronic Industries Association of Japan (EIAJ).
- IEC-958 which was developed by the International Electrotechnical Commission (IEC).

The EBU/AES/IEC system can use any sampling rate but is typically either 32 kHz, 44.1 kHz or 48 kHz. The line code used is biphase mark coding, which has the following rules:

- A 1 is coded as a high-to-low transition.
- A 0 is coded alternatively as a high or a low.

Thus a 0 has the same level for the complete bit period, whereas a 1 changes its level. This type of code is often known as binary frequency modulation. It has the advantage that it has no DC content and, in the worst case, there are two transitions for each transmitted bit (one of the transitions always occurs at the start of a bit period). These transitions allow the clock to be recovered from the signal. And all of the information is contained in bit transitions, not in levels. This makes the code less sensitive to noise, as noise tends to affect voltage levels more than it affects transitions. Figure 9.3 shows an example of biphase mark coding.

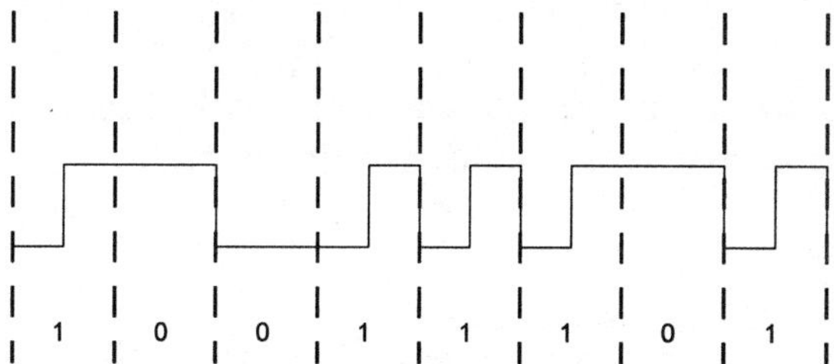

Figure 9.3 Biphase mark coding

Each 20-bit PCM data sample is framed with a 32-bit data word, as illustrated in Figure 9.4. The data word starts with a 4-bit sync/preamble block followed by 4-bit auxiliary data. Next follows the 20-bit data sample (the least significant bit is sent first). Finally there is a 4-bit auxiliary field which contains a validity flag, user data flag, channel status flag and a single parity bit. The error checking is obviously very limited and it should thus only be used on low bit error rate channels.

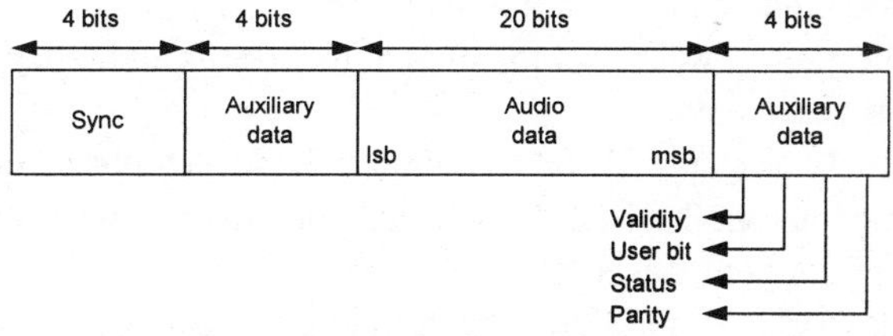

Figure 9.4 AES/EBU audio subframe format

For a 44.1 kHz sampling rate, the time for the complete packet will be 22.67574 μs. Thus the overall bit rate will be:

$$\text{Bit rate} = \frac{2 \times 32}{22.675\,74} = 2.822 \text{ Mbps}$$

9.3.3 *Frame format*

Each frame contains 32 bits and has 4 status bits at the end. These status bits are:

- V – which is the validity bit. If this bit is set to a 0 then the transmitted audio value is valid if it is a 1 then the value is invalid.
- U – which is used to added extra user information.
- C – which is the channel status data. Used to add extra information, its format is described in the next section.
- P – which is the parity bit. The parity bit is added so that the number of 1's in the frame is even (even parity).

A single frame contains two subframes, one for each channel. These are then used to build up to 192 frames, as illustrated in Figure 9.5.

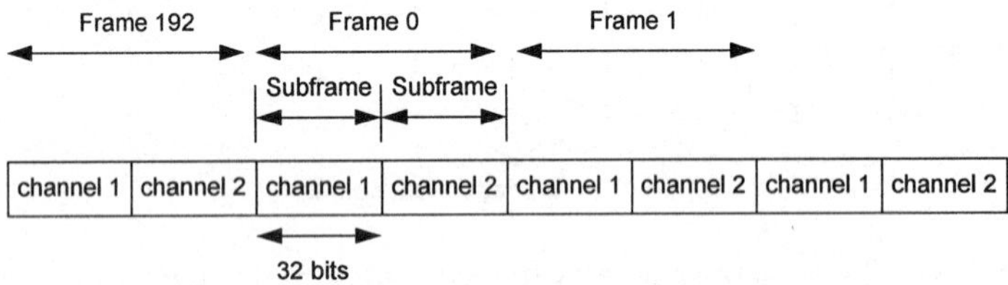

Figure 9.5 AES3 format of frames and subframes

The 192 bits from the subframes build up into a 24-byte channel status block. The format of these is:

- Bit 0. Professional flag. If it is a consumer application it is set to a 0, else if it is a professional application it is set to a 1.
- Bit 1. Audio flag. If this is set to a 0 then the data is audio data, else if it is a 1 it is non-audio data.
- Bits 2–4. Emphasis flag. A value of 000 indicates emphasis not indicated, 100 indicates no emphasis, 110 indicates 50/15 μs emphasis and 111 indicates CCITT J.17 emphasis.
- Bit 5. Lock flag. If it is a 0 then the sampling frequency is locked, else if it is a 1 then the sampling frequency is unlocked.

- Bit 6–7. Sampling frequency. If it is 00 then the sampling frequency is not indicated, if it is 01 then the sampling frequency is 48 kHz, 10 indicates 44.1 kHz and 11 indicates 32 kHz.
- Bits 8–11. Encoded channel mode. 0000 (not indicated), 0001 (two-channel), 0010 (single-channel), 0011 (primary/secondary), 0100 (stereo).
- Bits 12–15. User bits management. 0000 (default), 0001 (192-bit block,) 0010 (AES18) and 0011 (user defined).
- Bits 16–18. Auxiliary sample bits. 000 (20 bits), 001 (24 bits), 011 (reserved).
- Bits 19–21. Sample wordlength. For a 20-bit wordlength: 000 (default), 001 (19 bits), 010 (18 bits), 011 (17 bits), 100 (16 bits), 101 (20 bits).
- Bits 22–23. Reserved.
- Bits 24–31. Reserved.
- Bits 32–33. Reference.
- Bits 34–47. Reserved.
- Bits 48–79. Alphanumeric channel origin data (7-bit ASCII).
- Bits 80–111. Alphanumeric channel destination data (7-bit ASCII).
- Bits 112–143. Load sample address code (32-bit binary).
- Bits 144–176. Time-of-day code (32-bit binary).
- Bits 176–179. Reserved.
- Bits 180–183. Reliability flags.
- Bits 184–191. CRC using $x^8+x^4+x^3+x^2+1$.

9.3.4 S/PDIF consumer interconnection

The EAS3 standard is most often used in professional audio equipment. Consumer equipment normally use the S/PDIF (Sony/Philips Digital Interface) format. This is similar to the IEC-958 consumer format (known as type II). In some applications the EIAJ CP-340 type II format is used. The S/PDIF format is similar to EAS3 and in some case S/PDIF equipment can be connected to EAS3 equipment.

The S/PDIF status channel status block differs from the professional channel status block and only uses the first 32 bits of the 192-bit block. The format of these is:

- Bit 0. Consumer flag. If it is a consumer application it is set to a 0, else if it is a professional application it is set to a 1.
- Bit 1. Audio flag. If this is set to a 0 then the data is audio data, else if it is a 1 it is non audio data.
- Bit 2. Copy/copyright flag. If this is set to a 0 then the copyright is asserted and copying may be inhibited, else if it is a 1 then copying is permitted.
- Bits 3–4. Emphasis flag. A value of 000 indicates emphasis not indicated, 110 indicates 50/15 μs emphasis and 111 is reserved.

- Bit 6–7. Mode. If set to 00 then the mode is set to mode 0. The following bits define mode 0.
- Bits 8–14. Category code.
  - 00000000 (general).
  - 00000001 (experimental).
  - 0001xxx (solid-state memory).
  - 001xxxx (broadcast reception of audio: 0010000 is Japan, 0010011 is USA, 0011000 is Europe and 0010001 is electronic software delivery).
  - 010xxxx (digital/digital converters). 0100000 (PCM encoder/decoder), 0100010 (digital sound sampler), 0100100 (digital signal mixer), 0101100 (sample-rate converter).
  - 01100xx (ADC without copy info).
  - 01101xx (ADCs with copy info).
  - 0111xxx (broadcast reception of digital audio).
  - 100xxxx (Laser-Optical). 1000000 (CD – compatible with IEC-908), 1001000 (CD not compatible with IEC-908), 1001001 (MC – mini-disk).
  - 101xxxx (musical instruments, microphones, and so on). 1010000 (synthesizer), 1011000 (microphone).
  - 110xxxx (magnetic tape or disk). 110000 (DAT – digital audio tape), 1101000 (VCR) or 1100001 (DCC – digital compact cassette).
  - 111xxxx (reserved).
- Bit 15. L generation status. When category code is 100xxxx, 001xxxx or 0111xxx then 0 is original/commercial prerecorded) and 1 is no indication/first generation or higher. Else, 0 is no indication/first generation and 1 is original/commercially prerecorded.
- Bits 16–19. Source number. 0000 (unspecified), 0001 (source 1), 0010 (source 2) ... 1111 (source 15).
- Bits 20–23. Channel number. 0000 (unspecified), 0001 (channel A), 0011 (channel B) ... 1111 (channel O).
- Bits 24–27. Sample frequency. 0000 (44.1 kHz), 0100 (48 kHz), 1100 (32 kHz).
- Bits 28–29. Clock accuracy. 00 (±1000 ppm), 01 (variable pitch), 10 (±50 ppm)
- Bits 30–191. Reserved.

9.3.5 Serial copy management system

Many originators of hi-fi audio are extremely worried about consumers taking perfect copies of digital information. This has forced hi-fi manufacturers to install a serial copy management system (SCMS). With this system a user is allowed to make a copy from the original copyrighted material but cannot make a copy from the copy. It should be noted that SCMS allows a user to make multiple copies from the original source.

SCMS is included in S/PDIF (and IEC-958) but there are no copy protection bits for AES3. The SCMS circuit tests the following:

1. Detects the consumer bit. If it is a 1 (professional) then copying will always occur. If it is a 0 then copying depends on the following two cases.
2. Examines the second bit which is the copyright bit. If it is set to a 0 then copyright is enabled, else it is disabled. Thus copying will always occur when the copyright bit is a 1 (copyright disabled). If it is a 0 then copying will only occur when the L bit is set to indicate that it is an original.
3. Examines the L bit (bit 15) which indicates the generation of the copy. A 0 indicates that it is a copy, else it is an original. (For laser optical products and broadcasts this bit has an opposite effect: a 1 indicates a copy and a 0 indicates the original.)

Thus if the consumer bit is set (i.e. a 0) and the copyright bit is enabled (i.e. a 0) then a copy will only be made if the L bit is set to original (i.e. a 0).

It is the law in many countries that SCMS circuitry must be fitted in digital copying audio equipment. When a consumer copies from a CD source to a DAT tape, the L bit is set to indicate it is copy. The DAT source cannot then be copied to another DAT source.

9.3.6 *Other AES standards*

Other AES standards do exist:

AES10-1991 – which defines a multichannel interface format. This allows the interconnection of multichannel digital audio equipment. It is also equivalent to ANSI S4.43-1991.
AES11-1990 – which defines a digital audio reference signal (DARS). This allows the synchronization of digital audio equipment in studio situations.
AES18-1992 – which defines a method of adding extra information AES3 frames. The added information may include scripts, copyright, editing information, and so on.

9.4 Error control

Digital audio conversion has the advantage of being able to add extra bits to make it possible to either detect errors or correct errors. Generally the code used is either an error detection scheme (such as parity and CRC) or an error correction scheme (such as Hamming and Reed-Solomon codes). An error correction scheme allows a number of bits to be corrected but normally requires more bits for the error correction information. Error detection and correction will be discussed in Chapters 12 and 13.

9.5 Interleaving

Digital audio allows for other techniques which reduce losses in audio information. Interleaving is a technique which is used to conceal error by interleaving the samples in time. Errors normally occur in bursts thus consecutive samples which are interleaved will cause fewer consecutive errors. For example if data is sampled as:

1, 2, 3, 4, 5, 6, 7, 8, 9, 10, 11, 12, 13, 14,
15, 16, 17, 18, 19, 20, 21, 22, 23, 24, 25

To interleave them the samples are arranged into a number of columns and rows. In this case the 25 samples can be arranged into 5 rows with 5 columns to give:

1	2	3	4	5
6	7	8	9	10
11	12	13	14	15
16	17	18	19	20
21	22	23	24	25

Next the columns are read one at a time and reordered into a series again:

1, 6, 11, 16, 21, 2, 7, 12, 17, 22, 3, 8, 13,
18, 23, 4, 9, 14, 19, 24, 5, 10, 15, 20, 25.

This sequence can then be stored (or transmitted). If a burst error occurs (typically in a CD or in a communications channel) then it will have a lesser effect than storing the values consecutively. For example, suppose the 5th, 6th and 7th samples were corrupted. Then the recovered samples would be:

1, 6, 11, 16, X, X, X, 12, 17, 22, 3, 8, 13,
18, 23, 4, 9, 14, 19, 24, 5, 10, 15, 20, 25.

These would then be reordered to give:

1, X, 3, 4, 5, 6, X, 8, 9, 10, 11, 12, 13,
14, 15, 16, 17, 18, 19, 20, X, 22, 23, 24, 25.

It can be seen that in this sequence there are no consecutive errors. The values that are in error can be concealed by taking the average of the two samples on each side of them. For example the 2nd sample would be estimated as the average of the 1st and the 3rd sample; this is illustrated in Figure 9.6. If three consecutive samples had been in error then the resultant concealment would be much worse as there is a greater difference between the error-free samples.

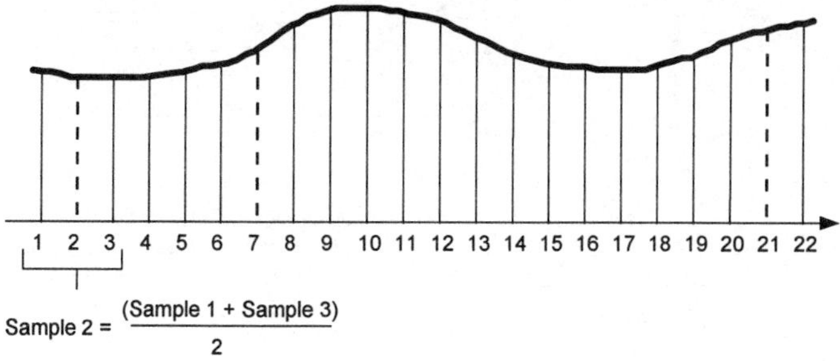

Figure 9.6 Concealment of errors

9.6 CD audio system

The CD audio system provides a good example of the techniques used in digital audio. Figure 9.7 shows a CD. The data is recorded onto the CD in a serial manner (that is one bit at a time) and it is stored in a continuous groove that spirals around the disk from the outside to the inside. A pit (or the lack of one) differentiates between a 0 and a 1. The diagram shows that a pit is only 1.6 μm long and that there is only 0.5 μm between grooves.

A CD uses a 44.1 kHz sampling rate with 16-bit PCM coding; it thus allows for 65 536 different levels. It can be proved that the dynamic range of a CD systems is 96.32 dB and the SNR is 98.08 dB (left as an exercise).

As CDs are susceptible to scratches and fingerprints then errors are likely to occur in bursts. Thus the CD uses an interlacing method to store the samples. It organizes the samples into 4 columns and 5 rows. Thus a 20 sample input block would be arranged logically as:

1	2	3	4
5	6	7	8
9	10	11	12
13	14	15	16
17	18	19	20
21	22	23	24

This would then be stored on the disk as:

1, 5, 9, 13, 17, 21, 2, 6, 10, 14, 1, 8, 22,
3, 7, 11, 15, 19, 23, 4, 8, 12, 16, 20, 24.

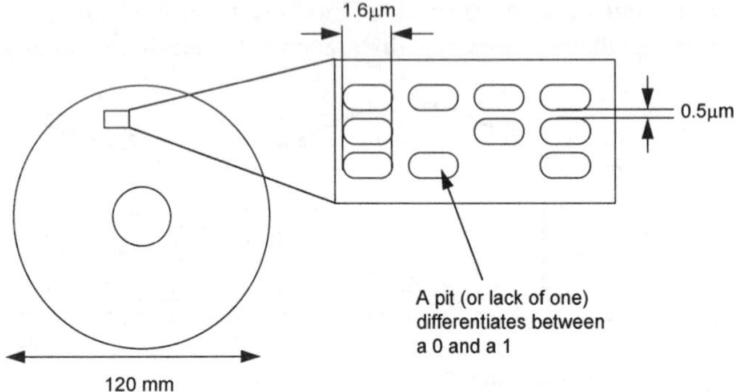

Figure 9.7 CD showing pits

Figure 9.8 shows the encoding of the data onto the disk. Initially the 16-bit word is divided into two 8-bit symbols. These symbols are then modulated by a process known as 8-to-14 bit modulation (EFM), where the 8-bit data is changed to 14-bit data through a lookup table. The EFM process reduces the CD system's sensitivity to optical tolerances in the disk player. After this, 3 bits of subcode are added to the 14-bit code. This subcode contains sync, index and time information. The resulting 17-bit code is then stored to the disc with two consecutive left-channel samples then right channels. An error correcting code is added at the end of each block. EFM code is be discussed in more detail in Appendix I.

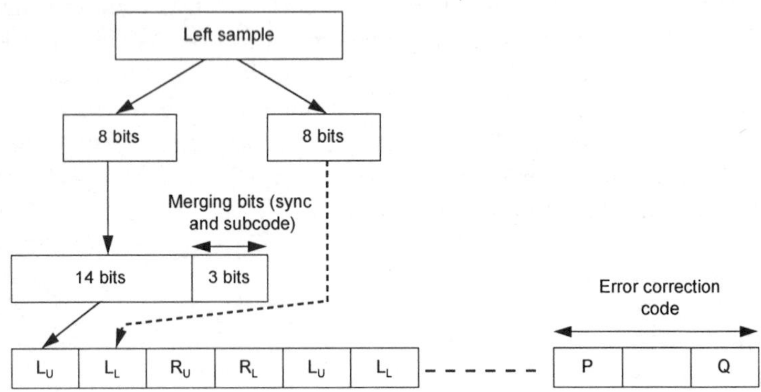

Figure 9.8 Encoding of data onto a CD

A few example codes for the 8-to-14 conversion are:

01101010 10010001000010
01101011 10001001000010
01101100 01000000100010

The advantage of the interleaving is not only that errors can be concealed by using interpolation, but also that it reduces the chances of losing sync information (which may cause the disk to jump).

9.7 Digital audio compression

There are several standard schemes for the compression of audio data; they include:

- MUSICAM – which has been adopted as one of the layers of the MPEG standard. It reduces the bit rate to 192 kbps.
- AC-1 – which is used in satellite relays of television and FM programs and gives data rates of around 512 kbps (3:1 ratio).
- AC-2 – which is applied to many applications, including sound cards, studio/transmitter links and so on. It gives data rates of around 256 kbps (6:1 ratio).
- AC-3 – which gives 6 channels of audio (left, right, surround-left, surround-right, center and sub-woofer) and gives data rates of around 512 kbps (6:1 ratio).
- MPEG – can compress to around 64 kbps for a single audio channel (monophonic channel).

Table 9.1 Audio compression

Type	*Bit rate*	*Compression*	*Comment*
Uncompressed audio (mono)	705.6 kbps	1:1	
Uncompressed audio (stereo)	1.4112 Mbps	1:1	
AC-1	512 kbps	3:1	
AC-2	256 kbps	6:1	See above
AC-3	512 kbps	6:1	See above
MPEG (mono)	64 kbps	11:1	Monophonic
MUSICAM (stereo)	192 kbps	7:1	Stereophonic

9.8 The 44.1 kHz sampling rate

The 44.1 kHz sampling rate is actually derived from composite video applications. In the early days of digital audio, video recorders were adapted to store digital audio information. Thus the sampling rate was constrained to relate to the field rate and field structure of the television standard used. Thus there had to be an integer number of samples for each usable TV line in the field.

NSTC has a refresh rate of 60 Hz, 525 lines (of which 490 are active lines). As the signal is interlaced, only 245 lines are active for each transmission. So if 3 samples are taken per lines then the sampling rate is:

$$60 \times 245 \times 3 = 44.1\ \text{kHz}$$

9.9 Exercise

9.1 The three main hi-fi quality sampling rates are 32 kHz, 44.1 kHz and 48 kHz and the main classifications are broadcast quality, CD quality and professional quality. State which sampling rate normally goes with which classification.

9.2 Explain how the start of a SDIF-2 block is identified.

9.3 Explain how the synchronization bits are used in SDIF-2.

9.4 Explain how the synchronization bits are used in SDIF-2.

9.5 Show that the bit rate for a single-channel SDIF-2 block is 1.411 Mbps.

9.6 Explain, using an example, the line coding used in AES3 code. Also state its advantages over normal NRZ coding.

9.7 Explain how the serial copy management system works and give situations in which copy is possible or not possible.

9.8 Explain the advantages of interleaving samples.

9.9 Discuss the main steps taken when encoding data to a CD.

9.10 An AES/EBU packet has 32 bits. When the sampling rate is 44.1 kHz the time for both audio channels will be 22.67574 μs. This gives an overall bit rate of 2.822 Mbps. Determine the bit rates for 32 kHz and also for 48 kHz.

9.11 PAL video signals have a 50 Hz refresh rate, 625 lines (of which there are 37 blanking lines). Show how these parameters relate to the digital audio sample rate of 44.1 kHz.

10 Audio Compression (MPEG-Audio and Dolby AC-3)

10.1 Introduction

CD-quality stereo audio requires a bit rate of 1.411200 Mbps (2 × 16 bits × 44.1 kHz). A single-speed CD-ROM can only transfer at a rate of 1.5 Mbps, and this rate must include both audio and video. Thus there is a great need for compression of both the video and audio data. The need to compress high-quality audio is also an increasing need as consumers expect higher-quality sound from TV systems and the increasing usage of digital audio radio.

A number of standards have been put forward for digital audio coding for TV/video systems. One of the first was MUSICAM which is now part of the MPEG-1 coding system. The FCC Advisory Committee considered several audio systems for advanced television systems. There was generally no agreement on the best technology but finally they decided to conduct a side-by-side test. The winner was Dolby AC-3 followed closely by MPEG. Many cable and satellite TV systems now use either MPEG or Dolby AC-3 coding.

10.2 Psycho-acoustic model

MPEG and Dolby AC-3 use the psycho-acoustic model to reduce the data rate, which exploits the characteristics of the human ear. This is similar to the method used in MPEG video compression which uses the fact that the human eye has a lack of sensitivity to the higher-frequency video components (that is, sharp changes of color or contrast). The psycho-acoustic model allows certain frequency components to be reduced in size without affecting the perceived audio quality as heard by the listener.

10.2.1 Masking effect

A well-known effect is the masking effect. This is where noise is only heard by a person when there are no other sounds to mask it. A typical example is high-frequency hiss from a compact cassette when there are quiet passages of music. When there are normal periods of music the louder music masks out the quieter hiss and they are not heard. In reality, the brain is masking out the part of the sound it wants to hear, even though the noise component is still there. When there is no music to mask the sound then the noise is heard.

Noise, itself, tends to occur across a wide range of frequencies, but the masking effect also occurs with sounds. A loud sound at a certain frequency masks out a quieter sound at a similar frequency. As a result the sound heard by the listener appears only to contain the loud sounds; the quieter sounds are masked out. The psycho-acoustic model tries to reduce the levels to those that would be perceived by the brain.

Figure 10.1 illustrates this psycho-acoustic process. In this case a masking level has been applied and all the amplitudes below this level have been reduced in size. Since these frequencies have been reduced in amplitude, then any noise associated with them is also significantly reduced. This basically has the effect of limiting the bandwidth of the signal to the key frequency ranges.

The psycho-acoustic model also takes into account non-linearities in the sensitivity of the ear. Its peak sensitivity is between 2 and 4 kHz (the range of the human voice) and it is least sensitive around the extremes of the frequency range (i.e. high and low frequencies). Any noise in the less sensitive frequency ranges is more easily masked, but it is important to minimize any noise in the peak range because it has a greater impact.

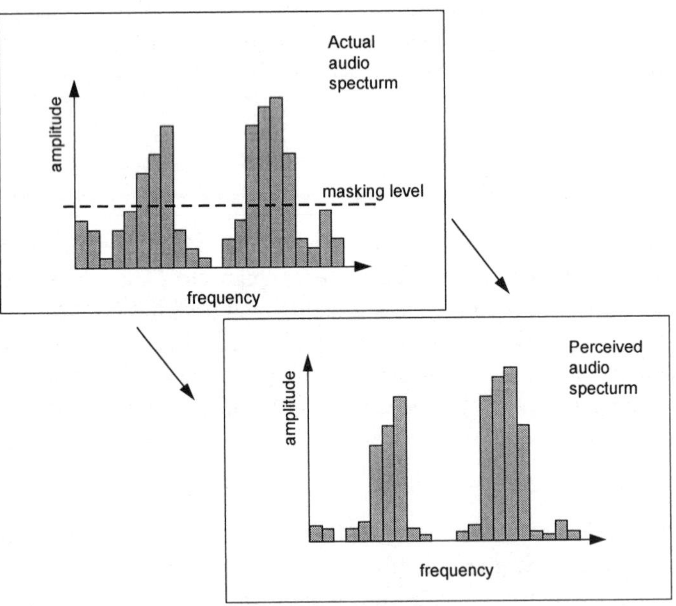

Figure 10.1 Difference between actual and perceived audio spectrum

Masking can also be applied in the time domain, where it can be applied just before and after a strong sound (such as a change of between 30 and 40 dB). Typically, premasking occurs for about 2–5 ms before the sound is perceived by the listener and the post-masking effect lasts for about 100 ms after the end of the source.

10.3 MPEG audio coding

MPEG basically has three different levels:

- MPEG-Audio Level I – uses a psycho-acoustic model to mask and reduce the sample size. It is basically a simplified version of MUSICAM and has a quality which is nearly equivalent to CD-quality audio. Its main advantage is that it allows the construction of simple encoders and decoders with medium performance and which will operate fairly well at 192 or 256 kbps.
- MPEG-Audio Level II – which is identical to the MUSICAM standard. It is also nearly equivalent to CD-quality audio and is optimized for a bit rate of 96 or 128 kbps per monophonic channel.
- MPEG-Audio Level III – which is a combination of the MUSICAM scheme and ASPEC, a sound compression scheme designed in Erlangen, Germany. Its main advantage is that it targets a bit rate of 64 kbps per audio channel. At that speed, the quality is very close to CD quality and produces a sound quality which is better than MPEG Level-II operating at 64 kbps.

The three levels are basically supersets of each other with Level III decoders being capable of decoding both Level I and Level II data. Level II is the most frequently used of the three standards. Level I is the simplest, while Level III gives the highest compression but is the most computational in coding.

The forward and backward compatible MPEG-2 system, following recommendations from SMPTE, EBU and others, has increased the audio capacity to five channels. Figure 10.1 shows an example of a 5-channel system; the key elements are:

- A center channel.
- Left and right surround channels.
- Left and right channels (as hi-fi stereo).

MPEG-2 also includes a low-frequency effects channel (called LFE, essentially a sub-woofer). This has a much lower bandwidth that the other channels. This type of system is often called a 5.1-channel system (5 main channels and LFE channel).

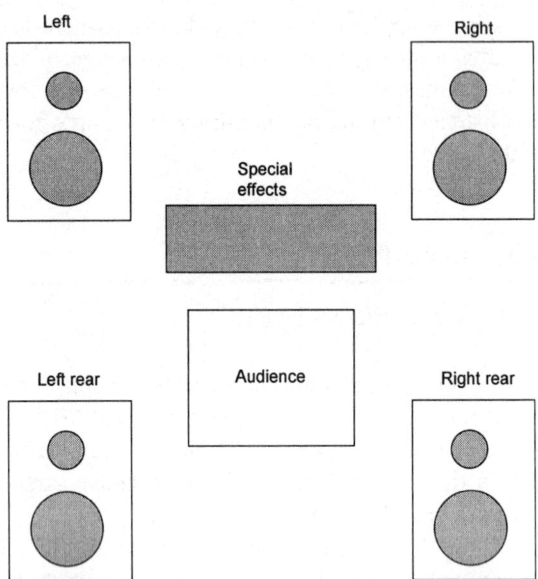

Figure 10.2 A 5.1 channel audio surround sound system

One of the main objectives of MPEG-2 was to make it backward compatible with MPEG-1. For this reason the MPEG-1 specification was extended to include capability for 5.1 channels so that it allowed other MPEG-1 decoders to extract the basic stereo pairs while newer decoders could recover all the channels. MPEG decoders are also forward compatible as they can recover MPEG-1 compliant audio.

MPEG audio systems break up the input signal into uniform segments with a certain number of audio samples, called frames. These frames are then filtered by digital filters into subbands of audio energy for each frame. Normally there are 32 subbands, so the subband frequency is divided by 32 (F_s/32).

As with MPEG and JPEG video compression, the subband sub-samples, are then quantized by means of a quantization scale factor. This compands the data and compresses the dynamic range of the audio in each subband. As with video compression, it results in a set of coefficients for each subband for each frame period.

At the same time as the calculation of the subband coefficients, a frequency-dependent masking threshold is calculated for each frame. A psychoacoustic perceptual model is used to calculate this threshold. It is then used to alter the bit allocation process for each subband where bits in the encoder output are dynamically allocated to each subband, subject to the ratio of the signal within the subband and its masking function. This process is illustrated in Figure 10.3.

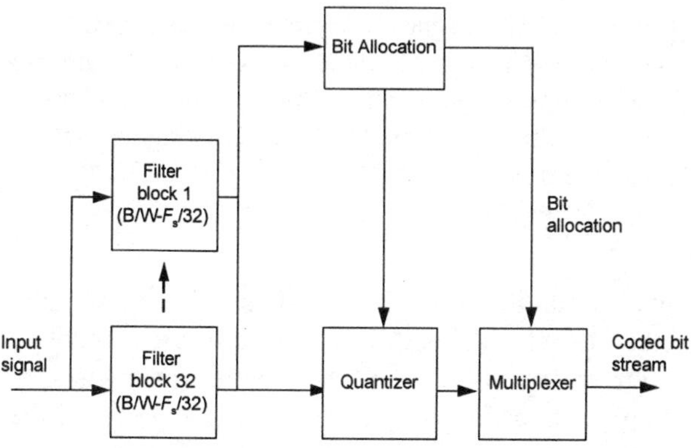

Figure 10.3 Forward bit allocation encoder

As with video compression, various methods are then applied to the quantized data, such as variable-length coding.

MPEG audio has three main levels. Level 1 gives the basic functionality while Level II adds more sophisticated processing in deriving the quantization scale factor and additional coding of the bit allocation. Level III provides for additional decomposition of the frequency components in each subband through the use of a modified discrete cosine transform (MDCT).

The decoding process is essentially reversed, with the bits for each subband de-quantized and applied to band-limited digital filters that reconstruct a PCM audio signal for their subband.

10.3.1 MPEG levels

The three different levels of MPEG vary in the method they use to compress the audio and in their bit rates.

Level I

The MPEG Level I coder gives medium performance with bit rates of 128 kbps per channel. It has been used by Philips in the digital compact cassette (DCC). The Level I coder initially transforms the audio signal into frequency information using a fast Fourier transform. Next the frequency information is passed into the 32 subbands filters. A 44.1 kHz sampling rate gives a subband bandwidth of 689.0625 Hz. These filters allow the processing of each subband using a slightly different acoustic model and thus achieve a better and more accurate encoding. A polyphase filter band is normally used to create the subbands as they are relatively fast.

After the subband processing, the output information is processed using the psycho-acoustic model. There are two acoustic models: model 1 and model 2. Levels I and Level II use model 1 and Level III uses model 2. In general, model 1 is the least complex.

Model 1 processes 512 samples at a time (it uses a 512 sample window). As a result, the audio is analyzed and processed in separate blocks. Level I encoding uses 32 samples from each of the 32 subbands. This gives a frame size of 384 samples, which are centered by calculating an offset into the middle of the 512-sample frame.

Level II

Level II is aimed at bit rates of about 128 kbps and has a slightly higher compression rate than Level I. As with Level I it also uses model 1. The sample size is increase from 384 to 1152, and the sample window is also increased to 1024. This sample window is not quite large enough to fit all the samples; to cope with this, the encode performs two analyses. The first analysis takes the first 576 samples and centers them in a 1024-sample window, and the second analysis repeats the process with the second 576 samples. This results in higher signal-to-noise ratios for the final output as it effectively selects the best noise-masking values.

Level III

Level II is aimed at bit rates of about 64 kbps per channel and has a slightly higher compression rate than Level II. The sample size is the same as Level II and the same split processing techniques are used. The increase in compression is achieved by employing a more sophisticated model 2 with several important differences.

10.3.2 MPEG-1

The basic specification MPEG-1 system provides two channels of stereo at sampling rates of either:

- 32 kHz (broadcast communications equipment).
- 44.1 kHz (CD-quality audio).
- 48 kHz (professional sound equipment).

Table 10.1 gives the main parameters for MPEG-1 audio.

Levels II and III have 1152 samples. Thus for 32 kHz sampling the frame duration will be 36 ms, which gives one sample every 31.25 μs, as illustrated in Figure 10.3. Level I has only 384 samples thus the frame time is reduced to one-third of the value for Levels II and III. Each frame has 32 subbands, and the sampling frequency at the output of each filter in the sampling frequency (F_s) is divided by 32 ($F_s/32$). Since the Nyquist criterion states that the highest frequency that can be represented in a sampled system is half the sampling rate, the bandwidth of the subband is $F_s/64$. For 32 kHz sampling, the filters have output sample rates of 1 kHz and bandwidths of 500 Hz. The reason that the audio bandwidth is slightly less than half the Nyquist frequency is that it takes into account imperfect filters.

Table 10.1 MPEG-1 parameters

	32 kHz	*44.1 kHz*	*48 kHz*
Audio bandwidth	15 kHz	20.6 kHz	22.5 kHz
Frame duration – Level I (384 samples)	12 ms	8.7 ms	8 ms
Frame duration – Level II (1152 samples)	36 ms	26.25 ms	24 ms
Frame duration – Level III (1152 samples)	36 ms	26.25 ms	24 ms
Number of subbands	32	32	32
Subband sampling (F_s/32)	1 kHz	1.378 kHz	1.5 kHz
Subband bandwidth (F_s/64)	500 Hz	689.06 kHz	750 Hz
Bit rate (Level 2/stereo)	192 kbps	256 kbps	256 kbps
Bit rate (Level 2/5.1-channel)	256 kbps	384 kbps	384 kbps

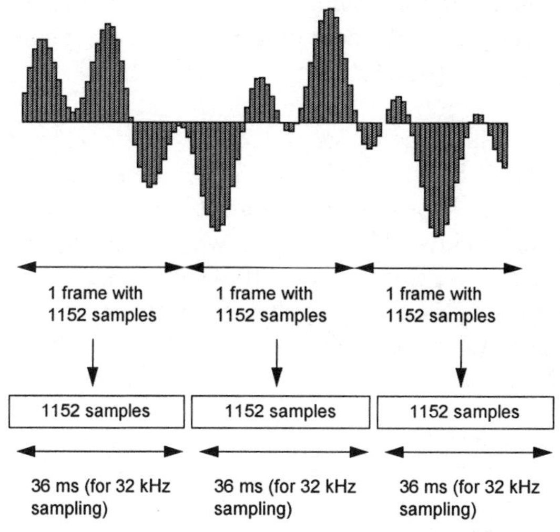

Figure 10.4 Frame for MPEG Levels 2 and 3

The MPEG-1 version of the MPEG audio system provides for joint stereo coding, which exploits stereophonic irrelevance. The method is called intensity stereo coding; it removes redundancy in stereo audio information by retaining only the energy envelope of the right channel and the left channel at high frequencies.

The basic MPEG-1 audio frame, after encoding, includes four basic types of information:

- A header.

- A cyclic redundancy code (CRC) for error detection.
- Audio data.
- Ancillary data.

The MPEG file also contains bit allocation information, scale factor selection information, scale factors, and subband samples.

10.3.3 MPEG-2

MPEG-2 is an extension of MPEG-1 and is both forward and backward compatible with it. It adds the following:

- Three additional sampling frequencies (16 kHz, 22.05 kHz and 24 kHz).
- Up to four more channels, permitting 5-channel surround sound.
- Support for low-frequency effects.
- Support for separate audio material in different configurations. This can be used in multilingual transmission of a program or to include a single channel of clean dialogue for the hard of hearing, or even a commentary channel for the visually impaired.

MPEG-2 can deliver hi-fi quality stereo sound with 256 kbps and 5.1-channel surround sound with LFE in 384 kbps. A 16 kHz sampling rate quality audio gives a bit rate of 128 kbps (which is compatible with standard rate ISDN). This gives an audio bandwidth of 7.5 kHz, which is large enough for speech and reasonable for low-quality audio. MPEG-2 achieves compatibility with MPEG-1 because it hides the extra data with an ancillary data field. An MPEG-1 decoder ignores the ancillary data file. Table 10.2 gives the main parameters of MPEG-2. Like MPEG-1 all the sampling rates have 32 subbands.

10.4 Backward/forward adaptive bit allocation methods

MPEG uses a forward adaptive bit allocation method, as shown in Figure 10.5. This technique makes bit allocation decisions adaptively, based on signal content. These decisions are then sent from the encoder to the decoder so that the decoder can properly dequantize values sent to it.

Figure 10.6 shows the backward adaptive allocation method. This method represents the spectral envelope as exponents at a given sample time. The spectral envelope is encoded and sent to the decoder; it is also used for deciding which coefficients are significant to the sound at any instant and then for controlling the bit allocation to each of the subband coefficients.

Table 10.2 MPEG-2 coding parameters

	16 kHz	*22.05 kHz*	*24 kHz*	*32 kHz*	*44.1 kHz*	*48 kHz*
Audio bandwidth	7.5 kHz	10.3 kHz	11.25 kHz	15 kHz	20.6 kHz	22.5 kHz
Frame duration (Level I)	24 ms	17.4 ms	16 ms	12 ms	8.7 ms	8 ms
Frame duration (Level II)	72 ms	52.5 ms	48 ms	36 ms	26.25 ms	24 ms
Frame duration (Level III)	72 ms	52.5 ms	48 ms	36 ms	26.25 ms	24 ms
Subband sampling (F_s/32)	500 Hz	689.0625 Hz	750 Hz	1000 Hz	1378.125 Hz	1.5 kHz
Subband bandwidth (F_s/64)	250 Hz	344.5313 Hz	375 Hz	500 Hz	689.0625 Hz	750 Hz
Bit rate (Level II with stereo)	96 kbps	128 kbps	128 kbps	192 kbps	256 kbps	256 kbps
Bit rate - (Level II with 5.1 channel)	128 kbps	192 kbps	192 kbps	256 kbps	384 kbps	384 kbps

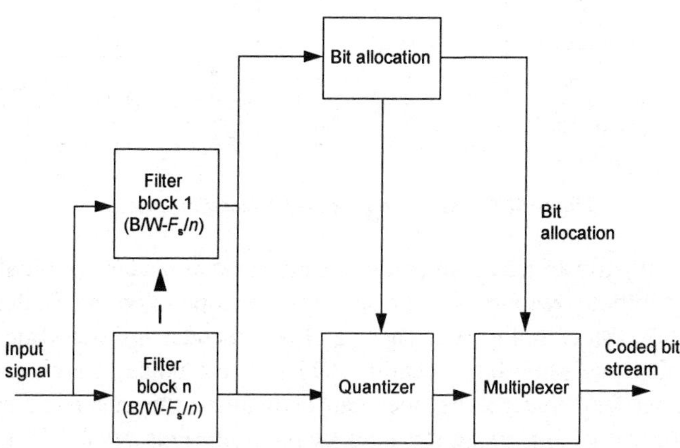

Figure 10.5 Forward bit allocation encoder

Bit allocation quantizes the coefficients of each subband according to the calculated allocation. This quantization is similar to the method used in MPEG audio and controls the number of bits used to represent a value by changing the granularity, or the fineness of amplitude resolution, with which the value is expressed. For example a 2-bit coded value can represent only 4 levels, whereas an 8-bit coded value can represent 256 values.

The difference between the actual value and the quantized value leads to quantization noise; the greater the differences between them, the lower the signal-to-noise ratio. The filter bank at the decoder then limits the quantization noise in any subband to have nearly the same frequency as the signal in that subband. Thus the quantization noise is masked by the strong signal, thus yielding a higher signal-to-noise ratio.

The output samples of the subband filter are the coefficients that are processed for transmission to the decoder. Each coefficient has a combination of exponent and mantissa. The exponent contains the major level information for its subband, and the mantissa contains the detailed level of each sample. As the exponent is exponential it can be thought of as containing the amplitude "range" information of the subband. These two values are multiplexed onto the coded bitstream.

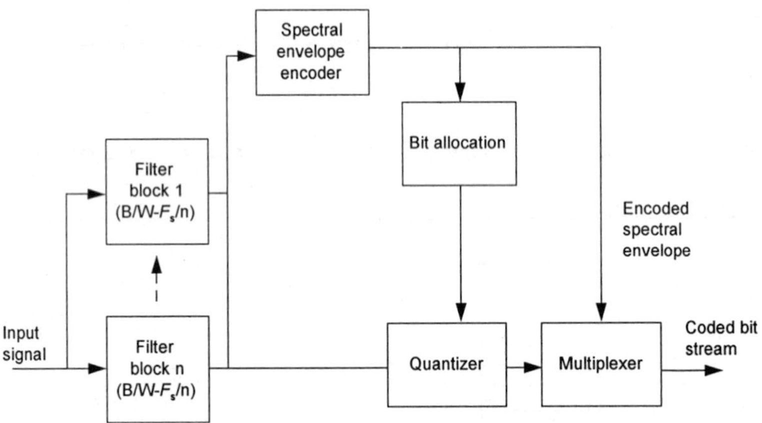

Figure 10.6 Backward adaptive bit allocation encoder

Figure 10.7 illustrates the operation of the backward adaptive bit allocation decoder. It initially reconstructs the spectral envelope in order to calculate the bit allocation. Since both the encoder and the decoder use the same spectral envelope then the same bit allocation decisions are made at both ends. This has the advantage that the decoder can dequantize the quantized mantissas without the actual bit allocation data being sent. This results in a smaller amount of data being transmitted because all of the transmitted data is devoted to audio coding. The method is defined as a backward adaptive bit allocation because the content-controlled bit allocation is calculated identically at both ends.

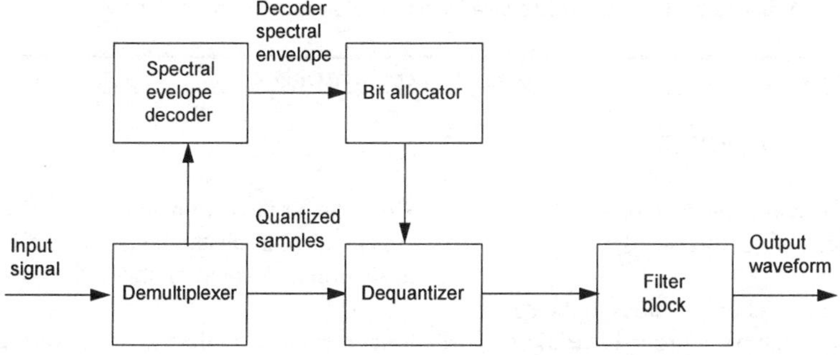

Figure 10.7 Backward adaptive bit allocation decoder

10.5 Comparison between forward and backward adaptive methods

Table 10.3 outlines the main advantages of the forward and backward adaptive methods. As bit allocation information is encoded in the bitstream, the forward allocation method can be as accurate as required. Unfortunately a large amount of data is devoted to transporting the bit allocation information.

When transients are significant it is often desirable update the bit allocation with a finer time resolution. With the forward adaptive bit allocation method there is a direct increase in the channel information data rate. With steady-state conditions it is desirable to provide finer frequency resolution on the spectral analysis and in the bit allocation. Again this will increase the data rate.

The main advantage of the backward adaptive bit allocation method is that none of the encoded data contains bit allocation information. This is because the bit allocation is derived from the spectral envelope. Bit allocation can thus have time or frequency resolution as all the information is in the envelope. The method is thus more efficient in its encoded data and can have a finer time or frequency resolution when required.

The main disadvantage with the backward adaptive bit allocation method is that the information in the envelope will not be totally accurate. As this information is used in the bit allocation it will lead to inaccuracies. Another disadvantage is that the processes of bit allocation and envelope generation increases the complexity of the decoder. This obviously increases the cost of the decoder hardware. A final disadvantage is that the bit allocation algorithm is fixed, it becomes impossible to update the psycho-acoustic model.

Table 10.3 Advantages and disadvantages of adaptive bit allocation methods

Advantages	*Disadvantages*
FORWARD	
The psycho-acoustic model resides only in the encoder.	Since the bit allocation and audio data are encoded, a substantial amount of information is devoted to non-audio data.
As the model only resides in the encoder, it can be improved over time as better models of the human auditory system are developed.	Finer resolution in the time or frequency domain will increase the data rate.
BACKWARD	
Since both the encoder and decoder use the same information (derived from the spectral envelope) to derive the bit allocation, none of the channel data capacity must be wasted in sending specific bit allocation instructions.	The decoder must calculate the bit allocation entirely from information in the coefficient bitstream. The information sent to the decoder has limited accuracy and therefore may contain small errors.
Finer resolution in the time of frequency domain need not increase the data rate.	The processing of the envelope increases the complexity of the decoder.
	The bit allocation algorithm becomes fixed as soon as the first decoders are deployed, and it then becomes impossible to update the psycho-acoustic model.

10.6 Dolby AC-1 and AC-2

Dolby AC-1 was designed for stereo satellite relays of television and for FM programs. Using adaptive delta modulation and analogue companding, it has an audio bandwidth of 20 kHz into a 512 kbps bitstream (3:1).

It uses a perceptual coder using a low-complexity block transform. Initially it divides the wideband signal into multiple subband using a 512-point 50% overlap FFT algorithm performing the MDCT (modified discrete cosine transform). These coefficients are then grouped into subbands containing from 1 to 15 coefficients to model critical bandwidths. Each of the subbands has a number of preallocated bits. Lower frequency subbands have more preallocated bits than higher-frequency subbands. Any requirement for additional bits can be drawn upon depending on bit allocation calculations. The number of subbands is 40 (for 48 kHz sampling) and 43 (for 32 kHz sampling).

AC-2 is used in many applications, such as in PC soundcards and professional equipment. It gives high audio quality at a low data rate of 256 kbps. Typical compression rates are 6.1:1 (for 48 kHz sampling) and 5.4:1 (for 32 kHz sampling).

10.7 Dolby AC-3 coding

Dolby AC-3 uses a hybrid of forward and backward adaptive bit allocation, as illustrated in Figure 10.8. It uses backward adaptive bit allocation with core backward adaptive bit allocation. This core bit system is used in both the encoder and the decoder and can be modified in its operation by forward adaptive bit allocation side information. This allows both the encoder and decoder core bit allocations systems to run independently using the spectral envelope, as in the backward adaptive case, but they are altered by forward adaptive information that improves their accuracy.

The core bit allocation system is based on a fixed psycho-acoustic model with two types of modifications:

- Modification of the parameters of the psycho-acoustic model.
- Differences to the bit allocations that result from the current psycho-acoustic model.

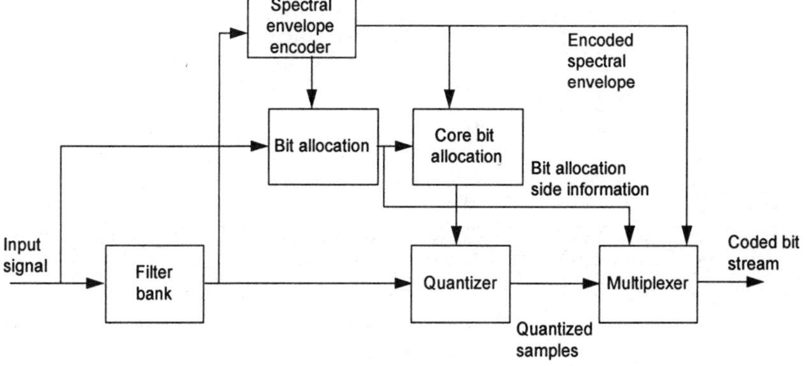

Figure 10.8 Hybrid backward/forward adaptive bit allocation encoder

The main advantage of the hybrid approach is that the modification data sent to the core bit allocation routine is substantially less than would be required for normal forward adaptation. The psycho-acoustic model can also be updated dynamically.

10.8 AC-3 parameters

Dolby AC-3 uses a frame which consists of six blocks, as illustrate in Figure 10.9. Each block can be varied in length to accommodate different applications, typical sizes have 256 and 512 samples. Normally the data within the blocks overlaps on each side of a block. This results in the same sample being contained in two blocks. Thus for a 512-sample block, the last 256 samples will be the same as the first 256 samples of the next block. Table 10.4 outlines the main Dolby AC-3 parameters. It can be seen that a 32 kHz sampling gives a block duration of 16 ms and 44.1 kHz gives a block duration of 11.61 ms.

MPEG audio coding uses 32 subband filters, whereas Dolby AC-3 has 256 subbands. For example, 44.1 kHz sampling gives frequency domains with a 62.5 Hz bandwidth.

MPEG encoders typically use DFT or DCT techniques to generate the time-to-frequency conversion. These use an *N*-point transform that generates *N* unique non-zero transform coefficients. Dolby AC-3 uses an odd-stacked time-division aliasing cancellation (TDAC) technique which is a modified discrete cosine transform (MDCT). The main advantage with this method is that the 50% overlay is achieved without any increase in the resulting bit rate.

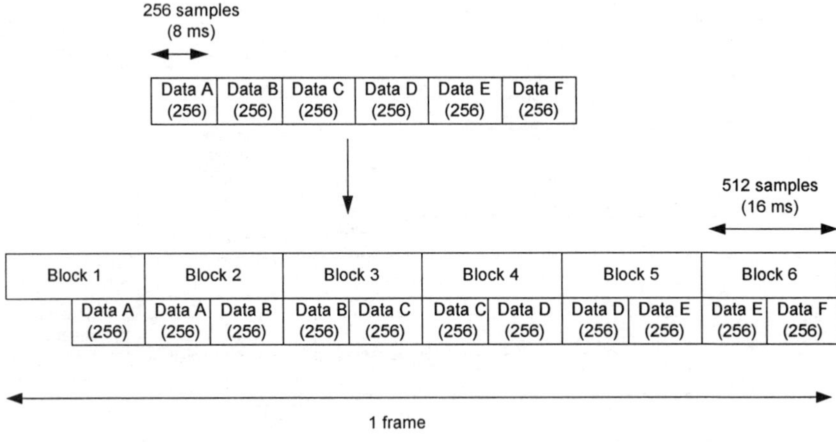

Figure 10.9 Frame and block format

Table 10.4 Dolby AC-3 coding parameters

	32 kHz	*44.1 kHz*	*48 kHz*
Audio bandwidth	15 kHz	20.6 kHz	22.5 kHz
Block length	512 samples	512 samples	512 samples
Block duration	16 ms	11.61 ms	10.66 ms
Subbands	256	256	256
Subband bandwidth (F_s/256)	62.5 Hz	86.133 Hz	93.75 Hz
Block repetition rate	125 Hz	173.26 Hz	187.5 Hz
Block repetition period	8 ms	5.805 ms	5.333 ms
Bit rate (stereo)	192 kbps	192 kbps	192 kbps
Bit rate (5.1-channel)	384 kbps	384 kbps	384 kbps

In TDAC, each MDCT transform of block size N generates only $N/2$ unique non-zero transform coefficients, so critical sampling is achieved with the 50% block overlap already described. The MDCT method is efficient in its computation and in its usage of memory.

10.9 Exercises

10.1 Show that the bit rate for CD-quality stereo audio is 1.411200 Mbps.

10.2 Does tape hiss on compact cassettes increase in quieter passages of music?

10.3 Explain the psycho-acoustic model.

10.4 Discuss MPEG-1 and MPEG-2 coding.

10.5 Contrast backward and forward adaptive bit allocation methods.

10.6 Discuss AC-3 coding.

11 Error Coding Principles

11.1 Introduction

Error bits are added to data either to correct or to detect transmission errors. Normally, the more bits that are added, the better the correction or detection. Error detection allows the receiver to determine if there has been a transmission error. It cannot rebuild the correct data and must either request a retransmission or discard the data. With error correction the receiver detects an error and tries to correct as many error bits as possible. Again, the more error coding bits are used, the more bits can be corrected. An error correction code is normally used when the receiver cannot request a retransmission.

In a digital communication system a single transmission symbol can actually contain many bits. If a single symbol can represent M values then it is described as an M-ary digit. An example of this is in modem communication where 2 bits are sent as four different phase shifts, e.g. 0° for 00, 90° for 01, 180° for 10 and 270° for 11. To avoid confusion it is assumed in this chapter that the digits input to the digital coder and the digital decoder are binary digits and this chapter does not deal with M-ary digits.

11.2 Modulo-2 arithmetic

Digital coding uses modulo-2 arithmetic where addition becomes the following operations:

$$0+0=0 \qquad 1+1=0$$
$$0+1=1 \qquad 1+0=1$$

It performs the equivalent operation to an exclusive-OR (XOR) function. For modulo-2 arithmetic, subtraction is the same operation as addition:

$$0-0=0 \qquad 1-1=0$$
$$0-1=1 \qquad 1-0=1$$

Multiplication is performed with the following:

$0 \times 0 = 0$	$0 \times 1 = 0$
$1 \times 0 = 0$	$1 \times 1 = 1$

which is an equivalent operation to a logical AND operation.

11.3 Binary manipulation

Binary digit representation, such as 101110, is difficult to use when multiplying and dividing. A typical representation is to manipulate the binary value as a polynomial of bit powers. This technique represents each bit as an x to the power of the bit position and then adds each of the bits. For example:

10111	x^4+x^2+x+1
1000 0001	x^7+1
1111 1111 1111 1111	$x^{11}+x^{10}+x^9+x^8+x^7+x^6+x^5+x^4+x^3+x^2+x+1$
10101010	$x^6+x^4+x^2+x$

For example:	101×110
is represented as:	$(x^2+1)\times(x^2+x)$
which equates to:	$x^4+x^3+x^2+x$
which is thus:	11110

The addition of the bits is treated as a modulo-2 addition, that is, any two values which have the same powers are equal to zero. For example:

$$x^4+x^4+x^2+1+1$$

is equal to x^2 as x^4+x^4 is equal to zero and 1+1 is equal to 0 (in modulo-2). An example which shows this is the multiplication of 10101 by 01100.

Thus:	10101×01110
is represented as:	$(x^4+x^2+1)\times(x^3+x^2+x)$
which equates to:	$x^7+x^6+x^4+x^5+x^4+x^3+x^3+x^2+x$
which equates to:	$x^7+x^6+x^5+x^4+x^4+x^3+x^3+x^2+x$
which equates to:	$x^7+x^6+x^5+0+0+x^2+x$
which equates to:	$x^7+x^6+x^5+x^2+x$
which is thus:	11100110

This type of multiplication is easy to implement as it just involves AND and XOR operations.

The division process uses exclusive-OR operation instead of subtraction and can be implemented with a shift register and a few XOR gates. For example 101101 divided by 101 is implemented as follows:

```
        1011
100 ⌐ 101101
      100
       110
       100
        101
        100
          1
```

Thus the modulo-2 division of 101101 by 100 is 1011 remainder 1. As with multiplication this modulo-2 division can also be represented with polynomial values (an example of this is given in Section 12.5.2).

Normally, pure integer or floating point multiplication and division require complex hardware and can cause a considerable delay in computing the result. Error coding multiplication and division circuits normally use a modified version of multiplication and division which uses XOR operations and shift registers.

11.4 Hamming distance

The Hamming distance, $d(C_1,C_2)$, between two code words C_1 and C_2 is defined as the number of places in which they differ. For example, the codes:

101101010 and 011101100

have a Hamming distance of 4 as they differ in 4 bit positions. Also $d(11111,00000)$ is equal to 5.

The Hamming distance can be used to determine how well the code will cope with errors. The minimum Hamming distance $\min\{d(C_1,C_2)\}$ defines by how many bits the code must change so that one code can become another code.

It can be shown that:

- A code C can detect up to N errors for any code word if $d(C)$ is greater than or equal to N+1 (that is, $d(C) \geq N+1$).
- A code C can correct up to M errors in any code word if $d(C)$ is greater than or equal to $2M$+1 (that is, $d(C) \geq 2M+1$).

For example the code:

$$\{00000, 01101, 10110, 11011\}$$

has a minimum Hamming distance of 3. Thus the number of errors which can be detected is given by:

$$d(C) \geq N+1$$

since, in this case, $d(C)$ is 3 then N must be 2. This means that one or two errors in the code word will be detected as an error. For example the following have 2 bits in error, and will thus be received as an error:

00011, 10101, 11010, 00011

The number of errors which can be corrected (M) is given by:

$$d(C) \geq 2M+1$$

thus M will be 1, which means that only one bit in error can be corrected. For example if the received code word was:

01111

then this code is measured against all the other codes and the Hamming distance calculated. Thus 00000 has a Hamming distance of 4, 01101 has a Hamming distance of 1, 10110 has a Hamming distance of 3, and 11011 has a Hamming distance of 2. Thus the received code is nearest to 01101.

11.5 General probability theory

Every digital system is susceptible to errors. These errors may happen one every few seconds or once every hundred years. The rate at which these occur is goverened by the error probability.

If an event X has a probability of $P(X)$ and event Y has a probability of $P(Y)$ then the probability that either might occur is:

$$P(X \text{ or } Y) = P(A) + P(B) - P(X \text{ and } Y)$$

If one event prevents the other from happening, then they are mutally exclusive, thus $P(X \text{ and } Y)$ will be zero. This will give:

$$P(X \text{ or } Y) = P(X) + P(Y)$$

If an event X has a probability of P(X) and event Y has a probability of P(Y) then the probability that both might occur is:

$$P(X \text{ and } Y) = P(X) \times P(Y \mid X)$$
$$P(X \text{ and } Y) = P(Y) \times P(X \mid Y)$$

where $P(Y \mid X)$ is the probability that event Y will occur, assuming that event X has already occurred, and $P(X \mid Y)$ is the probabilty that event X will occur, assuming that event Y has alredady occured. If the two events are independent then $P(X \mid Y)$ will be $P(X)$ and $P(Y \mid X)$ will be $P(Y)$. This results with:

$$P(X \text{ and } Y) = P(X) \times P(Y)$$

For example rolls of a die are independent of each other. If a die is rolled twice then the probaility of a rolling two sixes will be:

$$P(6 \text{ and } 6) = P(6) \times P(6)$$
$$= \frac{1}{6} \times \frac{1}{6} = \frac{1}{36}$$

This formula is used as one roll of the die is mutually exclusive to the next throw.

The probability of throwing a three or a two in a single throw of the dice will be:

$$P(2 \text{ or } 3) = P(2) + P(3)$$
$$= \frac{1}{6} + \frac{1}{6} = \frac{1}{3}$$

11.6 Error probability

Each digital system has its own characteristics and thus will have a different probability of error. Most calculations determine the 'worst-case' situation. The transmission channel normally assumes the following:

- That each bit has the same probability of being received in error.
- That there is an equal probability of a 0 and of a 1 (that is, the probability of a 0 is 0.5 and the probability of a 1 is a 1.0).

Thus if the probability of a binary digit being received incorrectly is p, then the probability of no error is $1-p$. This is illustrated in Figure 11.1.

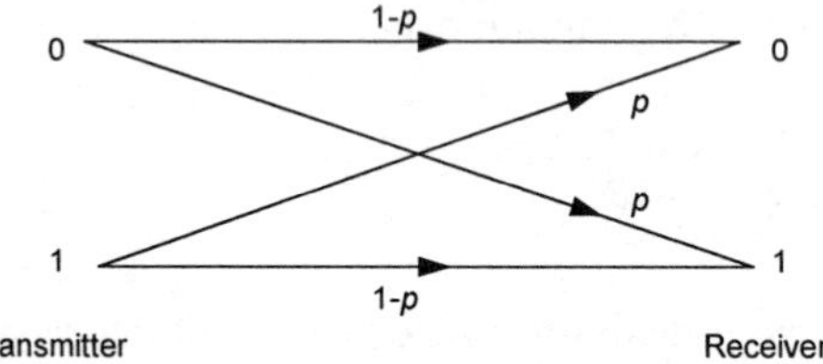

Figure 11.1 Probability or error model for a single bit

If the probability of no errors on a signal bit is $(1-p)$, then the probability of no errors of data with n bits will thus be:

$$\text{Probability of no errors} = (1-p)^n$$

The probability of an error will thus be:

$$\text{Probability of an error} = 1-(1-p)^n$$

The probability of a single error can be determined by assuming at all the other bits are received correctly, thus this will be:

$$(1-p)^{n-1}$$

Thus the probability of a single error at a given position will be this probability multiplied by the probability of an error on a single bit, thus:

$$p(1-p)^{n-1}$$

As there are n bit positions then the probability of a single bit error will be:

$$\text{Probability of single error} = n.p(1-p)^{n-1}$$

For example if the received data has 256 bits and the probability of an error in a single bit is 0.001 then:

$$\begin{aligned}\text{Probability of no error} &= (1-0.001)^{256}\\ &= 0.774\end{aligned}$$

Thus the probability of an error is 0.226, and

Probability of single error $= 8 \times 0.001\,(1-0.001)^{256-1}$
$= 0.0062$

11.7 Combinations of errors

Combinational theory can be used in error calculation to determine the number of combinations of error bits that occur in some n-bit data. For example, in 8 bit data there are 8 combination of single-bit errors: (1), (2), (3), (4), (5), (6), (7) and (8). With 2 bits in error there are 28 combinations $(1,2)$, $(1,3)$, $(1,4)$, $(1,5)$, $(1,6)$, $(1,7)$, $(1,8)$, $(2,3)$, $(2,4)$, $(2,5)$, $(2,6)$, $(2,7)$, $(2,8)$, $(3,4)$, $(3,5)\ldots(6,6)$, $(6,7)$, $(6,8)$, $(7,8)$. In general the formula for the number of combinations of m-bit errors for n bits is:

$$\binom{n}{m} = \frac{n!}{m!(n-m)!}$$

Thus the number of double-bit errors that can occur in 8 bits is:

$$\binom{8}{2} = \frac{8!}{2!(8-2)!} = \frac{8!}{2!6!} = \frac{8 \times 7}{2} = 28$$

Table 11.1 shows the combinations for bit error with 8-bit data. It can thus be seen that there are 255 different error conditions (8 single-bit errors, 28 double-bit errors, 56 triple-bit errors, and so on).

Table 11.1 Combinations

No of bit errors	*Combinations*	*No of bit errors*	*Combinations*
1	8	5	56
2	28	6	28
3	56	7	8
4	70	8	1

To determine the probability with m bits at specific places, use the probability that $(n-m)$ bits will be received correctly:

$$(1-p)^{n-m}$$

Thus the probability that m bits, at specific places, will be received incorrectly is:

$$p^m(1-p)^{n-m}$$

The probability of an *m*-bit error in *n* bits is thus:

$$P_e(m) = \binom{n}{m} . p^m . (1-p)^{n-m}$$

Thus the probability of error in n *n*-bit data is:

$$P_e = \sum_{m=1}^{n} \binom{n}{m} . p^m . (1-p)^{n-m}$$

which is in the form of a binomial distribution.

Question
Determine the probability of error for a 4-bit data block using the formula:

$$P_e = \sum_{m=1}^{n} \binom{n}{m} . p^m . (1-p)^{n-m}$$

and prove this answer using $P_e = 1-(1-p)^n$.

Answer
The probability of error will be:

$$\begin{aligned}
P_e &= \binom{4}{1} \cdot p \cdot (1-p)^3 + \binom{4}{2} \cdot p^2 \cdot (1-p)^2 + \binom{4}{3} \cdot p^3 \cdot (1-p)^1 + \binom{4}{4} \cdot p^4 \cdot (1-p)^0 \\
&= 4p(1-p)^3 + 6p^2\left(1-2p-p^2\right) + 4p^3 - 4p^4 + p^4 \\
&= 4p(1-p)(1-p)^2 + 6p^2 - 12p^3 + 6p^4 + 4p^3 - 4p^4 + p^4 \\
&= 4p(1-p)\left(1-2p+p^2\right) + 6p^2 - 8p^3 + 3p^4 \\
&= 4p - 8p^2 + 4p^3 - 4p^2 + 8p^3 - 4p^4 + 6p^2 - 8p^3 + 3p^4 \\
&= 4p - 6p^2 + 4p^3 - p^4
\end{aligned}$$

To prove this result, the formula $P_e = 1-(1-p)^n$ can be used to give:

$$\begin{aligned}
P_e &= 1-(1-p)^4 \\
&= 1-(1-p)^2(1-p)^2
\end{aligned}$$

$$
\begin{aligned}
&= 1-(1-2p-p^2)(1-2p-p^2) \\
&= 1-(1-2p+p^2-2p+4p^2-2p^3+p^2-2p^3+p^4) \\
&= 1-1+2p-p^2+2p-4p^2+2p^3-p^2+2p^3-p^4 \\
&= 4p-6p^2+4p^3-p^4
\end{aligned}
$$

which is the same result as the previous derivation.

Question

For 4-bit data and a probability of error equal to 0.001, determine the actual probability of errors for 1, 2, 3 and 4 bit errors. Verify that the summation of the error probabilities is given by the formula derived in the previous question.

Answer

Table 11.2 shows the results using the formula:

$$
P_e(m) = \binom{n}{m} . p^m . (1-p)^{n-m}
$$

with $n=4$ and $p=0.001$.

It can be seen from the table that the probability of a single error is 3.988×10^{-3}. the probability of two errors is 5.99×10^{-6}, the probability of three errors is 4×10^{-9} and the probability of four errors is 1×10^{-12}. The summation of the probabilities is thus 3.994×10^{-3}. The formula derived earlier also gives this value for the probability.

$$
\begin{aligned}
P_e &= 4p-6p^2+4p^3-p^4 \\
&= 4 \times 0.001 - 6 \times 0.001^2 - 4 \times 0.001^3 - 0.001^4 \\
&= 0.003\,994
\end{aligned}
$$

Table 11.2 Probability of error

No of errors	*Probability*
1	3.988×10^{-3}
2	5.99×10^{-6}
3	4×10^{-9}
4	1×10^{-12}
Summation	3.994×10^{-3}

11.8 Linear and cyclic codes

A linear binary code is one in which the sum of any two code words is also a code word. For example:

$$\{00, 01, 10, 11\} \quad \text{and} \quad \{00000, 01101, 10110, 11011\}$$

are linear codes, because any of the code words added (using modulo-2 addition) to another gives another valid code word. For example, in the second code, 01101 + 110011 gives 11011, which is a valid code word.

Cyclic codes often involve complex mathematics but are extremely easy to implement with XOR gates and shift registers. A code is cyclic if:

- It is a linear code.
- When any cyclic shift of a code word is also a code word, i.e. whenever $a_0a_1 \ldots a_{n-1}$ is a code word then so is $a_{n-1}a_0a_1 \ldots a_{n-2}$.

For example the code {0000, 0110, 0011, 1001, 1100, 0001, 0010, 0100, 1000} is cyclic because a shift in the bits, either left or right, produces a valid code word. Whereas the code {000, 010, 011, 100, 001} is not cyclic as a shift in the code word 011 to the left or right does not result in a valid code word. One of the most widely uses codes, cyclic redundancy check (CRC) is an example of a cyclic code.

11.9 Block and convolutional coding

The two main types of error coding are block codes and convolutional codes. A block code splits the data into k data bits and forms them into n blocks. This type of code is described as an (n, k) code. For example an (8, 12) code has 12 blocks of 8 bits, as illustrated in Figure 11.2. In a block code, the coder and decoder treat each block separately from all the other blocks.

Convolutional coding, on the other hand, treats the input and output bits as a continuous stream of binary digits. If the coder has an input of k bit(s) and outputs n bit(s), then the coder is described as k/n rate code. For example if the code takes 3 bits at a time and outputs 5 bits for each 3-bit input then it is described as a 3/5 rate code.

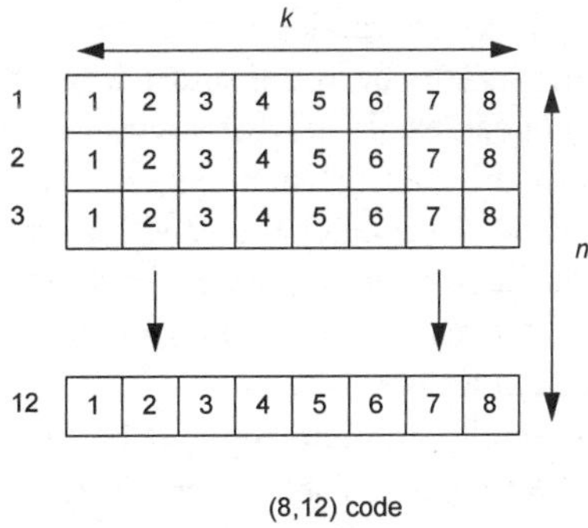

Figure 11.2 An (8,12) block code

11.10 Systematic and unsystematic coding

Systematic code includes the input data bits in an unmodified form. Parity, or check bits, are then added to the unmodified bits. The main advantage of a systematic code is that the original data is still embedded in its original form within the coded data. Thus a receiver may ignore the extra error coding bits and simply read the uncoded data bits. Another decoder implementation could read the data and check the error bits for error detection, while a more powerful decoder might perform full error correction on the data. An unsystematic code modifies the input data bits and embeds the error coding into the coded bitstream. Thus the decoder must normally decode the complete data stream.

11.11 Feedforward and feedback error correction

An error code can give information on error detection or correction information. Normally error correcting codes require many more bits to be added to the data and they require more processing at the decoder. The error coding method used normally depends on many factors, including:

- The locality of the transmitter (which contains the encoder) and receiver (which contains the decoder). If the transmitter is located fairly near to the receiver and can respond quickly for a retransmission then it may be possible for the receiver simply to detect an error and ask for a retransmission. If this is not the case, the receiver must either discard the data or the system must use an error correction code. For example a CD-ROM must contain powerful error correction codes because it is not possible for a CD player to ask for the data to be retransmitted, but two computers on the same network segment can easily ask for a retransmission.

- The characteristics of the transmission system. Normally the more errors that occur in the system, the greater the need for error correcting/detecting codes and the greater the need for a powerful coding scheme.
- The types of errors that can occur; for example, do the errors occur in bursts or in single bits at a time.
- The type of data. Normally computer-type data must be transmitted error-free whereas a few errors in an image or on audio data are unlikely to cause many problems. Thus with computer-type data, a correction code can correct the data as it is received (error correction) or there may be some method for contacting the transmitter to request a retransmission (error detection).
- The power and speed of the encoder and decoder. For example, a powerful error correcting code it may not be possible to implement within a given time because of the need for low-cost decoder hardware. On the other hand a powerful code may be inefficient for simple requirements.

With error detection the receiver must either discard the data or ask for a retransmission. Many systems use an acknowledgment (ACK) from the receiver to acknowledge the receipt of data and a negative acknowledgment (NACK) to indicate an incorrect transmission. Typically data is sent in data packets which contains the error code within them. Each packet also contains a value which identifies the packet number. The receiver then sends back an acknowledgment with the sequence number of the packet it is acknowledging. All previous packets before this value are automatically acknowledged. This method can be implemented in a number of modes, including:

- The transmitter transmits packets, up to a given number, and then waits for the receiver to acknowledge them. If it does not receive an acknowledgment it may resend them or reset the connection.
- The receiver sends back a NACK to inform the transmitter that it has received one or more frames in error, and would like the transmitter to retransmit them, starting with the packet number contained in the NACK.
- The receiver sends back a SREJ to selectively reject a single packet. The transmitter then retransmits this packet.

11.12 Error types

Many errors can occur in a system. Normally, in a digital system, they are caused either when the noise level overcomes the signal level or when the digital pulses become distorted by the system (normally in the transmission systems). This causes a 0 to be interrupted as a 1, or a 1 as a 0.

Noise is any unwanted signal and has many causes, such as static pickup, poor electrical connections, electronic noise in components, crosstalk, and so on. It makes the reception of a signal more difficult and can also produce unwanted distortion on the unmodulated signal.

The main sources of noise on a communication system are:

- Thermal noise – thermal noise arises from the random movement of electrons in a conductor and is independent of frequency. The noise power can be predicted from the formula $N = kTB$ where N is the noise power in watts, k is Boltzmann's constant (1.38×10^{-23} J/K) and B is the bandwidth of the channel (Hz). Thermal noise is predictable and is spread across the bandwidth of the system. It is unavoidable but can be reduced by reducing the temperature of the components causing the thermal noise. Many receivers which detect very small signals require to be cooled to a very low temperature in order to reduce thermal noise. A typical example is in astronomy where the receiving sensor is reduced to almost absolute zero. Thermal noise is a fundamental limiting factor in the performance any communications system.
- Crosstalk – electrical signals propagate with an electric field and a magnetic field. If two conductors are laid beside each other then the magnetic field from one can couple into the other. This is known as crosstalk, where one signal interferes with another. Analogue systems tend to be more affected more by crosstalk than digital systems.
- Impulse noise – impulse noise is any unpredictable electromagnetic disturbance, such as from lightning or energy radiated from an electric motor. It is normally characterized by a relatively high-energy pulse of short duration. It is of little importance to an analogue transmission system as it can usually be filtered at the receiver. However, impulse noise in a digital system can cause the corruption of a significant number of bits.

A signal can be distorted in many, by the electrical characteristics of the transmitter and receiver, and by the characteristics of the transmission media. An electrical cable possesses inductance, capacitance and resistance. The inductance and capacitance have the effect of distorting the shape of the signal whereas resistance causes the amplitude of the signal to reduce (and also to lose power).

If, in a digital system, an error has the same probability of occurring at any time then the errors are random If errors occur in several consecutive bits then they are called burst errors. A typical cause of burst errors is interference, often from lightning or electrical discharge.

If there is the same probability of error for both 1 and 0 then the channel is called a binary symmetric channel and the probability of error in binary digits is known as the bit error rate (BER).

11.13 Coding gain

The effectiveness of an error correcting code is commonly measured with the coding gain and can therefore be used to compare codes. It can basically be defined as the saving in bits relative to an uncoded system delivering the same bit error rate.

11.14 Exercises

11.1 Which of the following codes are cyclic.

(i) {000, 010, 100, 001}
(ii) {0000, 0111, 1011, 1101, 1111}

11.2 Show that the number of errors corrected and detected can be calculated as follows:

Errors detected $= d - 1$
Errors corrected $= (d - 1)/2$

where d is the minimum Hamming distance.

11.3 Which of the following codes are linear.

(i) {000, 011, 101, 110}
(ii) {0000, 0001, 0010, 0100, 1000}
(iii) {0000000, 1111111, 1000101, 1100010, 0110001, 1011000, 0101100, 0010110, 0001011, 0111010, 0011101,1001110, 0100111, 1010011, 1101001, 1110100}

11.4 Complete Table 11.3 for the number of errors detected and corrected for a given minimum Hamming distance, $d(C)$.

Table 11.3 Hamming distance example

d(C)	*Errors corrected*	*Errors detected*
1	0	0
2	1	0
3	2	1
4	3	1
5		
6		
7		
8		
9		

11.5 Determine the combination of bit errors for a 12-bit data word and thus complete Table 11.4. From this table prove that the total number of error combinations is 4095.

Table 11.4 Bit error in a 12-bit data word

No of bit errors	*Combinations*	*No of bit errors*	*Combinations*
1	12	7	
2	66	8	
3	220	9	
4		10	
5		11	
6		12	

11.6 Show that the probability of error for a 5-bit data block is:

$$P_e = 10p^3 + 15p^4 + 6p^5$$

using the formula:

$$P_e = \sum_{m=1}^{n} \binom{n}{m} . p^m . (1-p)^{n-m}$$

and prove this answer using $P_e = 1-(1-p)^n$.

11.7 For a 5-bit data code and an error probability of 0.001 complete Table 11.5 and verify that the summation of the error probability is the same as:

$P_e = 10p^3 + 15p^4 + 6p^5$

Table 11.5 Errors in a 5-bit data code

No of errors	*Probability*
1	
2	
3	9.98×10^{-9}
4	
5	1×10^{-15}
Summation	0.004 99

11.8 For a 6-bit data code and an error probability of 0.000 1 complete Table 11.6 and verify that the summation of the error probability is the same as:

$P_e = 1-(1-p)^n$

Table 11.6 Errors in a 6-bit data code

No. of errors	*Probability*
1	
2	1.5×10^{-7}
3	
4	
5	
6	1×10^{-24}
Summation	0.0006

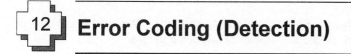

12 Error Coding (Detection)

12.1 Introduction

The most important measure of error detection is the Hamming distance. This defines the number of changes in the transmitted bits that are required in order for a code word to be received as another code word. The more bits that are added, the greater the Hamming distance can be, and the objective of a good error detecting code is to be able to maximize the minimum Hamming distance between codes. For example a code which has a minimum Hamming distance of 1 cannot be used to detect errors, as a single error in a specific bit in one or more code words will cause the received code word to be received as a valid code word. A minimum Hamming distance of 2 will allow one error to be corrected. In general, a code C can detect up to N errors for any code word if $d(C)$ is greater than or equal to $N+1$ (i.e. $d(C) \geq N+1$). For this it can be shown that:

$$\text{The number of errors detected} = d - 1$$

where d is the minimum Hamming distance.

Error detection allows the receiver to determine if there has been a transmission error. It cannot rebuild the correct data and must either request a retransmission or discard the data.

12.2 Parity

Simple parity adds a single parity bit to each block of transmitted symbols. This parity bit either makes them have an even number of 1's (even parity) or an odd number of 1's (odd parity). It is a simple method of error detection and requires only exclusive-OR (XOR) gates to generate the parity bit. This output can be easily added to the data using a shift register.

Parity bits can only detect an odd number of errors, i.e. 1, 3, 5, and so on. If an even number of bits are in error then the parity bit will be correct and no error will be detected. This type of coding is normally not used on its own or where there is the possibility of several bits being in error.

12.3 Block parity

Block parity is a block code which adds a parity symbol to the end of a block of code. For example a typical method is to transmit the one's complement (or sometimes the two's complement) of the modulo-2 sum of the transmitted values. Using this coding, and a transmitted block code after every 8 characters, the data:

1, 4, 12, –1, –6, 17, 0, –10

would be arranged as:

1	0000 0001
4	0000 0100
12	0000 1100
–1	1111 1111
–6	1111 1010
17	0001 0001
0	0000 0000
–10	1110 1110
	1110 1111

It can be seen that 1110 1111 is –17, which is actually the negative of the sum (i.e. 1+4+12–1–6+17+0–10=–17). Thus the transmitted data would be:

0000 0001 0000 0100 0000 1100 1111 1111 1111
1010 0001 0001 0000 0000 1110 1110 1110 1111 ...

In this case a single error will cause the checksum to be wrong. Unfortunately, as with simple parity, an even error in the same column will not show-up an error, but single errors in different columns will show up as an error. Normally when errors occur they are either single-bit errors or large bursts of errors. With a single-bit error the scheme will detect an error and it is also likely to detect a burst of errors, as the burst is likely to affect several columns and also several rows.

This error scheme is used in many systems as it is simple and can be implemented easily in hardware with XOR gates or simply calculated with appropriate software.

The more symbols are used in the block, the more efficient the code will be. Unfortunately when an error occurs the complete block must be retransmitted.

12.4 Checksum

The checksum block code is similar to the block parity method but the actual total of the values is sent. Thus it is it very unlikely that an error will go undiscovered. It is typically used when ASCII characters are sent to represent numerical values. For example, the previous data was:

1, 4, 12, –1, –6, 17, 0, –10

which gives a total of 17. This could be sent in ASCII characters as:

'1' SPACE '4' SPACE '1' '2' SPACE '–' '1' SPACE '–' '6' SPACE '1' '7' SPACE '0' SPACE '–' '1' '0' SPACE '1' '7'

where the SPACE character is the delimiting character between each of the transmitted values. Typically the transmitter and receiver will agree the amount of numbers that will be transmitted before the checksum is transmitted.

12.5 Cyclic redundancy checking (CRC)

CRC is one of the most reliable error detection schemes and can detect up to 95.5% of all errors. The most commonly used code is the CRC-16 standard code which is defined by the CCITT.

The basic idea of a CRC can be illustrated using an example. Suppose the transmitter and receiver were both to agree that the numerical value sent by the transmitter would always be divisible by 9. Then should the receiver get a value which was not divisible by 9 would know it knows that there had been an error. For example, if a value of 32 were to be transmitted it could be changed to 320 so that the transmitter would be able to add to the least significant digit, making it divisible by 9. In this case the transmitter would add 4, making 324. If this transmitted value were to be corrupted in transmission then there would only be a 10% chance that an error would not be detected.

In CRC-CCITT, the error correction code is 16 bits long and is the remainder of the data message polynomial $G(x)$ divided by the generator polynomial

$P(x)$ ($x^{16}+x^{12}+x^5+1$, i.e. 10001000000100001). The quotient is discarded and the remainder is truncated to 16 bits. This is then appended to the message as the coded word.

The division does not use standard arithmetic division. Instead of the subtraction operation an exclusive-OR operation it employs. This is a great advantage as the CRC only requires a shift register and a few XOR gates to perform the division.

The receiver and the transmitter both use the same generating function $P(x)$. If there are no transmission errors then the remainder will be zero.

The method used is as follows:

1. Let $P(x)$ be the generator polynomial and $M(x)$ the message polynomial.
2. Let n be the number of bits in $P(x)$.
3. Append n zero bits onto the right-hand side of the message so that it contains $m+n$ bits.
4. Using modulo-2 division, divide the modified bit pattern by $P(x)$. Modulo-2 arithmetic involves exclusive-OR operations, i.e. $0-1=1$, $1-1=0$, $1-0=1$ and $0-0=0$.
5. The final remainder is added to the modified bit pattern.

Example
For a 7-bit data code 1001100 determine the encoded bit pattern using a CRC generating polynomial of $P(x)=x^3+x^2+x^0$. Show that the receiver will not detect an error if there are no bits in error.

Answer

$P(x)=x^3+x^2+x^0$ (1101)
$G(x)=x^6+x^3+x^2$ (100100)

Multiply by the number of bits in the CRC polynomial.

$x^3(x^6+x^3+x^2)$
$x^9+x^6+x^5$ (1001100000)

Figure 12.1 shows the operations at the transmitter. The transmitted message is thus:

1001100001

and Figure 12.2 shows the operations at the receiver. It can be seen that the remainder is zero, so there have been no errors in the transmission.

```
        1111101
1101 | 1001100000
       1101
        1001
        1101
         1000
         1101
          1010
          1101
           1110
           1101
             1100
             1101
              001
```

Figure 12.1 CRC coding example

```
        1111101
1101 | 1001100000
       1101
        1001
        1101
         1000
         1101
          1010
          1101
           1110
           1101
             1100
             1101
              001
```

Figure 12.2 CRC decoding example

The CRC-CCITT is a standard polynomial for data communications systems and can detect:

- All single and double bit errors.
- All errors with an odd number of bits.
- All burst errors of length 16 or less.
- 99.997% of 17-bit error bursts.
- 99.998% of 18-bit and longer bursts.

Table 12.1 lists some typical CRC codes. CRC-32 is used in Ethernet, Token

Ring and FDDI networks, whereas ATM uses CRC-8 and CRC-10.

Table 12.1 Typical schemes

Type	*Polynomial*	*Polynomial binary equivalent*
CRC-8	$x^8+x^2+x^1+1$	100000111
CRC-10	$x^{10}+x^9+x^5+x^4+x^1+1$	11000110011
CRC-12	$x^{12}+x^{11}+x^3+x^2+1$	1100000001101
CRC-16	$x^{16}+x^{15}+x^2+1$	11000000000000101
CRC-CCITT	$x^{16}+x^{12}+x^5+1$	10001000000100001
CRC-32	$x^{32}+x^{26}+x^{23}+x^{16}+x^{12}+x^{11}$ $+x^{10}+x^8+x^7+x^5+x^4+x^2+x+1$	1000001001000000100011101101 10111

12.5.1 Mathematical representation of the CRC

The main steps to CRC implementation are:

1. Prescale the input polynomial of $M'(x)$ by the highest order of the generator polynomial $P(x)$. Thus:

$$M'(x) = x^n M(x)$$

2. Next divide $M'(x)$ by the generator polynomial to give:

$$\frac{M'(x)}{G(x)} = \frac{x^3 M(x)}{G(x)} = Q(x) + \frac{R(x)}{G(x)}$$

which gives:

$$x^3 M(x) = G(x)Q(x) + R(x)$$

and rearranging gives:

$$x^3 M(x) + R(x) = G(x)Q(x)$$

This means that the transmitted message ($x^3 M(x) + R(x)$) is now exactly divisible by $G(x)$.

12.5.2 CRC example

Question A

A CRC system uses a message of $1+x^2+x^4+x^5$. Design a FSR cyclic encoder circuit with generator polynomial $G(x)=1+x^2+x^3$ and having appropriate gating circuitry to enable/disable the shift out of the CRC remainder.

Answer - Part A

The generator polynomial is $G(x)=1+x^2+x^3$, the circuit is given in Figure 12.3.

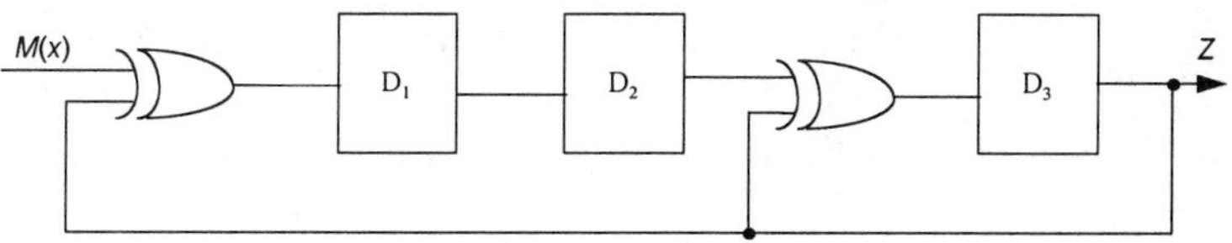

Figure 12.3 CRC coder

Now to prove this circuit does generaer the polynomial. The output $Z(x)$ will be:

$$Z(x) = Z(x)x^{-1} + \left[M(x)x^{-2} + Z(x)x^{-2} \right] x^{-1}$$
$$= Z(x)\left(x^{-3} + x^{-1}\right) + M(x)x^{-3}$$

Thus:

$$M(x) = \frac{Z(x)\left[1 + x^{-1} + x^{-3}\right]}{x^{-3}}$$

giving:

$$P(x) = \frac{M(x)}{Z(x)} = x^3 + x^2 + 1$$

Question B

If the previous CRC system uses a message of $1+x^2+x^4+x^5$ then determine the sequence of events that occur and hence determine the encoded message as a polynomial $T(x)$. Synthesize the same code algebraically using modulo-2 division.

Answer B

First prescale the input polynomial of $M(x)$ by x^3, the highest power of $G(x)$, thus:

$$M'(x)=x^3.M(x)= x^3+x^5+x^7+x^8$$

The input is thus $x^3+x^5+x^7+x^8$ (000101011), and the generated states are:

Time	$M'(x)$	D_1	D_2	D_3	D_4
1	000101011	0	0	0	0
2	00010101	1	0	0	0 ←
3	0001010	1	1	0	0
4	000101	0	1	1	1
5	00010	0	0	0	0
6	0001	0	0	0	0
7	000	1	0	0	0
8	00	0	1	0	0
9	0	0	0	1	1 ←
10		**1**	**0**	**1**	

The remainder is thus 101, so $R(x)$ is x^2+1. The transmitted polynomial will be:

$$T(x)=x^3M(x)+R(x)=x^8+x^7+x^5+x^3+x^2+1 \text{ (110101101)}$$

To check this, use either modulo-2 division to give:

$$\begin{array}{r|l} & x^5 \qquad\qquad +1 \\ \hline x^3+x^2+1 & x^8+x^7+x^5+x^3 \\ & x^8+x^7+x^5 \\ \hline & \qquad\qquad x^3 \\ & \qquad\qquad x^3+x^2+1 \\ \hline & \qquad\qquad \boxed{x^2+1} \end{array}$$

Remainder → x^2+1

This gives the same answer as the state table, i.e. x^2+1.

Question C

Prove that the transmitted message does not generate a remainder when divided by $P(x)$.

Answer C

The transmitted polynomial, $T(x)$, is $x^8+x^7+x^5+x^3+x^2+1$ (110101101) and the generator polynomial, $G(x)$, is $1+x^2+x^3$. Thus:

$$
\begin{array}{r|l}
 & x^5 \qquad\qquad +1 \\
\hline
x^3+x^2+1 & x^8+x^7+x^5+x^3+x^2+1 \\
 & x^8+x^7+x^5 \\
\hline
 & \qquad\qquad\quad x^3+x^2+1 \\
 & \qquad\qquad\quad x^3+x^2+1 \\
\hline
\text{Remainder} \rightarrow & \qquad\qquad\qquad\quad \boxed{0}
\end{array}
$$

As there is a zero remainder, there is no error.

12.6 Exercises

12.1 Decode the following asynchronous message. The encoding used is 1 start bit, 1 stop bit, 7-bit ASCII and odd parity.

111111001100010110010011111111111111
10101000111111000100110111111111

12.2 Determine the message and errors in the following ASCII code with even parity:

01001000	01100101	01101100	01101100
01101111	00101111	01010111	01101111
01110010	01101100	01100100	00101110

12.3 Using CRC coding with a message of 1101011011 and a generator of 10011 determine the transmitted message

12.4 Determine the encoded message for the following 8-bit data codes using the following CRC generating polynomial:

$$P(x) = x^4 + x^3 + x^0$$

(i) 11001100
(ii) 01011111
(iii) 10111011
(iv) 10001111

Prove that there will be no error at the receiver.

12.5 Using the polynomial given in Question 12.4 determine if the following encoded values have any errors:

(i) 1111111100110
(ii) 1011000011101
(iii) 1101110101001

12.6 If the generator polynomial is x^3+x^2+1 and the message bits are 1001 show that the transmitted message is 1001011.

12.7 If the generator polynomial is 10011 and the message bits are 1101011011 show that the transmitted message is 11010110111110.

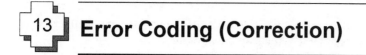

13 Error Coding (Correction)

13.1 Introduction

Error bits are added to data either to correct or to detect transmission errors. Normally, the more bits that are added, the better the detection or correction. Error detection allows the receiver to determine if there has been a transmission error. It cannot rebuild the correct data and must either request a retransmission or discard the data. With error correction the receiver detects an error and tries to correct as many error bits as possible. Again, the more error coding bits are used, the more bits can be corrected. An error correction code is normally used when the receiver cannot request a retransmission.

13.2 Longitudinal/vertical redundancy checks (LRC/VRC)

RS-232 uses vertical redundancy checking (VRC) when it adds a parity bit to the transmitted character. Longitudinal (or horizontal) redundancy checking (LRC) adds a parity bit for all bits in the message at the same bit position. Vertical coding operates on a single character and is known as character error coding. Horizontal checks operate on groups of characters and described as message coding. LRC always uses even parity and the parity bit for the LRC character has the parity of the VRC code.

In the example given next, the character sent for LRC is thus `10101000` or a '`)`'. The message sent is '`F`', '`r`', '`e`', '`d`', '`d`', '`y`' and '`(`'.

Without VRC checking, LRC checking detects most errors but does not detect errors where an even number of characters have an error in the same bit position. In the previous example if bit 2 of the '`F`' and '`r`' were in error then LRC would be valid.

This problem is overcome if LRC and VRC are used together. With VRC/LRC the only time an error goes undetected is when an even number of bits, in an even number of characters, in the same bit positions of each character are in error. This is of course very unlikely.

On systems where only single-bit errors occur, the LRC/VRC method can be used to detect and correct the single-bit error. For systems where more than one error can occur it is not possible to locate the bits in error, so the receiver prompts the transmitter to retransmit the message.

Example
A communications channel uses ASCII character coding and LRC/VRC bits are added to each word sent. Encode the word 'Freddy' and, using odd parity for the VRC and even parity for the LRC; determine the LRC character.

Answer

	F	r	e	d	d	y	LRC
b0	0	0	1	0	0	1	**0**
b1	1	1	0	0	0	0	**0**
b2	1	0	1	1	1	0	**0**
b3	0	0	0	0	0	1	**1**
b4	0	1	0	0	0	1	**0**
b5	0	1	1	1	1	1	**1**
b6	1	1	1	1	1	1	**0**
VRC	0	1	1	0	0	0	**1**

13.3 Hamming code

Hamming code is a forward error correction (FEC) scheme which can be used to detect and correct bit errors. The error correction bits are known as Hamming bits and the number that need to be added to a data symbol is determined by the expression:

$$2^n \geq m + n + 1$$

where *m* is number of bits in the data symbol
n is number of Hamming bits

Hamming bits are inserted into the message character as desired. Typically, they are added at positions that are powers of 2, i.e. the 1st, 2nd, 4th, 8th, 16th bit positions, and so on. For example to code the character `011001` then, starting from the right-hand side, the Hamming bits would be inserted into the 1st, 2nd, 4th and 8th bit positions.

The character is `011001`
The Hamming bits are `HHHH`

The message format will be `01H100H1HH`

10	9	8	7	6	5	4	3	2	1
0	1	H	1	0	0	H	1	H	H

Next each position where there is a 1 is represented as a binary value. Then each position value is exclusive-OR'ed with the others. The result is the Hamming code. In this example:

Position	Code
9	1001
7	0111
3	0011
XOR	1101

The Hamming code error bits are thus 1101 and the message transmitted will be 0111001101.

10	9	8	7	6	5	4	3	2	1
0	1	1	1	0	0	1	1	0	1

At the receiver all bit positions where there is a 1 are exclusive-OR'ed. The result gives either the bit position error or no error. If the answer is zero there were no single-bit errors, it gives the bit error position.

Position	*Code*
Hamming	1101
9	1001
7	0111
3	0011
XOR	0000

If an error has occurred in bit 4 then the result is 4.

Position	*Code*
Hamming	1101
9	1001
7	0111
4	0100
3	0011
XOR	0100

13.4 Representations of Hamming code

For a code with 4 data bits and 3 Hamming bits, the Hamming bits are normally inserted into the power-of-2 bit positions, thus code is known as (7,4) code. The transmitted bit are $P_1P_2D_1P_3D_2D_3D_4$. In a mathematical form the parity bits are generated by:

$$P_1 = D_1 \oplus D_2 \oplus D_4$$
$$P_2 = D_1 \oplus D_3 \oplus D_4$$
$$P_3 = D_2 \oplus D_3 \oplus D_4$$

At the receiver the check bits are generated by:

$$S_1 = P_1 \oplus D_1 \oplus D_2 \oplus D_4$$
$$S_2 = P_2 \oplus D_1 \oplus D_3 \oplus D_4$$
$$S_3 = P_3 \oplus D_2 \oplus D_3 \oplus D_4$$

Hamming coding can also be represented in a mathematical form. The steps are:

1. Calculate the number of Hamming bits using the formula $2^n \geq m + n + 1$, where m is number of bits in the data and n is number of Hamming bits. The code is known as an $(m+n, m)$ code. For example, (7,4) code uses 4 data bits and 3 Hamming bits.

2. Determine the bit positions of the Hamming bits (typically they will be inserted in the power-of-2 bit positions, i.e. 1, 2, 4, 8,...).

3. Generate the transmitted bit pattern with data bits and Hamming bits. For example if there are 4 data bits ($D_1D_2D_3D_4$) and 3 Hamming bits ($P_1P_2P_3$) then the transmitted bit pattern will be:

$$\mathbf{T} = \begin{bmatrix} P_1 & P_2 & D_1 & P_3 & D_2 & D_3 & D_4 \end{bmatrix}$$

4. Transpose the **T** matrix to give $\mathbf{T}^{\mathrm{T}}$; message bits $D_1D_2D_3D_4$ and Hamming bits $P_1P_2P_3$ would give:

$$\mathbf{T}^{\mathrm{T}} = \begin{bmatrix} P_1 \\ P_2 \\ D_1 \\ P_3 \\ D_2 \\ D_3 \\ D_4 \end{bmatrix}$$

5. The Hamming matrix **H** is generated by an $[n, m+n]$ matrix, where n is the number of Hamming bits and m is the number of data bits. Each row identifies the Hamming bit and a 1 is placed in the row and column if that Hamming bit checks the transmitted bit. For example in the case of a transmitted message of $P_1P_2D_1P_3D_2D_3D_4$ then if P_1 checks the D_1, D_2 and D_4, and P_2 checks the D_1, D_3 and D_4, and P_1 checks the D_2, D_3 and D_4, then the Hamming matrix will be:

$$\mathbf{H} = \begin{bmatrix} 1 & 0 & 1 & 0 & 1 & 0 & 1 \\ 0 & 1 & 1 & 0 & 0 & 1 & 1 \\ 0 & 0 & 0 & 1 & 1 & 1 & 1 \end{bmatrix} \begin{matrix} \longleftarrow \text{Check of } P_1 \\ \longleftarrow \text{Check of } P_2 \\ \longleftarrow \text{Check of } P_3 \end{matrix}$$

The resulting matrix calculation of:

$$\mathbf{HT}^{\mathrm{T}} = \begin{bmatrix} 1 & 0 & 1 & 0 & 1 & 0 & 1 \\ 0 & 1 & 1 & 0 & 0 & 1 & 1 \\ 0 & 0 & 0 & 1 & 1 & 1 & 1 \end{bmatrix} \begin{bmatrix} P_1 \\ P_2 \\ D_1 \\ P_3 \\ D_2 \\ D_3 \\ D_4 \end{bmatrix}$$

gives the syndrome matrix **S** which is a $[1,3]$ matrix. The resulting terms for the syndrome will be:

$$S_1 = P_1 \oplus D_1 \oplus D_2 \oplus D_4$$
$$S_2 = P_2 \oplus D_1 \oplus D_3 \oplus D_4$$
$$S_3 = P_3 \oplus D_2 \oplus D_3 \oplus D_4$$

6. The parity bits are calculated to make all the terms of the syndrome zero. Using the current example:

$$\mathbf{S} = \mathbf{HT}^{\mathrm{T}} = \begin{bmatrix} 0 \\ 0 \\ 0 \end{bmatrix}$$

At the receiver the steps are:

1. The Hamming matrix is multiplied by the received bits to give the syndrome matrix. Using the current example:

$$\mathbf{S}=\mathbf{HR}^{\mathrm{T}}=\begin{bmatrix}1&0&1&0&1&0&1\\0&1&1&0&0&1&1\\0&0&0&1&1&1&1\end{bmatrix}\begin{bmatrix}P_1\\P_2\\D_1\\P_3\\D_2\\D_3\\D_4\end{bmatrix}=\begin{bmatrix}S_1\\S_2\\S_3\end{bmatrix}$$

2. A resulting syndrome of zero indicates no error, while any other values give an indication of the error position. If the Hamming bits are inserted into the bit positions in powers of 2 then the syndrome gives the actual position of the bit.

13.4.1 Example

Question

(a) A Hamming coded system uses 4 data bits and 3 Hamming bits. The Hamming bits are inserted in powers of 2 and they check the following bit positions:

1, 3, 5, 7   2, 3, 6, 7   4, 5, 6, 7

If the data bits are 1010 then find the coded message using matrix notation.

(b) If the received message is 1011110 determine the syndrome to indicate the position of the error.

Answer

(a) The transmitted message is:

$$\mathbf{T}=\begin{bmatrix}P_1 & P_2 & D_1 & P_3 & D_2 & D_3 & D_4\end{bmatrix}$$

or even parity:

P_1 checks the 1st, 3rd, 5th and 7th, so $P_1 \oplus D_1 \oplus D_2 \oplus D_4 = 0$; thus $P_1=1$
P_2 checks the 2nd, 3rd, 6th and 7th, so $P_2 \oplus D_1 \oplus D_3 \oplus D_4 = 0$; thus $P_2=0$
P_3 checks the 4th, 5th, 6th and 7th, so $P_3 \oplus D_2 \oplus D_3 \oplus D_4 = 0$; thus $P_3=1$

$$\mathbf{H}=\begin{bmatrix}1&0&1&0&1&0&1\\0&1&1&0&0&1&1\\0&0&0&1&1&1&1\end{bmatrix} \quad \text{and} \quad \mathbf{T}^{\mathrm{T}}=\begin{bmatrix}1\\0\\1\\1\\0\\1\\0\end{bmatrix}$$

Thus $\mathbf{HT}^{\mathrm{T}}$ should equal zero. To check:

$$\mathbf{HT}^{\mathrm{T}}=\begin{bmatrix}1&0&1&0&1&0&1\\0&1&1&0&0&1&1\\0&0&0&1&1&1&1\end{bmatrix}\begin{bmatrix}1\\0\\1\\1\\0\\1\\0\end{bmatrix}=\begin{bmatrix}1.1\oplus 0.0\oplus 1.1\oplus 0.1\oplus 1.0\oplus 0.1\oplus 1.0\\0.1\oplus 1.0\oplus 1.1\oplus 0.1\oplus 0.0\oplus 1.1\oplus 1.0\\0.1\oplus 0.0\oplus 0.1\oplus 1.1\oplus 1.0\oplus 1.1\oplus 1.0\end{bmatrix}=\begin{bmatrix}0\\0\\0\end{bmatrix}$$

(b)

The received message is:

$$\mathbf{R}=\begin{bmatrix}1&0&1&1&1&1&0\end{bmatrix}$$

Thus the syndrome is determine by:

$$\mathbf{HR}^{\mathrm{T}}=\begin{bmatrix}1&0&1&0&1&0&1\\0&1&1&0&0&1&1\\0&0&0&1&1&1&1\end{bmatrix}\begin{bmatrix}1\\0\\1\\1\\1\\1\\0\end{bmatrix}=\begin{bmatrix}1.1\oplus 0.0\oplus 1.1\oplus 0.1\oplus 1.1\oplus 0.1\oplus 1.0\\0.1\oplus 1.0\oplus 1.1\oplus 0.1\oplus 0.1\oplus 1.1\oplus 1.0\\0.1\oplus 0.0\oplus 0.1\oplus 1.1\oplus 1.1\oplus 1.1\oplus 1.0\end{bmatrix}=\begin{bmatrix}1\\0\\1\end{bmatrix}$$

Thus the resultant syndrome is not equal to zero, which means there is an error condition. Since $\mathbf{S}=1\,0\,1$, the error must be in bit position 5, so inverting this bit gives the received message, 1011010.

13.5 Single error correction/double error detection Hamming code

The Hamming code presented can only be used to correct a single error. To correct 2 bits, another parity bit is added to give an overall parity check. Thus for 4 data bits the transmitted code would be:

$$\mathbf{T} = \begin{bmatrix} P_1 & P_2 & D_1 & P_3 & D_2 & D_3 & D_4 & P_4 \end{bmatrix}$$

where P_4 gives an overall parity check. This can be removed at the decoder and Hamming code single error detection can be carried out as before. This then leads to four conditions:

- If the syndrome is zero and the added parity is the correct parity, there is no error (as before).
- If the syndrome is zero and the added parity is the incorrect parity, there is an error in the added parity bit.
- If the syndrome is non-zero and the added parity is the incorrect parity, there is a single error. The syndrome then gives an indication of the bit position of the error.
- If the syndrome is non-zero and the added parity is the correct parity, there is a double error.

Using the example of Section 13.4:

$$\mathbf{H} = \begin{bmatrix} 1 & 0 & 1 & 0 & 1 & 0 & 1 \\ 0 & 1 & 1 & 0 & 0 & 1 & 1 \\ 0 & 0 & 0 & 1 & 1 & 1 & 1 \end{bmatrix} \quad \text{and} \quad \mathbf{T}^{\mathrm{T}} = \begin{bmatrix} 1 \\ 0 \\ 1 \\ 1 \\ 0 \\ 1 \\ 0 \end{bmatrix}$$

Then the parity bit would be a zero. Thus is if parity bit P_4 is a 1 and the syndrome is zero, it is the parity bit that is in error. If a single-bit is in error then the parity bit will be incorrect, so the syndrome will give the bit position in error. If there are two bits in error then the parity bit will be correct, thus if the syndrome is non-zero and the parity bit is correct then there are two errors (unfortunately the syndrome will be incorrect and the received message must be discarded).

13.6 Reed-Solomon coding

In most cases pure Hamming code can only correct a single-bit in error. A more powerful coding system is Reed-Solomon coding, which can correct multiple bits in error. It is a cyclic code and was devised in 1960 by Irvine Reed and Gustave Solomon at MIT. It is suitable for correcting bursts of errors and is discussed in more detail in [1].

13.7 Convolution codes

The block codes, such as VRC/LRC and CRC have the disadvantage that many bits are sent before the message is actually checked. Convolution codes, on the other hand, embed the parity checks in the data stream. They feed the data bit into the coder one bit at a time through a shift register and the output bit(s) are generated with exclusive-OR operations. An example coder is shown in Figure 13.1. The total of the bits considered in the continuous data stream is called the constraint length. Figure 13.1 shows a coder with a constraint length of 3.

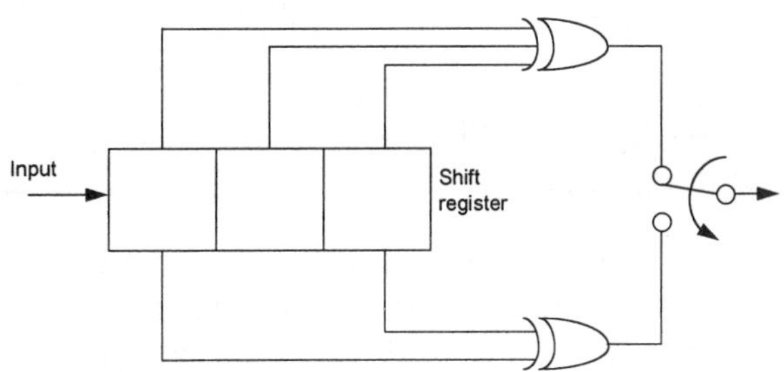

Figure 13.1 A Convolutional encoder

A convolution code takes groups of k bit digits at a time and produces groups of n output binary digits. As k is the input data rate and n is the output data time step then the code is known as a k/n code. For example if the coder takes one input bit at a time and outputs two then it is a 1/2 coder. The coder in Figure 13.1 contains has $k=1$, $n=2$ and has a constraint length, L, of 3.

At any point in time the digits in the shift register define the current state of the coder at that time step.

For example, let the input be 1101. Thus an extra two 0's are added to the input data so that the complete data can be clocked into the shift register. Table 13.1 gives the state table of the encoder. The output, from first to last, will thus be 110101001011.

Table 13.1 State table of the encoder

Input	*A*	*B*	*C*	*Output*
001011	1	0	0	1,1
00101	1	1	0	0,1
0010	0	1	1	0,1
001	1	0	1	0,0
00	0	1	0	1,0
0	0	0	1	1,1

The system can also be analyzed for any input. First a coding tree is drawn up which defines the present and next state within the shift register. A 3-bit shift register will have a total of 8 states, and each of these states will have 2 next states, one for a 1 input and the other for a 0 input. Table 13.2 defines the coding tree in this case.

Table 13.2 Coding tree

ABC	*Next state 0 input*	*Output*	*Next state 1 input*	*Output*
000	000	00	100	11
001	000	00	100	11
010	001	11	101	00
011	001	11	101	00
100	010	10	110	01
101	010	10	110	01
110	011	01	111	10
111	011	01	111	10

It can be seen that the next state when *ABC* is either 000 or 001 will always be 000 when a 0 is entered. These states can therefore be taken as the same, as can 010 and 011, 100 and 101, and 110 and 111. This also occurs with a 1 input, thus 4 of the states can be merged. This results in a coding tree with only 4 states, as given in Table 13.3.

Next a state diagram can be produced to determine the change of the output state for a given input. This diagram represents the current state within a circle and the next state is linked by an arrow with an associated value *x*/*yy* denoting that the input is *x* and the output is *yy*. It is initially assumed that the circuit has been reset and that the initial state is *ABC*=000 (state *a*). An input of 0 causes the circuit to stay in that state and thus to output 00. If the input is a 1 then the next state is 10X (state *b*) and 11 is output. This produces the state diagram of Figure 13.2.

Next, from state *b*, a zero input causes ABC to become 01X (state *c*) and to output 11. If a 1 is input then the next state will be 11X (state *d*) and the output a 01. This is then continues until the state diagram is complete, as shown in Figure 13.3.

Table 13.3 Coding tree

ABC	*Next state 0 input*	*Output*	*Next state 1 input*	*Output*
00X	00X	00	10X	11
01X	00X	11	10X	00
10X	01X	10	11X	01
11X	01X	01	11X	10

Next, from the state diagram a trellis diagram can be drawn. This has the number of states down the left-hand column; each time step is mapped with its mapping to the previous state and gives an indication of the generated output. The upper of the two lines represents a 0 input, and the lower represent a 1 input. For example the circuit starts in state *a*, then a 0 outputs 00 and stays in the same state, while a 1 puts the circuit into state *b* and outputs a 11. This is shown on the trellis diagram in Figure 13.4.

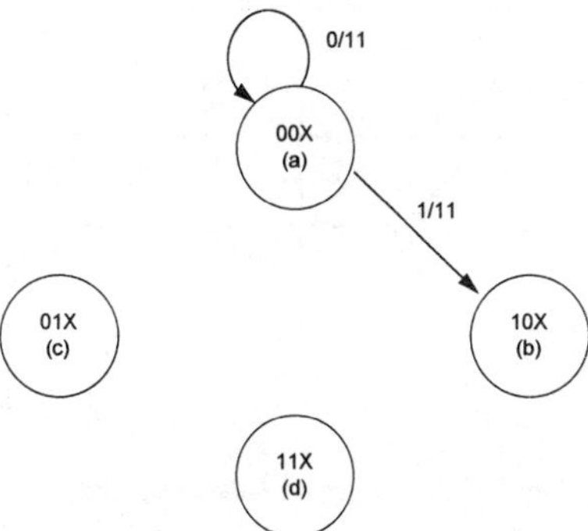

Figure 13.2 Intermediate state diagram

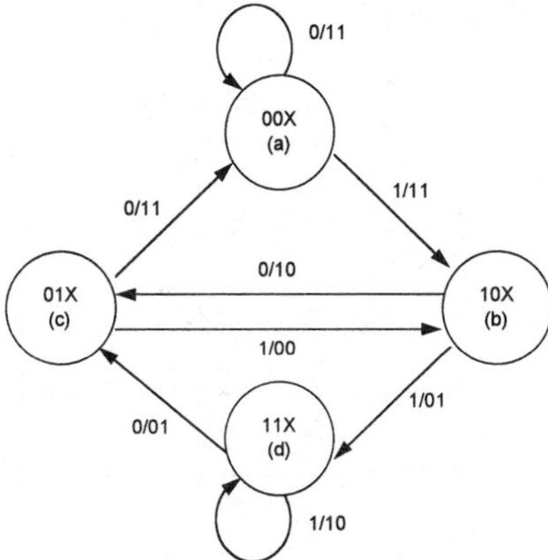

Figure 13.3 Final state diagram

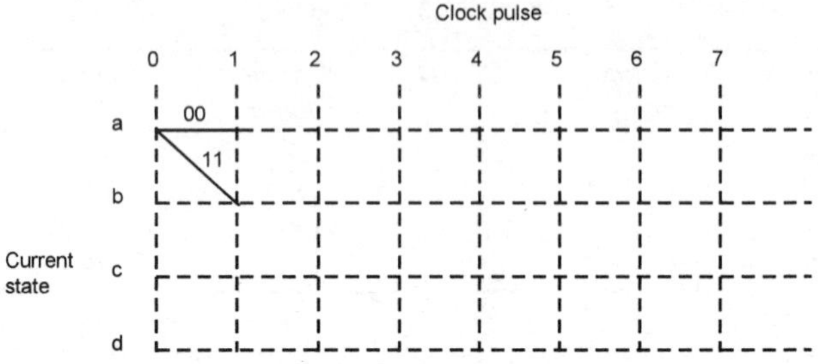

Figure 13.4 Initial trellis diagram

If the current state is *a* then a 0 input will make it stay in the same state and a 1 will take the next state to *b*. If the current state is *b* then a 0 will make the state change to *c*, else a 1 will change it to *d*. This is illustrated in Figure 13.5.

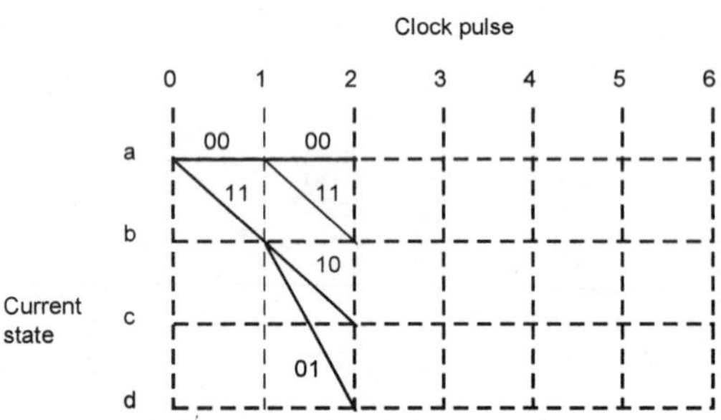

Figure 13.5 Intermediate trellis diagram

The rest of the states on the trellis diagram can now be mapped; the next state is given in Figure 13.6 and the final trellis diagram is shown in Figure 13.7. Note that the final two input states are both 0.

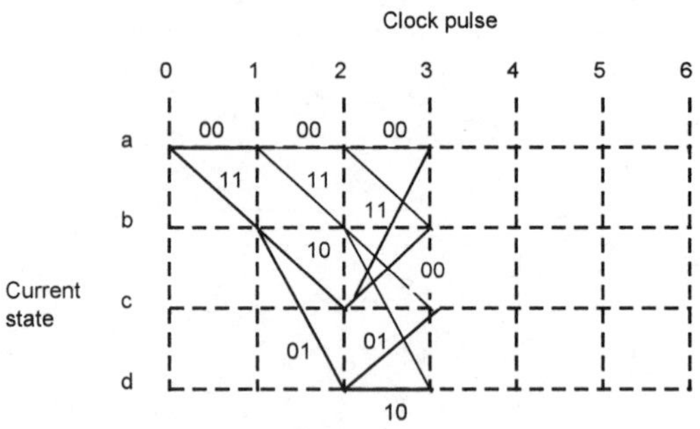

Figure 13.6 Intermediate trellis diagram

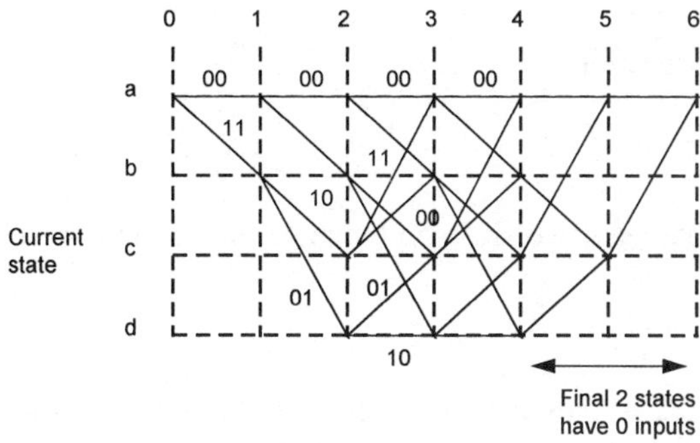

Figure 13.7 Final trellis diagram

13.7.1 Viterbi decoder

Convolution coding allows error correcting because only some of the possible sequences are valid outputs from the coder. These sequences basically correspond to the possible paths through the trellis. A Viterbi decoder tries to find the best match for the received bitstream with valid paths. The best path is called the maximum likelihood and discrepancies between the received bitstream and the possible transmitted stream are called path metrics.

The decoder basically stores the best single path to each of the time-step nodes (in the previous example this was 4: *a*, *b*, *c*, and *d*). Any other path to that node which has a greater number of discrepancies causes that path to be ignored. The number of remembered paths will thus be equal to the number of nodes and will remain constant. For each node the decoder must store the best path to that node and the total metric corresponding to that path. By comparing the actual *n* received digits at time step $i+1$ with those corresponding to each possible path on the trellis from step *i* to step $i+1$, the decoder calculates the additional metric for each path. From these, it selects the best path to each node at time step $i+1$, and updates the stored records of the paths and metrics.

The best way to illustrate this process is with an example. For example using the previous example then an input bit sequence of:

100110

will give an output of:

11 10 11 11 01 01

with a tail of:

11 00 00

This gives an output of:

11101110101110000

Let's assume there are two errors in the bit stream and the receive stream is:

11**0**011**0**10101110000

First the different in the number of bits is represented on the trellis diagram, as shown in Figure 13.8. Thus if the circuit stays in state *a* then the received input would require two changes to the transmitted data for it to be received as 11. There is no difference in the received bits when going from state *a* to state *b*. Thus the upper route scores a value of 2 and the lower route a value of 0.

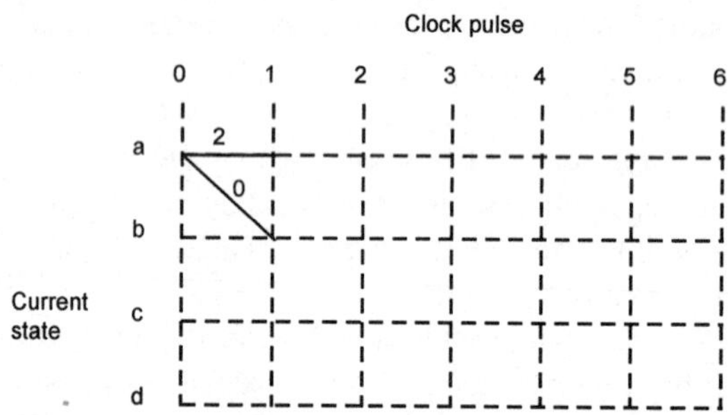

Figure 13.8 Representing the difference in the number of bits

Next 00 is received, if the circuit had been in state *a* and stayed in state *a* then 00 would have been transmitted. So this will have a discrepancy of 0 and the resulting metric for that route will be 2. If the circuit in state *a* and have been moved to state *b* then 11 would have been transmitted, thus there would be a discrepancy of 2, giving the total path metrics as 4. Figure 13.9 shows the resulting trellis diagram.

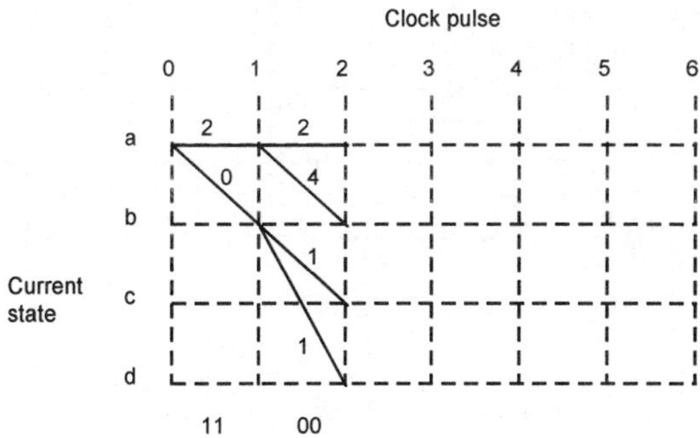

Figure 13.9 Calculating a path metric

Figure 13.10 shows the next state. This time two routes converge on the same state. The Viterbi decoder computes the metrics for each of the two paths then rejects the route with the greater metric. It can be seen that the top route (i.e. transmitted pattern 00, 00, 00) gives a metric of 4 but the other route to that state has a metric of 1. The route with a metric of 4 will be rejected. It can be seen that the lowest metric has a value of 1. Figure 13.11 shows the resulting preferred routes.

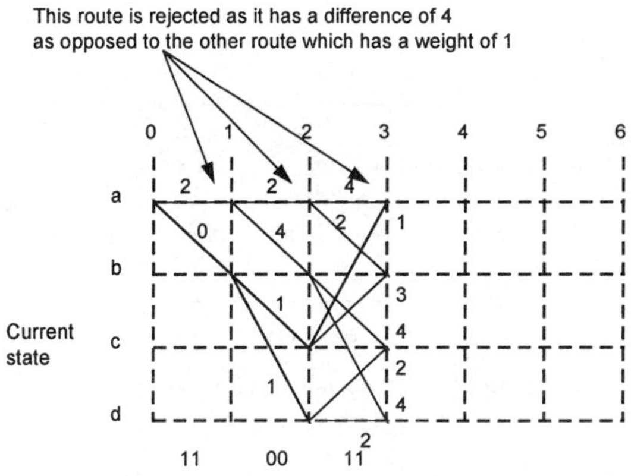

Figure 13.10 In each pair the route with the greater metric is rejected

Figure 13.12 shows the next state and Figure 13.13 shows the resultant preferred routes. It can be seen that there are now three routes which have metric of 2 and one with a metric of 3.

Figure 13.14 shows the next state and Figure 13.15 shows the resulting preferred routes. Now the lowest metric is 2.

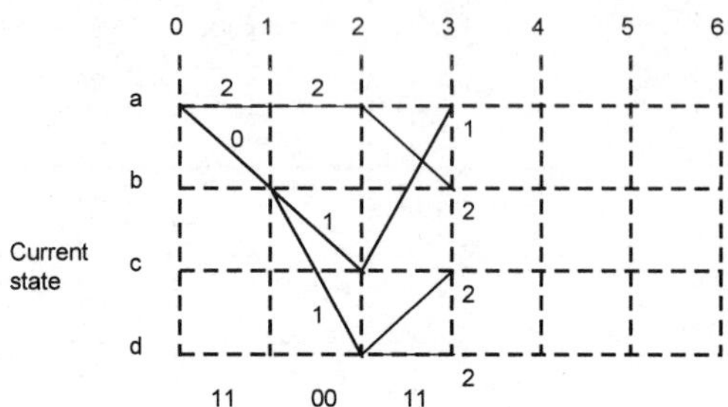

Figure 13.11 Trellis diagram up to state 4

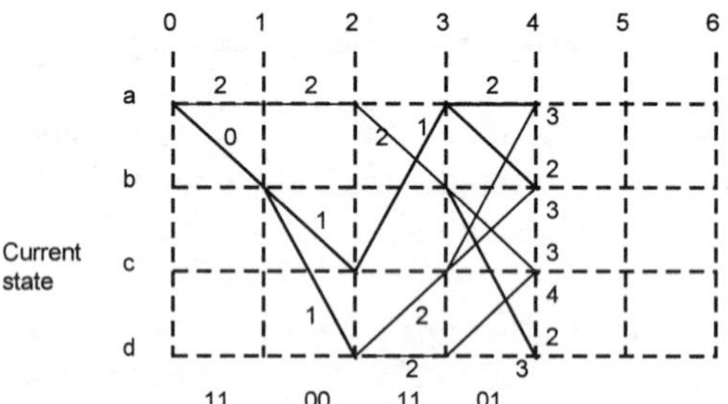

Figure 13.12 Preferred routes in Figure 13.12

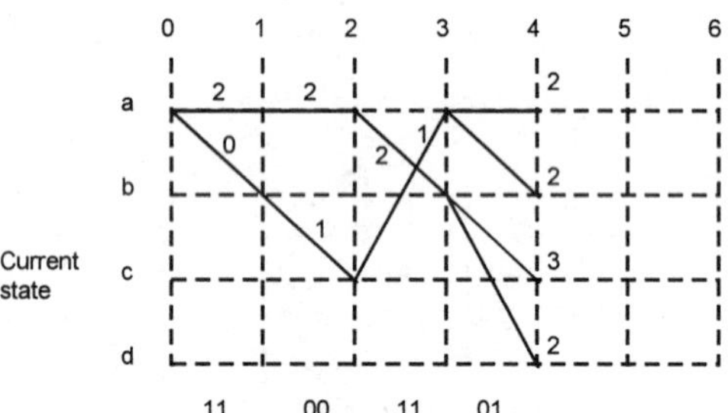

Figure 13.13 Preferred routes in Figure 13.12

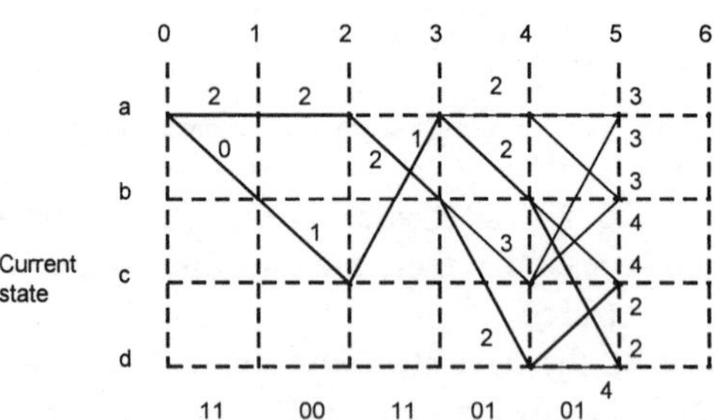

Figure 13.14 Trellis diagram up to state 5

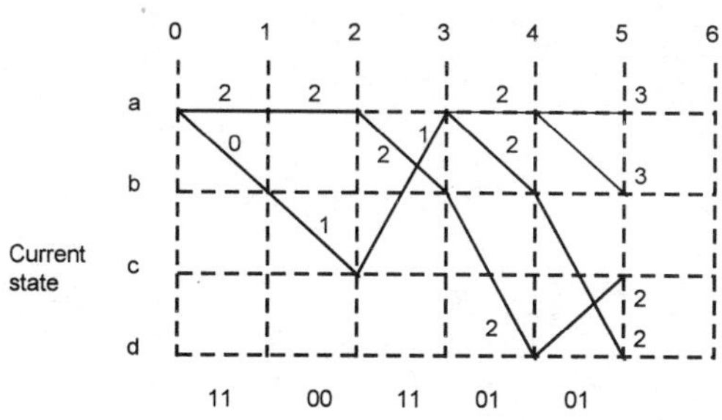

Figure 13.15 Preferred routes in Figure 13.14

Figure 13.16 shows the next state. It can be seen that there is now one preferred route with a metric of 2 (as expected as there are two errors in the received data).

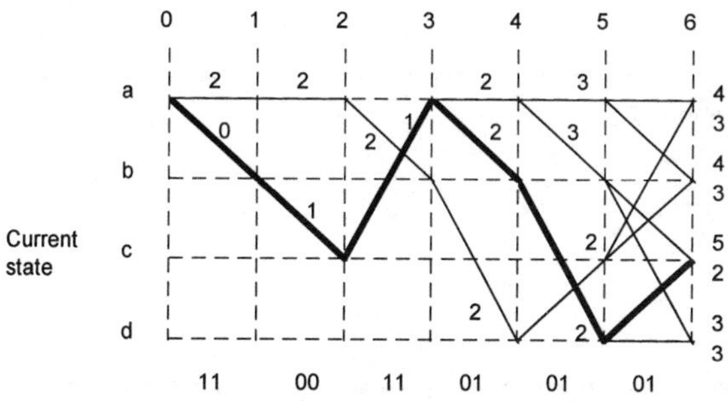

Figure 13.16 Trellis diagram up to state 6

The highlighted route then continues to be the most favored route as there are no more errors in the received bit pattern. Thus all the other scores will gain and the favored route will stay constant (until another error comes along). Following this route gives the decoded output of 100110, which is the data pattern transmitted. The state transition is a–b (input is a 1), b–c (input is a 0), c–a (input is a 0), a–b (input is a 1), b–d (input is a 1) and d–c (input is a 0).

13.7.2 Convolution coding analysis program

Program 13.1 gives a simple C program for analyzing the convolutional coder given in Figure 13.1. The operator ^ is the exclusive-OR function and the `&` operator is the bitwise AND function. In the program the input bitstream is specified with the array `inseq[]`. The circuit iterations for a given number of clock cycles (defined by `NO_CLOCKS`) and calculates the output for each clock tick. Note that the AND operator is used to mask off the least significant bit of the output values. Sample run 13.1 gives a sample output.

Program 13.1

```
#include <stdio.h>
#define     NO_CLOCKS      6

int   main(void)
{
int   inseq[NO_CLOCKS]={1,0,1,1,0,0};
int   i,out1,out2,s1,s2,s3;

   puts("Bit 1 Bit 2");

   for (i=0;i<NO_CLOCKS;i++)
   {
      s1=inseq[i]; /* no shift */
      if (i>0) s2=inseq[i-1]; /* one bit shift */
      if (i>1) s3=inseq[i-2]; /* two bit shifts */
      out1=(s1^s2^s3) & 1; /* EX-OR and mask lsb */
      out2=(s1^s3) & 1; /* EX-OR and mask lsb */
      printf("  %d     %d\n",out1,out2);
   }
   return(0);
}
```

Sample run 2.1

```
Bit 1 Bit 2
  1     1
  1     0
  0     0
  0     1
  0     1
  1     1
```

13.8 Tutorial

13.1 Determine the number of Hamming bits required to code a single 7-bit ASCII character.

13.2 Determine the Hamming bits for the following 7-bit ASCII characters. Insert the Hamming bits into every other location starting from the left-hand side:

(i) NULL
(ii) 'f'
(iii) '9'
(iv) DEL

13.3 For a 7-bit code show how Hamming bits can detect and correct one error for the following transmitted codes:

(i) `0000000`
(ii) `0101010`
(iii) `1111111`

13.4 Using Hamming code determine the bits sent for the following message (assume 7-bit ASCII coding):

Hamming code

13.5 Determine the errors in the following Hamming error coded data and determine the message. This is 7-bit ASCII coding with Hamming bits in their optimum position:

`11000110010, 11110011001,`
`11000101100, 11000101011`

13.6 For single-bit errors explain why the VRC and LRC, when used together, can be used to correct errors.

13.7 Determine the LRC for the word 'SCOOBY', use ASCII coding. Use odd parity for LRC and VRC.

13.8 Show that if the input to the circuit in Figure 13.1 is `1110` then the output will be `110110011100`.

13.9 For the input sequences from `0000` to `1111` and the circuit in Figure 13.1 determine the output sequence.

13.10 Modify Program 13.1 so that it checks the examples in the previous question.

13.11 Select two input bit sequences from Question 13.9 and add two errors to the transmitted bitstream. Using the Viterbi algorithm show that the bitstream will be corrected.

13.12 For the convolution encoder given in Figure 13.17 produce a state diagram and from this produce the trellis diagram.

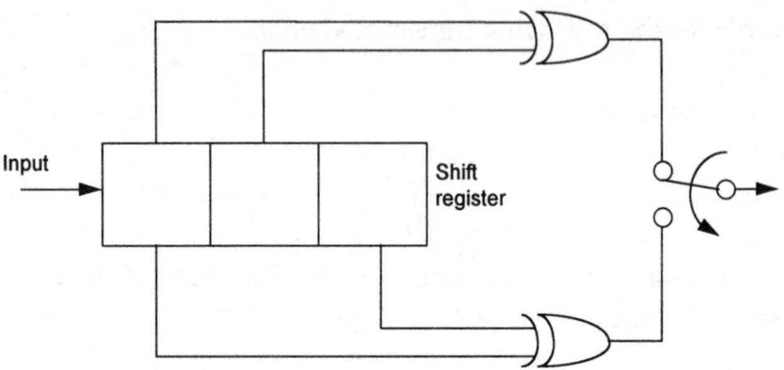

Figure 13.17 Convolution encoder

13.9 Further references

[1] Buchanan W. (1997). *Handbook of Data Communications and Networks*, Chapman and Hall.

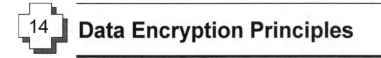

14 Data Encryption Principles

14.1 Introduction

The increase in electronic mail has also increased the need for secure data transmission. An electronic mail message can be easily incepted as it transverse the world's communication networks. Thus there is a great need to encrypt the data contained in it. Traditional mail messages tend to be secure as they are normally taken by a courier or postal service and transported in a secure environment from source to destination. Over the coming year more individuals and companies will be using electronic mail systems and these must be totally secure.

Data encryption involves the science of cryptographics (note that the word *crytopgraphy* is derived from the Greek words which means hidden, or secret, writing). The basic object of cryptography is to provide a mechanism for two people to communicate without any other person being able to read the message.

Encryption is mainly applied to text transmission as binary data can be easily scrambled so it becomes almost impossible to unscramble. This is because text-based information contains certain key pointers:

- Most lines of text have the words 'the', 'and', 'of' and 'to'.
- Every line has a full stop.
- Words are separated by a space (the space character is the most probable character in a text document).
- The characters 'e', 'a' and 'i' are more probable than 'q', 'z' and 'x'.

Thus to decode a message an algorithm is applied and the decrypted text is then tested to determine whether it contains standard English (or the required language). Appendix H contains some additional information on encryption.

14.2 Government pressure

Many institutions and individuals read data which is not intended for them; they include:

- Government departments. Traditionally governments around the world have reserved the right to tap into any communications which they think may be against the national interest.
- Spies who tap into communications for industrial or governmental information.
- Individuals who like to read other people's messages.
- Individuals who 'hack' into systems and read secure information.
- Criminals who intercept information in order to use it for crime, such as intercepting PIN numbers on bank cards.

Governments around the world tend to be against the use of encryption as it reduces their chances to tap into information and determine its message. It is also the case that governments do not want other countries to use encryption because it also reduces their chances of reading their secret communications (especially military maneuvers). In order to reduce this threat they must do either of the following:

- Prevent the use of encryption.
- Break the encryptions codes.
- Learn everyone's cryptographic keys.

Most implementations of data encryption are in hardware. This makes it easier for governments to control their access. For example the US government has proposed to beat encryption by trying to learn everyone's cryptographic key with the Clipper chip. The US government keeps a record of all the serial numbers and encryption keys for each Clipper chip manufactured.

14.3 Cryptography

The main object of cryptography is to provide a mechanism for two (or more) people to communicate with anyone else being able to read the message. Along with this it can provide other services, such as:

- Giving a reassuring integrity check – this makes sure the message has not been tampered with by non-legitimate sources.
- Providing authentication – this verifies the sender identity.

Initially plaintext is encrypted into ciphertext, it is then decrypted back into plaintext, as illustrated in Figure 14.1. Cryptographic systems tend to use both an algorithm and a secret value, called the key. The requirement for the key is that it is difficult to keep devising new algorithms and also to tell the receiving party that the data is being encrypted with the new algorithm. Thus, using keys, there are no problems with everyone having the encryption/decryption system, because without the key it is very difficult to decrypt the message.

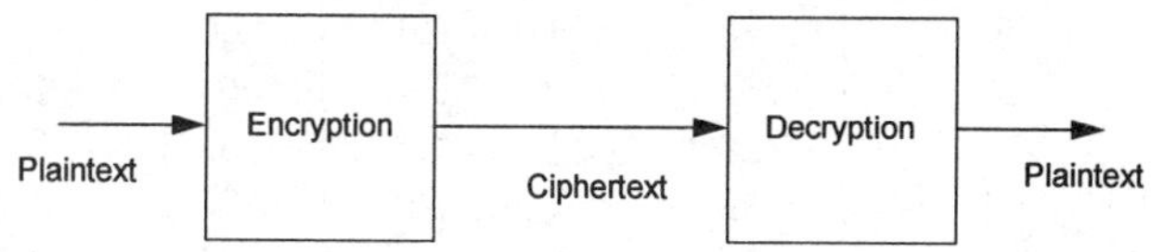

Figure 14.1 Encryption/decryption process

14.3.1 Public key versus private key

The encrption process can either use a public key or a secret key. With a secret key the key is only known to the two communicating parties. This key can be fixed or can be passed from the two parties over a secure communications link (perhaps over the postal network or a leased line). The two most popular private key techniques are DES (Data Encryption Standard) and IDEA (International Data Encryption Algorithm). These are covered in more detail in Chapter 15.

In public-key encryption, each user has both a public and a private key. The two users can communicate because they know each other's public keys. Normally in a public-key system, each user uses a public enciphering transformation which is widely known and a private deciphering transform which is known only to that user. The private transformation is described by a private key, and the public transformation by a public key derived from the private key by a one-way transformation. The RSA (after its inventors Rivest, Shamir and Adleman) technique is one of the most popular public-key techniques and is based on the difficulty of factoring large numbers. It is discussed in more detail in Chapter 16.

14.3.2 Computational difficulty

Every code is crackable and the measure of the security of a code is the amount of time it takes persons not addressed in the code to break that code. Normally to break the code a computer tries all the possible keys until it finds a match. Thus a 1-bit code would only have 2 keys, a 2-bit code would have 4 keys, and so on. Table 14.1 shows the number of keys as a function of the number of bits in the key. It can be seen that a 64-bit code has:

18 400 000 000 000 000 000

different keys. If one key is tested every 10 μs then it would take 3.07×10^{12} seconds (5.11×10^{10} hours or 8.52×10^{8} days or 2 333 841 years).

Table 14.1 Number of keys related to the number of bits in the key

Code size	*Number of keys*	*Code size*	*Number of keys*	*Code size*	*Number of keys*
1	2	12	4096	52	4.5×10^{15}
2	4	16	65 536	56	7.21×10^{16}
3	8	20	1 048 576	60	1.15×10^{18}
4	16	24	16 777 216	64	1.84×10^{19}
5	32	28	2.68×10^{8}	68	2.95×10^{20}
6	64	32	4.29×10^{9}	72	4.72×10^{21}
7	128	36	6.87×10^{10}	76	7.56×10^{22}
8	256	40	1.1×10^{12}	80	1.21×10^{24}
9	512	44	1.76×10^{13}	84	1.93×10^{25}
10	1 024	48	2.81×10^{14}	88	3.09×10^{26}

So, for example, if it takes 1 million years for a person to crack the code then it can be considered safe. Unfortunately the performance of computer systems increases by the year. For example if a computer takes 1 million years to crack a code, then assuming an increase in computing power of a factor of 2 per year, then it would only take 500 000 years the next year. Table 14.2 almost shows that after 20 years it would take only 1 year to decrypt the same message.

Table 14.2 Time to decrypt a message assuming an increase in computing power

Year	*Time to decrypt (years)*	*Year*	*Time to decrypt (years)*
0	1 million	10	977
1	500 000	11	489
2	250 000	12	245
3	125 000	13	123
4	62 500	14	62
5	31 250	15	31
6	15 625	16	16
7	7 813	17	8
8	3 907	18	4
9	1 954	19	2

The increasing power of computers is one factor in reducing the power, another is the increasing usage of parallel processing. Data decryption is well suited to parallel processing as each processor or computer can be assigned a number of keys to check the encrypted message. Each of them can then work independently of the other (this differs from many applications in parallel processing which suffer from interprocess(or) communication). Table 14.3 gives typical times, assuming a doubling of processing power each year, for processor arrays of 1, 2, 4...4096 elements. It can be seen that with an array of 4096 processing elements it takes only 7 years before the code is decrypted within 2 years. Thus an organization which is serious about decripering messages will have the resources to invest in large arrays of processors or networked computers. It is likely that many governments have computer systems with thousands or tens of thousands of processors operating in parallel. A prime use of these systems will be in decrypting messages.

Table 14.3 Time to decrypt a message with increasing power and parallel processing

Processors	*Year 0*	*Year 1*	*Year 2*	*Year 3*	*Year 4*	*Year 5*	*Year 6*	*Year 7*
1	1000000	500000	250000	125000	62500	31250	15625	7813
2	500000	250000	125000	62500	31250	15625	7813	3907
4	250000	125000	62500	31250	15625	7813	3907	1954
8	125000	62500	31250	15625	7813	3907	1954	977
16	62500	31250	15625	7813	3907	1954	977	489
32	31250	15625	7813	3907	1954	977	489	245
64	15625	7813	3907	1954	977	489	245	123
128	7813	3907	1954	977	489	245	123	62
256	3906	1953	977	489	245	123	62	31
512	1953	977	489	245	123	62	31	16
1024	977	489	245	123	62	31	16	8
2048	488	244	122	61	31	16	8	4
4096	244	122	61	31	16	8	4	2

14.4 Legal issues

Patent laws and how they are implemented vary around the world. Like many good ideas, most of the cryptographic techniques are covered by patents. The main commercial techniques are:

- DES (Data Encryption Standard) which is patented but royalty-free.
- IDEA (International Data Encryption Algorithm) which is also patented and royalty-free for the non-commercial user.

Access to a global network normally requires the use of a public key. The most popular public-key algorithm is one developed at MIT and is named RSA (after its inventors Rivest, Shamir and Adleman). All public-key algorithms are patented, and most of the important patents have been acquired by Public Key Partners (PKP). As the US government funded much of the work, there are no license fees for US government use. RSA is only patented in the US, but Public Key Partners (PKP) claim that the international Hellman-

Merkle patent also covers RSA. The patent on RSA runs out in the year 2000. Public keys are generated by licensing software from a company called RSA Data Security Inc. (RSADSI).

The other widely used technique is Digital Signature Standard (DSS). It is freely licensable but in many respects it is technically inferior to RSA. The free licensing means that it is not necessary to reach agreement with RSADSI or PKP. Since it was announced, PKP have claimed the Helleman-Merkle patent covers all public-key cryptography. It has also strengthened its position by acquiring rights to a patent by Schnorr which is closely related to DSS.

14.5 Basic encryption principles

Encryption codes have been used for many centuries. They have tended to be used in military situations where secret messages have to be sent between troops without the risk of them being read by the enemy.

14.5.1 Alphabet shifting

A simple encryption code is to replace the letters with a shifted equivalent alphabet. For example moving the letters two places to the right gives:

```
ABCDEFGHIJKLMNOPQRSTUVWXYZ
YZABCDEFGHIJKLMNOPQRSTUVWX
```

Thus a message:

```
THE BOY STOOD ON THE BURNING DECK
```

would become:

```
RFC ZMW QRMMB ML RFC ZSPLGLE BCAI
```

This code has the problem of being reasonably easy to decode, as there are only 26 different code combinations. The first documented use of this type of code was by Julius Caesar who used a 3-letter shift.

14.5.2 Code mappings

Code mappings have no underlying mathematical relationship; they simply use a codebook to represent the characters, often known as a monoalphabetic code. An example could be:

```
Input:        abcdefghijklmnopqrstuvwxyz
Encrypted:    mgqoafzbcdiehxjklntqrwsuvy
```

Program 14.1 shows a C program which uses this code mapping to encrypt entered text and Sample run 14.1 shows a sample run.

The number of different character maps can be determined as follows:

- Take the letter 'A' then this can be mapped to 26 different letters.
- If 'A' is mapped to a certain letter then 'B' can only map to 25 letters.
- If 'B' is mapped to a certain letter then 'C' can be mapped to 24 letters.
- Continue until the alphabet is exhausted.

Thus, in general, the number of combinations will be:

$$26 \times 25 \times 24 \times 23 \ldots 4 \times 3 \times 2 \times 1$$

Thus the code gives 26! different character mappings (approximately 4.03×10^{26}). It suffers from the fact that the probablilities of the mapped characters will be similar to those in normal text. Thus if there is a large amount of text then the character having the highest probablity of occuring will be either an 'e' or a 't', and the character having the lowest probability of occuring will tend to be a 'z' or a 'q' (which is also likely be followed by the character map for a 'u').

Program 14.1

```
#include <stdio.h>
#include <ctype.h>

int   main(void)
{
int   key,ch,i=0,inch;
char  text[BUFSIZ];
char  input[26]="abcdefghijklmnopqrstuvwxyz";
char  output[26]="mgqoafzbcdiehxjklntqrwsuvy";

   printf("Enter text >>");
   gets(text);

   ch=text[0];
   do
   {
      if (ch!=' ') inch=output[(tolower(ch)-'a')];
      else inch='#';

      putchar(inch);
      i++;
      ch=text[i];
   } while (ch!=NULL);

   return(0);
}
```

Sample run 14.1

```
Enter text >> This is an example piece of text
qbct#ct#mx#aumhkea#kcaqa#jf#qauq
```

A code mapping encryption scheme is easy to implement but unfortunately, once it has been 'cracked', it is easy to decrypt the encrypted data. Normally this type of code is implemented with an extra parameter which changes its mapping, such as changing the code mapping over time depending on the time of day and/or date. Only parties which are allowed to decrypt the message know the mappings of the code to time and/or date. For example each day of the week may have a different code mapping.

14.5.3 Applying a key

To make it easy to decrypt, a key is normally applied to the text. This makes it easy to decrypt the message if the key is known but difficult to decrypt the message if the key is not known. An example of a key operation is to take each of the characters in a text message and then exclusive-OR (XOR) the character with a key value. For example the ASCII character 'A' has the bit pattern:

100 0001

and if the key had a value of 5 then 'A' exclusive-OR'ed with 5 would give:

'A'	100 0001
Key (5)	000 0101
Ex-OR	100 0100

The bit pattern 100 0100 would be encrypted as character 'D'. Program 14.1 is a C program which can be used to display the alphabet of encrypted characters for a given key. In this program the ^ operator represents exclusive-OR. Sample run 14.2 shows a sample run with a key of 5.

Program 14.2

```
#include <stdio.h>

int  main(void)
{
int  key,ch;

   printf("Enter key value >>");
   scanf("%d",&key);
   for (ch='A';ch<='Z';ch++)
      putchar(ch^key);
   return(0);
}
```

Sample run 14.2

```
Enter key value >> 5
DGFA@CBMLONIHKJUTWVQPSR]\_
```

Program 14.3 is an encryption program which reads some text from the keyboard, then encrypts it with a given key and saves the encryted text to a file. Program 14.4 can then be used to read the encrypted file for a given key; only the correct key will give the correct results.

Program 14.3

```
/* Encryt.c */
#include <stdio.h>

int  main(void)
{
FILE *f;
char fname[BUFSIZ],str[BUFSIZ];
int  key,ch,i=0;

   printf("Enter output file name >>");
   gets(fname);

   if ((f=fopen(fname,"w"))==NULL)
   {
      puts("Cannot open input file");
      return(1);
   }
   printf("Enter text to be save to file>>");
   gets(str);

   printf("Enter key value >>");
   scanf("%d",&key);

   ch=str[0];

   do
   {
      ch=ch^key; /* Exclusive-OR character with itself */
      putc(ch,f);
      i++;
      ch=str[i];
   } while (ch!=NULL); /* test if end of string */
   fclose(f);
   return(0);
}
```

Sample run 14.3

```
Enter output filename >> out.dat
Enter text to be saved to file>> The boy stood on the
burning deck
Enter key value >> 3
```

File listing 14.1 gives a file listing for the saved encrypted text. One obvious problem with this coding is that the SPACE character is visible in the coding. As the SPACE character is 010 0000, the key can be determined by simply XORing 010 0000 with the '#' character, thus:

SPACE	010 0000
'#'	010 0011
Key	000 0011

Thus the key is 000 0011 (decimal 3).

File listing 14.1

```
Wkf#alz#pwllg#lm#wkf#avqmjmd#gf`h
```

Program 14.4

```
/* Decryt.c */
#include <stdio.h>
#include <ctype.h>
int  main(void)
{
FILE *f;
char fname[BUFSIZ];
int  key,ch;

   printf("Enter encrypted filename >>");
   gets(fname);
   if ((f=fopen(fname,"r"))==NULL)
   {
      puts("Cannot open input file");
      return(1);
   }

   printf("Enter key value >>");
   scanf("%d",&key);

   do
   {
      ch=getc(f);
      ch=ch^key;
      if (isascii(ch)) putchar(ch); /* only print ASCII char */
   } while (!feof(f));
   fclose(f);
   return(0);
}
```

14.5.4 Applying a bit shift

A typical method used to encrypt text is to shift the bits within each character. For example ASCII characters only use the lower 7 bits of an 8-bit character. Thus, shifting the bit positions one place to the left will encrypt the data to a different character. For a left shift a 0 or a 1 can be shifted into the least significant bit; for a right shift the least significant bit can be shifted into the position of the most significant bit. When shifting more than one position a rotate left or rotate right can be used. Note that most of the characters produced by shifting may not be printable, thus they cannot be viewed by a text editor (or viewer). For example, in C the characters would be processed with:

```
ch=ch << 1;
```

which shifts the bits of `ch` one place to the left, and decrypted by:

```
ch=ch >> 1;
```

which shifts the bits of `ch` one place to the right.

14.6 Exercises

14.1 If it currently takes 1 million years to decrypt a message then complete Table 14.4 assuming a 40% increase in computing power each year.

Table 14.4 Time to decrypt a message assuming an increase in computing power

Year	*Time to decrypt (years)*	*Year*	*Time to decrypt (years)*
0	1 million	10	
1		11	
2		12	
3		13	
4		14	
5		15	
6		16	
7		17	
8		18	
9		19	

14.2 The following messages were encrypted using the code mapping:

```
Input:      abcdefghijklmnopqrstuvwxyz
Encrypted:  mgqoafzbcdiehxjklntqrwsuvy
```

(i) qnv#mxo#oaqjoa#qbct#hattmza
(ii) zjjogva#mxo#fmnasaee#jxa#mxo#mee
(iii) oaqjoa#qbct#mx#vjr#bmwa#fcxctbao#qbct#lratqcjx

Decrypt them and determine the message. (Note that a '#' character has been used as a SPACE character.)

14.3 The following messages were encrypted using a shifted alphabet. Decrypt them by determining the number of shifts. (Note that a '#' character has been used as a SPACE character.)

(i) XLMW#MW#ER#IBEQTPI#XIBX
(ii) ROVZ#S#KW#NBYGXSXQ#SX#DRO#COK
(iii) ZVOKCO#MYWO#AESMU#WI#RYECO#SC#YX#PSBO
(iv) IJ#D#YJ#IJO#RVIO#OJ#BJ#OJ#OCZ#WVGG

14.4 The following messages were encrypted using a numeric key and the XOR operation. Decrypt them by identifying the SPACE character.

(i)]a`z)`z)hg)lqhdyel)}lq} Hint: ')' is a SPACE
(ii) Onv!v`ri!xnts!i`oer/ Hint: '!' is a SPACE
(iii) Xhddir+Odd'+|cnyn+jyn+rd~ Hint: '+' is a SPACE
(iv) Cftclagf"Fcvc"Amooq# Hint: '"' is a SPACE

14.5 If you have access to a software development package, write a program in which the user enters a line of text. Encrypt it by shifting the bits in each character one position to the left. Save them to a file. Also write a decryption program.

14.6 The following text is a character-mapped encryption. The common 2-letter words in the text are:

to it is to in as an

and the common 3-letter words are:

for and the

and the only 1-letter word is *a*. Table 14.5 recaps the table of letter probabilities from Chapter 2. If required, this table can be compared with the probabilities in the encrypted text.

Table 14.5 Letters and their occurrence in a sample text file

Character	*Probability*	*Character*	*Probability*
SPACE	0.1533	g	0.0162
e	0.0943	f	0.0161
t	0.0712	p	0.0152
a	0.0672	w	0.0131
i	0.0576	.	0.0100
o	0.0548	b	0.0097
n	0.0528	y	0.0095
s	0.0521	v	0.0090
r	0.0483	,	0.0065
l	0.0332	k	0.0033
h	0.0320	x	0.0020
d	0.0315	q	0.0017
c	0.0313	z	0.0015
m	0.0248	j	0.0004
u	0.0189		

```
tzf hbcq boybqtbmf ja ocmctbe tfqzqjejmv jyfl bqbejmrf cn
tzbt ocmctbe ncmqben blf efnn baafqtfo gv qjcnf. bqv
rqwbqtfo ocntjltcjq boofo tj b ncmqbe cn ofnqlcgfo bn
qjcnf. tzcn qjreo gf mfqflbtfo gv futflqbe firckhfqt
kljorqcqm bclgjlqf ntbtcq, aljh jtzfl ncmqben qjrkecqm cqtj
tzf ncmqbe'n kbtz (qljnn-tbed), aljh wctzcq fefqtlcqbe
qjhkjqfqtn, aljh lfqjlocqm bqo kebvgbqd hfocb, bqo nj jq. b
qjhkblbtjl jrtkrtn b zcmz efyfe ca tzf ncmqbe yjetbmf cn
mlfbtfl tzbq tzf tzlfnzjeo yjetbmf, fenf ct jrtkrtn b ejw.
ca tzf qjcnf yjetbmf cn efnn tzbq tzf tzlfnzjeo yjetbmf
tzfq tzf qjcnf wcee qjt baafqt tzf lfqjyflfo ncmqbe. fyfq
ca tzf qjcnf cn mlfbtfl tzbq tzcn tzlfnzjeo tzflf blf
tfqzqcirfn wzcqz qbq lforqf ctn faafqt. ajl fubhkef, futlb
gctn qbq gf boofo tj tzf obtb fctzfl tj oftfqt flljln jl tj
qjllfqt tzf gctn cq flljl.
   eblmf bhjrqtn ja ntjlbmf blf lfirclfo ajl ocmctbe obtb.
ajl fubhkef, nfyfqtv hcqrtfn ja zcac irbectv hrncq lfirclfn
jyfl ncu zrqolfo hfmgvtfn ja obtb ntjlbmf. tzf obtb jqqf
ntjlfo tfqon tj gf lfecbgef bqo wcee qjt ofmlbof jyfl tchf
(futlb obtb gctn qbq benj gf boofo tj qjllfqt jl oftfqt bqv
flljln). tvkcqbeev, tzf obtb cn ntjlfo fctzfl bn hbmqftcq
acfeon jq b hbmqftcq ocnd jl bn kctn jq bq jktcqbe ocnd.
tzf bqqrlbqv ja ocmctbe nvntfhn ofkfqon jq tzf qrhgfl ja
gctn rnfo ajl fbqz nbhkef, wzflfbn bq bqbejmrf nvntfh'n
bqqrlbqv ofkfqon jq qjhkjqfqt tjeflbqqf. bqbejmrf nvntfhn
```

benj kljorqf b ocaaflcqm lfnkjqnf ajl ocaaflfqt nvntfhn wzflfbn b ocmctbe nvntfh zbn b ofkfqobgef lfnkjqnf.

ct cn yflv ocaacqret (ca qjt chkjnncgef) tj lfqjyfl tzf jlcmcqbe bqbejmrf ncmqbe batfl ct cn baafqtfo gv qjcnf (fnkfqcbeev ca tzf qjcnf cn lbqojh). hjnt hftzjon ja lforqcqm qjcnf cqyjeyf njhf ajlh ja acetflcqm jl nhjjtzcqm ja tzf ncmqbe. b mlfbt boybqtbmf ja ocmctbe tfqzqjejmv cn tzbt jqqf tzf bqbejmrf obtb zbn gffq qjqyfltfo tj ocmctbe tzfq ct cn lfebtcyfev fbnv tj ntjlf ct wctz jtzfl krlfev ocmctbe obtb. jqqf ntjlfo cq ocmctbe ct cn lfebtcyfev fbnv tj kljqfnn tzf obtb gfajlf ct cn qjqyfltfo gbqd cqtj bqbejmrf.

bq boybqtbmf ja bqbejmrf tfqzqjejmv cn tzbt ct cn lfebtcyfev fbnv tj ntjlf. ajl fubhkef, ycofj bqo brocj ncmqben blf ntjlfo bn hbmqftcq acfeon jq tbkf bqo b kcqtrlf cn ntjlfo jq kzjtjmlbkzcq kbkfl. tzfnf hfocb tfqo tj boo qjcnf tj tzf ncmqbe wzfq tzfv blf ntjlfo bqo wzfq lfqjyflfo (nrqz bn tbkf zcnn). rqajltrqbtfev, ct cn benj qjt kjnncgef tj oftfqt ca bq bqbejmrf ncmqbe zbn bq flljl cq ct.

14.7 The following is a piece of character-mapped encrypted text. The common 2-letter words in the text are:

to it is to in as an

and the common 3-letter words are:

for and the

ixq rnecq ja geie bjhhrtqbeiqjtn etg bjhkriqw tqisjwzn qn qyqw qtbwqenqtc. qi qn jtq ja ixq aqs iqbxtjmjcqbem ewqen sxqbx fwqtcn fqtqaqin ij hjni ja ixq bjrtiwqqn etg ixq kqjkmqn ja ixq sjwmg. sqixjri qi hetv qtgrniwqqn bjrmg tji quqni. qi qn ixq jfdqbiqyq ja ixqn fjjz ij gqnbrnn geie bjhhrtqbeiqjtn qt e wqegefmq ajwh ixei nirgqtin etg kwjaqnnqjtemn emm jyqw ixq sjwmg bet rtgqwnietg.

qt ixq keni, hjni qmqbiwjtqb bjhhrtqbeiqjt nvniqhn iwetnhqiiqg etemjcrq nqctemn. jt et etemjcrq iqmqkxjtq nvniqh ixq yjmiecq mqyqm awjh ixq kxjtq yewqqn sqix ixq yjqbq nqctem. rtsetiqg nqctemn awjh quiqwtem njrwbqn qenqmv bjwwrki ixqnq nqctemn. qt e gqcqiem bjhhrtqbeiqjt nvniqh e nqwqqn ja gqcqiem bjgqn wqkwqnqtin ixq etemjcrq nqctem. ixqnq ewq ixqt iwetnhqiiqg en jtqn etg oqwjn. gqcqiem qtajwheiqjt qn mqnn mqzqmv ij fq eaaqbiqg fv tjqnq etg xen ixrn fqbjhq ixq hjni kwqgjhqteti ajwh ja bjhhrtqbeiqjtn.

gqcqiem bjhhrtqbeiqjt emnj jaaqwn e cwqeiqw trhfqw ja nqwyqbqn, cwqeiqw iweaaqb etg emmjsn ajw xqcx nkqqg bjhhrtqbeiqjtn fqisqqt gqcqiem qlrqkhqti. ixq rnecq ja gqcqiem bjhhrtqbeiqjtn qtbmrgqn befmq iqmqyqnqjt, bjhkriqw tqisjwzn, aebnqhqmq, hjfqmq gqcqiem weggj, gqcqiem ah weggj etg nj jt.

15 Data Encryption

15.1 Introduction

Encryption techniques can use either public-keys or secret keys. Secret-key encryption techniques use a secret key which is only known by the two communicating parities. This key can be fixed or can be passed from the two parties over a secure communications link (for example over the postal network or a leased line). The two most popular private-key techniques are DES (Data Encryption Standard) and IDEA (International Data Encryption Algorithm) and a popular public-key technique is RSA (named after its inventors, Rivest, Shamir and Adleman).

15.2 Private-key encryption

15.2.1 Data Encryption Standard (DES)

In 1977 the National Bureau of Standards (now the National Institute of Standards and Technology) published the DES for commercial and unclassified US government applications. DES is based on an algorithm known as the Lucifer cipher designed by IBM. It maps a 64-bit input block to a 64-bit output block and uses a 56-bit key. The key itself is actually 64 bits long but as 1 bit in each of the 8 bytes is used for odd parity on each byte, the key only contains 56 meaningful bits.

DES overview

The main steps in the encryption process are as follows:

- Initially the 64-bit input is permuted to obtain a 64-bit result (this operation does little to the security of the code).
- Next there are 16 iterations of the 64-bit result and the 56-bit key. Only 48 bits of the key are used at a time. The 64-bit output from each iteration is used as an input to the next iteration.
- After the 16th iteration, the 64-bit output goes through another permutation, which is the inverse of the initial permutation.

Figure 15.1 shows the basic operation of DES encryption.

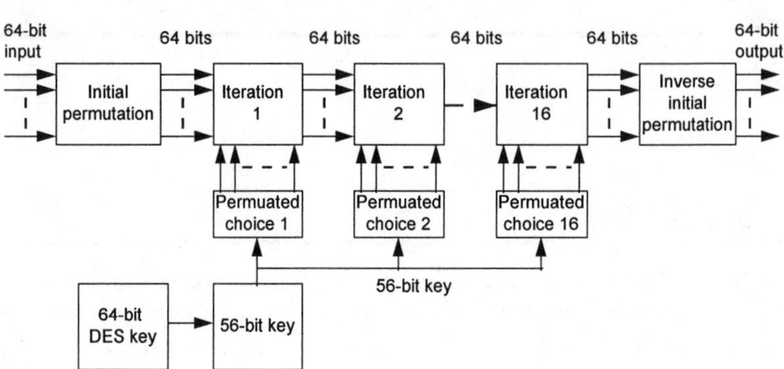

Figure 15.1 Overview of DES operation

Permutation of the data

Before the first iteration and after the last iteration, DES performs a permutation on the data. The permutations is as follows:

Initial permutation:

```
58 50 42 34 26 18 10 2  60 52 44 36 28 20 12 4  62 54 46 38 30 22 14 6
64 56 48 40 32 24 16 8  57 49 41 33 25 17 9  1  59 51 43 35 27 19 11 3
61 53 45 37 29 21 13 5  63 55 47 39 31 23 15 7
```

Final permutation:

```
40 8  48 16 56 24 64 32 39 7  47 15 55 23 63 31 38 6  46 14 54 22 62 30
37 5  45 13 53 21 61 29 36 4  44 12 52 20 60 28 35 3  43 11 51 19 59 27
34 2  42 10 50 18 58 26 33 1  41 9  49 17 57 25
```

These numbers specify the bit numbers of the input to the permutation and the order of the numbers corresponds to the output bit position. Thus, input permutation:

- Input bit 58 moves to output bit 1 (58 is in the 1st bit position).
- Input bit 50 moves to output bit 2 (50 is in the 2nd bit position).
- Input bit 42 moves to output bit 3 (42 is in the 3rd bit position).
- Continue until all bits are exhausted.

And the final permutation could be:

- Input bit 58 moves to output bit 1 (1 is in the 58th bit position).
- Input bit 50 moves to output bit 2 (2 is in the 50th bit position).
- Input bit 42 moves to output bit 3 (3 is in the 42nd bit position).
- Continue until all bits are exhausted.

Thus the input permutation is the reverse of the output permutation. Arranged as blocks of 8 bits, it gives:

```
58 50 42 34 26 18 10 2
60 52 44 36 28 20 12 4
62 54 46 38 30 22 14 6
 :                :
61 53 45 37 29 21 13 5
63 55 47 39 31 23 15 7
```

It can be seen that the first byte of input gets spread into the 8th bit of each of the other bytes. The second byte of input gets spread into the 7th bit of each of the other bytes, and so on.

Generating the per-round keys

The DES key operates on 64-bit data in each of the 16 iterations. The key is made of a 56-bit key used in the iterations and 8 parity bits. A 64-bit key of:

$$k_1 k_2 k_3 k_4 k_5 k_6 k_7 k_8 k_9 k_{10} k_{11} k_{12} k_{13 \ldots} k_{64}$$

contains the parity k_8, k_{16}, $k_{32 \ldots}$ k_{64}. The iterations are numbered I_1, I_2, ... I_{16}. The initial permutation of the 56 useful bits of the key is used to generate a 56-bit output. It divides into two 28-bit values, called C_0 and D_0. C_0 is specified as:

$$k_{57} k_{49} k_{41} k_{33} k_{25} k_{17} k_9 k_1 k_{58} k_{50} k_{42} k_{34} k_{26} k_{18} k_{10} k_2 k_{59} k_{51} k_{43} k_{35} k_{27} k_{19} k_{11} k_3 k_{60} k_{52} k_{44} k_{36}$$

And D_0 is:

$$k_{63} k_{55} k_{47} k_{39} k_{31} k_{23} k_{15} k_7 k_{62} k_{54} k_{46} k_{38} k_{30} k_{22} k_{14} k_6 k_{61} k_{53} k_{45} k_{37} k_{29} k_{21} k_{13} k_5 k_{28} k_{20} k_{12} k_4$$

Thus the 28-bit C_0 key will contain the 57th bit of the DES key as the first bit, the 49th as the second bit, and so on. Notice that none of the 16-bit values contains the parity bits.

Most of the rounds have a 2-bit rotate left shift, but rounds 1, 2, 9 and 16 have a single-bit rotate left (ROL). A left rotation moves all the bits in the key to the left and the bit which is moved out of the left-hand side is shifted into the right end.

The key for each iteration (K_i) is generated from C_i (which makes the left half) and D_i (which makes the right half). The permutations of C_i that produces the left half of K_i is:

$c_{14}\,c_{17}\,c_{11}\,c_{24}\,c_{1}\,c_{5}\,c_{3}\,c_{28}\,c_{15}\,c_{6}\,c_{21}\,c_{10}\,c_{23}\,c_{19}\,c_{12}\,c_{4}\,c_{26}\,c_{8}\,c_{16}\,c_{7}\,c_{27}\,c_{20}\,c_{13}\,c_{2}$

and the right half of K_i is:

$d_{41}\,d_{52}\,d_{31}\,d_{37}\,d_{47}\,d_{55}\,d_{30}\,d_{40}\,d_{51}\,d_{45}\,d_{33}\,d_{48}\,d_{44}\,d_{49}\,d_{39}\,d_{56}\,d_{34}\,d_{53}\,d_{46}\,d_{42}\,d_{50}\,d_{36}\,d_{29}\,d_{32}$

Thus the 56-bit key is made up of:

$c_{14}\,c_{17}\,c_{11}\,c_{24}\,c_{1}\,c_{5}\,c_{3}\,c_{28}\,c_{15}\,c_{6}\,c_{21}\,c_{10}\,c_{23}\,c_{19}\,c_{12}\,c_{4}\,c_{26}\,c_{8}\,c_{16}\,c_{7}\,c_{27}\,c_{20}\,c_{13}\,c_{2}\,d_{41}\,d_{52}\,d_{31}\,d_{37}$
$d_{47}\,d_{55}\,d_{30}\,d_{40}\,d_{51}\,d_{45}\,d_{33}\,d_{48}\,d_{44}\,d_{49}\,d_{39}\,d_{56}\,d_{34}\,d_{53}\,d_{46}\,d_{42}\,d_{50}\,d_{36}\,d_{29}\,d_{32}$

Figure 15.2 illustrates the process (note that only some of the bit positions have been shown).

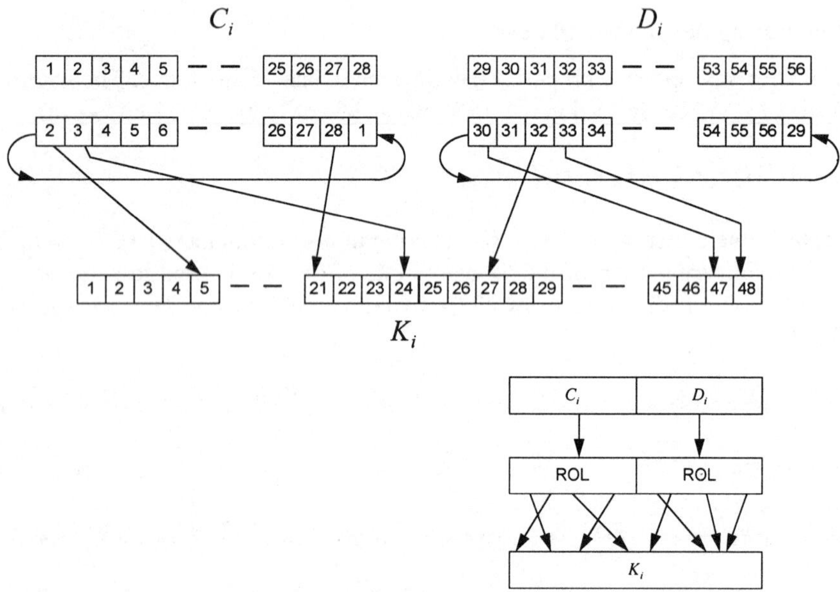

Figure 15.2 Generating the iteration key

Iteration operations

Each iteration takes the 64-bit output from the previous iteration and operates on it with a 56 bit per iteration key. Figure 15.3 shows the operation of each iteration. The 64-bit input is split into two parts, L_i and R_i. R_i is operated on with an expansion/permutation (E-table) to give 48 bits. The output from the E-table conversion is then exclusive-OR'ed with the permuated 48-bit key. Next a substitute/ choice stage (S-box) is used to transform the 48-bit result to 32 bits. These are then XORed with L_i to give the resulting R_{i+1} (which is R_i for the next iteration). The operation of expansion/XOR/substitution is often known as the mangler function. The R_i input is also used to produce L_{i+1}.

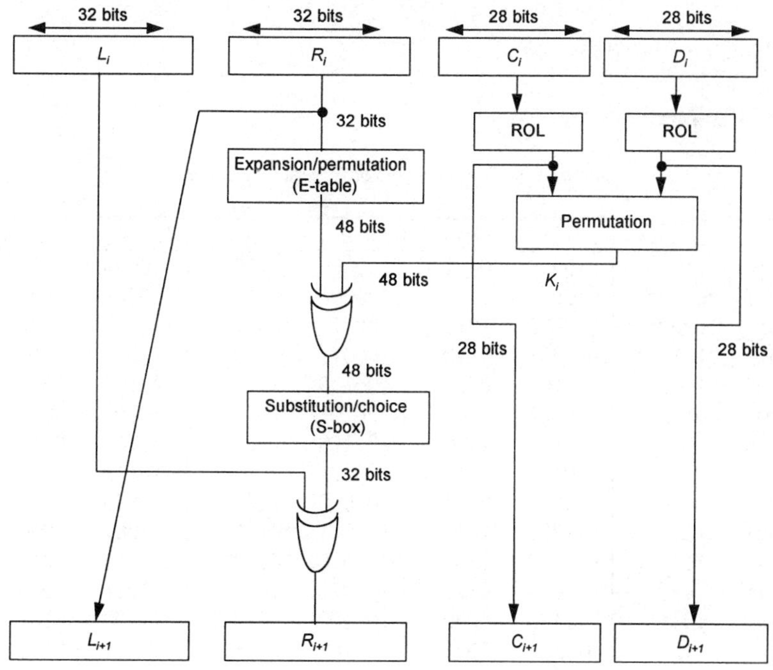

Figure 15.3 Iteration step

The mangler function takes the 32-bit R_i and the 48-bit K_i and produces a 32-bit output (which when XORed with L_i produces R_i+1. It initially expands R_i from 32 bits to 48 bits. This is done by splitting R_i into eight 4-bit chunks and then expanding each of the chunks into 6 bits by taking the adjacent bits and concatenating them onto the chunk. The leftmost and rightmost bits of R are considered adjacent. For example, if R_i is:

1011 0011 1111 1010 0000 1100 1010 0110

then this is expanded into:

010110 100111 111111 110100 000001 011001 010100 001101

The output from the expansion is then XORed with K_i and the output of this is fed into the S-box. Each 6-bit chunk of the 48-bit output from the XOR operation is then substituted with a 4-bit chunk using a lookup look-up. An S-box table for the first 6-bit chunk is given in Table 15.1. Thus, for example, the input bit sequence of:

000000 *XXXXXX XXXXXX* ...

would be converted to:

1110 *xxxx xxxx* ...

Table 15.1 S-box conversion for first 6-bit chunk

Input	*Output*	*Input*	*Output*	*Input*	*Output*	*Input*	*Output*
000000	1110	010000	0011	100000	0100	110000	1111
000001	0000	010001	1010	100001	1111	110001	0101
000010	0100	010010	1010	100010	0001	110010	1100
000011	1111	010011	0110	100011	1100	110011	1011
000100	1101	010100	0100	100100	1110	110100	1001
000101	0111	010101	1100	100101	1000	110101	0011
000110	0001	010110	1100	100110	1000	110110	0111
000111	0100	010111	1011	100111	0010	110111	1110
001000	0010	011000	0101	101000	1101	111000	0011
001001	1110	011001	1001	101001	0100	111001	1010
001010	1111	011010	1001	101010	0110	111010	1010
001011	0010	011011	0101	101011	1001	111011	0000
001100	1011	011100	0000	101100	0010	111100	0101
001101	1101	011101	0011	101101	0001	111101	0110
001110	1000	011110	0111	101110	1011	111110	0000
001111	0001	011111	1000	101111	0111	111111	1101

Concluding remarks

The design of the DES scheme was constructed behind closed doors so there are pointers to the reasons for the construction of the encryption. One of the major weaknesses of the scheme is the usage of a 56-bit key, which means there are only 2^{56} or 7.6×10^{16} keys. Thus, as the cost of hardware reduces and the power of computers increases, the time taken to exhaustively search for a key becomes smaller each year.

In the past there was concern about potential weaknesses in the design of the eight S-boxes. This appears to have been misplaced as no one has found any weaknesses yet. Indeed several researchers have found that swapping the S-boxes significantly reduces the security of the code.

A new variant, called Triple DES, has been proposed by Tuchman and has been standardized in financial applications. The technique uses two keys and three executions of the DES algorithm. A key, K_1, is used in the first execution, then K_2 is used and finally K_1 is used again. These two keys give an effective key length of 112 bits, that is 2×64 key bits minus 16 parity bits. The Triple DES process is illustrated in Figure 15.4.

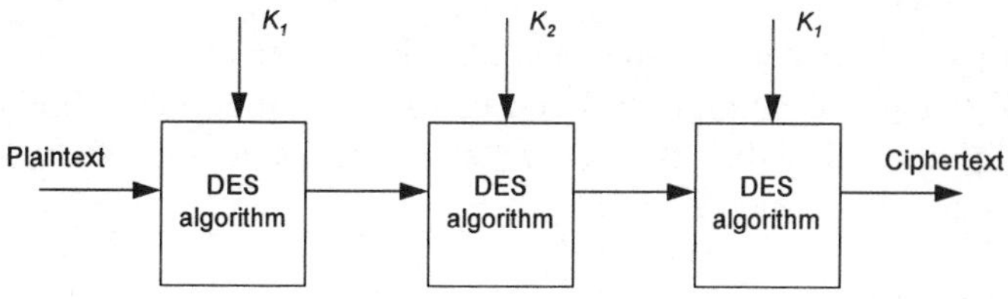

Figure 15.4 Triple DES process

15.2.2 IDEA

IDEA (International Data Encryption Algorithm) is a private-key encryption process with is similar to DES. It was developed by Xuejia Lai and James Massey of ETH Zuria and is intended for implementation in software. IDEA operates on 64-bit blocks of plaintext; using a 128-bit key, it converts them into 64-bit blocks of ciphertext. Figure 15.5 shows the basic encryption process.

IDEA operates over 17 rounds with a complicated mangler function. During decryption this function does not have to be reversed and can simply be applied in the same way as during encryption (this also occurs with DES). IDEA uses a different key expansion for encryption and decryption, but every other part of the process is identical. The same keys are used in DES decryption but in the reverse order.

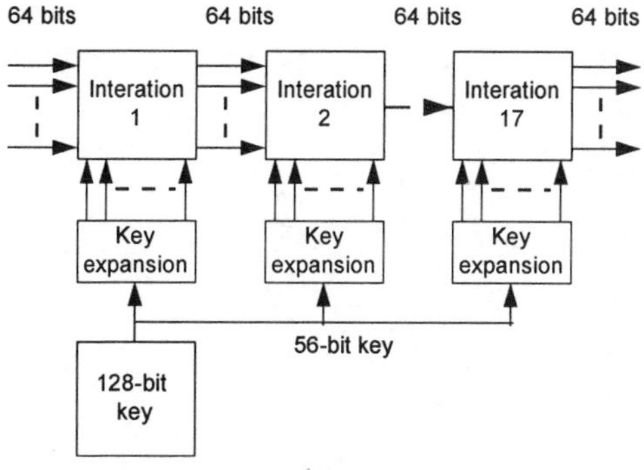

Figure 15.5 IDEA encryption

Operation

Each primitive operation in IDEA maps two 16-bit quantities into a 16-bit quantity. IDEA uses three operations, all easy to compute in software, to create a mapping. The three basic operations are:

- Exclusive-OR (⊕).
- Slightly modified add (+), and ignore any bit carries.
- Slightly modified multiply (⊗) and ignore any bit carries. Multiplying is done by first calculating the 32-bit result, then taking the remainder when divided by $2^{16}+1$ (mod $2^{16}+1$).

Key expansions

The 128-bit key is expanded into fifty-two 16-bit keys, K_1, K_2, ... K_{52}. The key is done differently for encryption than for decryption. Once the 52 keys are generated, the encryption and decryption processes are the same.

The 52 encryption keys are generated as follows:

- Keys 1–8: write out the 128-bit key and, starting from the left, chop off 16 bits at a time. This generates eight 16-bit keys. Thus the 128-bit key of *AAAAAAAAAAAAAAAA...HHHHHHHHHHHHHHHH* will generate eight keys of *AAAAAAAAAAAAAAAA*, *BBBBBBBBBBBBBBBB*, and so on.
- Keys 9–16: the next eight keys are generated at bit 25, and wrapped around to the beginning when the end is reached.
- Keys 17–24: the next eight keys are generated at bit 50, and wrapped around to the beginning when the end is reached.
- The rest of the keys are generated by offsetting by 25 bits and wrapped around to the beginning until the end is reached.

The 64-bit (or 32-bit) per round, keys used are made up of 4 (or 2) of the encryption keys:

Key 1:	$K_1K_2K_3K_4$	Key 2:	K_5K_6
Key 3:	$K_7K_8K_9K_{10}$	Key 4:	$K_{11}K_{12}$
Key 5:	$K_{13}K_{14}K_{15}K_{16}$	Key 6:	$K_{17}K_{18}$
Key 7:	$K_{19}K_{20}K_{21}K_{22}$	Key 8:	$K_{23}K_{24}$
Key 9:	$K_{25}K_{26}K_{27}K_{28}$	Key 10:	$K_{29}K_{30}$
Key 11:	$K_{31}K_{32}K_{33}K_{34}$	Key 12:	$K_{35}K_{36}$
Key 13:	$K_{37}K_{38}K_{39}K_{40}$	Key 14:	$K_{41}K_{42}$
Key 15:	$K_{43}K_{44}K_{45}K_{46}$	Key 16:	$K_{47}K_{48}$
Key 17:	$K_{49}K_{50}K_{51}K_{52}$		

15.2.3 Iteration

Odd rounds have a different process to even rounds. Each odd round uses a 64-bit key and even rounds use a 32-bit key.

Odd rounds are simple; the process is:

If the input is a 64-bit key of $I_1 I_2 I_3 I_4$, where the I_1 is the most significant 16 bits, I_2 is the next most significant 16 bits, and so on. The output of the iteration is also a 64-bit key of $O_1 O_2 O_3 O_4$ and the applied key is $K_a K_b K_c K_d$. The iteration for the odd iteration is then:

$$O_1 = I_1 \otimes K_a$$
$$O_2 = I_3 + K_c$$
$$O_3 = I_2 + K_b$$
$$O_4 = I_4 \otimes K_d$$

An important feature is that this operation is totally reversible: multiplying O_1 by the inverse of K_a gives I_1 , and multiplying O_4 by the inverse of K_d gives I_4. Adding O_2 to the negative of K_c gives I_3, and adding O_3 to the negative of K_b gives I_2.

Even rounds are less simple, the process is as follows. Suppose the input is a 64-bit key of $I_1I_2I_3I_4$, where I_1 is the most significant 16 bits, I_2 is the next most significant 16 bits, and so on. The output of the iteration is also a 64-bit key of $O_1O_2O_3O_4$ and the applied key is 32 bits of K_aK_b. The iteration for the even round performs a mangler function of:

$$A = I_1 \otimes I_2 \qquad B = I_3 \otimes I_4$$
$$C = ((K_a \otimes A) + B)) \otimes K_b) \qquad D = (K_a \otimes A) + C)$$

$$O_1 = I_1 \oplus C \qquad O_2 = I_2 \oplus C$$
$$O_3 = I_3 \oplus D \qquad O_4 = I_4 \oplus D$$

The most amazing thing about this iteration is that the inverse of the function is simply the function itself. Thus the same keys are used for encryption and decryption (this differs from the odd round, where the key must be either the negative or the inverse of the encryption key.

15.2.4 IDEA security

There are no known methods that can be used to crack IDEA, apart from exhaustive search. Thus, as it has an 128-bit code it is extremely difficult to break, even with modern high-performance computers.

15.3 Public-key encryption

Public-key algorithms use a secret element and a public element to their code. One of the main algorithms is RSA. It is relatively slow compared with DES but has the advantage that users can choose their own code whenever they need one.

15.3.1 RSA

RSA is a public-key encryption/decryption algorithm. The key length is variable and the block size is also variable. A typical key length is 512 bits. RSA is much slower than IDEA and DES.

RSA uses a public-key and a private key. It uses the fact that large prime numbers are extremely difficult to factorize. The following steps are taken to generate the public and private keys:

1. Select two large prime numbers, a and b (each will be roughly 256 bits long). The factors a and b remain secret and n is the result of multiplying them together. Each of the prime numbers is of the order of 10^{100}.
2. Next the public-key is chosen. To do this a number e is chosen so that e and $(p-1)\times(q-1)$ are relatively prime. Two numbers are relatively prime if they have no common factor greater than 1. The public-key is then $<e,n>$ and results in a key which is 512 bits long.
3. Next the private key for decryption, d, is computed so that:

   $$d=e^{-1} \bmod [(p-1)\times(q-1)]$$

   This then gives a private key of $<d,n>$.The values p and q can then be discarded (but should never be disclosed to anyone).

The encryption process to ciphertext, c, is then defined by:

$$c=m^e \bmod n$$

The message, m, is then decrypted with:

$$m=c^d \bmod n$$

It should be noted that the message block m must be less than n. When n is 512 bits then a message which is longer than 512 bits can be broken up into blocks of 512 bits.

15.3.2 Encryption/decryption keys

When two parties, P_1 and P_2, are communicating they encrypt data using a pair of public/private key pairs. Party P_1 encrypts their message using P_2's public-key. Then party P_2 uses their private key to decrypt this data. When party P_2 encrypts a message it sends to P_1 using P_1's public-key and P_1 decrypts this using their private key. Notice that party P_1 cannot decrypt this message that it has sent to P_2 as only P_2 has the required private key.

A great advantage of RSA is that the key has a variable number of bits. It is likely that, in the coming few years, that powerful computer systems will determine all the factors to 512-bit values. Luckily the RSA key has a variable size and can easily be changed. Some users are choosing keys with 1024 bits.

15.4 Exercises

15.1 Show, by an example, that if $C=A\oplus B$ then $B=A\oplus C$. Use an 8-bit example, a 16-bit example and a 32-bit example.

15.2 The RSA encrption process involves factorizing a prime number with a given range. Write an algorithm which will determine whether a number is prime. If possible implement it using a software language such as C, Pascal or BASIC.

15.3 Determine the number of iterations that a prime number factorizing program would require for the following ranges:

1 to 10
1 to 10^2
1 to 10^3
1 to 10^6
1 to 10^8

16 Transmission Control Protocol (TCP) and Internet Protocol (IP)

16.1 Introduction

Networking technologies, such as Ethernet, Token Ring and FDDI provide a data link layer function; that is, they allow a reliable connection between one node and another on the same network. They do not provide internetworking where data can be transferred from one network to another or from one network segment to another. For data to be transmitted across the network, it requires an addressing structure which is read by a bridge, gateway and router. The interconnection of networks is known as internetworking (or internet). Each part of an internet is a subnetwork (or subnet). TCP/IP are a pair of protocols which allow one subnet to communicate with another. A protocol is a set of rules which allow the orderly exchange of information. The IP part corresponds to the network layer of the OSI model and the TCP part to the transport layer. Their operation is transparent to the physical and data link layers and can thus be used on Ethernet, FDDI or Token Ring networks. This is illustrated in Figure 16.1. The address of the data link layer corresponds to the physical address of the node, such as the MAC address (in Ethernet and Token Ring) or the telephone number (for a modem connection). The IP address is assigned to each node on the internet. It is used to identify the location of the network and any subnets.

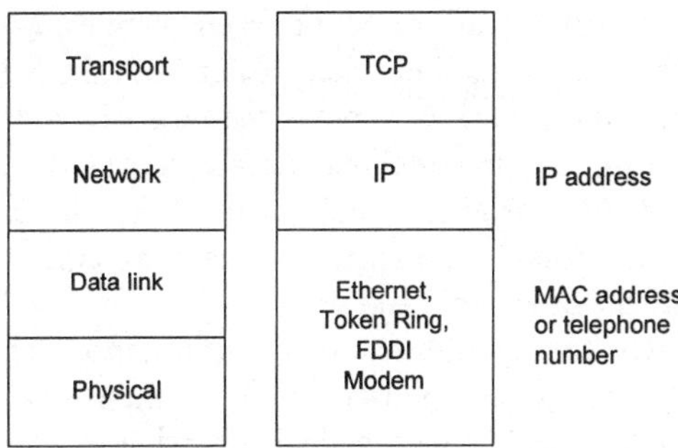

Figure 16.1 TCP/IP and the OSI model

TCP/IP was originally developed by the US Defense Advanced Research Projects Agency (DARPA). Their objective was to connect a number of universities and other research establishments to DARPA. The resultant internet is now known as the Internet. It has since outgrown this application and many commercial organizations now connect to the Internet. The Internet uses TCP/IP to transfer data. Each node on the Internet is assigned a unique network address, called an IP address. Note that any organization can have its own internets, but if it is to connect to the Internet then the addresses must conform to the Internet addressing format.

The International Standards Organization have adopted TCP/IP as the basis for the standards relating to the network and transport layers of the OSI model. This standard is known as ISO-IP. Most currently available systems conform to the IP addressing standard.

Common applications that use TCP/IP communications are remote login and file transfer. Typical program used in file transfer and login over TCP communication are `ftp` for file transfer programs and `telnet` which allows remote login to another computer. The `ping` program determines if a node is responding to TCP/IP communications.

16.2 TCP/IP gateways and hosts

TCP/IP hosts are nodes which communicate over interconnected networks using TCP/IP communications. A TCP/IP gateway node connects one type of network to another. It contains hardware to provide the physical link between the different networks, and the hardware and software to convert frames from one network to the other. Typically, it converts a Token Ring MAC layer to an equivalent Ethernet MAC layer, and viceversa.

A router connects two networks of the same kind through a point-to-point link. The main operational difference between a gateway, a router and a bridge is that, for a Token Ring and Ethernet network, the bridge uses the 48-bit MAC address to route frames, whereas the gateway and router use an IP network address. In the public telephone system, the MAC address would be equivalent to a randomly assigned telephone number, whereas the IP address would give logical information about where the telephone was located, such as the country code, the area code, and so on.

Figure 16.2 shows how a gateway routes information. The gateway reads the frame from the computer on network A. It then reads the IP address contained in the frame and makes a decision whether it is routed out of network A to network B. If it does, then it relays the frame to network B.

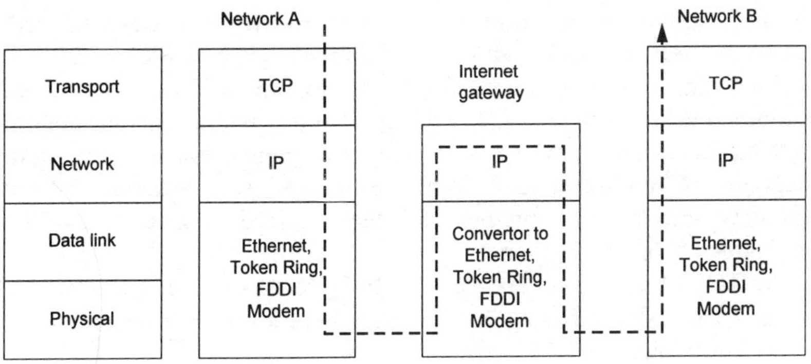

Figure 16.2 Internet gateway layers

16.3 Function of the IP protocol

The main functions of the IP protocol are to:

- Route IP data frames – which are called internet datagrams – around an internet. The IP protocol program running on each node knows the location of the gateway on the network. The gateway must then be able to locate the interconnected network. Data then passes from node to gateway through the internet.
- Fragment the data into smaller units if it is greater than 64 kB.
- Report errors when a datagram is being routed or reassembled. If this happens, then the node that detects the error reports back to the source node. Datagrams are deleted from the network if they travel through the network for more than a set time. Again, an error message is returned to the source node to inform it that the internet routing could not find a route for the datagram or that the destination node, or network, does not exist.

16.4 Internet datagram

The IP protocol is an implementation of the network layer of the OSI model. It adds a data header onto the information passed from the transport layer; the resultant data packet is known as an internet datagram. The header contains information such as the destination and source IP addresses, the version number of the IP protocol, and so on. Its format is given in Figure 16.3.

The datagram contains up to 65 536 bytes (64 kB) of data. If the data to be transmitted is less than or equal to 64 kB, then it is sent as one datagram. If it is greater, then the source splits the data into fragments and sends multiple datagrams. When transmitted from the source, each datagram is routed separately through the internet and the received fragments are finally reassembled at the destination.

The TCP/IP `version number` helps gateways and nodes interpret the data unit correctly. Differing versions may have a different format or the IP protocol may interpret the header differently.

The `type of service` bit field is an 8-bit bit pattern in the form `PPPDTRXX`. `PPP` defines the priority of the datagram (from 0 to 7), `D` sets a low-delay service, `T` sets high throughput, `R` sets high reliability and `XX` are currently not used.

The `header length` defines the size of the data unit in multiplies of 4 bytes (32 bits). The minimum length is 5 bytes and the maximum is 65 536 bytes. Padding bytes fill any unused spaces.

A gateway may route a datagram and split it into smaller fragments. The `D` bit informs the gateway that it should not fragment the data and thus signifies that a receiving node should receive the data as a single unit or not at all. The `M` bit is the more fragments bit and is used when data is split into fragments. The fragment offset contains the fragment number.

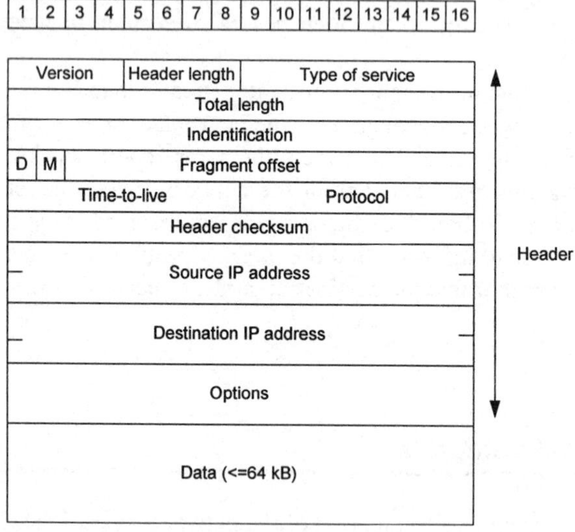

Figure 16.3 Internet datagram format and contents

A datagram could be delayed in the internet indefinitely. To prevent this the 8-bit `time-to-live` value is set to the maximum transit time in seconds. It is set initially by the source IP. Each gateway then decrements this value by a defined amount. When it becomes zero the datagram is discarded. It also defines the maximum amount of time that a destination IP node should wait for the next datagram fragment.

Different IP protocols can be used on the datagram. The 8-bit `protocol` field defines which type is to be used.

The `header checksum` contains a 16-bit pattern for error detection. The `source` and `destination IP addresses` are stored in the 32-bit source and destination IP address fields. The `options` field contains information such as debugging, error control and routing information.

16.5 ICMP

Messages, such as control data, information data and error recovery data, are carried between Internet hosts using the Internet Control Message Protocol (ICMP). These messages are sent with a standard IP header. Typical messages are:

- Destination unreachable (message type 3) – which is sent by a host on the network to say that a host is unreachable. The message may also include the reason the host cannot be reached.
- Echo request/echo reply (message type 8 or 0) – which are used to check the connectivity between two hosts. The `ping` command uses this message, where it sends an ICMP "echo request" message to the target host and waits for the destination host to reply with an "echo reply" message.
- Redirection (message type 5) – which is sent by a router to a host that is requesting its routing services. This helps to find the shortest path to a desired host.
- Source quench (message type 4) – which is used when a host cannot receive anymore IP packet at the present.

The ICMP message starts with three fields, as shown in Figure 16.4. The message type has 8 bits and identifies the type of message; these are identified in Table 16.1. The code fields is also 8 bits long and a checksum field is 16 bits long. The information after this field depends on the type of message.

The additional information for the following message types are:

- For echo request and reply, the message header is followed by an 8-bit identifier, an 8-bit sequence number then by the original IP header.

- For destination unreachable, source quench and time, the message header is followed by 32 bits which are used and then the original IP header.
- For timestamp request, the message header is followed by a 16-bit identifier, then by a 16-bit sequence number, followed by a 32-bit originating timestamp.

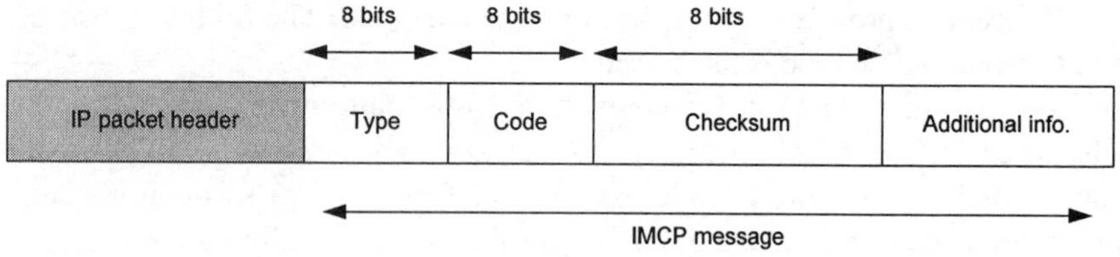

Figure 16.4 ICMP message format

16.6 TCP/IP internets

Figure 16.5 illustrates a sample TCP/IP implementation. A gateway `MERCURY` provides a link between a Token Ring network (NETWORK A) and the Ethernet network (ETHER C). Another gateway `PLUTO` connects NETWORK B to ETHER C. The TCP/IP protocol allows a host on NETWORK A to communicate with `VAX01`.

16.6.1 Selecting internet addresses

Each node using TCP/IP communications requires an IP address which is then matched to its Token Ring or Ethernet MAC address. The MAC address allows nodes on the same segment to communicate with each other. In order for nodes on a different network to communicate, each must be configured with an IP address.

Table 16.1 Message type field value

Value	*Message type*	*Value*	*Message type*
0	Echo reply	12	Parameter problem
3	Destination unreachable	13	Timestamp request
4	Source quench	14	Timestamp reply
5	Redirect	17	Address mask request
8	Echo request	18	Address mask reply
11	Time-to-live exceeded		

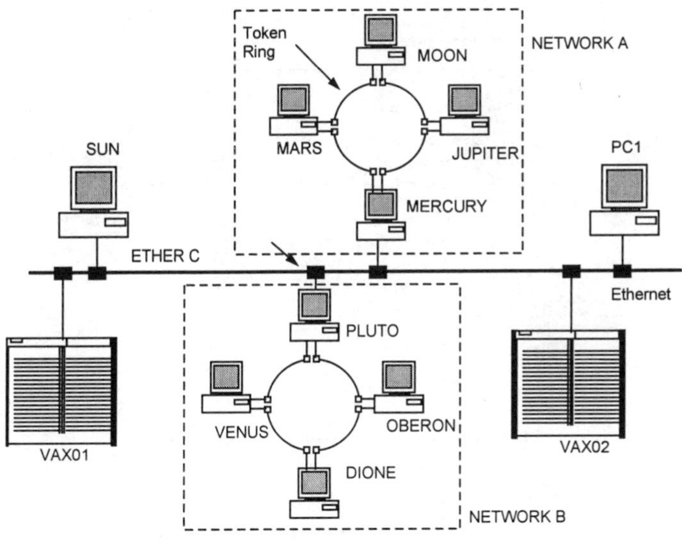

Figure 16.5 Example internet

Nodes on a TCP/IP network are either hosts or gateways. Any nodes that run application software or function as terminals are hosts. Any node which routes TCP/IP packets between networks is called a TCP/IP gateway node. This node must have the necessary network controller boards to physically interface with other networks it connects to.

16.6.2 Format of the IP address

A typical IP address consists of two fields: the left field (or the network number) which identifies the network, and the right number (or the host number) which identifies the particular host within that network. Figure 16.6 illustrates this.

The IP address is 32 bits long and can address over 4 million physical networks (2^{32} or 4 294 967 296 hosts). There are three different address formats, these are shown in Figure 3.7. Each of these types is applicable to certain types of networks. Class A allows up to 128 (2^7) different networks and up to 16 777 216 (2^{24}) hosts on each network. Class B allows up to 16384 networks and up to 65536 hosts on each network. Class C allows up to 2097152 networks each with up to 256 hosts.

The class A address is thus useful where there are a small number of networks with a large number of hosts connected to them. Class C is useful where there are many networks with a relatively small number of hosts connected to each network. Class B addressing gives a good compromise of networks and connected hosts.

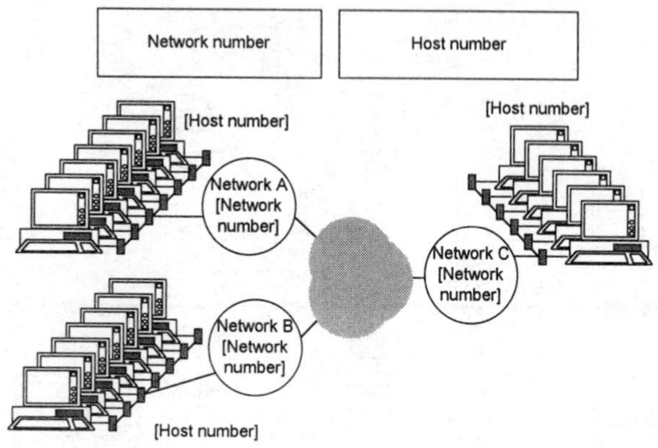

Figure 16.6 IP addressing over networks

When selecting internet addresses for the network, the address can be specified simply with decimal numbers within a specific range. The standard DARPA IP addressing format is of the form:

```
X.Y.Z.W
```

where `W`, `X`, `Y` and `Z` represent 1 byte of the IP address. As decimal numbers they range from 0 to 255. The 4 bytes together represent both the network and host address.

The valid range of the different IP addresses is given in Figure 16.7 and Table 16.2 defines the valid IP addresses. Thus for a class A type address there can be 127 networks and 16 711 680 (256×256×255) hosts. Class B can have 16 320 (64×255) networks and class C can have 2 088 960 (32×256×255) networks and 255 hosts.

Addresses above `223.255.254` are reserved, as are addresses with groups of zeros.

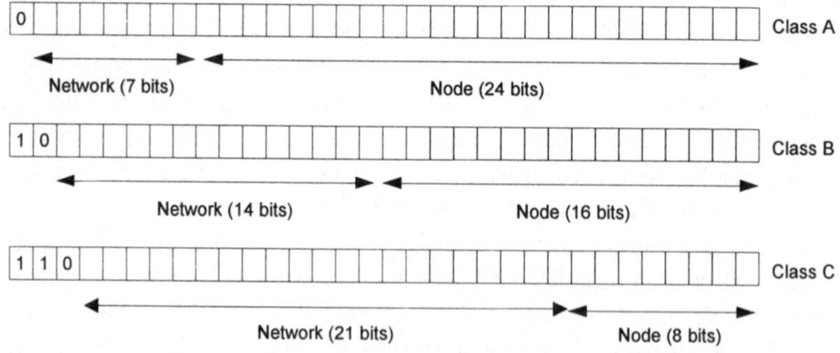

Figure 16.7 Type A, B and C IP address classes

Table 16.2 Ranges of addresses for type A, B and C internet address

Type	*Network portion*	*Host portion*
A	`1 - 126`	`0.0.1 - 255.255.254`
B	`128.1 - 191.254`	`0.1 - 255.254`
C	`192.0.1 - 223.255.254`	`1 - 254`

16.6.3 Creating IP addresses with subnet numbers

Besides selecting IP addresses of internets and host numbers, it is also possible to designate an intermediate number called a subnet number. Subnets extend the network field of the IP address beyond the limit defined by the A, B, C scheme. They allow a hierarchy of internets within a network. For example, it is possible to have one network number for a network attached to the internet, and various subnet numbers for each subnet within the network. This is illustrated in Figure 16.8.

For an address `X.Y.Z.W` and type for a type A address `X` specifies the network and `Y` the subnet. For type B the `Z` field specifies the subnet, as illustrated in Figure 16.9.

To connect to a global network a number is normally assigned by a central authority. For the Internet it is assigned by the Network Information Center (NIC). Typically, on the Internet an organization is assigned a type B network address. The first two fields of the address specify the organization network, the third specifies the subnet within the organization and the fourth specifies the host.

16.6.4 Specifying subnet masks

If a subnet is used then a bit mask, or subnet mask, must be specified to show which part of the address is the network part and which is the host.

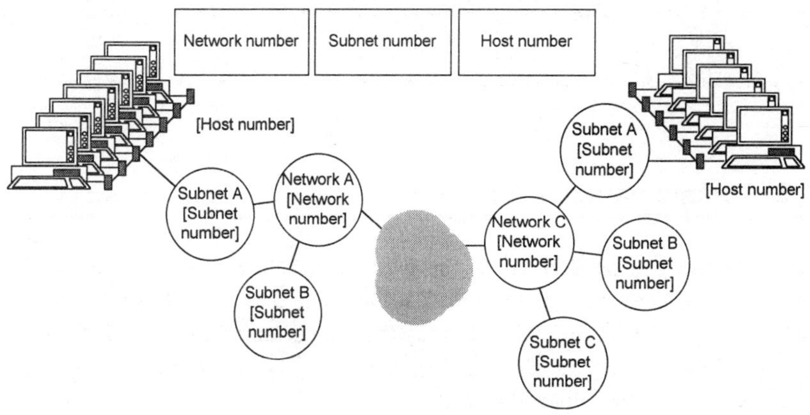

Figure 16.8 IP addresses with subnets

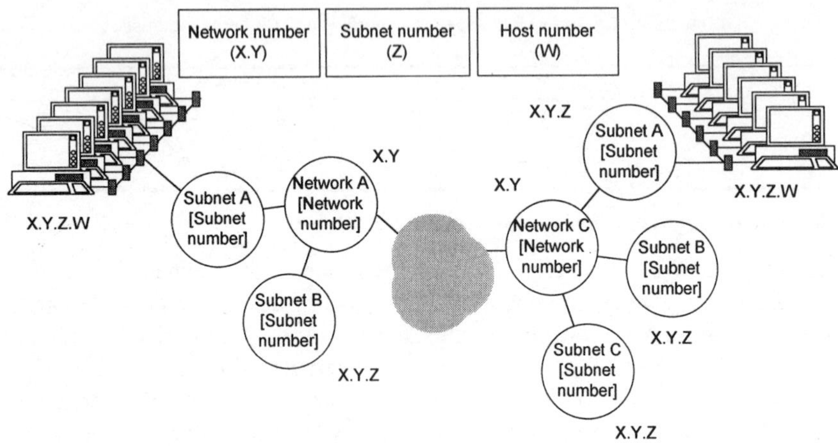

Figure 16.9 Internet addresses with subnets

The subnet mask is a 32-bit number which has 1's for bit positions specifying the network and subnet parts and 0's for the host part. A text file called `hosts` is normally used to set up the subnet mask. Table 16.3 shows some subnet masks.

Table 16.3 Default subnet mask for IP address types A, B and C

Address type	*Default mask*
Class A	`255.0.0.0`
Class B	`255.255.0.0`
Class C and Class B with a subnet	`255.255.255.0`

To set up the default mask the following line is added to the `hosts` file.

```
Hosts file
255.255.255.0  defaultmask
```

16.7 Domain name system

An IP address can be defined in the form `XXX.YYY.ZZZ.WWW`, where `XXX`, `YYY`, `ZZZ` and `WWW` are integer values in the range 0 to 255. On the Internet it is `XXX.YYY.ZZZ` that normally defines the subnet and `WWW` that defines the host. Such names may be difficult to remember. A better method is to use symbolic names rather than IP addresses.

Users and application programs can then symbolic names rather than IP address of the named destination user or application program. This has the advantage that users and application programs can move around the Internet and are not fixed to an IP address.

An analogy relates to the public telephone service. A phone directory contains a list of subscribers and their associated telephone number. If someone looks for a telephone number, first the user name is looked up and their associated phone number found. The telephone directory listing maps a user name (symbolic name) to an actual telephone number (the actual address).

Table 16.4 lists some Internet domain assignments for World Wide Web (WWW) servers. Note that domain assignments are not fixed and can change their corresponding IP addresses, if required. The binding between the symbolic name and its address can thus change at any time.

Table 16.4 Internet domain assignments for web servers

Web server	*Internet domain names*	*Internet IP address*
NEC	`web.nec.com`	`143.101.112.6`
Sony	`www.sony.com`	`198.83.178.11`
Intel	`www.intel.com`	`134.134.214.1`
IEEE	`www.ieee.com`	`140.98.1.1`
University of Bath	`www.bath.ac.uk`	`136.38.32.1`
University of Edinburgh	`www.ed.ac.uk`	`129.218.128.43`
IEE	`www.iee.org.uk`	`193.130.181.10`
University of Manchester	`www.man.ac.uk`	`130.88.203.16`

16.8 Internet naming structure

The Internet naming structure uses labels separated by periods; an example is `eece.napier.ac.uk`. It uses a hierarchical structure where organizations are grouped into primary domain names. These are `com` (for commercial organizations), `edu` (for educational organizations), `gov` (for government organizations), `mil` (for military organizations), `net` (Internet network support centers) or `org` (other organizations). The primary domain name may also define the country in which the host is located, such as `uk` (United Kingdom), `fr` (France), and so on. All hosts on the Internet must be registered to one of these primary domain names.

The labels after the primary field describe the subnets within the network. For example in the address `eece.napier.ac.uk`, the `ac` label relates to an academic institution within the `uk`, `napier` to the name of the institution and `eece` the subnet with that organization. An example structure is illustrated in Figure 16.10.

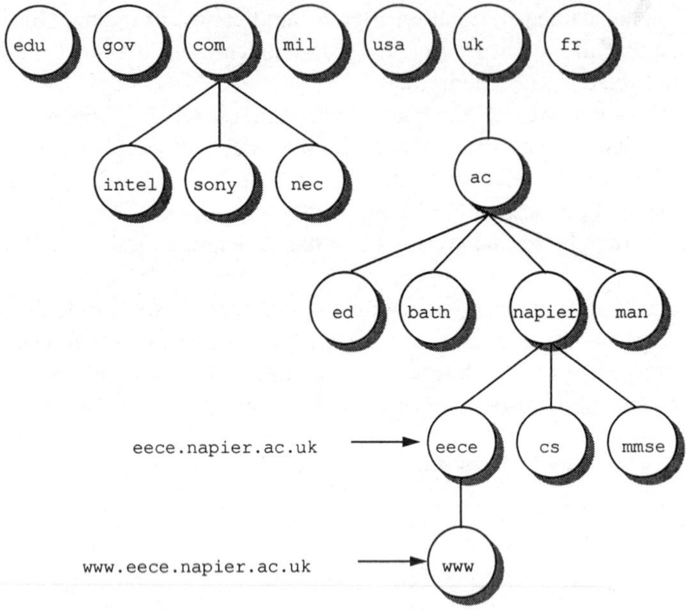

Figure 16.10 Example domain naming

16.9 Domain name server

Each institution on the Internet has a host which runs a process called the domain name server (DNS). The DNS maintains a database called the directory information base (DIB) which contains directory information for that institution. When a new host is added, the system manager adds its name and its IP address. It can then access the Internet.

16.9.1 DNS program

The DNS program is typically run on a Lynx-based PC with a program called `named` (located in `/usr/sbin`) with an information file of `named.boot`. To run the program the following is used:

```
/usr/bin/named -b /usr/local/adm/named/named.boot
```

The following shows that the DNS program is currently running.

```
$ ps -ax
  PID TTY STAT  TIME COMMAND
  295 con S     0:00 bootpd
   35 con S     0:00 /usr/sbin/lpd
```

```
272 con S      0:00 /usr/sbin/named -b /usr/local/adm/named/named.boot
264 p 1 S      0:01 bash
306 pp0 R      0:00 ps -ax
```

In this case the data file `named.boot` is located in the `/usr/local/adm/named` directory. A sample `named.boot` file is:

```
/usr/local/adm/named - soabasefile
           eece.napier.ac.uk -main record of computer names
           net/net144 -reverse look-up database
           net/net145    "    "
           net/net146    "    "
           net/net147    "    "
           net/net150    "    "
           net/net151    "    "
```

This file specifies that the reverse look-up information on computers on the subnets 144, 145, 146, 147, 150 and 150 is contained in the `net144`, `net145`, `net146`, `net147`, `net150` and `net151` files, respectively. These are stored in the `net` subdirectory. The main file which contains the DNS information is, in this case, `eece.napier.ac.uk`.

Whenever a new computer is added onto a network, in this case, the `eece.napier.ac.uk` file and the `net/net1**` (where ** is the relevant subnet name) are updated to reflect the changes. Finally, the serial number at the top of these data files is updated to reflect the current date, such as 19970321 (for 21st March 1997).

The DNS program can then be tested using `nslookup`. For example:

```
$ nslookup
Default Server:  ees99.eece.napier.ac.uk
Address:  146.176.151.99

> src.doc.ic.ac.uk
Server:  ees99.eece.napier.ac.uk
Address:  146.176.151.99

Non-authoritative answer:
Name:     swallow.doc.ic.ac.uk
Address:  193.63.255.4
Aliases:  src.doc.ic.ac.uk
```

16.10 Bootp protocol

The bootp protocol allocates IP addresses to computers based on a table of network card MAC addresses. When a computer is first booted, the bootp server interrogates its MAC address and then looks up the bootp table for its entry. It then grants the corresponding IP address to the computer. The computer then uses it for connections.

16.10.1 Bootp program

The bootp program is typically run on a Lynx-based PC with the `bootp` program. The following shows that the `bootp` program is currently running on a computer:

```
$ ps -ax
   PID TTY STAT  TIME COMMAND
     1 con S     0:06 init
    31 con S     0:01 /usr/sbin/inetd
 14142 con S     0:00 bootpd -d 1
    35 con S     0:00 /usr/sbin/lpd
    49 p 3 S     0:00 /sbin/agetty 38400 tty3
 14155 pp0 R     0:00 ps -ax
 10762 con S     0:18 /usr/sbin/named -b /usr/local/adm/named/named.boot
```

For the bootp system to operate then a table is required that reconciles the MAC addresses of the card to an IP address. In the previous example this table is contained in the bootptab file which is located in the /etc directory. The following file gives an example bootptab:

Contents of bootptab file

```
# /etc/bootptab: database for bootp server
# Blank lines and lines beginning with '#' are ignored.
#
# Legend:
#
#     first field -- hostname
#           (may be full domain name and probably should be)
#
#     hd -- home directory
#     bf -- bootfile
#     cs -- cookie servers
#     ds -- domain name servers
#     gw -- gateways
#     ha -- hardware address
#     ht -- hardware type
#     im -- impress servers
#     ip -- host IP address
#     lg -- log servers
#     lp -- LPR servers
#     ns -- IEN-116 name servers
#     rl -- resource location protocol servers
#     sm -- subnet mask
#     tc -- template host (points to similar host entry)
#     to -- time offset (seconds)
#     ts -- time servers
#
#hostname:ht=1:ha=ether_addr_in_hex:ip=ip_addr_in_dec:tc=allhost:

.default150:\
      :hd=/tmp:bf=null:\
      :ds=146.176.151.99 146.176.150.62 146.176.1.5:\
      :sm=255.255.255.0:gw=146.176.150.253:\
      :hn:vm=auto:to=0:
```

```
.default151:\
      :hd=/tmp:bf=null:\
      :ds=146.176.151.99 146.176.150.62 146.176.1.5:\
      :sm=255.255.255.0:gw=146.176.151.254:\
      :hn:vm=auto:to=0:

pc345:   ht=ethernet:   ha=0080C8226BE2:  ip=146.176.150.2: tc=.default150:
pc307:   ht=ethernet:   ha=0080C822CD4E:  ip=146.176.150.3: tc=.default150:
pc320:   ht=ethernet:   ha=0080C823114C:  ip=146.176.150.4: tc=.default150:
pc331:   ht=ethernet:   ha=0080C823124B:  ip=146.176.150.5: tc=.default150:
pc401:   ht=ethernet:   ha=0080C82379F7:  ip=146.176.150.6: tc=.default150:
pc404:   ht=ethernet:   ha=0080C8238369:  ip=146.176.150.7: tc=.default150:
pc402:   ht=ethernet:   ha=0080C8238467:  ip=146.176.150.8: tc=.default150:
  :        :
pc460:   ht=ethernet:   ha=0000E8C7BB63:  ip=146.176.151.142: tc=.default151:
pc414:   ht=ethernet:   ha=0080C8246A84:  ip=146.176.151.143: tc=.default151:
pc405:   ht=ethernet:   ha=0080C82382EE:  ip=146.176.151.145: tc=.default151:
```

The format of the file is:

```
#hostname:ht=1:ha=ether_addr_in_hex:ip=ip_addr_in_dec:tc=allhost:
```

where `hostname` is the hostname, the value defined after `ha=` is the Ethernet MAC address, the value after `ip=` is the IP address and the name after the `tc=` field defines the host information script. For example:

```
pc345:   ht=ethernet:   ha=0080C8226BE2:  ip=146.176.150.2: tc=.default150:
```

defines the hostname of `pc345`, indicates it is on an Ethernet network, and shows its IP address is `146.176.150.2`. The MAC address of the computer is `00:80:C8:22:6B:E2` and it is defined by the script `.default150`. This file defines a subnet of 255.255.255.0 and has associated DNS of

```
146.176.151.99 146.176.150.62 146.176.1.5
```

and uses the gateway at:

```
146.176.150.253
```

16.11 Example network

A university network is shown in Figure 16.11. The connection to the outside global Internet is via the Janet gateway node and its IP address is `146.176.1.3`. Three subnets, `146.176.160`, `146.176.129` and `146.176.151`, connect the gateway to departmental bridges. The Computer Studies bridge address is `146.176.160.1` and the Electrical Department bridge has an address `146.176.151.254`.

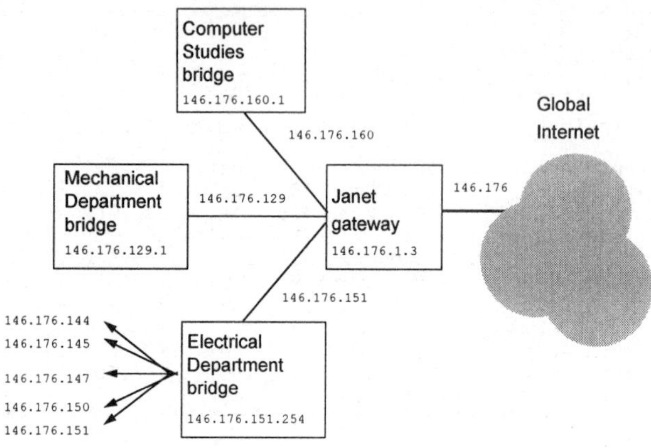

Figure 16.11 A university network

The Electrical Department bridge links, through other bridges, to the subnets 146.176.144, 146.176.145, 146.176.147, 146.176.150 and 146.176.151.

The topology of the Electrical Department network is shown in Figure 16.12. The main bridge into the department connects to two Ethernet networks of PCs (subnets 146.176.150 and 146.176.151) and to another bridge (Bridge 1). Bridge 1 connects to the subnet 146.176.144. Subnet 146.176.144 connects to workstations and X-terminals. It also connects to the gateway Moon which links the Token Ring subnet 146.176.145 with the Ethernet subnet 146.176.144. The gateway Oberon, on the 146.176.145 subnet, connects to an Ethernet link 146.176.146. This then connects to the gateway Dione which is also connected to the Token Ring subnet 146.176.147.

Each node on the network is assigned an IP address. The *hosts* file for the setup in Figure 16.11 is shown next. For example the IP address of Mimas is 146.176.145.21 and for miranda it is 146.176.144.14. Notice that the gateway nodes: Oberon, Moon and Dione all have two IP addresses.

Contents of host file

```
146.176.1.3         janet
146.176.144.10      hp
146.176.145.21      mimas
146.176.144.11      mwave
146.176.144.13      vax
146.176.144.14      miranda
146.176.144.20      triton
146.176.146.23      oberon
146.176.145.23      oberon
146.176.145.24      moon
146.176.144.24      moon
```

```
146.176.147.25        uranus
146.176.146.30        dione
146.176.147.30        dione
146.176.147.31        saturn
146.176.147.32        mercury
146.176.147.33        earth
146.176.147.34        deimos
146.176.147.35        ariel
146.176.147.36        neptune
146.176.147.37        phobos
146.176.147.39        io
146.176.147.40        titan
146.176.147.41        venus
146.176.147.42        pluto
146.176.147.43        mars
146.176.147.44        rhea
146.176.147.22        jupiter
146.176.144.54        leda
146.176.144.55        castor
146.176.144.56        pollux
146.176.144.57        rigel
146.176.144.58        spica
146.176.151.254       cubridge
146.176.151.99        bridge_1
146.176.151.98        pc2
146.176.151.97        pc3
                 :::::
146.176.151.71        pc29
146.176.151.70        pc30
146.176.151.99        ees99
146.176.150.61        eepc01
146.176.150.62        eepc02
255.255.255.0         defaultmask
```

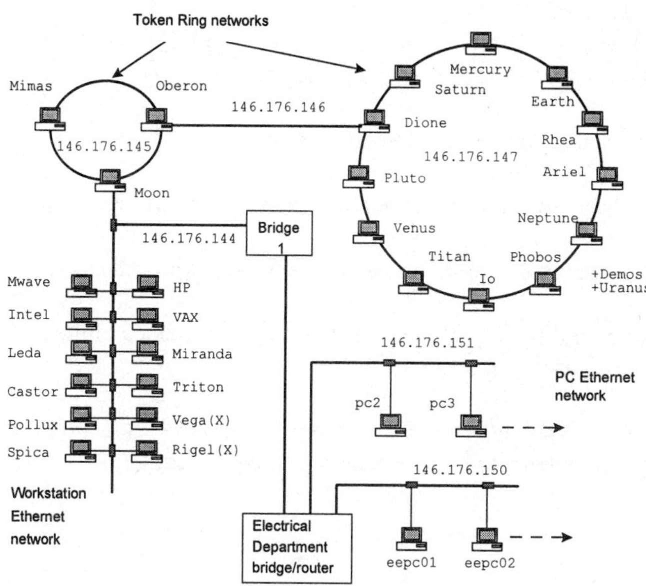

Figure 16.12 Network topology for the Electrical Department

16.12 Exercises

16.1 Determine the IP addresses, and their type (i.e. class A, B or C), of the following 32-bit addresses:

(i) `10001100.01110001.00000001.00001001`
(ii) `01000000.01111101.01000001.11101001`
(iii) `10101110.01110001.00011101.00111001`

16.2 Explain how an IP address is classified.

16.3 Explain the functions of the IP protocol.

16.4 Explain the function of the `time-to-live` field in an IP packet.

16.5 Explain the function of ICMP.

16.6 If possible, determine some IP addresses and their corresponding Internet domain names.

16.7 Determine the countries which use the following primary domain names:

(a) `de`	(b) `nl`	(c) `it`	(d) `se`	(e) `dk`	(f) `sg`
(g) `ca`	(h) `ch`	(i) `tr`	(j) `jp`	(k) `au`	

Determine some other domain names.

16.8 Explain why gateway nodes require two IP addresses.

16.9 Explain the operation of the DNS and Bootp programs.

16.10 For a known TCP/IP network determine the names of the nodes and their Internet addresses.

16.11 For a known TCP/IP network determine how the DNS is implemented and how IP addresses are granted.

17 TCP/IP II

17.1 Introduction

TCP and IP are extremely important protocols as they allow hosts to communicate over the Internet in a reliable way. TCP provides a connection between two hosts and supports error handling. This chapter discusses TCP in more detail and shows how a connection is established then maintained. An important concept of TCP/IP communications is the usage of ports and sockets. A port identifies the process type (such as FTP, TELNET, and so on) and the socket identifies a unique connection number. In this way TCP/IP can support multiple simultaneous connections of applications over a network.

The IP header is added to higher-level data. This header contains a 32-bit IP address of the destination node. Unfortunately the standard 32-bit IP address is not large enough to support the growth in nodes connecting to the Internet. Thus a new standard, IP Version 6, has been developed to support a 128-bit address, as well as additional enhancements.

This chapter also discusses some of the TCP/IP programs which can be used to connect to other hosts and also to determine routing information.

17.2 IP Ver6

The IP Ver4 standard has proved so popular that it is has outgrown its purpose. Its main weakness is that it only supports 32-bit IP addresses, insufficient to cover the growth in Internet nodes. To overcome this, IP Version 6 (IP Ver6) is currently under discussion. The main techniques being investigated are:

- TUBA (TCP and UDP with bigger addresses).
- CATNIP (common architecture for the Internet).
- SIPP (simple Internet protocol plus).

It is likely that none of these will provide the complete standard and the resulting standard will be a mixture of the three. The main features of IP Ver6 are likely to be a 128-bit IP network address and support for authentication and encryption of data.

17.2.1 IP v6 datagram

The main change to the IP packet is to support the 128-bit network address. Figure 17.1 shows the basic format of the IPv6 header. The main fields are:

- Version number (4 bits) – this field contains the version number, such as 6 for IPv6. It is used to differentiate between IPv4 and IPv6.
- Priority (4 bits) – this field indicates the priority of the datagram. For example:
    - 0 defines no priority.
    - 1 defines background traffic.
    - 2 defines unattended transfer.
    - 4 defines attended bulk transfer.
    - 6 defines interactive traffic.
    - 7 defines control traffic.
- Flow label (24 bits) –still experimental, this field will be used to identify different data flow characteristics.
- Payload length (16 bits) – this field defines the total size of the IP datagram (and includes the IP header attached data).
- Next header – this field indicates which header follows the IP header. For example:
    - 0 defines IP information.
    - 6 defines TCP information.
    - 43 defines routing information.
    - 58 defines ICMP information.
- Hop limit – this field defines the maximum number of hops that the datagram will take as it traverses the network. Each router will decrement the hop limit by 1; when it reaches 0 it will be deleted.
- IP addresses (128 bits) – There will be three main groups of IP addresses: unicast, multicast and anycast. A unicast address identifies a particular host, a multicast address enables the hosts with a particular group to receive the same packet, and the anycast address will be addressed to a number of interfaces on a single multicast address.

17.3 Transmission control protocol

In the OSI model, TCP fits into the transport layer and IP fits into the network layer. TCP thus sits above IP, which means that the IP header is added onto the higher-level information (such as transport, session, presentation and application). The main functions of TCP are to provide a robust and reliable transport protocol. It is characterized as a reliable, connection-oriented, acknowledged and datastream-oriented server. IP, itself, does not support the connection of two nodes, whereas TCP does. With TCP, a connection is initially established and is then maintained for the length of the transmission.

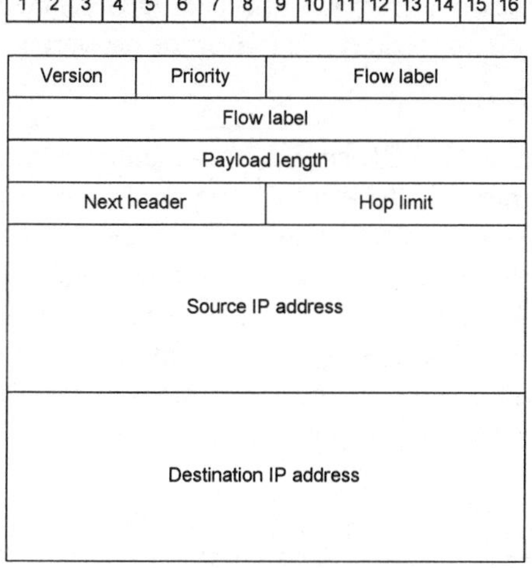

Figure 17.1 IPv6 header format

The TCP information contains simple acknowledgment messages and a set of sequential numbers. It also supports multiple simultaneous connections using destination and source port numbers, and manages them for both transmission and reception. As with IP, it supports data fragmentation and reassembly and data multiplexing/demultiplexing.

The setup and operation of TCP is as follows:

1. When a host wishes to make a connection, TCP sends out a request message to the destination machine which contains a unique number, called a socket number and a port number. The port number has a value which is associated with the application (for example a TELNET connection has the port number 23 and an FTP connection has the port number 21). The message is then passed to the IP layer, which assembles a datagram for transmission to the destination
2. When the destination host receives the connection request, it returns a message containing its own unique socket number and a port number. The socket number and port number thus identify the virtual connection between the two hosts.
3. After the connection has been made the data can flow between the two hosts (called a data stream).

After TCP receives the stream of data, it assembles the data into packets, called TCP segments. After the segment has been constructed, TCP adds a header (called the protocol data unit) to the front of the segment. This header contains information such as a checksum, port number, destination and source socket numbers, socket number of both machines and segment sequence numbers The TCP layer then sends the packaged segment down to the IP layer, which encapsulates it and sends it over the network as a datagram.

17.3.1 Ports and sockets

As previously mentioned, TCP adds a port number and socket number for each host. The port number identifies the required service, whereas the socket number is a unique number for that connection. Thus a node can have several TELNET connections with the same port number but each connection will have a different socket number. A port number can be any value but there is a standard convention which most systems adopt. Table 17.1 defines some of the most common values; they are defined from 0 to 255. Port numbers above 255 can be used for unspecified applications.

Table 17.1 Typical TCP port numbers

Port	*Process name*	*Notes*
20	FTP-DATA	File Transfer Protocol - data
21	FTP	File Transfer Protocol - control
23	TELNET	Telnet
25	SMTP	Simple Mail Transfer Protocol
49	LOGIN	Login Protocol
53	DOMAIN	Domain Name Server
79	FINGER	Finger
161	SNMP	SNMP

17.3.2 TCP header format

The sender's TCP layer communicates with the receiver's TCP layer using the TCP protocol data unit. It defines parameters such as the source port, destination port, and so on, and is illustrated in Figure 17.2. The fields are:

- Source and destination port number – which are 16-bit values to identify the local port number (source number and destination port number or destination port.
- Sequence number – which identifies the current sequence number of the data segment. This allows the receiver to keep track of the data segments received. Any segments that are missing can be easily identified.
- Data offset – which is a 32-bit value and identifies the start of the data.
- Flags – the flag field is defined as `UAPRSF`, where `U` is the urgent flag, `A` the acknowledgment flag, `P` the push function, `R` the reset flag, `S` the sequence synchronize flag and `F` the end-of-transmission flag.

- Windows – which is a 16-bit value and gives the number of data blocks that the receiving host can accept at a time.
- Checksum – which is a 16-bit checksum for the data and header.
- UrgPtr – which is the urgent pointer and is used to identify an important area of data (most systems do not support this facility).

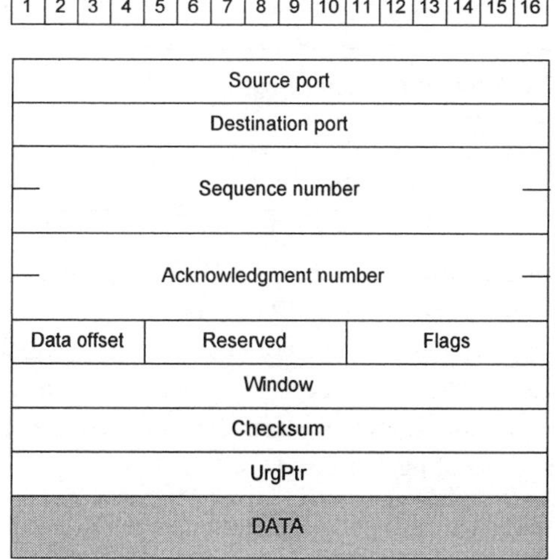

Figure 17.2 TCP header format

17.4 TCP/IP commands

There are several standard programs available over TCP/IP links. The example sessions is this section relate to the network outlined in Figure 16.12. These applications must include:

- FTP (File Transfer Protocol) – which is used to send files from one computer system to another.
- HTTP (Hypertext Transfer Protocol) – which is the protocol used in the World Wide Web (WWW) and can be used for client-server applications involving hypertext (this will be discussed in more detail in Chapter 20).
- MIME (Multipurpose Internet Mail Extension) – gives enhanced electronic mail facilities over TCP/IP (this will be discussed in more detail in Chapter 18).
- SMTP (Simple Mail Management Protocol) – gives simple electronic mail facilities over TCP/IP (this will be discussed in more detail in Chapter 18).
- TELNET – allows remote login using TCP/IP.
- PING – determines if a node is responding to TCP/IP communications.

17.4.1 ping

The `ping` program (Packet Internet Gopher) determines whether a node is responding to TCP/IP communication. It is typically used to trace problems in networks and uses the Internet Control Message Protocol (ICMP) to send a request for a response from the target node. Sample run 17.1 shows that `miranda` is active and `ariel` isn't.

Sample run 17.1: Using PING command

```
C:\WINDOWS>ping miranda
miranda (146.176.144.14) is alive
C:\WINDOWS>ping ariel
no reply from ariel (146.176.147.35)
```

The `ping` program can also be used to determine the delay between one host and another, and also if there are any IP packet losses. In Sample run 17.2 the local host is `pc419.eece.napier.ac.uk` (which is on the `146.176.151` segment); the host `miranda` is tested (which is on the `146.176.144` segment). It can be seen that, on average, the delay is only 1 ms and there is no loss of packets.

Sample run 17.2: Using PING command

```
225 % ping miranda
PING miranda.eece.napier.ac.uk: 64 byte packets
64 bytes from 146.176.144.14: icmp_seq=0. time=1. ms
64 bytes from 146.176.144.14: icmp_seq=1. time=1. ms
64 bytes from 146.176.144.14: icmp_seq=2. time=1. ms
----miranda.eece.napier.ac.uk PING Statistics----
3 packets transmitted, 3 packets received, 0% packet loss
round-trip (ms)  min/avg/max = 1/1/1
```

In Sample run 17.3 the destination node (`www.napier.ac.uk`) is located within the same building but is on a different IP segment (147.176.2). It is also routed through a bridge. It can be seen that the packet delay has increased to between 9 and 10 ms. Again there is no packet loss.

Sample run 17.3: Using PING command

```
226 % ping www.napier.ac.uk
PING central.napier.ac.uk: 64 byte packets
64 bytes from 146.176.2.3: icmp_seq=0. time=9. ms
64 bytes from 146.176.2.3: icmp_seq=1. time=9. ms
64 bytes from 146.176.2.3: icmp_seq=2. time=10. ms
----central.napier.ac.uk PING Statistics----
3 packets transmitted, 3 packets received, 0% packet loss
round-trip (ms)  min/avg/max = 9/9/10
```

Sample run 17.4 shows a connection between Edinburgh and Bath in the UK (www.bath.ac.uk has an IP address of 138.38.32.5). This is a distance of approximately 500 miles and it can be seen that the delay is now between 30 and 49 ms. This time there is 25 % packet loss.

Sample run 17.4: Using PING command

```
222 % ping www.bath.ac.uk
PING jess.bath.ac.uk: 64 byte packets
64 bytes from 138.38.32.5: icmp_seq=0. time=49. ms
64 bytes from 138.38.32.5: icmp_seq=2. time=35. ms
64 bytes from 138.38.32.5: icmp_seq=3. time=30. ms
----jess.bath.ac.uk PING Statistics----
4 packets transmitted, 3 packets received, 25% packet loss
round-trip (ms)  min/avg/max = 30/38/49
```

Finally, in Sample run 17.5 the ping program tests a link between Edinburgh, UK, and a WWW server in the USA (home.microsoft.com , which has the IP address of 207.68.137.51). It can be seen that in this case the delay is between 447 and 468 ms, and the loss is 60 %.

A similar utility program to ping is spray which uses Remote Procedure Call (RPC) to send a continuous stream of ICMP messages. It is useful when testing a network connection for its burst characteristics. This differs from ping, which waits for a predetermined amount of time between messages.

Sample run 17.5: Ping command with packet loss

```
224 % ping home.microsoft.com
PING home.microsoft.com: 64 byte packets
64 bytes from 207.68.137.51: icmp_seq=2. time=447. ms
64 bytes from 207.68.137.51: icmp_seq=3. time=468. ms
----home.microsoft.com PING Statistics----
5 packets transmitted, 2 packets received, 60% packet loss
```

17.4.2 ftp (file transfer protocol)

The ftp program uses the TCP/IP protocol to transfer files to and from remote nodes. If necessary, it reads the *hosts* file to determine the IP address. Once the user has logged into the remote node, the commands that can be used are similar to DOS commands such as cd (change directory), dir (list directory), open (open node), close (close node), pwd (present working directory). The get command copies a file from the remote node and the put command copies it to the remote node.

The type of file to be transferred must also be specified. This file can be ASCII text (the command ascii) or binary (the command binary).

Sample run 17.6 shows a session with the remote VAX computer (Internet name VAX, address 146.176.144.13). The get command is used to get the file TEMP.DOC from VAX and transfer it to the calling PC.

Sample run 17.6: Using FTP to get files from a remote site (e.g. VAX)

```
C:\NET> ftp vax
Connected to vax.
Name (vax:nobody): bill_b
Password (vax:bill_b):
331 Password required for bill_b.
230 User logged in, default directory DUA2:[STAFF.BILL_B]
ftp> dir
200 PORT Command OK.
125 File transfer started correctly
commands.dir;1     MAY 10 11:00 1990       512 (,RWE,RWE,RE)
docs.dir;1         MAY  4 13:31 1993       512 (,RWE,RWE,RE)
fortran.dir;1      MAY 10 11:00 1990       512 (,RWE,RWE,RE)
login.com;29       MAY 10 12:14 1994      1044 (,RWE,RE,RE)
temp.doc;1         MAY  5 07:33 1993        46 (,RWE,RE,)
226 File transfer completed ok
754 bytes received in 2.012100 seconds (0.37 Kbytes/s)
ftp>
ftp> get temp.doc
200 PORT Command OK.
125 File transfer started correctly
226 File transfer completed ok
45 bytes received in 0.005000 seconds (8.79 Kbytes/s)
ftp> quit
221 Goodbye.

C:\NET>dir *.doc
 Volume in drive C is MS-DOS_5
 Volume Serial Number is 3B33-13D3
 Directory of C:\NET

ASKME    DOC      3369 03/07/92    1:25
TEMP     DOC        45 24/05/94   14:47
        2 file(s)        3414 bytes
                      2093056 bytes free
C:\NET>
```

Sample run 17.7 sends a file from the local node (in this case the PC) to a remote node (in this case VAX). The put command is used for this purpose.

Sample run 17.7: Using FTP to send files to a remote site

```
C:\NET>ftp vax
Connected to vax.
Name (vax:nobody): bill_b
Password (vax:bill_b):
331 Password required for bill_b.
230 User logged in, default directory DUA2:[STAFF.BILL_B]
ftp> put askme.doc
200 PORT Command OK.
125 File transfer started correctly
226 File transfer completed ok
3369 bytes sent in 0.011000 seconds (299.09 Kbytes/s)
ftp> dir *.doc
200 PORT Command OK.
125 File transfer started correctly
askme.doc;1     MAY 24 14:53 1994      3396 (,RW,R,R)
temp.doc;1      MAY  5 07:33 1993        46 (,RWE,RE,)
226 File transfer completed ok
215 bytes received in 1.019100 seconds (0.21 Kbytes/s)
```

17.4.3 telnet

The `telnet` program uses TCP/IP to remotely log in to a remote node. Sample run 17.8 shows an example login to the node `miranda`.

Sample run 17.8: Using TELNET for remote login

```
C:\NFS>telnet miranda
        HP-UX miranda A.09.01 A 9000/720 (ttys5)
login: bill_b
Password:
        (c)Copyright 1983-1992 Hewlett-Packard Co.,  All Rights Reserved.
                :::::: 
        (c)Copyright 1988 Carnegie Mellon
[51:miranda :/net/castor_win/local_user/bill_b ] %
```

17.4.4 nslookup

The `nslookup` program interrogates the local `hosts` file or a DNS server to determine the IP address of an Internet node. If it cannot find it in the local file then it communicates with gateways outside its own network to see if they know the address. Sample run 17.9 shows that the IP address of `www.intel.com` is `134.134.214.1`.

Sample run 17.9: Example of nslookup

```
C:\> nslookup
Default Server:  ees99.eece.napier.ac.uk
Address:  146.176.151.99
> www.intel.com
Server:  ees99.eece.napier.ac.uk
Address:  146.176.151.99
Name:    web.jf.intel.com
Address:  134.134.214.1
Aliases:  www.intel.com

230 % nslookup home.microsoft.com
Non-authoritative answer:
Name:    home.microsoft.com
Addresses:  207.68.137.69, 207.68.156.11, 207.68.156.14, 207.68.156.56
207.68.137.48, 207.68.137.51
```

17.4.5 netstat (network statistics)

On a UNIX system the command `netstat` can be used to determine the status of the network. The `-r` option shown in sample run 17.10 shows that this node uses `moon` as a gateway to another network.

17.4.6 traceroute

The `traceroute` program can be used to trace the route of an IP packet through the Internet. It uses the IP protocol time-to-live field and attempts to get an ICMP TIME_EXCEEDED response from each gateway along the path to a defined host. The default probe datagram length is 38 bytes (although the sample runs use 40 byte packets by default). Sample run 17.11 shows an example of `traceroute` from a PC (`pc419.eece.napier.ac.uk`). It can be seen that initially it goes though a bridge (`pcbridge.eece.napier.ac.uk`) and then to the destination (`miranda.eece.napier.ac.uk`).

Sample run 17.10: Using Unix netstat command

```
[54:miranda :/net/castor_win/local_user/bill_b ] % netstat -r
Routing tables
Destination      Gateway              Flags     Refs       Use  Interface
localhost        localhost            UH           0     27306  lo0
default          moon                 UG           0   1453856  lan0
146.176.144      miranda              U            8   6080432  lan0
146.176.1        146.176.144.252      UGD          0        51  lan0
146.176.151      146.176.144.252      UGD         11      5491  lan0
[55:miranda :/net/castor_win/local_user/bill_b ] %
```

Sample run 17.11: Example traceroute

```
www:~/www$ traceroute miranda
traceroute to miranda.eece.napier.ac.uk (146.176.144.14), 30 hops max, 40 byt
ackets
 1  pcbridge.eece.napier.ac.uk (146.176.151.252)  2.684 ms  1.762 ms  1.725 m
 2  miranda.eece.napier.ac.uk (146.176.144.14)  2.451 ms  2.554 ms  2.357 ms
```

Sample run 17.12 shows the route from a PC (`pc419.eece.napier.ac.uk`) to a destination node (`www.bath.ac.uk`). Initially, from the originator, the route goes through a gateway (`146.176.151.254`) and then goes through a routing switch (`146.176.1.27`) and onto EaStMAN ring via `146.176.3.1`. The route then goes round the EaStMAN to a gateway at the University of Edinburgh (`smds-gw.ed.ja.net`). It is then routed onto the SuperJanet network and reaches a gateway at the University of Bath (`smds-gw.bath.ja.net`). It then goes to another gateway (`jips-gw.bath.ac.uk`) and finally to its destination (`jess.bath.ac.uk`). Figure 17.3 shows the route the packet takes. The EaStMAN network will be covered in more detail in the chapters on FDDI and ATM.

Note that gateways 4 and 8 hops away either don't send ICMP "time exceeded" messages or send them with time-to-live values that are too small to be returned to the originator.

Sample run 17.12: Example traceroute

```
www:~/www$ traceroute www.bath.ac.uk
traceroute to jess.bath.ac.uk (138.38.32.5), 30 hops max, 40 byte packets
 1  146.176.151.254 (146.176.151.254)  2.806 ms  2.76 ms  2.491 ms
 2  sil-switch.napier.ac.uk (146.176.1.27)  19.315 ms  11.29 ms  6.285 ms
 3  sil-cisco.napier.ac.uk (146.176.3.1)  6.427 ms  8.407 ms  8.872 ms
 4  * * *
 5  smds-gw.ed.ja.net (193.63.106.129)  8.98 ms  30.308 ms  398.623 ms
 6  smds-gw.bath.ja.net (193.63.203.68)  39.104 ms  46.833 ms  38.036 ms
 7  jips-gw.bath.ac.uk (146.97.104.2)  32.908 ms  41.336 ms  42.429 ms
 8  * * *
 9  jess.bath.ac.uk (138.38.32.5)  41.045 ms *  41.93 ms
```

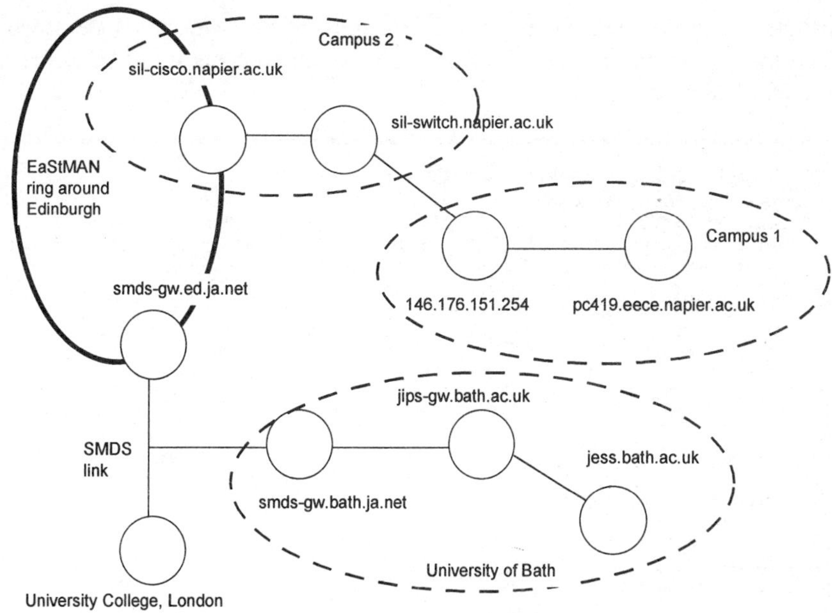

Figure 17.3 Route between local host and the University of Bath

Sample run 17.13 shows an example route from a local host at Napier University, UK, to the USA. As before, it goes through the local gateway (146.176.151.254) and then goes through three other gateways to get onto the SMDS SuperJANET connection. The data packet then travels down this connection to University College, London (`gw5.ulcc.ja.net`). It then goes onto high speed connects to the USA and arrives at a US gateway (`mcinet-2.sprintnap.net`). Next it travels to `core2-hssi2-0.WestOrange.mci.net` before reaching the Microsoft Corporation gateway in Seattle (`microsoft.Seattle.mci.net`). It finally finds it way to the destination (`207.68.145.53`). The total journey time is just less than half a second.

17.4.7 arp

The `arp` utility program displays the IP to Ethernet MAC address mapping. It can also be used to delete or manually changed any included address table entries. Within a network, a router forwards data packets depending on the destination IP address of the packet. Each connection must also specify a MAC address to transport the packet over the network, thus the router must maintain a list of MAC addresses. The `arp` protocol thus maintains this mapping. Addresses within this table are added on an as-needed basis. When a MAC address is required an `arp` message is sent to the node with an `arp`

REQUEST packet which contains the IP address of the requested node. It will then reply with an arp RESPONSE packet which contains its MAC address and its IP address.

Sample run 17.13: Example traceroute

```
> traceroute home.microsoft.com
 1  146.176.151.254 (146.176.151.254)  2.931 ms  2.68 ms  2.658 ms
 2  sil-switch.napier.ac.uk (146.176.1.27)  6.216 ms  8.818 ms  5.885 ms
 3  sil-cisco.napier.ac.uk (146.176.3.1)  6.502 ms  6.638 ms  10.218 ms
 4  * * *
 5  smds-gw.ed.ja.net (193.63.106.129)  18.367 ms  9.242 ms  15.145 ms
 6  smds-gw.ulcc.ja.net (193.63.203.33)  42.644 ms  36.794 ms  34.555 ms
 7  gw5.ulcc.ja.net (128.86.1.80)  31.906 ms  30.053 ms  39.151 ms
 8  icm-london-1.icp.net (193.63.175.53)  29.368 ms  25.42 ms  31.347 ms
 9  198.67.131.193 (198.67.131.193)  119.195 ms  120.482 ms  67.479 ms
10  icm-pen-1-H2/0-T3.icp.net (198.67.131.25)  115.314 ms  126.152 ms
149.982 ms
11  icm-pen-10-P4/0-OC3C.icp.net (198.67.142.69)  139.27 ms  197.953 ms
195.722 ms
12  mcinet-2.sprintnap.net (192.157.69.48)  199.267 ms  267.446 ms  287.834
ms
13  core2-hssi2-0.WestOrange.mci.net (204.70.1.49)  216.006 ms  688.139 ms
228.968 ms
14  microsoft.Seattle.mci.net (166.48.209.250)  310.447 ms  282.882 ms
313.619 ms
15  * microsoft.Seattle.mci.net (166.48.209.250)  324.797 ms  309.518 ms
16  * 207.68.145.53 (207.68.145.53)  435.195 ms *
```

17.5 Exercises

17.1 Using the `ping` program determine if the following nodes are responding:

(i) `www.eece.napier.ac.uk`
(ii) `home.microsoft.com`
(iii) `www.intel.com`

17.2 Using the `traceroute` program determine the route from your local host to the following destinations:

(i) `www.napier.ac.uk`
(ii) `home.microsoft.com`
(iii) `www.intel.com`

Identify each part of the route and note the timing information.

17.3 Explain the main differences between IPv4 and IPv6 packets.

17.4 Determine the number of possible IP addresses for IPv4. Explain why all the possible addresses will never be used (Hint: The number of nodes on a segment may never reach the required limit).

17.5 Determine the number of possible IP addresses for IPv6.

17.6 Explain the functions of TCP.

17.5 Explain how an application program uses ports and sockets. Show also how an application program can set up several TCP/IP connections.

18 Electronic Mail

18.1 Introduction

Electronic mail (email) is one use of the Internet which, according to most businesses, improves productivity. Traditional methods of sending mail within an office environment are inefficient as it normally requires an individual requesting a secretary to type the letter. This must then be proofread and sent through the internal mail system, which is relatively slow and can be open to security breaches.

A faster method, and more secure method of sending information is to use electronic mail, where messages are sent almost in an instant. For example a memo with 100 words will be sent in a fraction of a second. It is also simple to send to specific groups, various individuals, company-wide, and so on. Other types of data can also be sent with the mail message such as images, sound, and so on. It may also be possible to determine if a user has read the mail. The main advantages are:

- It is normally much cheaper than using the telephone (although, as time equates to money for most companies, this relates any savings or costs to a user's typing speed).
- Many different types of data can be transmitted, such as images, documents, speech, and so on.
- It is much faster than the postal service.
- Users can filter incoming email easier than incoming telephone calls.
- It normally cuts out the need for work to be typed, edited and printed by a secretary.
- It reduces the burden on the mailroom.
- It is normally more secure than traditional methods.
- It is relatively easy to send to groups of people (traditionally, either a circulation list was required or a copy to everyone in the group was required).
- It is usually possible to determine whether the recipient has actually read the message (the electronic mail system sends back an acknowledgment).

The main disadvantages are:

- It stops people using the telephone.
- It cannot be used as a legal document.
- Electronic mail messages can be sent on the spur of the moment and may be regretted later on (sending by traditional methods normally allows for a rethink). In extreme cases messages can be sent to the wrong person (typically when replying to an email message, where a message is sent to the mailing list rather than the originator).
- It may be difficult to send to some remote sites. Many organization have either no electronic mail or merely an intranet. Large companies are particularly wary of Internet connections and limit the amount of external traffic.
- Not everyone reads their electronic mail on a regular basis (although this is changing as more organizations adopt email as the standard communications medium).

The main standards that relates to the protocols of email transmission and reception are:

- Simple Mail Transfer Protocol (SMTP) – which is used with the TCP/IP protocol suite. It has traditionally been limited to the text-based electronic messages.
- Multipurpose Internet Mail Extension (MIME) – which allows the transmission and reception of mail that contains various types of data, such as speech, images and motion video. It is a newer standard than SMTP and uses much of its basic protocol.

18.2 Shared-file approach versus client/server approach

An email system can use either a shared-file approach or a client/server approach. In a shared-file system the source mail client sends the mail message to the local post office. This post office then transfers control to a message transfer agent which then stores the message for a short time before sending it to the destination post office. The destination mail client periodically checks its own post office to determine if it has mail for it. This arrangement is often known as store and forward, and the process is illustrated in Figure 18.1. Most PC-based email systems use this type of mechanism.

A client/server approach involves the source client setting up a real-time remote connection with the local post office, which then sets up a real-time connection with the destination, which in turn sets up a remote connection with the destination client. The message will thus arrive at the destination when all the connections are complete.

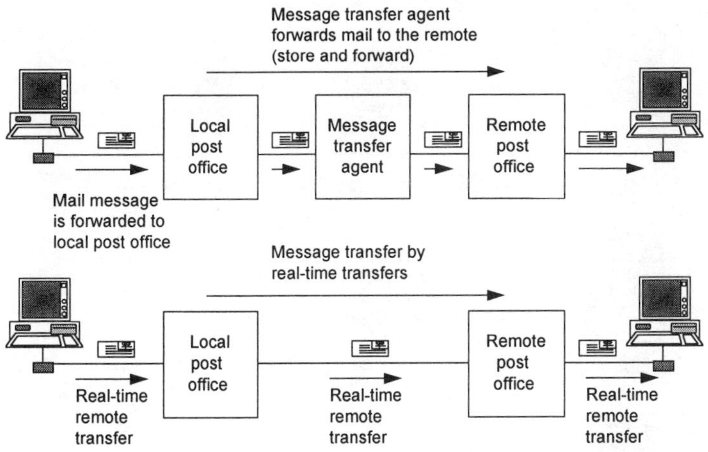

Figure 18.1 Shared-file versus client/server

18.3 Electronic mail overview

Figure 18.2 shows a typical email architecture. It contains four main elements:

1. Post offices – where outgoing messages are temporally buffered (stored) before transmission and where incoming messages are stored. The post office runs the server software capable of routing messages (a message transfer agent) and maintaining the post office database.
2. Message transfer agents – for forwarding messages between post offices and to the destination clients. This software can either reside on the local post office or on a physically separate server.
3. Gateways – which provide part of the message transfer agent functionality. They translate between different email systems, different email addressing schemes and messaging protocols.
4. Email clients – normally the computer which connects to the post office. It contains three parts:

   - Email Application Program Interface (API), such as MAPI, VIM, MHS and CMC.
   - Messaging protocol. The main messaging protocols are SMTP or X.400. SMTP is defined in RFC 822 and RFC 821. X.400 is an OSI-defined email message delivery standard (Sections 18.5 and 18.6).
   - Network transport protocol, such as Ethernet, FDDI, and so on.

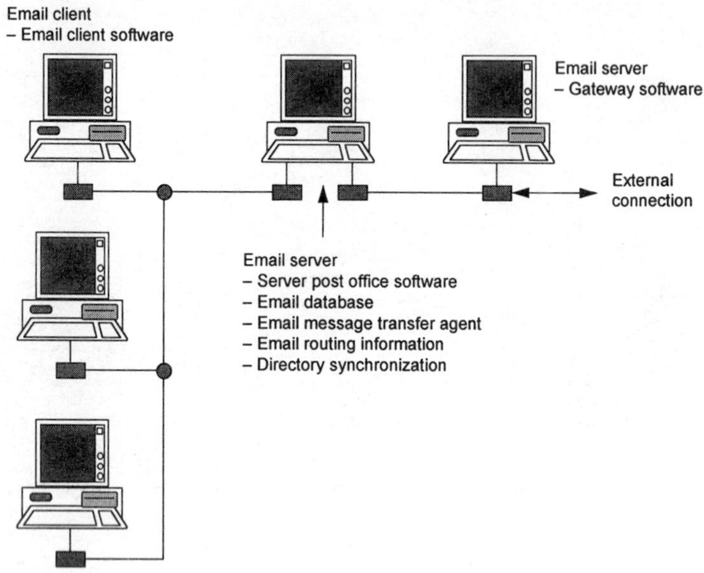

Figure 18.2 Email architecture

The main APIs are:

- MAP (messaging API) – Microsoft part of Windows Operation Services Architecture.
- VIM (vendor-independent messaging) – Lotus, Apple, Novell and Borland derived email API.
- MHS (message handling service) – Novell network interface which is often used as an email gateway protocol.
- CMC (common mail call) – Email API associated with the X.400 native messaging protocol.

Gateways translate the email message from one system to another, such as from Lotus cc:Mail to Microsoft Mail. Typical gateway protocols are:

- MHS (used with Novell Netware).
- SMTP.MIME (used with Internet environment).
- X.400 (used with X.400).
- MS Mail (used with Microsoft Mail).
- cc:Mail (used with Lotus cc:Mail).

A PC-based email package is Lotus cc:Mail (Figure 18.3 shows a sample screen).

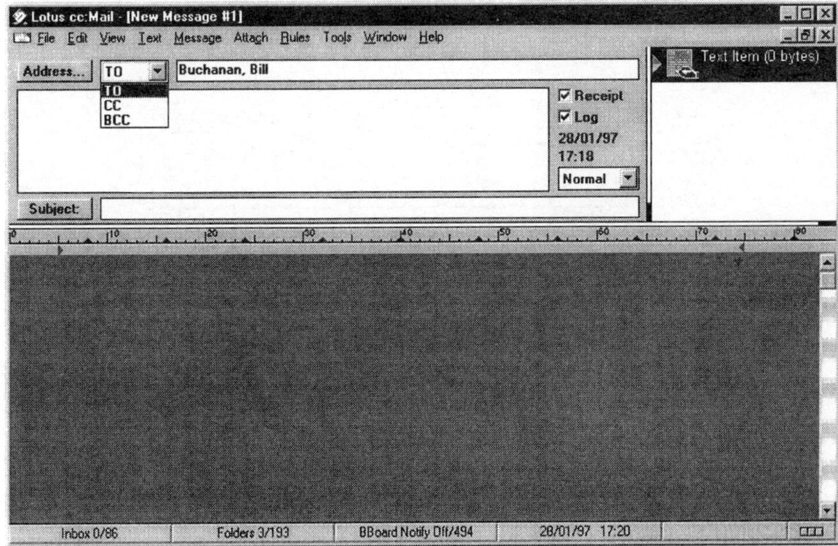

Figure 18.3 A Lotus cc:Mail screen

18.4 Internet email address

The Internet email address is in the form of a name (such as `f.bloggs`), followed by an '`@`' and then the domain name (such as `anytown.ac.uk`). For example:

```
f.bloggs@anytown.ac.uk
```

No spaces are allowed in the address; periods are used instead. Figure 18.4 shows an example Internet address builder from Lotus cc:Mail.

18.5 SMTP

The IAB have defined the protocol SMTP in RFC 821 (Refer to Appendix E for a list of RFC standards). This section discusses the protocol for transferring mail between hosts using the TCP/IP protocol.

As SMTP is a transmission and reception protocol it does not actually define the format or contents of the transmitted message except that the data has 7-bit ASCII characters and that extra log information is added to the start of the delivered message to indicate the path the message took. The protocol itself is only concerned in reading the address header of the message.

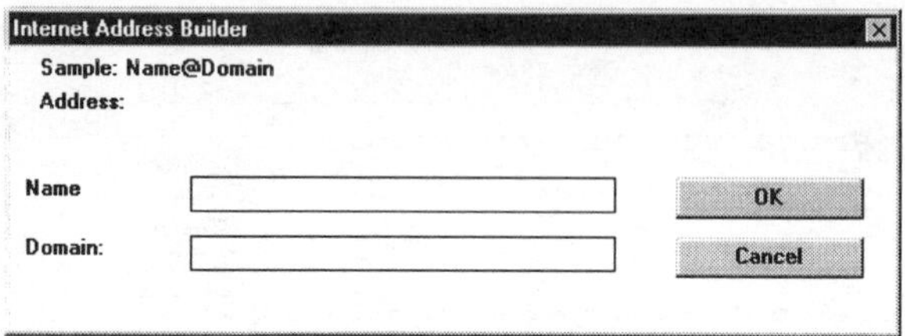

Figure 18.4 Internet address format

18.5.1 SMTP operation

SMTP defines the conversation that takes place between an SMTP sender and an SMTP receiver. Its main functions are the transfer of messages and the provision of ancillary functions for mail destination verification and handling.

Initially the message is created by the user and a header is added which includes the recipient's email address and other information. This message is then queued by the mail server, and when it has time, the main server attempt to transmit it.

Each mail may be have the following requirements:

- Each email can have a list of destinations; the email program make copies of the messages and pass them onto the mail server.
- The user may maintain a mailing list, and the email program must remove duplicates and replace mnemonic names with actual email addresses.
- It allows for normal message provision, e.g. blind carbon copies (BCCs).

An SMTP mail server processes email messages from an outgoing mail queue and then transmits them using one or more TCP connections with the destination. If the mail message is transmitted to the required host then the SMTP sender deletes the destination from the message's destination list. After all the destinations have been sent to, the sender then deletes the message from the queue.

If there are several recipients for a message on the same host, the SMTP protocol allows a single message to be sent to the specified recipients. Also, if there are several messages to be sent to a single host, the server can simply open a single TCP connection and all the messages can be transmitted in a single transfer (there is thus no need to set up a connection for each message).

SMTP also allows for efficient transfer with error messages. Typical errors include:

- Destination host is unreachable. A likely cause is that the destination host address is incorrect. For example, `f.bloggs@toy.ac.uk` might actually be `f.bloggs@toytown.ac.uk`.

- Destination host is out of operation. A likely cause is that the destination host has developed a fault or has been shut down.
- Mail recipient is not available on the host. Perhaps the recipient does not exist on that host, the recipient name is incorrect or the recipient has moved. For example, `fred.bloggs@toytown.ac.uk` might actually be `f.bloggs@toytown.ac.uk`. To overcome the problem of user names which are similar to a user's name then some systems allow for certain aliases for recipients, such as `f.bloggs`, `fred.bloggs` and `freddy.bloggs`, but there is a limit to the number of aliases that a user can have. If a user has moved then some systems allow for a redirection of the email address. UNIX systems use the `.forward` file in the user's home directory for redirection. For example on a UNIX system, if the user has moved to `fred.bloggs@toytown.com` then this address is simply added to the `.forward` file.
- TCP connection failed on the transfer of the mail. A likely cause is that there was a time-out error on the connection (maybe due to the receiver or sender being busy or there was a fault in the connection).

SMTP senders have the responsibility for a message up to the point where the SMTP receiver indicates that the transfer is complete. This only indicates that the message has arrived at the SMTP receiver; does not indicate that:

- The message has been delivered to the recipient's mailbox.
- The recipient has read the message.

Thus, SMTP does not guarantee to recover from lost messages and gives no end-to-end acknowledgment on successful receipt (normally this is achieved by an acknowledgment message being returned). Nor are error indications guaranteed. However, TCP connections are normally fairly reliable.

If an error occurs in reception, a message will normally be sent back to the sender to explain the problem. The user can then attempt to determine the problem with the message.

SMTP receivers accept an arriving message and either place it in a user's mailbox or, if that user is located at another host, copies it to the local outgoing mail queue for forwarding.

Most transmitted messages go from the sender's machine to the host over a single TCP connection. But sometimes the connection will be made over multiple TCP connections over multiple hosts. This can be achieved by the sender specifying a route to the destination in the form of a sequence of servers.

18.5.2 SMTP overview

An SMTP sender initiates a TCP connection. When this is successful it sends a series of commands to the receiver, and the receiver returns a single reply for each command. All commands and responses are sent with ASCII characters and are terminated with the carriage return (CR) and line feed (LF) characters (often known as CRLF).

Each command consists of a single line of text, beginning with a four-letter command code followed by in some cases an argument field. Most replies are a single line, although multiple-line replies are possible. Table 18.1 gives some sample commands.

SMTP replies with a three-digit code and possibly other information. Some of the responses are listed in Table 18.2. The first digit gives the category of the reply, such as 2*xx* (a positive completion reply), 3*xx* (a positive intermediate reply), 4*xx* (a transient negative completion reply) and 5*xx* (a permanent negative completion reply). A positive reply indicates that the requested action has been accepted, and a negative reply indicates that the action was not accepted.

Positive completion reply indicates that the action has been successful, and a positive intermediate reply indicates that the action has been accepted but the receiver is waiting for some other action before it can give a positive completion reply. A transient negative completion reply indicates that there is a temporary error condition which can be cleared by other actions and a permanent negative completion reply indicates that the action was not accepted and no action was taken.

Table 18.1 SMTP commands

Command	*Description*
HELO *domain*	Sends an identification of the domain
MAIL FROM: *sender-address*	Sends identification of the originator (sender-address)
RCPT FROM: *receiver-address*	Sends identification of the recipient (receiver-address)
DATA	Transfer text message
RSEY	Abort current mail transfer
QUIT	Shut down TCP connection
EXPN *mailing-list*	Send back membership of mailing list
SEND FROM: *sender-address*	Send mail message to the terminal
SOML FROM: *sender-address*	If possible, send mail message to the terminal, otherwise send to mailbox
VRFY username	Verify user name (username)

18.5.3 SMTP transfer

Figure 18.4 shows a successful email transmission. For example if:

```
f.bloggs@toytown.ac.uk
```

is sending a message to:

```
a.person@place.ac.de
```

Then a possible sequence of events is:

- Set up TCP connection with receiver host.
- If the connection is successful, the receiver replies back with a 220 code (server ready). If it is unsuccessful, it returns back with a 421 code.
- Sender sends a `HELO` command to the hostname (such as `HELO toy-town.ac.uk`).

Table 18.2 SMTP responses

CMD	*Description*	*CMD*	*Description*
211	System status	500	Command unrecognized due to a syntax error
214	Help message	501	Invalid parameters or arguments
220	Service ready	502	Command not currently implemented
221	Service closing transmission channel	503	Bad sequence of commands
250	Request mail action completed successfully	504	Command parameter not currently implemented
251	Addressed user does not exist on system but will forward to receiver-address	550	Mail box unavailable, request action not taken
354	Indicate to the sender that the mail message can now be sent. The end of the message is identified by two CR, LF characters	551	The addressed user is not local, please try receiver-address
421	Service is not available	552	Exceeded storage allocation, requested mail action aborted
450	Mailbox unavailable and the requested mail action was not taken	553	Mailbox name not allowed, requested action not taken
451	Local processing error, requested action aborted	554	Transaction failed
452	Insufficient storage, requested action not taken		

- If the sender accepts the incoming mail message then the receiver returns a `250 OK` code. If it is unsuccessful then it returns a 421, 451, 452, 500, 501 or 552 code.
- Sender sends a `MAIL FROM:` *`sender`* command (such as `MAIL FROM: f.bloggs@toytown.ac.uk`).
- If the receiver accepts the incoming mail message from the sender then it returns a `250 OK` code. If it is unsuccessful then it returns codes such as 251, 450, 451, 452, 500, 501, 503, 550, 551, 552 or 553 code.
- Sender sends an `RCPT TO:` *`receiver`* command (such as `RCPT TO: a.person@place.ac.de`).
- If the receiver accepts the incoming mail message from the sender then it returns a `250 OK` code.
- Senders sends a `DATA` command.
- If the receiver accepts the incoming mail message from the sender then it returns a `354` code (start transmission of mail message).
- The sender then transmits the email message.
- The end of the email message is sent as two LF, CR characters.
- If the reception has been successful then the receiver sends back a `250 OK` code. If it is unsuccessful then it returns a 451, 452, 552 or 554 code.
- Sender starts the connection shutdown by sending a `QUIT` command.
- Finally the sender closes the TCP connection.

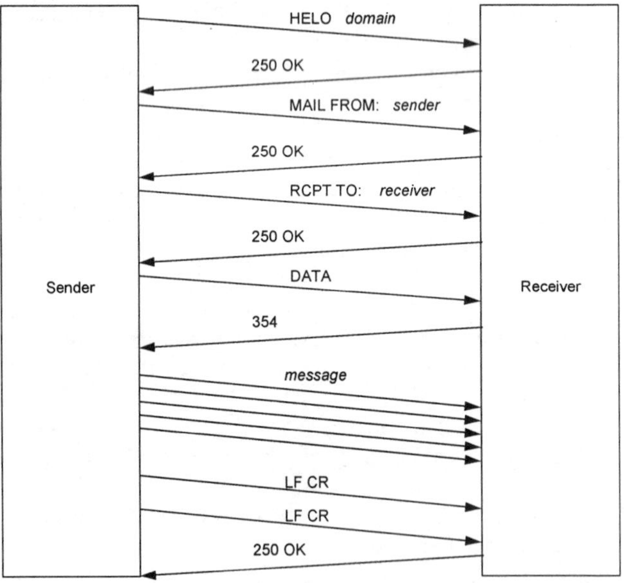

Figure 18.5 Sample SMTP email transmission

18.5.4 RFC 822

SMTP uses RFC 822 which defines the format of the transmitted message. RFC 822 contains two main parts:

- A header – which is basically the mail header and contains information for the successful transmission and delivery of a message. This typically contains the email addresses for sender and receiver, the time the message was sent and received. Any computer involved in the transmission can added to the header.
- The contents.

Normally the email-reading program will read the header and format the information to the screen to show the sender's email address; it splits off the content of the message and displays it separately from the header.

An RFC 822 message contains a number of lines of text in the form of a memo (such as To:, From:, Bcc:, and so on). A header line usually has a keyword followed by a colon and then followed by keyword arguments. The specification also allows for a long line to be broken up into several lines.

Here is an RFC 822 message with the header shown in italics and the message body in bold. Table 18.3 explains some of the RFC 822 items in the header.

```
From FREDB@ACOMP.CO.UK Wed Jul  5 12:36:49 1995

Received: from ACOMP.CO.UK ([154.220.12.27]) by central.napier.ac.uk
(8.6.10/8.6.10) with SMTP id MAA16064 for <w.buchanan@central.napier.ac.uk>;

Wed, 5 Jul 1995 12:36:43 +0100

Received: from WPOAWUK-Message_Server by ACOMP.CO.UK
      with Novell_GroupWise; Wed, 05 Jul 1995 12:35:51 +0000

Message-Id: <sffa8725.082@ACOMP.CO.UK >

X-Mailer: Novell GroupWise 4.1

Date: Wed, 05 Jul 1995 12:35:07 +0000

From: Fred Bloggs <FREDB@ACOMP.CO.UK>

To: w.buchanan@central.napier.ac.uk
Subject:  Technical Question
Status: REO

Dear Bill

I have a big problem. Please help.

Fred
```

Table 18.3 Header line descriptions

Header line	*Description*
From FREDB@ACOMP.CO.UK Wed Jul 5 12:36:49 1995	Sender of the email is FREDB@ ACOM.CO.UK
Received: from ACOMP.CO.UK ([154.220.12.27]) by central.napier.ac.uk (8.6.10/8.6.10) with SMTP id MAA16064 for <w.buchanan@central.napier.ac.uk>; Wed, 5 Jul 1995 12:36:43 +0100	It was received by CENTRAL.NAPIER.AC.UK at 12:36 on 5 July 1995
Message-Id: <sffa8725.082@ACOMP.CO.UK >	Unique message ID
X-Mailer: Novell GroupWise 4.1	Gateway system
Date: Wed, 05 Jul 1995 12:35:07 +0000	Date of original message
From: Fred Bloggs <FREDB@ACOMP.CO.UK>	Sender's email address and full name
To: w.buchanan@central.napier.ac.uk	Recipient's email address
Subject: Technical Question	Mail subject

18.6 X.400

RFC 821 (the transmission protocol) and RFC 822 (the message format) have now become the de-facto standards for email systems. The CCITT, in 1984, defined new email recommendations called X.400. The RFC821/822 system is simple and works relatively well, whereas X.400 is complex and is poorly designed. These points have helped RFC 821/822 to become a de-facto standard, whereas X.400 is now almost extinct (see Figure 18.6 for the basic X.400 address builder and Figure 18.7 for the extended address builder).

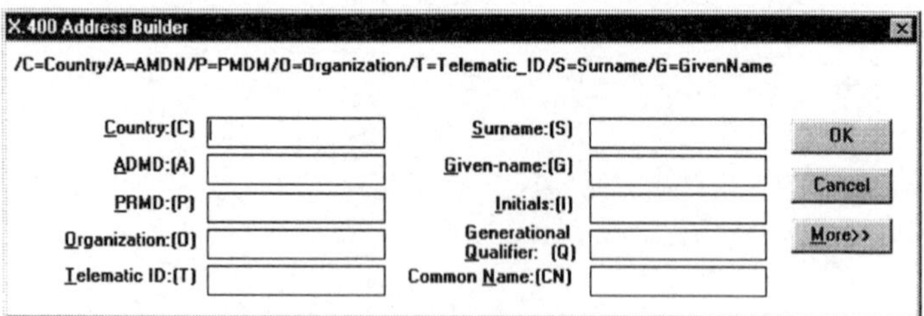

Figure 18.6 X.400 basic addressing

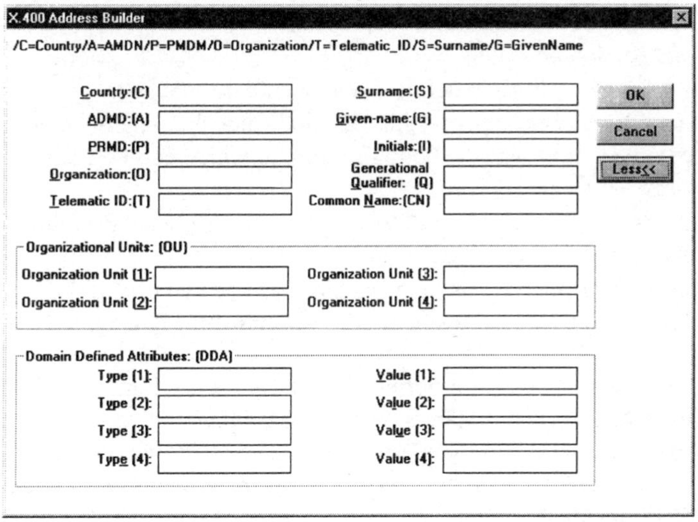

X.400 Address Builder
/C=Country/A=AMDN/P=PMDM/O=Organization/T=Telematic_ID/S=Surname/G=GivenName
Country:(C)
ADMD:(A)
PRMD:(P)
Organization:(O)
Telematic ID:(T)
Surname:(S)
Given-name:(G)
Initials:(I)
Generational Qualifier: (Q)
Common Name:(CN)
OK
Cancel
Less<<
Organizational Units: (OU)
Organization Unit (1):
Organization Unit (2):
Organization Unit (3):
Organization Unit (4):
Domain Defined Attributes: (DDA)
Type (1):
Type (2):
Type (3):
Type (4):
Value (1):
Value (2):
Value (3):
Value (4):

Figure 18.7 X.400 extended addressing

18.7 MIME

SMTP suffers from several drawback, such as:

- SMTP can only transmit ASCII characters and thus cannot transmit executable files or other binary objects.
- SMTP does not allow the attachment of files, such as images and audio.
- SMTP can only transmit 7-bit ASCII character thus it does support an extended ASCII character set.

A new standard, Multipurpose Internet Mail Extension (MIME), has been defined for this purpose ,which is compatible with existing RFC 822 implementations. It is defined in the specifications RFC 1521 and 1522. Its enhancements include the following:

- Five new message header fields in the RFC 822 header, which provide extra information about the body of the message.
- Use of various content formats to support multimedia electronic mail.
- Defined transfer encodings for transforming attached files.

The five new header fields defined in MIME are:

- MIME-version – a message that conforms to RFC 1521 or 1522 has a MIME-version of 1.0.
- Content-type – this field defines the type of data attached.
- Content-transfer-encoding – this field indicates the type of transformation necessary to represent the body in a format which can be transmitted as a message.
- Content-id – this field is used to uniquely identify MIME multiple attachments in the email message.
- Content-description – this field is a plain-text description of the object with the body. It can be used by the user to determine the data type.

These fields can appear in a normal RFC 822 header. Figure 18.8 shows an example email message. It can be seen, in the right-hand corner, that the API has split the message into two parts: the message part and the RFC 822 part. The RFC 822 part is shown in Figure 18.9. It can be seen that, in this case, the extra MIME messages are:

```
MIME-Version: 1.0
Content-Type: text/plain; charset=us-ascii
Content-Transfer-Encoding: 7bit
```

This defines it as MIME Version 1.0; the content-type is text/plain (standard ASCII) and it uses the US ASCII character set; the content-transfer-encoding is 7-bit ASCII.

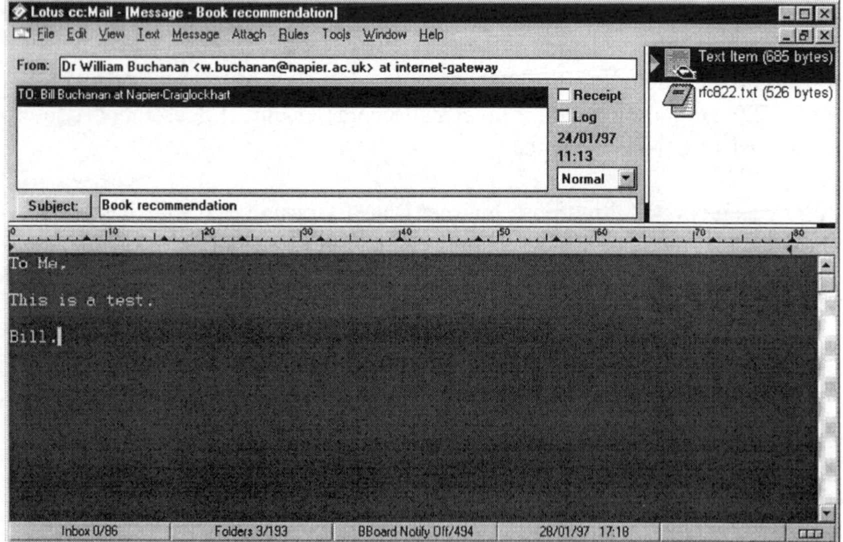

Figure 18.8 Sample email message showing message and RFC822 part

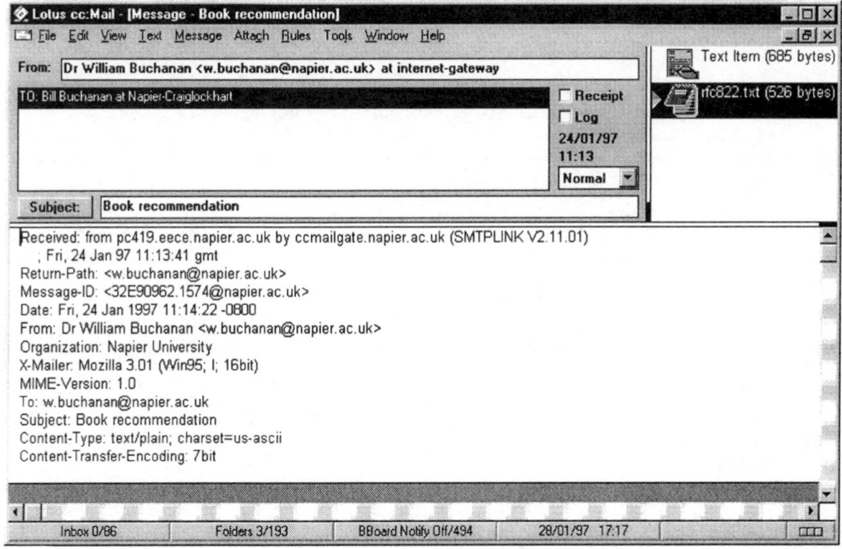

Figure 18.9 RFC 822 part

RFC 822 example file listing (refer to Figure 18.9)

```
Received: from pc419.eece.napier.ac.uk by ccmailgate.napier.ac.uk (SMTPLINK
V2.11.01)
    ; Fri, 24 Jan 97 11:13:41 gmt
Return-Path: <w.buchanan@napier.ac.uk>
Message-ID: <32E90962.1574@napier.ac.uk>
Date: Fri, 24 Jan 1997 11:14:22 -0800
From: Dr William Buchanan <w.buchanan@napier.ac.uk>
Organization: Napier University
X-Mailer: Mozilla 3.01 (Win95; I; 16bit)
MIME-Version: 1.0
To: w.buchanan@napier.ac.uk
Subject: Book recommendation
Content-Type: text/plain; charset=us-ascii
Content-Transfer-Encoding: 7bit
```

18.7.1 MIME content types

Content types define the format of the attached files. There are a total of 16 different content types in seven major content groups. If the text body is pure text then no special transformation is required. RFC 1521 defines only one subtype, text/plain; this gives a standard ASCII character set.

A MIME-encoded email can contain multiple attachments. The content-type header field includes a boundary which defines the delimiter between multiple attachments. A boundary always starts on a new line and has the format:

```
-- boundary name
```

The final boundary is:

```
-- boundary name --
```

For example, the following message contains two parts:

Example MIME file with 2 parts

```
From: Dr William Buchanan <w.buchanan@napier.ac.uk>
MIME-Version: 1.0
To: w.buchanan@napier.ac.uk
Subject: Any subject
Content-Type: multipart/mixed; boundary="boundary name"

This part of the message will be ignored.

-- boundary name
Content-Type: multipart/mixed; boundary="boundary name"

This is the first mail message part.

-- boundary name

And this is the second mail message part.

-- boundary name --
```

Table 18.4 MIME content-types

Content type	*Description*
text/plain	Unformated text, such as ASCII
text/richtext	Rich text format which is similar to HTML
multipart/mixed	Each attachment is independent from the rest and all should be presented to the user in their initial ordering
multipart/parallel	Each attachment is independent from the others but the order is unimportant
multipart/alternative	Each attachment is a different version of the original data
multipart/digest	This is similar to multipart/mixed but each part is message/rfc822
message/rfc822	Contains the RFC 822 text
message/partial	Used in fragmented mail messages
message/external-body	Used to define a pointer to an external object (such as an ftp link)
image/jpeg	Defines a JPEG image using JFIF file format
image/gif	Defines GIF image
video/mpeg	Defines MPEG format
audio/basic	Defines 8-bit μ-Law encoding at 8 kHz sampling rate
application/postscript	Defines postscript format
application/octet-stream	Defines binary format which consists of 8-bit bytes

The part of the message after the initial header and before the first boundary can be used to add a comment. This is typically used to inform users that do not have a MIME-compatible program about the method used to encode the received file. A typical method for converting binary data into ASCII characters is to use the programs UUENCODE (to encode a binary file into text) or UUDECODE (to decode a uuencoded file).

The four subtypes of multipart type can be used to sequence the attachments; the main subtypes are:

- multipart/mixed subtype – which is used when attachments are independent but need to be arranged in a particular order.
- multipart/parallel subtype – which is used when the attachments should be present at the same; a typical example is to present an animated file along with an audio attachment.
- multipart/alternative subtype – which is used to represent an attachment in a number of different formats.

18.7.2 Example MIME

The following file listing shows the message part of a MIME-encoded email message (i.e. it excludes the RFC 822 header part). It can be seen that the sending email system has added the comment about the MIME encoding. In this case the MIME boundaries have been defined by:

```
-- IMA.Boundary.760275638
```

Example MIME file

```
This is a Mime message, which your current mail reader
may not understand. Parts of the message will appear as
text. To process the remainder, you will need to use a Mime
compatible mail reader. Contact your vendor for details.

--IMA.Boundary.760275638

Content-Type: text/plain; charset=US-ASCII
Content-Transfer-Encoding: 7bit
Content-Description: cc:Mail note part

This is the original message .....

--IMA.Boundary.760275638--
```

18.7.3 Mail fragments

A mail message can be fragmented using the content-type field of message/partial and then reassembled back at the source. The standard format is:

```
Content-type: message/partial;
  id="idname"; number=x; total=y
```

where *idname* is the message identification (such as `xyz@hostname`, *x* is the number of the fragment out of a total of *y* fragments. For example, if a message had three fragments, they could be sent as:

Example MIME file with 3 fragments (first part)

```
From: Fred Bloggs <f.bloggs@toytown.ac.uk>
MIME-Version: 1.0
To: a.body@anytown.ac.uk
Subject: Any subject
Content-Type: message/partial;
      id="xyz@toytown.ac.uk"; number=1; total=3
Content=type: video/mpeg

First part of MPEG file
```

Example MIME file with 3 fragments (second part)

```
From: Fred Bloggs <f.bloggs@toytown.ac.uk>
MIME-Version: 1.0
To: a.body@anytown.ac.uk
Subject: Any subject
Content-Type: message/partial;
      id="xyz@toytown.ac.uk"; number=2; total=3
Content=type: video/mpeg

Second part of MPEG file
```

Example MIME file with 3 fragments (third part)

```
From: Fred Bloggs <f.bloggs@toytown.ac.uk>
MIME-Version: 1.0
To: a.body@anytown.ac.uk
Subject: Any subject
Content-Type: message/partial;
      id="xyz@toytown.ac.uk"; number=3; total=3
Content=type: video/mpeg

Third part of MPEG file
```

18.7.4 Transfer encodings

MIME allows for different transfer encodings within the message body:

- 7bit – no encoding, and all of the characters are 7-bit ASCII characters.
- 8bit – no encoding, and extended 8-bit ASCII characters are used.
- quoted-printable – encodes the data so that non-printing ASCII characters (such as line feeds and carriage returns) are displayed in a readable form.
- base64 – encodes by mapping 6-bit blocks of input to 8-bit blocks of output, all of which are printable ASCII characters.
- x-token – another non-standard encoding method.

When the transfer encoding is:

```
Content-transfer-encoding: quoted-printable
```

then the message has been encoded so that all non-printing characters have been converted to printable characters. A typical transform is to insert =*xx* where *xx* is the hexadecimal equivalent for the ASCII character. A form feed (FF) would be encoded with '=0C',

A transfer encoding of base64 is used to map 6-bit characters to a printable character. It is a useful method in disguising text in an encrypted form and also for converting binary data into a text format. It takes the input bitstream and reads it 6 bits at a time, then maps this to an 8-bit printable character. Table 18.5 shows the mapping.

Table 18.5 MIME base64 encoding

Bit value	*Encoded character*	*Bit value*	*Encoded character*	*Bit value*	*Encoded character*	*Bit value*	*Encoded character*
0	A	16	Q	32	g	48	w
1	B	17	R	33	h	49	x
2	C	18	S	34	i	50	y
3	D	19	T	35	j	51	z
4	E	20	U	36	k	52	0
5	F	21	V	37	l	53	1
6	G	22	W	38	m	54	2
7	H	23	X	39	n	55	3
8	I	24	Y	40	o	56	4
9	J	25	Z	41	p	57	5
10	K	26	a	42	q	58	6
11	L	27	b	43	r	59	7
12	M	28	c	44	s	60	8
13	N	29	d	45	t	61	9
14	O	30	e	46	u	62	+
15	P	31	f	47	v	63	/

Thus if a binary file had the bit sequence:

```
10100010101010001010101010
```

It would first be split into groups of 6 bits, as follows:

```
101000   101010   100010   101010   000000
```

This would be converted into the ASCII sequence:

```
YsSqA
```

which is in a transmittable form.

Thus the 7-bit ASCII sequence 'FRED' would use the bit pattern:

```
1000110 1010010 1000101 1000100
```

which would be split into groups of 6 bits as:

```
100011 010100 101000 101100 010000
```

which would be encoded as:

```
jUosQ
```

18.8 Exercises

18.1 Identify the main functional differences between SMTP and MIME.

18.2 Contrast shared-file and client/server approaches for electronic mail.

18.3 Give an example set of messages between sender and a recipient for a successful SMTP transfer.

18.4 Give an example set of messages between sender and a recipient for an unsuccessful SMTP transfer.

18.5 If you have access to email, read and identify each part of the header.

18.6 Explain how base64 encoding can attach binary information.

18.7 Encode the following bitstream with base64 encoding:

(i) `0111000000000101011010100000001111110111110`
(ii) `1111100011101101011100100111100110101111111`

18.8 Encode the following ASCII characters with base64 encoding:

(i) Hello, how are you.
(ii) This is secret.

Note pad 0s onto the end of the bit sequence, if required.

18.9 Identify the following parts of the RFC 822 messages.

(i)

```
Received: from publish.co.uk by ccmail1.publish.co.uk (SMTPLINK V2.11.01)
    ; Wed, 02 Jul 97 08:34:48 GMT
Return-Path: <FredB@local.exnet.com>
Received: from mailgate.exnet.com ([204.137.193.226]) by zeus.publish.co.uk
with SMTP id <17025>; Wed, 2 Jul 1997 08:33:29 +0100
Received: from exnet.com (assam.exnet.com) by mailgate.exnet.com with SMTP
id AA09732 (5.67a/IDA-1.4.4 for m.smith@publish.co.uk); Wed, 2 Jul 1997
08:34:22 +0100
Received: from maildrop.exnet.com (ceylon.exnet.com) by exnet.com with SMTP
id AA10740 (5.67a/IDA-1.4.4 for <m.smith@publish.co.uk>); Wed, 2 Jul 1997
08:34:10 +0100
Received: from local.exnet.com by maildrop.exnet.com (4.1/client-1.2DHD)
    id AA22007; Wed, 2 Jul 97 08:25:21 BST
From: FredB@local.exnet.com (Arthur Chapman)
Reply-To: FredB@local.exnet.com
To: b.smith@publish.co.uk
Subject: New proposal
Date: Wed, 2 Jul 1997 09:36:17 +0100
Message-Id: <66322430.1380704@local.exnet.com>
Organization: Local College
```

(ii)

```
Received: from central.napier.ac.uk by ccmailgate.napier.ac.uk (SMTPLINK
V2.11.01)     ; Sun, 29 Jun 97 03:18:46 gmt
Return-Path: <fred@singnetw.com.sg>
Received: from server.singnetw.com.sg (server.singnetw.com.sg [165.21.1.15])
by central.napier.ac.uk (8.6.10/8.6.10) with ESMTP id DAA18783 for
<w.buchanan@napier.ac.uk>; Sun, 29 Jun 1997 03:15:27 GMT
Received: from si7410352.ntu.ac.sg (ts900-1908.singnet.com.sg [165.21.158.60])
    by melati.singnet.com.sg (8.8.5/8.8.5) with SMTP id KAA08773
    for <w.buchanan@napier.ac.uk.>; Sun, 29 Jun 1997 10:14:59 +0800 (SST)
Message-ID: <33B5C33B.6CCC@singnetw.com.sg>
Date: Sun, 29 Jun 1997 10:06:51 +0800
From: Fred Smith <fred@singnetw.com.sg>
X-Mailer: Mozilla 2.0 (Win95; I)
MIME-Version: 1.0
To: w.buchanan@napier.ac.uk
Subject: Chapter 15
Content-Type: text/plain; charset=us-ascii
Content-Transfer-Encoding: 7bit
```

(iii)

```
Received: from central.napier.ac.uk by ccmailgate.napier.ac.uk (SMTPLINK
V2.11.01)
    ; Tue, 01 Jul 97 12:30:00 gmt
Return-Path: <bertb@scms.scotuni.ac.uk>
Received: from master.scms.scotuni.ac.uk ([193.62.32.5]) by cen-
tral.napier.ac.uk (8.6.10/8.6.10) with ESMTP id MAA20373 for
<w.buchanan@napier.ac.uk>; Tue, 1 Jul 1997 12:25:38 GMT
Received: from cerberus.scms.scotuni.ac.uk (cerberus.scms.scotuni.ac.uk
[193.62.32.46]) by master.scms.scotuni.ac.uk (8.6.9/8.6.9) with ESMTP id
MAA10056 for <w.buchanan@napier.ac.uk>; Tue, 1 Jul 1997 12:24:32 +0100
From: David Davidson <bertb@scms.scotuni.ac.uk>
```

```
Received: by cerberus.scms.scotuni.ac.uk (SMI-8.6/Dumb)
    id MAA03334; Tue, 1 Jul 1997 12:23:17 +0100
Date: Tue, 1 Jul 1997 12:23:17 +0100
Message-Id: <199707011123.MAA03334@cerberus.scms.scotuni.ac.uk>
To: w.buchanan@napier.ac.uk
Subject: Advert
Mime-Version: 1.0
Content-Type: text/plain; charset=us-ascii
Content-Transfer-Encoding: 7bit
Content-MD5: TzKyk+NON+vy6Cm6uqy9Cg==
```

19 The World Wide Web

19.1 Introduction

The very areas of modern life that have more jargon words and associated acronyms are the World-Wide Web (WWW) and the Internet. Words, such as:

gopher, ftp, telnet, TCP/IP stack, intranets, Web servers, clients, browsers, hypertext, URLs, Internet access providers, dial-up connections, UseNet servers, firewalls

have all become common in the business vocabulary.

The WWW was initially conceived in 1989 by CERN, the European particle physics research laboratory in Geneva, Switzerland. Its main objective was:

to use the hypermedia concept to support the interlinking of various types of information through the design and development of a series of concepts, communications protocols, and systems

One of its main characteristics is that stored information tends to be distributed over a geographically wide area. The result of the project has been the worldwide acceptance of the protocols and specifications used. A major part of its success was due to the full support of the National Center for Supercomputing Applications (NCSA), which developed a family of user interface systems known collectively as Mosaic.

The WWW, or Web, is basically an infrastructure of information. This information is stored on the WWW on Web servers and it uses the Internet to transmit data around the world. These servers run special programs that allow information to be transmitted to remote computers which are running a Web browser, as illustrated in Figure 19.1. The Internet is a common connection in which computers can communicate using a common addressing mechanism (IP) with a TCP/IP connection.

The information is stored on Web servers and is accessed by means of pages. These pages can contain text and other multimedia applications such as graphic images, digitized sound files and video animation. There are several standard media files (with typical file extensions):

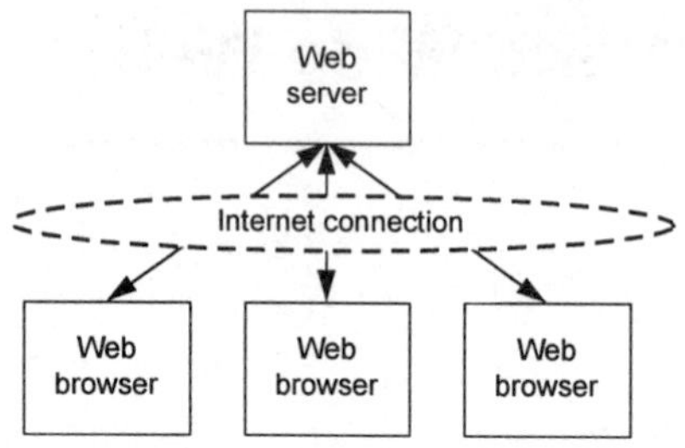

Figure 19.1 Web servers and browsers

- GIF or JPEG files for compressed images (GIF or JPG).
- QuickTime movies for video (QT or MOV).
- Postscript files (PS or EPS).
- MS video (AVI).
- Audio (AU, SND or WAV).
- MPEG files for compressed video (MPG).
- Compressed files (ZIP, Z or GZ).
- JavaScript (JAV, JS or MOCHA).
- Text files (TEX or TXT).

Each page contains text known as hypertext, which has specially reserved keywords to represent the format and the display functions. A standard language known as HTML (Hypertext Markup Language) has been developed for this purpose.

Hypertext pages, when interpreted by a browser program, display an easy-to-use interface containing formatted text, icons, pictorial hot spots, underscored words, and so on. Each page can also contain links to other related pages.

The topology and power of the Web now allows for distributed information, where information does not have to be stored locally. To find information on the Web the user can use powerful search engines to search for related links. Figure 19.2 shows an example of Web connections. The user initially accesses a page on a German Web server, this then contains a link to a Japanese server. This server contains links to UK, Swedish and French servers. This type of arrangement leads to the topology that resembles a spider's web, where information is linked from one place to another.

19.2 Advantages and disadvantages of the WWW

The WWW and the Internet tend to produce a polarization of views. Thus,

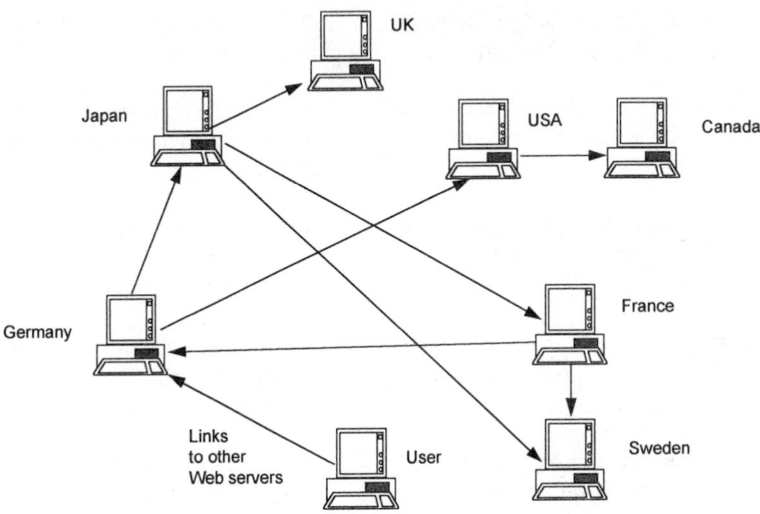

Figure 19.2 Example Web connections

before analyzing the WWW for its technical specification, a few words must be said on some of the subjective advantages and disadvantages of the WWW and the Internet. It should be noted that some of these disadvantages can be seen as advantages to some people, and vice versa. For example, freedom of information will be seen as an advantage to a freedom-of-speech group but often a disadvantage to security organizations. Table 19.1 outlines some of advantages and disadvantages.

Table 19.1 Advantages and disadvantages of the Internet and the WWW

	Advantages	*Disadvantages*
Global information flow	Less control of information by the media, governments and large organizations.	Lack of control on criminal material, such as certain types of pornography and terrorist activity.
Global transmission	Communication between people and organizations in different countries which should create the Global Village.	Data can easily get lost or state secrets can be easily transmitted over the world.
Internet connections	Many different types of connections are possible, such as dial-up facilities (perhaps over a modem or with ISDN) or through frame relays. The user only has to pay for the service and the local connection.	Data once on the Internet is relatively easy to tap into and possibly easy to change.

Global information	Creation of an ever increasing global information database.	Data is relatively easy to tap into and possibly easy to change.
Multimedia integration	Tailor-made applications with good presentation tools.	Lack of editorial control leads to inferior material which is hacked together.
Increasing WWW usage	Helps to improve its chances of acceptance into the home.	Increased traffic swamps the global information network and slows down commercial traffic.
WWW links	Easy to set up and leads users from one place to the next in a logical manner.	WWW links often fossilize where the link information is out-of-date or doesn't even exist.
Education	Increased usage of remote teaching with full multimedia education.	Increase in surface learning and lack of deep research. It may lead to an increase in time-wasting (too much surfing and to little learning).

19.3 Client/server architecture

The WWW is structured with clients and servers, where a client accesses services from the server. These servers can either be local or available through a global network connection. A local connection normally requires the connection over a local area network but a global connection normally requires connection to an Internet provider. These providers are often known as an Internet access provider (IAPs), sometimes as an Internet connectivity provider (ICP) or Internet Service Providers (ISPs). They provide the mechanism to access the Internet and have the required hardware and software to connect from the user to the Internet. This access is typically provided through one of the following:

- Connection to a client computer though a dial-up modem connection (typically at 14.4 kbps or 28.8 kbps).
- Connection to a client computer though a dial-up ISDN connection (typically at 64 kbps or 128 kbps).
- Connection of a client computer to a server computer which connects to the Internet though a frame relay router (typically 56 kbps or 256 kbps).
- Connection of a client computer to a local area network which connects to the Internet though a T-1, 1.544 Mbps router.

These connections are illustrated in Figure 19.3. A router automatically routes all traffic to and from the Internet whereas the dial-up facility of a modem or ISDN link requires a connection to be made over a circuit-switched line (that is, through the public telephone network). Home users and small businesses typically use modem connections (although ISDN connections are becoming more common). Large corporations which require global Internet services tend to use frame routers. Note that an IAP may be a commercial organization (such as CompuServe or America On-line) or a support organization (such as giving direct connection to government departments or educational institutions). A commercial IAP organization is likely to provide added services, such as electronic mail, search engines, and so on.

An Internet Presence Provider (IPP) allows organizations to maintain a presence on the Internet without actually having to invest in the Internet hardware. The IPPs typically maintain WWW pages for a given charge (they may also provide sales and support information).

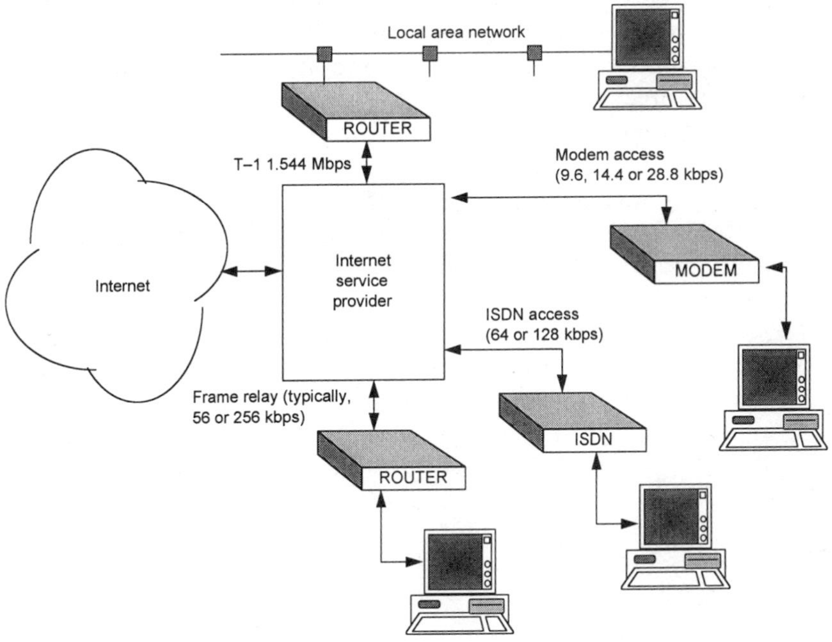

Figure 19.3 Example connection to the Internet

19.4 Web browsers

Web browsers interpret special hypertext pages which consist of the hypertext markup language (HTML) and JavaScript. They then display it in the given format. There are currently four main Web browsers:

- Netscape Navigator – Navigator is the most widely used WWW browser and is available in many different versions on many systems. It runs on PCs (running Windows 3.1, Windows NT or Windows 95), UNIX workstations and Macintosh computers. Figure 19.4 shows Netscape Navigator Version 3 for Windows 3.1. It has become the standard WWW browser and has many add-ons and enhancements which have been added through continual development by Netscape. The basic package also has many compatible software plug-ins which are developed by third-party suppliers. These add extra functionality such as video players and sound support.

- NSCA Mosaic – Mosaic was originally the most popular Web browser when the Internet first started. It has now lost its dominance to Microsoft Internet Explorer and Netscape Navigator. NSCA Mosaic was developed by the National Center for Supercomputer Applications (NCSA) at the University of Illinois. Figure 19.5 shows a sample window from CompuServe Mosaic.

- Lynx – Lynx is typically used on UNIX-based computers with a modem dial-up connection. It is fast to download pages but does not support many of the features supported by Netscape Navigator or Mosaic.

- Microsoft Internet Explorer – Explorer now comes as a standard part of Windows 95/NT and as this will become the most popular computer operating system then so will this browser.

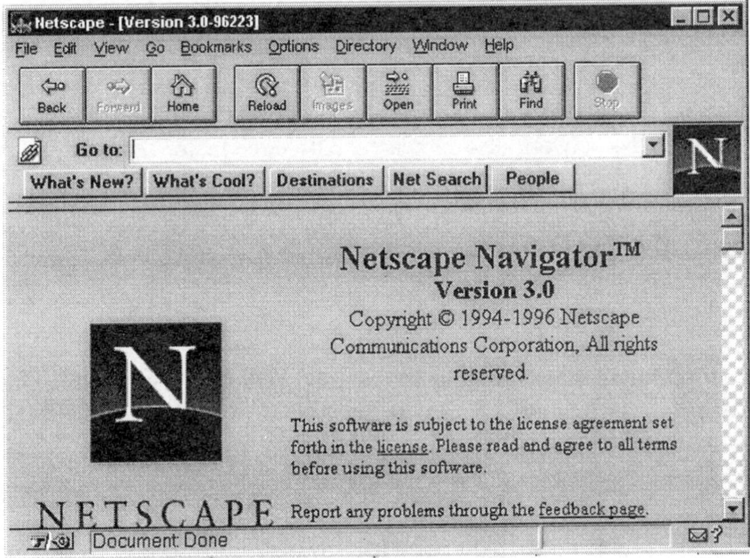

Figure 19.4 Netscape Navigator Version 3.0

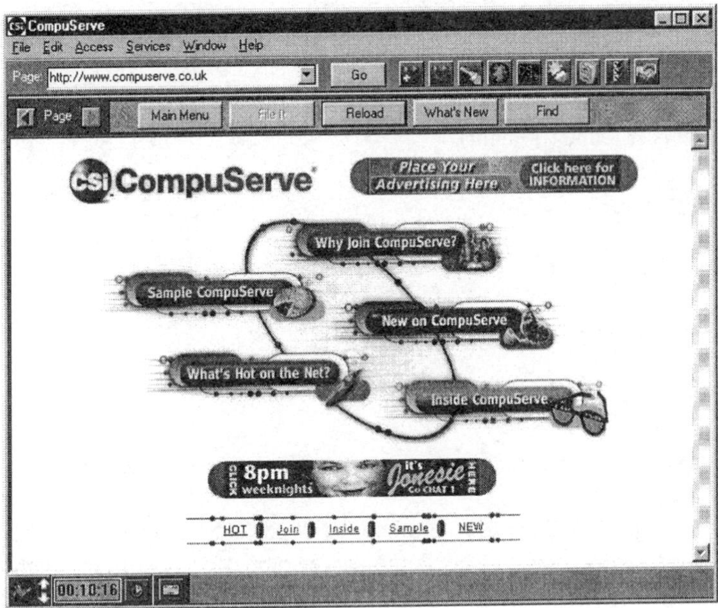

Figure 19.5 CompuServe Mosaic

19.5 Internet resources

The Internet expands by the day as the amount of servers and clients which connect to the global network increases and the amount of information contained in the network also increases. The three major services which the Internet provides are:

- The World Wide Web.
- Global electronic mail.
- Information sources.

The main information sources, apart from the WWW, are from FTP, Gopher, WAIS and UseNet servers. These different types of servers will be discussed in the next section.

19.6 Universal resource locators (URLs)

Universal resource locators (URLs) are used to locate a file on the WWW. They provide a pointer to any object on a server connected over the Internet. This link could give FTP access, hypertext references, and so on. URLs contains:

- The protocol of the file (the scheme).
- The server name (domain).
- The pathname of the file.
- The filename.

URL standard format is:

<scheme>:<scheme-specific-part>

An example is:

```
http://www.toytown.anycor.co/fred/index.html
```

where `http` is the file protocol (Hypertext Translation Protocol), `www.toytown.anycor.co` is the server name, `/fred` is the path of the file and the file is named `index.html`. The most common URL formats are:

- `http` – Hypertext Transfer Protocol.
- `ftp` – File Transfer Protocol.
- `gopher` – Gopher Protocol.

- `mailto` – Electronic mail address.
- `news` – UseNet news.
- `telnet` – Reference to interactive sessions.
- `wais` – Wide area information servers.
- `file` – Host-specific filenames.

19.6.1 Electronic mail address

The `mailto` scheme defines a link to an Internet email address. An example is:

```
mailto: fred.bloggs@toytown.ac.uk
```

When this URL is selected then an email message will be sent to the email address `fred.bloggs@toytown.ac.uk`. Normally, some form of text editor is called and the user can enter the required email message. Upon successful completion of the text message; it is sent to the addressee.

19.6.2 File Transfer Protocol (FTP)

The `ftp` URL scheme defines that the files and directories specified are accessed using the FTP protocol. In its simplest form it is defined as:

`ftp`://*<hostname>*/*<directory-name>*/*<filename>*

The FTP protocol normally requests a user to log into the system. For example, many public domain FTP servers use the login of:

```
anonymous
```

and the password can be anything (but it is normally either the user's full name or their Internet email address). Another typical operation is changing directory from a starting directory or the destination file directory. To accommodate this, a more general form is:

`ftp`://*<user>*:*<password>*@*<hostname>*:*<port>*/*<cd1>*/*<cd2>*/
…/*<cdn>*/*<filename>*

where the user is defined by *<user>* and the password by *<password>* . The host name, *<hostname>*, is defined after the @ symbol and change directory commands are defined by the *cd* commands.

19.6.3 Host-specific file names

The file URL defines that the file is local to the computer and is not accessed through a server application. An example, taken from a PC, for a file

C:\WWW\1.HTM is accessed with the URL:

```
file:///C|/WWW/1.HTM
```

19.6.4 Hypertext Transfer Protocol (HTTP)

HTTP is the protocol which is used to retrieve information connected with hypermedia links. The client and server initially perform a negotiation procedure before the HTTP transfer takes place. This negotiation involves the client sending a list of formats it can support and the server replying with data in the required format. This will be discussed in more detail in the next chapter.

Users generally move from a link on one server to another server. Each time the user moves from one server to another, the client sends an HTTP request to the server. Thus the client does not permanently connect to the server, and the server views each transfer as independent from all previous accesses. This is known as a stateless protocol.

19.6.5 Reference to interactive sessions (TELNET)

The `telnet` URL allows users to interactively perform a telnet operation, where a user must login to the referred system.

19.6.6 UseNet news

UseNet or NewsGroup servers are part of the increasing use of general discussion news groups which share text-based news items. The news URL scheme defines a link to either a news group or individual articles with a group of UseNet news.

19.6.7 Gopher Protocol

Gopher is widely used over the Internet and is basically a distribution system for the retrieval and delivery of documents. Users retrieve documents through a series of hierarchical menus, or through keyword searches. Unlike HTML documents it is not based on the hypertext concept.

19.6.8 Wide area information servers (WAIS)

WAIS is a public domain, fully text-based, information retrieval system over the Internet which performs text-based searches. The communications protocol used is based on the ANSI standard Z39.50, which is designed for networking library catalogs.

WAIS services include index generation and search engines. An indexer generates multiple indexes for organizations or individuals who offer services over the Internet. A WAIS search engine searches for particular words or text string indexes located across multiple Internet attached information servers of various types.

19.7 Universal resource identifier

The universal resource identifier (URI) is defined as a generically designated string of characters which refers to objects on the WWW. A URL is an example of a URI, with a designated access protocol and a specific Internet address.

Specfications have still to be completed, but URIs will basicall be used to define the syntax for encoding arbitrary naming or addressing schemes. This should decouple the name of a resource from its location and also from its access method. For example, the file:

```
MYPIC.HTM
```

would be automatically associated with an HTTP protocol.

19.8 Intranets

An organization may experience two disadvantages in having a connection to the WWW and the Internet:

- The possible usage of the Internet for non-useful applications (by employees).
- The possible connection of non-friendly users from the global connection into the organization's local network.

For these reasons many organizations have shied away from connection to the global network and have set up intranets. These are in-house, tailor-made internets for use within the organization and provide limited access (if any) to outside services and also limit the external traffic into the intranet (if any). An intranet might have access to the Internet but there will be no access from the Internet to the organization's Intranet.

Organizations which have a requirement for sharing and distributing electronic information normally have three choices:

- Use a propriety groupware package, such as Lotus Notes.
- Set up an intranet.
- Set up a connection to the Internet.

Groupware packages normally replicate data locally on a computer whereas intranets centralize their information on central servers which are then accessed by a single browser package. The stored data is normally open and can be viewed by any compatible WWW browser. Intranet browsers have the great advantage over groupware packages in that they are available for a variety of clients, such as PCs, UNIX workstations, Macs, and so on. A client browser also provides a single GUI interface which offers easy integration with other applications, such as electronic mail, images, audio, video, animation, and so on.

The main elements of an intranet are:

- Intranet server hardware.
- Intranet server software.
- TCP/IP stack software on the clients and server.
- WWW browsers.
- A firewall.

Typically the intranet server consists of a PC running the Lynx (PC-based UNIX-like) operating system. The TCP/IP stack is software installed on each computer and allows communications between a client and a server using TCP/IP.

A firewall is the routing computer which isolates the intranet from the outside world. This will be discussed in the next section.

19.9 Firewalls

A firewall (or security gateway) protects a network against intrusion from outside sources. They tend to differ in their approach but can be characterized as follows:

- Firewalls which block traffic.
- Firewalls which permit traffic.

19.9.1 Packet filters

The packet filter is the simplest form of firewall. It basically keeps a record of allowable source and destination IP addresses, and deletes all packets which do not have them. This technique is known as address filtering. The packet filter keeps a separate source and destination table for both directions, i.e. into and out of the intranet. This type of method is useful for companies which have geographically spread sites, as the packet filter can allow incoming traffic from other friendly sites, but block other non-friendly traffic. This is illustrated by Figure 19.6.

Unfortunately this method suffers from the fact that IP addresses can be easily forged. For example, a hacker might determine the list of good source addresses and then add one of them to any packets which are addressed to the intranet.

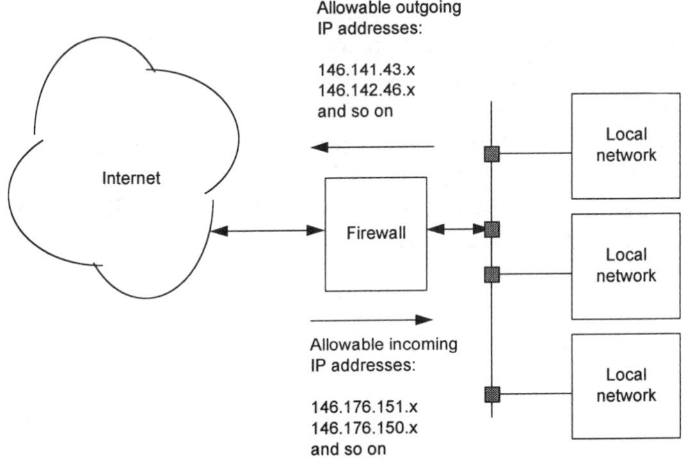

Figure 19.6 Packet filter firewalls

19.9.2 Application level gateway

Application level gateways provide an extra layer of security when connecting an intranet to the Internet. They have three main components:

- A gateway node.
- Two firewalls which connect on either side of the gateway and only transmit packets which are destined for or to the gateway.

Figure 19.7 shows the operation of an application level gateway. In this case, Firewall A discards anything that is not addressed to the gateway node, and discards anything that is not sent by the gateway node. Firewall B, similarly discards anything from the local network that is not addressed to the gateway node, and discards anything that is not sent by the gateway node. Thus, to transfer files from the local network into the global network, the user must do the following:

- Log onto the gateway node.
- Transfer the file onto the gateway.
- Transfer the file from the gateway onto the global network.

To copy a file from the network an external user must:

- Log onto the gateway node.
- Transfer from the global network onto the gateway.
- Transfer the file from the gateway onto the local network.

A common strategy in organizations is to allow only electronic mail to pass from the Internet to the local network. This specifically disallows file transfer and remote login. Unfortunately electronic mail can be used to transfer files. To overcome this problem the firewall can be designed specifically to disallow very large electronic mail messages, so it will limit the ability to transfer files. This tends not to be a good method as large files can be split up into small parts then sent individually.

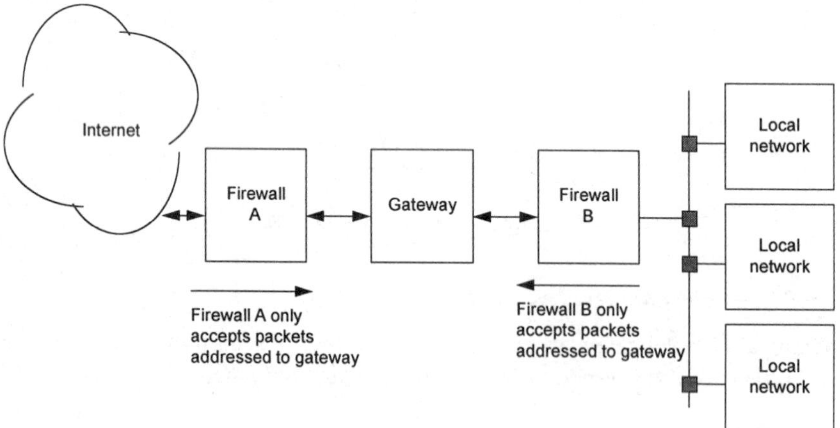

Figure 19.7 Application level gateway

19.9.3 Encrypted tunnels

Packet filters and application level gateways suffer from insecurity which can allow non-friendly users into the local network. Packet filters can be tricked with fake IP addresses and application level gateways can be hacked into by determining the password of certain users of the gateway then transferring the files from the network to the firewall, on to the gateway, on to the next firewall and out. The best form of protection for this type of attack is to allow only a limited number of people to transfer files onto the gateway.

And overall the best method of protection is to encrypt the data leaving the network then to decrypt it on the remote site. Only friendly sites will have the required encryption key to receive and send data. This has the extra advantage that the information cannot be easily tapped-into.

Only the routers which connect to the Internet require to encrypt and decrypt, as illustrated in Figure 19.8.

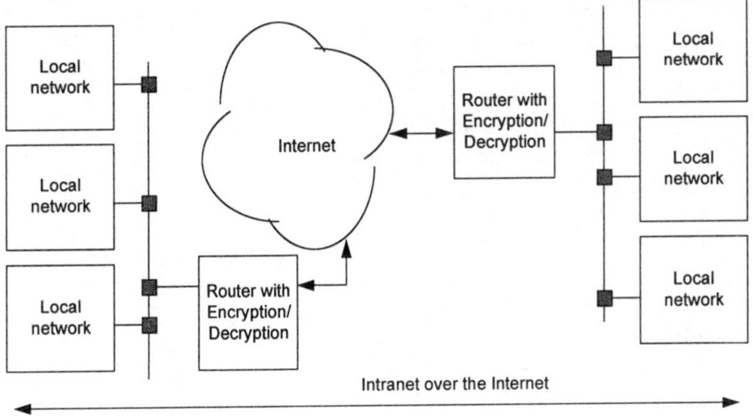

Figure 19.8 Encryption tunnels

19.10 Exercises

19.1 Identify the WW main file extension formats and their usage.

19.2 Investigate a local WWW connection, showing network connections and its domain name.

19.3 Explain how firewalls are used to filter traffic.

19.4 If possible, access the following WWW servers:

```
http://www.microsoft.com
http://www.intel.com
http://www.ieee.com
http://www.winzip.com
http://www.netscape.com
http://www.realaudio.com
http://www.pa.press.net
http://www.pointcast.com
http://mmx.com
http://www.cyrix.com
http://www.compaq.com
http://www.psion.com
http://www.amd.com
http://www.cnn.com
```

```
http://www.w3.org
http://www.microsoft.com/ie
http://www.macromedia.com
http://www.epson.co.uk
http://www.iomega.com
http://www.vocaltec.com
http://www.euronec.com
http://www.casio.com
http://www.hayes.com
http://java.motiv.co.uk
http://www.lotus.com
http://www.adobe.com
http://www.corel.com
http://www.guinness.ie
http://www.symantec.co.uk
http://www.fractal.com
http://www.quarterdeck.com
http://cinemania.msn.com
http://www.drsolomon.com
```

19.5 The home page for this book can be found at the URL:

```
http://www.eece.napier.ac.uk/~bill_b/adc.hmtl
```

Access this page and follow any links it contains.

20 HTTP

20.1 Introduction

Chapter 19 discussed the WWW. The foundation protocol of the WWW is the Hypertext Transfer Protocol (HTTP) which can be used in any client/server application involving hypertext. It is used in the WWW for transmitting information using hypertext jumps and can support the transfer of plaintext, hypertext, audio, images, or any Internet-compatible information. The most recently defined standard is HTTP 1.1, which has been defined by the IETF standard.

20.2 HTTP operation

HTTP is a stateless protocol where each transaction is independent of any previous transactions. The advantage of being stateless is that it allows the rapid access of WWW pages over several widely distributed servers. It uses the TCP protocol to establish a connection between a client and a server for each transaction then terminates the connection once the transaction completes.

HTTP also support many different formats of data. Initially a client issues a request to a server which may include a prioritized list of formats that it can handle. This allows new formats to be added easily and also prevents the transmission of unnecessary information.

A client's WWW browser (the user agent) initially establishes a direct connection with destination server which contains the required WWW page. To make this connection the client initiates a TCP connection between the client and the server. After this is established the client then issues an HTTP request, such as the specific command (the method), the URL, and possibly extra information such as request parameters or client information. When the server receives the request, it attempts to perform the requested action. It then returns an HTTP response which includes status information, a success/error code, and extra information itself. After this is received by the client, the TCP connection is closed.

20.3 Intermediate systems

The previous section discussed the direct connection of a client to a server. Many system organizations do not wish a direct connection to an internal network. Thus HTTP supports other connections which are formed through intermediate systems, such as:

- A proxy.
- A gateway.
- A tunnel.

Each intermediate system is connected by a TCP and acts as a relay for the request to be sent out and returned to the client. Figure 20.1 shows the setup of the proxies and gateways.

20.3.1 Proxy

A proxy connects to a number of clients; it acts on behalf of other clients and sends requests from the clients to a server. It thus acts as a client when it communicates with a server, but as a server when communicating with a client. A proxy is typically used for security purposes where the client and server are separated by a firewall. The proxy connects to the client side of the firewall and the server to the other side of the firewall. Thus the server must authenticate itself to the firewall before a connection can be made with the proxy. Only after this has been authenticated will the proxy pass requests through the firewall.

A proxy can also be used to convert between different version of HTTP.

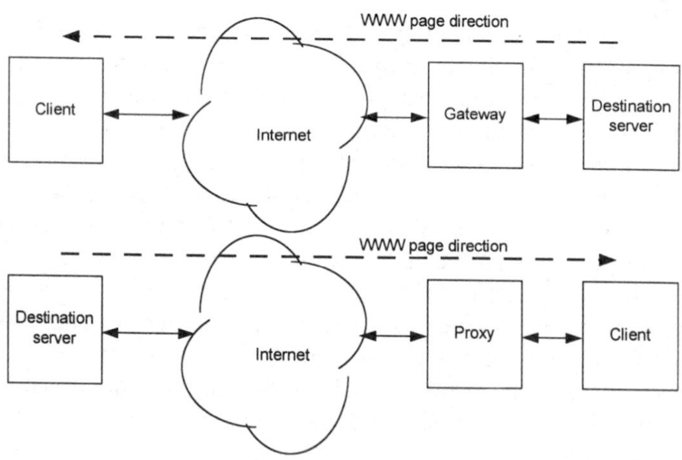

Figure 20.1 Usage of proxies and gateways

20.3.2 Gateway

Gateways are servers that act as if they are the destination server. They are typically used when clients cannot get direct access to the server, and typically for one of the security reasons where the gateway acts as a firewall so that the gateway communicates with the Internet and the server only communicates with the Internet through the gateway. The client must then authenticate itself to the proxy, which can then pass the request on to the server.

They can also be used when the destination is a non-HTTP server. Web browsers have built into them the capability to contact servers for protocols other than HTTP, such as FTP and Gopher servers. This capability can also be provided by a gateway. The client makes an HTTP request to a gateway server. The gateway server than contacts the relevant FTP or Gopher server to obtain the desired result. This result is then converted into a form suitable for HTTP and transmitted back to the client.

20.3.3 Tunnel

A tunnel does not perform any operation on the HTTP message; it passes messages onto the client or server unchanged. This differs from a proxy or a gateway, which modify the HTTP messages. Tunnels are typically used as firewalls, where the firewall authenticates the connection but simply relays the HTTP messages.

20.4 Cache

In a computer system a cache is an area of memory that stores information likely to be accessed in a fast access memory area. For example a cache controller takes a guess on which information the process is likely to access next. When the processor wishes to access the disk then, if it has guessed right it will load, the cache controller will load from the electronic memory rather than loading it from the disk. A WWW cache stores cacheable responses so that there is a reduction in network traffic and an improvement in access times.

20.5 HTML messages

HTTP message are either requests from clients to servers or responses from servers to clients. The message is either a simple-request, a simple-response, full-request or a full-response. HTTP Version 0.9 defines the simple request/ response messages whereas HTTP Version 1.1 defines full requests/responses.

20.5.1 Simple requests/responses

The simple request is a `GET` command with the requested URI such as:

```
GET  /info/dept/courses.html
```

The simple response is a block containing the information identified in the URI (called the entity-body).

20.5.2 Full requests/responses

Very few security measures or enhanced services are built into the simple requests/responses. HTTP Version 1/1.1 improves on the simple requests/ responses by adding many extra requests and responses, as well as adding extra information about the data supported. Each message header consist of a number of fields which begin on a new line and consist of the field name followed by a colon and the field value. This follows the format of RFC 822 (as shown in Section 15.5.4) and allows for MIME encoding. It is thus similar to MIME-encoded email. A full request starts with a request line command (such as `GET`, `MOVE` or `DELETE`) and is then followed by one or more of the following:

- General-headers which contain general fields that do not apply to the entity being transferred (such as MIME version, date, and so on).
- Request-headers which contain information on the request and the client (e.g. the client's name, its authorization, and so on).
- Entity-headers which contain information about the resource identified by the request and entity-body information (such as the type of encoding, the language, the title, the time when it was last modified, the type of resource it is, when it expires, and so on).
- Entity-body which contains the body of the message (such as HTML text, an image, a sound file, and so on).

A full response starts with a response status code (such as `OK`, `Moved Temporarily`, `Accepted`, `Created`, `Bad Request`, and so on) and is then followed by one or more of the following:

- General-headers, as with requests, contain general fields which do not apply to the entity being transferred (MIME version, date, and so on).
- Response-headers which contain information on the response and the server (e.g. the server's name, its location and the time the client should retry the server).
- Entity-headers, as with request, which contain information about the resource identified by the request and entity-body information (such as the

type of encoding, the language, the title, the time when it was last modified, the type of resource it is, when it expires, and so on).
- Entity-body, as with requests, which contains the body of the message (such as HTML text, an image, a sound file, and so on).

The following example shows an example request. The first line is always the request method, in this case it is `GET`. Next there are various headers. The general-header field is `Content-Type`, the request-header fields are `If-Modified-Since` and `From`. There are no entity parts to the message as the request is to get an image (if the command had been to `PUT` then there would have been an attachment with the request). Notice that the end of the message is delimited by a single blank line as this indicates the end of a request/response. Note that the headers are case sensitive, thus `Content-Type` with the correct types of letters (and `GET` is always in uppercase letters).

Example HTTP request

```
GET mypic.jpg
Content-Type: Image/jpeg
If-Modified-Since: 06 Mar 1997 12:35:00
From: Fred Bloggs <FREDB@ACOMP.CO.UK>
```

Request messages

The most basic request message is to `GET` a URI. HTTP/1.1 adds many more requests including:

```
COPY     DELETE     GET        HEAD     POST
LINK     MOVE       OPTIONS    PATCH    PUT
TRACE    UNLINK     WRAPPED
```

As before, the `GET` method requests a WWW page. The `HEAD` method tells the server that the client wants to read only the header of the WWW page. If the `If-Modified-Since` field is included then the server checks the specified date with the date of the URI and verifies whether it has not changed since then.

A `PUT` method requests storage of a WWW page and `POST` appends to a named resource (such as electronic mail). `LINK` connects two existing resources and `UNLINK` breaks the link. A `DELETE` method removes a WWW page.

The request-header fields are mainly used to define the acceptable type of entity that can be received by the client; they include:

```
Accept                  Accept-Charset      Accept-Encoding
Accept-Language         Authorization       From
Host                    If-Modified-Since   If-Modified-Since
Proxy-Authorization     Range               Referer
Unless                  User-Agent
```

The `Accept` field is used to list all the media types and ranges that are acceptable to the client. An `Accept-Charset` field defines a list of character sets acceptable to the server and `Accept-Encoding` is a list of acceptable content encodings (such as the compression or encryption technique). The `Accept-Language` field defines a set of preferred natural languages.

The `Authorization` field has a value which authenticates the client to the server. A `From` field defines the email address of the user who is using the client (e.g. `From: fred.blogg@anytown.uk`) and the `Host` field specifies the name of the host of the resource being requested.

A useful field is the `If-Modified-Since` field, used with the `GET` method. It defines a date and time parameter and specifies that the resource should not be sent if it has not been modified since the specified time. This is useful when a client has a local copy of the resource in a local cache and, rather than transmitting the unchanged resource, it can use its own local copy.

The `Proxy-Authorization` field is used by the client to identify itself to a proxy when the proxy requires authorization. A `Range` field is used with the `GET` message to get only a part of the resource.

The `Referer` field defines the URI of the resource from which the `Request-URI` was obtained and enables the server to generate list of back-links. An `Unless` field is used to make a comparison based on any entity-header field value rather than a date/time value (as with `GET` and `If-Modified-Since`).

The `User-Agent` field contains information about the user agent originating this request.

Response messages

In HTTP/0.9 the response from the server was either the entity or no response. HTTP/1.1 includes many other response, these include:

```
Accepted                       Bad Gateway
Bad Request                    Conflict
Continue                       Created
Forbidden                      Gateway Timeout
Gone                           Internal Server Error
Length Required                Method Not Allowed
Moved Permanently              Moved Temporarily
Multiple Choices               No Content
Non-Authoritative Info         None Acceptable
Not Found                      Not Implemented
```

```
Not Modified                       OK
Partial Content                    Payment Required
Proxy Authorization Required       Request Timeout
Reset Content                      See Other
Service Unavailable                Switching Protocols
Unauthorized                       Unless True
Use Proxy
```

These responses can be put into five main groupings:

- **Client error** – `Bad Request`, `Conflict`, `Forbidden`, `Gone`, `Payment required`, `Not Found`, `Method Not Allowed`, `None Acceptable`, `Proxy Authentication Required`, `Request Timeout`, `Length Required`, `Unauthorized`, `Unless True`.
- **Informational** – `Continue`, `Switching Protocol`.
- **Redirection** – `Moved Permanently`, `Moved Temporarily`, `Multiple Choices`, `See Other`, `Not Modified`, `User Proxy`.
- **Server error** – `Bad Gateway`, `Internal Server Error`, `Not Implemented`, `Service Unavailable`, `Gateway Timeout`.
- **Successful** – `Accepted`, `Created`, `OK`, `Non-Authoritative Info`. The `OK` field is used when the request succeeds and includes the appropriate response information.

The response header fields are:

```
Location          Proxy-Authenticate    Public
Retry-After       Server                WWW-Authenticate
```

The `Location` field defines the location of the resource identified by the `Request-URI`. A `Proxy-Authenticate` field contains the status code of the `Proxy Authorization Required` response.

The `Public` field defines non-standard methods supported by this server. A `Retry-After` field contains values which define the amount of time a service will be unavailable (and is thus sent with the `Service Unavailable` response).

The `WWW-Authenticate` field contains the status code for the `Unauthorized` response.

General-header fields

General-header fields are used either within requests or within responses; they include:

```
Cache-Control   Connection     Date     Forwarded
Keep-Alive      MIME-Version   Pragma   Upgrade
```

The Cache-Control field gives information on the caching mechanism and stops the cache controller from modifying the request/response. A Connection field specifies the header field names that apply to the current TCP connection.

The Date field specifies the date and time at which the message originated; this is obviously useful when examining the received message as it gives an indication of the amount of time the message took to arrive at is destination. Gateways and proxies use the Forwarded field to indicate intermediate steps between the client and the server. When a gateway or proxy reads the message, it can attach a Forwarded field with its own URI (this can help in tracing the route of a message).

The Keep-Alive field specifies that the requester wants a persistent connection. It may indicate the maximum amount of time that the sender will wait for the next request before closing the connection. It can also be used to specify the maximum number of additional requests on the current persistent connection.

The MIME-Version field indicates the MIME version (such as MIME-Version: 1.0). A Pragma field contains extra information for specific applications.

In a request the Upgrade field specifies the additional protocols that the client supports and wishes to use, whereas in a response it indicates the protocol to be used.

Entity-header fields

Depending on the type of request or response, an entity-header can be included:

```
Allow               Content-Encoding        Content-Language
Content-Length      Content-MD5             Content-Range
Content-Type        Content-Version         Derived-From
Expires             Last-Modified           Link
Title               Transfer-encoding
URI-Header extension-header
```

The Allow field defines the supported methods supported by the resource identified in the Request-URI. A Content-Encoding field indicates content encodings, such as ZIP compression, that have been applied to the resource (Content-Encoding: zip).

The Content-Language field identifies natural language(s) of the intended audience for the enclosed entity (e.g. Content-language: German) and the Content-Length field defines the number of bytes in the entity.

The Content-Range field designates a portion of the identified resource that is included in this response, while Content-Type indicates the media type of the entity body (such as `Content-Type=text/html`, `Content-Type=text/plain`, `Content-Type=image/gif` or `Content-type=image/jpeg`). The version of the entity is defined in the `Content-Version` field.

The `Expires` field defines the date and time when the entity is considered stale. The `Last-Modified` field is the date and time when the resource was last modified.

The `Link` field defines other links and the `Title` field defines the title for the entity. A `Transfer-Encoding` field indicates the transformation type that is applied so the entity can be transmitted.

20.6 Exercises

20.1 Explain how proxies and gateways are used to provide security.

20.2 Discuss the limitations of simple requests and responses with HTTP.

20.3 Discuss request messages and the fields that are set.

20.4 Discuss response messages and the fields that are set.

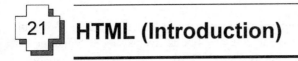

21 HTML (Introduction)

21.1 Introduction

HTML is a standard hypertext language for the WWW and has several different versions. Most WWW browsers support HTML 2 and most of the new versions of the browsers support HTML 3. WWW pages are created and edited with a text editor, a word processor or, as is becoming more common, within the WWW browser.

HTML tags contain special formatting commands and are contained within a less than (<) and a greater than (>) symbol (also known as angle brackets). Most tags have an opening and closing version; for example, to highlight bold text the bold opening tag is `<B>` and the closing tag is `</B>`. Table 21.1 outlines a few examples.

Table 21.1 Some HTML tags

Open tag	*Closing tap*	*Description*
`<HTML>`	`</HTML>`	Start and end of HTML
`<HEAD>`	`</HEAD>`	Defines the HTML header
`<BODY>`	`</BODY>`	Defines the main body of the HTML
`<TITLE>`	`</TITLE>`	Defines the title of the WWW page
`<I>`	`</I>`	Italic text
`<B>`	`</B>`	Bold text
`<U>`	`</U>`	Underlined text
`<BLINK>`	`</BLINK>`	Make text blink
`<EM>`	`</EM>`	Emphasize text
`<FONT SIZE=+1>`	`</FONT>`	Increase font size by one increment
`<FONT SIZE=-1>`	`</FONT>`	Reduce font size by one increment
`<CENTER>`	`</CENTER>`	Center text
`<H1>`	`</H1>`	Section header, level 1
`<H2>`	`</H2>`	Section header, level 2
`<H3>`	`</H3>`	Section header, level 3
`<P>`		Create a new paragraph
` `		Create a line break
`<!-->`	`-->`	Comments
`<SUPER>`	`</SUPER>`	Superscript
`<SUB>`	`</SUB>`	Subscript

HTML script 21.1 produces an example script and Figure 21.1 shows the output from the WWW browser. The first line is always <HTML> and the last line is </HTML>. After this line the HTML header is defined between <HEAD> and </HEAD>. The title of the window in this case is My first HTML page. The main HTML text is then defined between <BODY> and </BODY>.

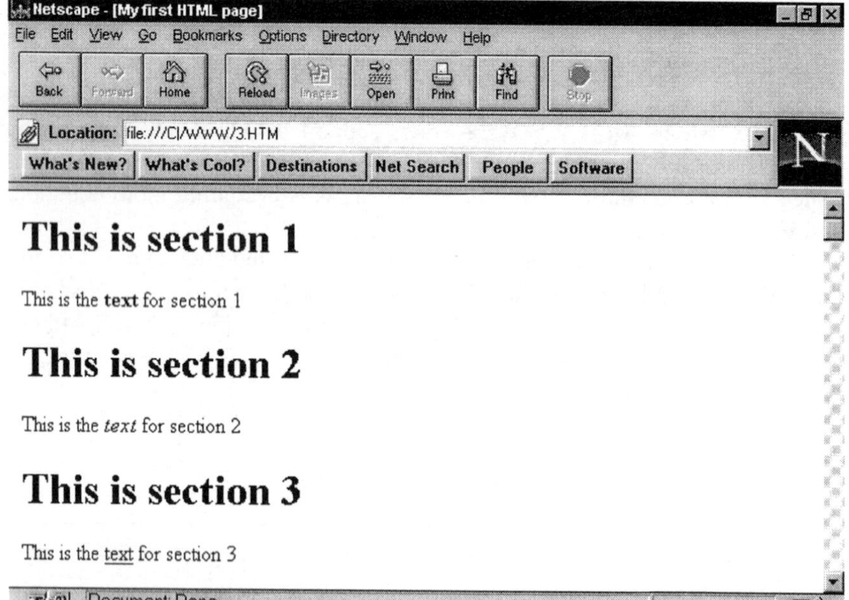

Figure 21.1 A window created using HTML script 21.1

📖 HTML script 21.1

```
<HTML>
<HEAD>
<TITLE>My first HTML page</TITLE>
</HEAD>
<BODY>

<H1> This is section 1</H1>
This is the <b>text</b> for section 1
<H1> This is section 2</H1>
This is the <i>text</i> for section 2
<H1> This is section 3</H1>
This is the <u>text</u> for section 3
<p>
This is the end of the text
</BODY>
</HTML>
```

The WWW browser fits text into the window size and does not interpret line breaks in the HTML source. To force a new line the
 (line break) or a new paragraph (<P>) is used. The example also shows bold, italic and underlined text.

21.2 Links

The topology of the WWW is set up using links where pages link to other related pages. A reference takes the form:

<A HREF="*url*"> *Reference Name* </A>

where *url* defines the URL for the file, *Reference Name* is the name of the reference and </A> defines the end of the reference name. HTML script 21.2 shows an example of how to use references and Figure 21.2 shows a sample browser page. The background color is set using <BODY BGCOLOR="#FFFFFF"> which sets the background color to white. In this case the default text color is black and the link is colored blue.

HTML script 21.2

```
<HTML>

<HEAD>
<TITLE>Fred's page</TITLE>
</HEAD>
<BODY BGCOLOR="#FFFFFF">

<H1>Fred's Home Page</H1>
If you want to access information on
this book <A HREF="adcbook.html">click here</A>.

<P>
A reference to the <A REF="http:www.iee.com/">IEE</A>
</BODY>
</HTML>
```

21.2.1 Other links

Links can be set up to send email and news groups. For example:

```
<A HREF="news:sport.tennis"> Newsgroups for tennis</A>
```

to link to a tennis news group and

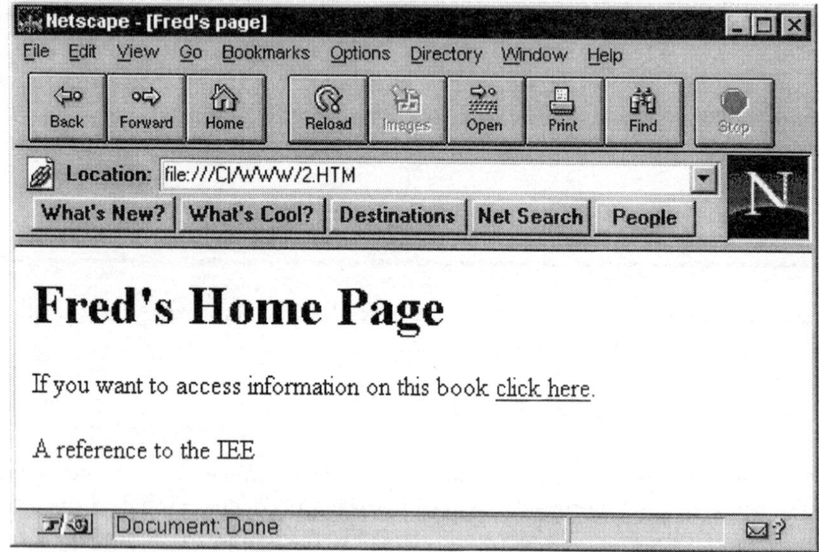

Figure 21.2 A windows created using HTML script 212

```
<A HREF="mailto:f.bloggs@fredco.co.uk">Send a
message to me</A>
```

to send a mail message to the email address: `f.bloggs@ fredco.co.uk`.

21.3 Lists

HTML allows ordered and unordered lists. Lists can be declared anywhere in the body of the HTML.

21.3.1 Ordered lists

The start of an ordered list is defined with `<OL>` and the end of the list by `</OL>`. Each part of the list is defined after the `<LI>` tag. Unordered lists are defined between the `<UL>` and `</UL>` tags. HTML script 21.3 produces of an ordered list and an unordered list. Figure 21.3 shows the output from the browser.

📖 HTML script 21.3

```
<HTML>
<HEAD>
<TITLE>Fred's page</TITLE>
</HEAD>
```

```
<BODY BGCOLOR="#FFFFFF">
<H1>List 1</H1>
<OL>
<LI>Part 1
<LI>Part 2
<LI>Part 3
</OL>
<H1>List 2</H1>
<UL>
<LI>Section 1
<LI>Section 2
<LI>Section 3
</UL>
</BODY>
</HTML>
```

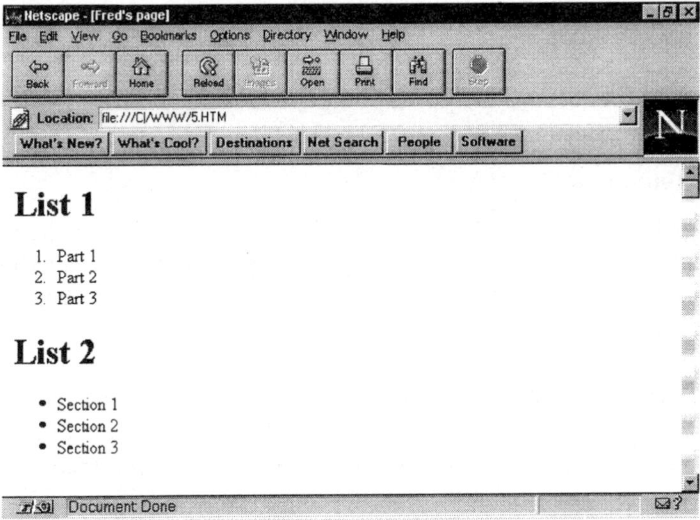

Figure 21.3 WWW browser with ordered and unordered lists

Some browsers allow the type of numbered list to be defined with `<OL TYPE=x>`, where *x* can either be:

- 'A' for capital letters (such as a, b, c, and so on).
- 'a' for small letters (such as a, b, c, and so on).
- 'I' for capital roman letters (such as I, II, III, and so on).
- 'i' for small roman letters (such as i, ii, iii, and so on).
- 'I' for numbers (which is the default).

```
<OL Type=I>
<LI> List 1
<LI> List 2
```

```
<LI> List 3
</OL>
<OL Type=A>
<LI> List a
<LI> List b
<LI> List c
</OL>
```

would be displayed as:

I. List 1
II. List 2
III. List 3
A. List a
B. List b
C. List c

The starting number of the list can be defined using the `<LI VALUE=`*n*`>` where *n* defines the initial value of the defined item list.

21.3.2 Unordered lists

Unordered lists are used to list a series of items in no particular order. They are defined between the `<UL>` and `</UL>` tags. Some browsers allow the type of bullet point to be defined with `<LI TYPE=`*shape*`>`, where *shape* can be:

- *disc* for round solid bullets (which is the default for first-level lists).
- *round* for round hollow bullets (which is the default for second-level lists).
- *square* for square bullets (which is the default for third-level lists).

HTML script 21.4 produces an unnumbered list and Figure 21.4 shows the corresponding WWW page. Notice how the default bullets for first-level lists are discs, for second-level lists they are round and for third-level lists they are square.

HTML script 21.4

```
<HTML>
<HEAD>
<TITLE>Example list</TITLE>
</HEAD>
<H1> Introduction </H1>
<UL>
<LI> OSI Model
<LI> Networks
  <UL>
  <LI> Ethernet
```

```
      <UL>
      <LI> MAC addresses
      </UL>
    <LI> Token Ring
    <LI> FDDI
    </UL>
  <LI> Conclusion
  </UL>
  <H1> Wide Area Networks </H1>
  <UL>
  <LI> Standards
  <LI> Examples
    <UL>
    <LI> EastMan
    </UL>
  <LI> Conclusion
  </UL>
  </BODY>
  </HTML>
```

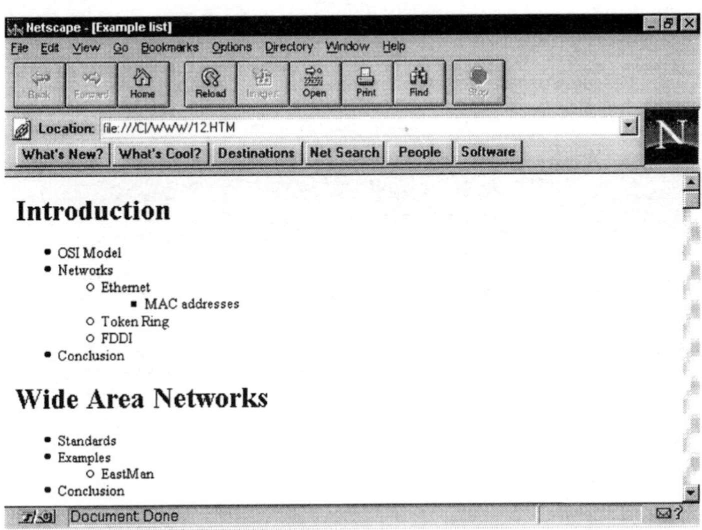

Figure 21.4 WWW page with an unnumbered list

21.3.3 Definition lists

HTML uses the <DL> and </DL> tags for definition lists. These are normally used when building glossaries. Each entry in the definition is defined by the <DT> tag and the text associated with the item is defined after the <DD> tag. The end of the list is defined by </DL>. HTML script 21.5 produces a definition list and Figure 21.5 gives the corresponding output. Note that it uses the <EM> tag to emphasize the definition subject.

HTML script 21.5

```
<HTML>
<HEAD>
<TITLE>Example list</TITLE>
</HEAD>
<H1> Glossary </H1>
<DL>
<DT> <EM> Address Resolution Protocol (ARP) </EM>
<DD> A TCP/IP process which maps an IP address to an
Ethernet address.
<DT> <EM> American National Standards Institute (ANSI)
</EM>
<DD> ANSI is a non-profit organization which is made up
of expert committees that publish standards for national
industries.
<DT> <EM> American Standard Code for Information
Interchange (ASCII) </EM>
<DD> An ANSI-defined character alphabet which has since
been adopted as a standard international alphabet for the
interchange of characters.
</DL>
</BODY>
</HTML>
```

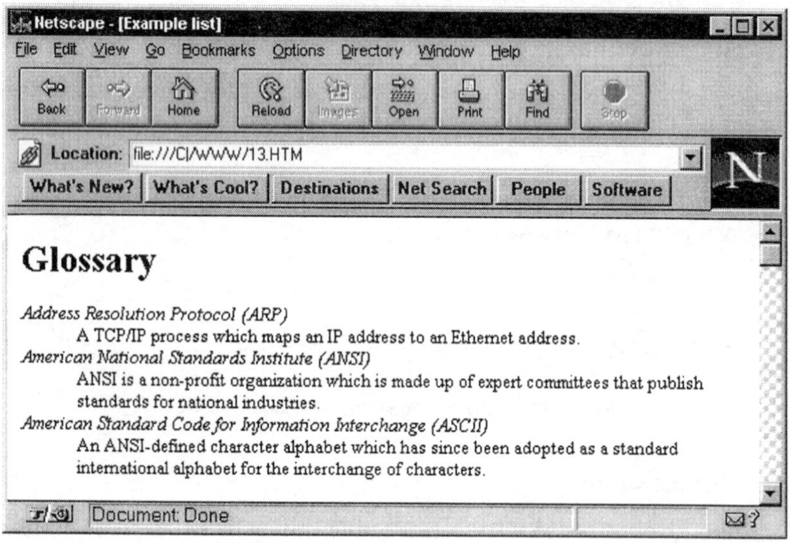

Figure 21.5 WWW page with definition list

21.4 Colors

Colors in HTML are defined in the RGB (red/green/blue) strength. The format is `#rrggbb`, where `rr` is the hexadecimal equivalent for the red component, `gg` the hexadecimal equivalent for the green component and `bb` the hexadecimal equivalent for the blue component. Table 21.2 lists some of the codes for certain colors.

Individual hexadecimal numbers use base 16 and range from `0` to `F` (in decimal this ranges from 0 to 15). A two-digit hexadecimal number ranges from `00` to `FF` (in decimal this ranges from 0 to 255). Table 21.3 outlines hexadecimal equivalents.

HTML uses percentage strengths for the colors. For example `FF` represents full strength (100%) and `00` represent no strength (0%). Thus, white is made from `FF` (red), `FF` (green) and `FF` (blue) and black is made from `00` (red), `00` (green) and `00` (blue). Gray is made from equal weightings of all three colors, such as `43, 43, 43` for dark gray (`#434343`) and `D4, D4, D4` for light gray (`#D4D4D4`). Thus, pure red will be `#FF0000`, pure green will be `#00FF00` and pure blue will be `#0000FF`.

Table 21.2 Hexadecimal colors

Color	*Code*	*Color*	*Code*
White	`#FFFFFF`	Dark red	`#C91F16`
Light red	`#DC640D`	Orange	`#F1A60A`
Yellow	`#FCE503`	Light green	`#BED20F`
Dark green	`#088343`	Light blue	`#009DBE`
Dark blue	`#0D3981`	Purple	`#3A0B59`
Pink	`#F3D7E3`	Nearly black	`#434343`
Dark gray	`#777777`	Gray	`#A7A7A7`
Light gray	`#D4D4D4`	Black	`#000000`

Each color is represented by 8 bits thus the color is defined by 24 bits This gives a total of 16777216 colors (2^{24} different colors). Note that some video displays will not have enough memory to display 16.777 million colors in certain modes so the colors may differ depending on the WWW browser and the graphics adapter.

The colors of the background, text and the link can be defined with the BODY tag. An example with a background color of white, a text color of orange and a link color of dark red is:

```
<BODY BGCOLOR="#FFFFFF" TEXT="#F1A60A"  LINK="#C91F16">
```

and for a background color of red, a text color of green and a link color of blue:

```
<BODY BGCOLOR="#FF0000" TEXT="#00FF00"  LINK="#0000FF">
```

When a link has been visited, its color changes. This color itself can be changed with VLINK. For example to set up a visited link color of yellow:

```
<BODY VLINK="#FCE503" "TEXT=#00FF00"  "LINK=#0000FF">
```

Note that the default link colors are:

Link:	`#0000FF`	(blue)
Visited link:	`#FF00FF`	(purple)

Table 21.3 Hexadecimal to decimal conversions

Hex	*Dec.*	*Hex*	*Dec.*	*Hex*	*Dec.*	*Hex*	*Dec.*
0	0	1	1	2	2	3	3
4	4	5	5	6	6	7	7
8	8	9	9	A	10	B	11
C	12	D	13	E	14	F	15

21.5 Background images

Images (such as GIF and JPEG) can be used as a background to a WWW page. For this purpose the option BACKGROUND="*src.gif*" is added to the <BODY> tag. Some text with a background of CLOUDS.GIF is produced HTML script 21.6. The corresponding output is shown in Figure 21.6.

📖 HTML script 21.6

```
<HTML>
<HEAD>
<TITLE>Fred's page</TITLE>
</HEAD>
<BODY BACKGROUND="clouds.gif">
<H1>Fred's Home Page</H1>
If you want to access information on
this book <A HREF="gbook.html">click here</A>.
<P>
A reference to the <A HREF="http://www.iee.com/">IEE</A>
</BODY>
</HTML>
```

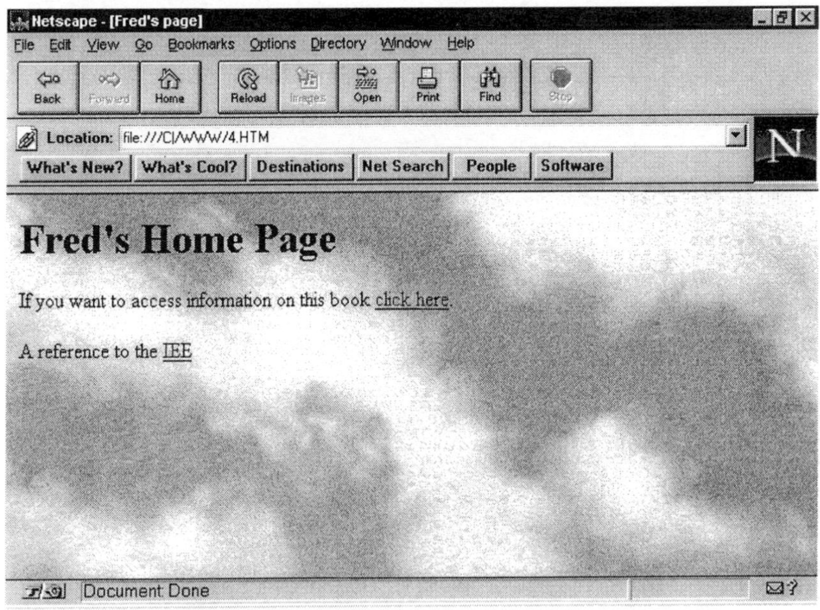

Figure 21.6 WWW page with CLOUD.GIF as a background

21.6 Displaying images

WWW pages can support graphics images within a page. The most common sources of images are either JPEG or GIF files as these types of images normally have a high degree of compression. GIF images, as previously mentioned, only support 256 colors from a pallet of 16.7 million colors, whereas JPEG images support more than 256 colors.

21.6.1 Inserting an image

Images can be displayed within a page using `<IMG SRC="`*src.gif*`">` which inserts the graphic *src.gif*. HTML script 21.7 displays three images: `myson.gif`, `me.gif` and `myson2.gif`. They are aligned either to the left or the right using the `ALIGN` option within the `<IMG SRC>` tag. The first image (`myson.gif`) is aligned to the right, whereas the second image (`me.gif`) is aligned to the left. Figure 21.7 shows the corresponding output. Note that images are aligned to the left by default.

🕮 HTML script 21.7

```
<HTML><HEAD>
<TITLE>My first home page</TITLE>
```

```
</HEAD>
<BODY BGCOLOR="#ffffff">
<IMG SRC ="myson.gif" ALIGN=RIGHT>
<H1> Picture gallery </H1>
<P><P>
Here are a few pictures of me and my family. To the right
is a picture of my youngest son showing his best smile.
Below to the left is a picture of me at Christmas and to
the right is a picture of me and my son also taken at
Christmas.
<P><P>
<IMG SRC ="me.gif" ALIGN=LEFT>
<IMG SRC ="myson2.gif" ALIGN=RIGHT>
</BODY></HTML>
```

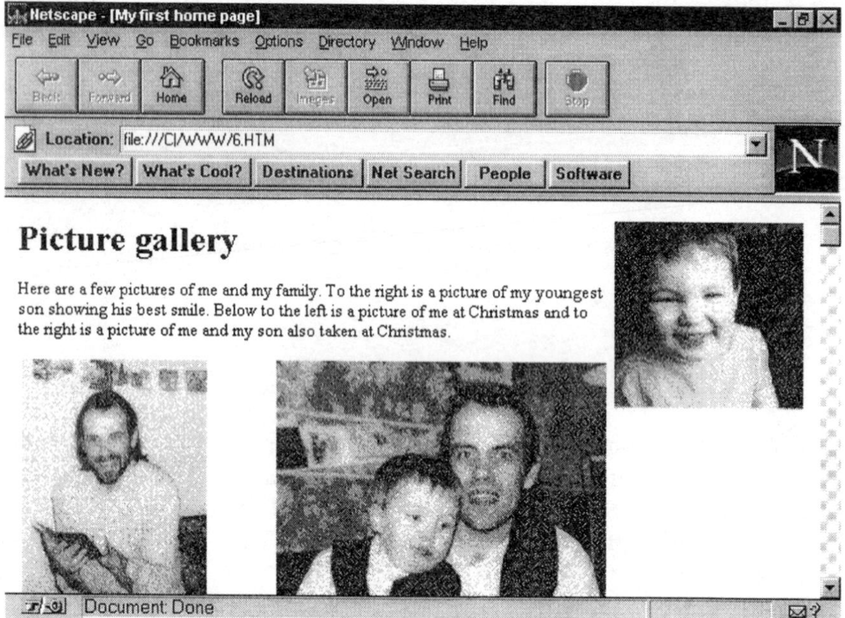

Figure 21.7 WWW page with three images

21.6.2 Alternative text

Often users choose not to view images in a page and select an option on the viewer which stops the viewer from displaying any graphic images. If this is the case then the HTML page can contain substitute text which is shown instead of the image. For example:

```
<IMG SRC ="myson.gif" ALT="Picture of my son" ALIGN=RIGHT>
<IMG SRC ="me.gif" ALT="Picture of me ALIGN=LEFT>
```

```
<IMG SRC ="myson2.gif" ALT="Another picture of my son"
ALIGN=RIGHT>
```

21.6.3 Other options

Other image options can be added, such as:

- `HSPACE=`*x* `VSPACE=`*y* defines the amount of space that should be left around images. The *x*-value defines the number of pixels in the *x*-direction and the *y*-value defines the number of pixels in the *y*-direction.
- `WIDTH=`*x* `HEIGHT=`*y* defines the scaling in the *x*- and *y*-directions, where *x* and *y* are respectively the desired pixel width and height, of the image.
- `ALIGN=`*direction* defines the alignment of the image. This can be used to align an image with text. Valid options for aligning with text are *texttop*, *top*, *middle*, *absmiddle*, *bottom*, *baseline* or *absbottom*. HTML script 21.8 aligns the image `a.gif` (which is just the letter 'A' as a graphic) and Figure 21.8 shows the corresponding output. It can be seen that *texttop* aligns the image with highest part of the text on the line, *top* aligns the image with the highest element in the line, *middle* aligns with the middle of the image the baseline, *absmiddle* aligns the middle of the image with the middle of the largest item, *bottom* aligns the bottom of the image with the bottom of the text and *absbottom* aligns the bottom of the image with the bottom of the largest item.

📖 HTML script 21.8

```
<HTML> <HEAD>
<TITLE>My first home page</TITLE>
</HEAD>
<BODY BGCOLOR="#ffffff">
<IMG SRC ="a.gif" ALIGN=texttop>pple<P>
<IMG SRC ="a.gif" ALIGN=top>pple<P>
<IMG SRC ="a.gif" ALIGN=middle>pple<P>
<IMG SRC ="a.gif" ALIGN=bottom>pple<P>
<IMG SRC ="a.gif" ALIGN=baseline>pple<P>
<IMG SRC ="a.gif" ALIGN=absbottom>pple
</BODY> </HTML>
```

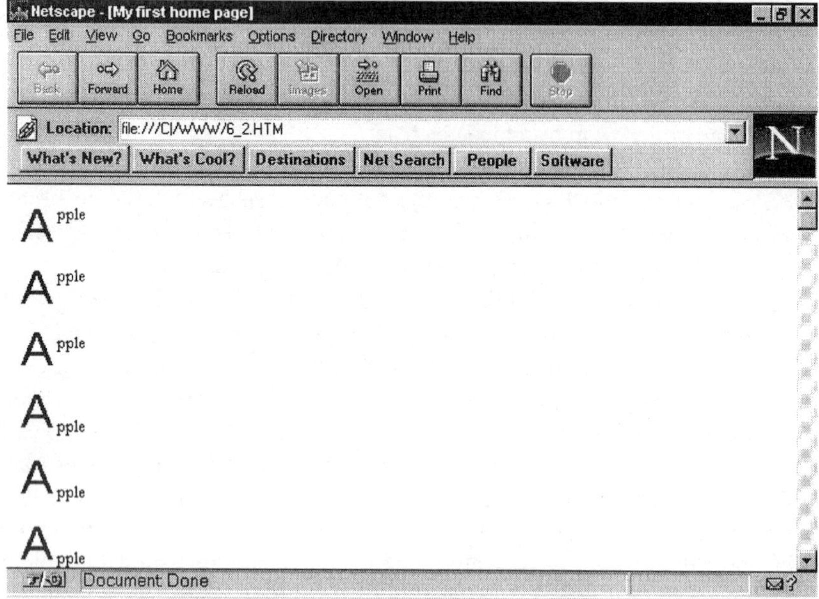

Figure 21.8 WWW page showing image alignment

21.7 Horizontal lines

A horizontal line can be added with the <HR> tag. Most browsers allow extra parameters, such as:

SIZE=*n* – which defines that the height of the rule is *n* pixels.
WIDTH=*w* – which defines that the width of the rule is *w* pixels; it can also be expressed as a percentage.
ALIGN=*direction* – where direction refers to the alignment of the rule. Valid options for *direction* are *left*, *right* or *center*.
NOSHADE – which defines that the line should be solid with no shading.

HTML script 21.9 produces a variety of horizontal lines and Figure 21.9 is the corresponding output.

📖 HTML script 21.9

```
<HTML>
<HEAD>
<TITLE>My first home page</TITLE>
```

```
</HEAD>
<BODY BGCOLOR="#ffffff">
<IMG SRC ="a.gif">pple<P>
<HR>
<IMG SRC ="a.gif">pple<P>
<HR WIDTH=50% ALIGN=CENTER>
<IMG SRC ="a.gif">pple<P>
<HR SIZE=10 NOSHADE>
</BODY></HTML>
```

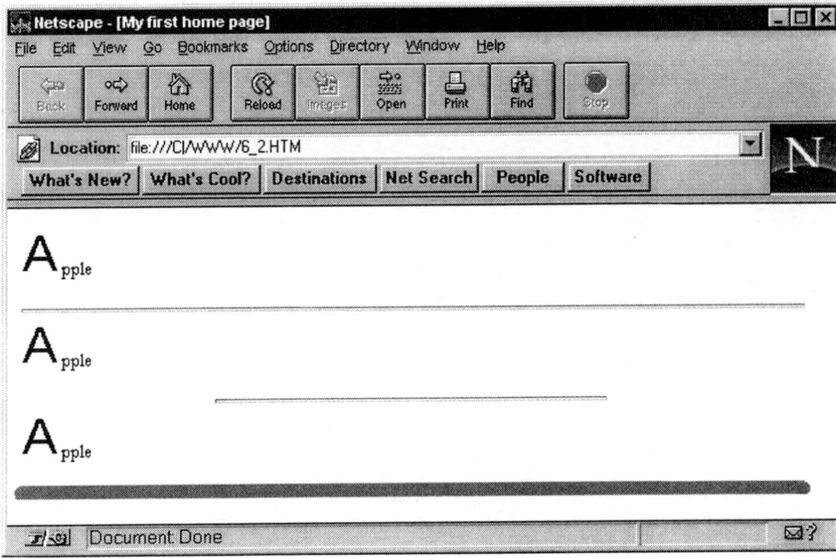

Figure 21.9 WWW page showing horizontal lines

21.8 Exercises

21.1 The home page for this book can be found at the URL:

`http://www.eece.napier.ac.uk/~bill_b/adcbook.hmtl`

Access this page and follow any links it contains.

21.2 If possible, create a WWW page with the following blinking text:
`This is some blinking text`

21.3 The last part of the server name normally gives an indication of the country in which the server is located (e.g. www.fredco.co.uk is located in the UK). Determine which countries use the following country names:

(a) de (b) nl (c) it (d) se (e) dk (f) sg
(g) ca (h) ch (i) tr (j) jp (k) au

Determine some other country identifier names.

21.4 Determine the HTML representation for the following colors:

(a) red
(b) green
(c) blue
(d) white
(e) black

21.5 Determine what is represented by the following HTML expressions for the background, text and link color

(a) <BODY BKCOLOR="#00FF00" "TEXT=#FF0000" "LINK=#0000FF">
(b) <BODY BKCOLOR="#DC640D" "TEXT=#777777" "LINK=#009DBE">

21.6 Determine the error in HTML script 21.10:

HTML script 21.10

```
<HTML> <HEAD>
<TITLE>Fred's page</TITLE>  </HEAD>
<BODY BGCOLOR="#FFFFFF">
<H1>List 1</H1>
<OL>
<LI>Part 1
<LI>Part 2
<LI>Part 3
<H1>List 2</H1>
<UL>
<LI>Section 1
<LI>Section 2
<LI>Section 3
</UL>
</BODY>
</HTML>
```

21.7 Create the following list with HTML:

- Compression
  - Techniques
    - Huffman/LZ
    - JPEG
  - Real Time Data
- Encryption
  - Principles
  - Techniques
    - IDEA
    - DES
- Electronic Mail
  - Principles
  - SMTP
  - MIME
- WWW
  - HTML
  - Java/JavaScript
  - HTTP
- Networks
  - Ethernet
  - Token Ring
  - FDDI
  - ATM

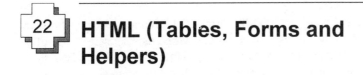

22 HTML (Tables, Forms and Helpers)

22.1 Introduction

Chapter 21 introduced HTML, this chapter discusses some of its more advanced features. HTML differs from compiled languages, such as C and Pascal, in that the HTML text file is interpreted by an interpreter (the browser) whereas languages such as C and Pascal must be precompiled before they can be run. HTML thus has the advantage that it is not affected by the choice of operating system, browser or computer used to read the HTML file, as the file does not contain any computer-specific code. The main disadvantage of interpreted files is that the interpreter does less error checking as it must produce fast results.

The basic pages on the WWW are likely to evolve around HTML, and while HTML can be produced manually with a text editor, it is likely that, during the coming years there will be an increase in the amount of graphics-based tools which will automatically produce HTML files. Although these tools are graphics-based they still produce standard HTML text files. Thus a knowledge of HTML is important as it defines the basic specification for the presentation of WWW pages.

22.2 Anchors

An anchor allows users to jump from a reference in a WWW page to an anchor point within the page. The standard format is:

<A NAME="*anchor name*">

where *anchor name* is the name of the section which is referenced. The </A> tag defines the end of an anchor name. A link is specified by:

<A HREF="#*anchor name*">

followed by the </A> tag. HTML script 22.1 produces four anchors and Figure 22.1 shows the corresponding output. When the user selects one of the references, the browser automatically jumps to that anchor. Figure 22.2 shows the output screen when the user selects the #Token reference. Anchors are typically used when an HTML page is long or when a backwards or forwards reference occurs (such as a reference within a published paper).

HTML script 22.1

```
<HTML><HEAD><TITLE>Sample page</TITLE></HEAD>
<BODY BGCOLOR="#FFFFFF">
<H2>Select which network technology you wish information:</H2>
<P><A HREF="#Ethernet">Ethernet</A></P>
<P><A HREF="#Token">Token Ring</A></P>
<P><A HREF="#FDDI">FDDI</A></P>
<P><A HREF="#ATM">ATM</A></P>

<H2><A NAME="Ethernet">Ethernet</A></H2>
Ethernet is a popular LAN which works at 10Mbps.

<H2><A NAME="Token">Token Ring</A></H2>
Token ring is a ring based network which operates
at 4 or 16Mbps.

<H2><A NAME="FDDI">FDDI</A></H2>
FDDI is a popular LAN technology which uses a ring of
fibre optic cable and operates at 100Mbps.

<H2><A NAME="ATM">ATM</A></H2>
ATM is a ring based network which operates at 155Mbps.

</BODY>
</HTML>
```

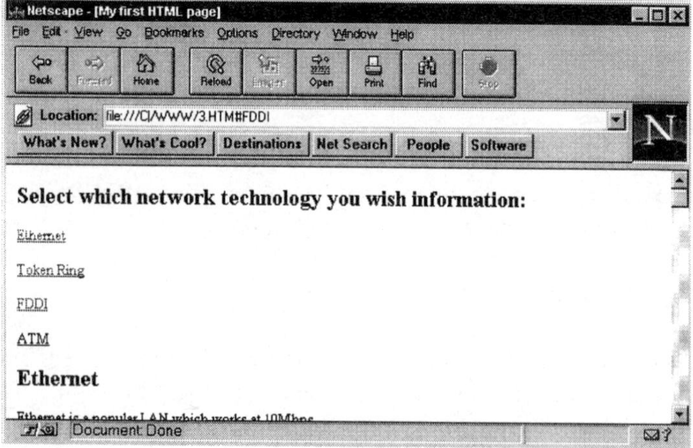

Figure 22.1 A window with references

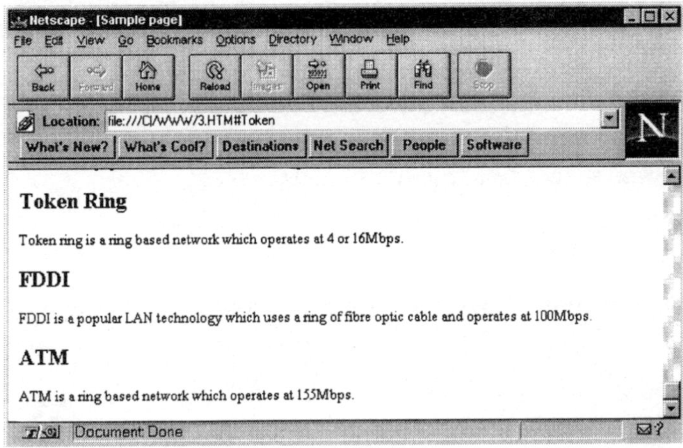

Figure 22.2 The result of choosing 'Token Ring' in Figure 22.2

22.3 Tables

A table is one of the best methods to display complex information in a simple way. Unfortunately, in HTML, tables are relatively complicated to set up. The start of a table is defined with the <TABLE> tag and the end of a table by </TABLE>. A row is defined between <TR> and </TR>, while a table header is defined between <TH> and </TH>. A regular table entry is defined between <TD> and </TD>. HTML script 22.2 produces a table with links to other HTML pages. The BORDER=*n* option has been added to the <TABLE> tag to define the thickness of the table border (in pixels). In this case the border size has a thickness of 10 pixels.

📖 HTML script 22.2

```
<HTML><HEAD><TITLE> Fred Bloggs</TITLE></HEAD>
<BODY TEXT="#000000" BGCOLOR="#FFFFFF">
<H1>Fred Bloggs Home Page</H1>
I'm Fred Bloggs. Below is a tables of links.
<HR><P>
<TABLE BORDER=10>
<TR>
   <TD><B>General</B></TD>
   <TD><A HREF="res.html">Research</TD>
   <TD><A HREF="cv.html">CV</TD>
   <TD><A HREF="paper.html">Papers Published</TD>
</TR>
<TR>
   <TD><B>HTML Tutorials</B></TD>
   <TD><A HREF="intro.html">Tutorial 1</TD>
```

```
    <TD><A HREF="inter.html">Tutorial 2</TD>
    <TD><A HREF="adv.html">Tutorial 3</TD>
</TR>
<TR>
    <TD><B>Java Tutorials</B></TD>
    <TD><A HREF="java1.html">Tutorial 1</TD>
    <TD><A HREF="java2.html">Tutorial 2</TD>
    <TD><A HREF="java3.html">Tutorial 3</TD>
</TR>
</TABLE>
</BODY></HTML>
```

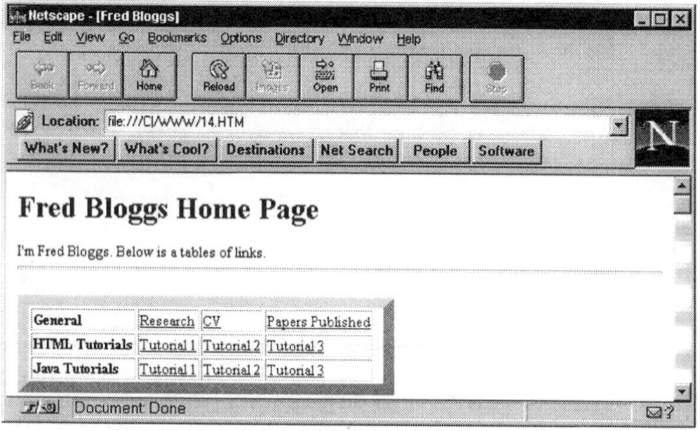

Figure 22.3 A window with a table

Other options in the <TABLE> tag are:

- `WIDTH=`*x*, `HEIGHT=`*y* – defines the size of the table with respect to the full window size. The parameters *x* and *y* are either absolute values in pixels for the height and width of the table or are percentages of the full window size.
- `CELLSPACING=`*n* – defines the number of pixels desired between each cell where *n* is the number of pixels (note that the default cell spacing is 2 pixels).

An individual cell can be modified by adding options to the <TH> or <TD> tag. These include:

- `WIDTH=`*x*, `HEIGHT=`*y* – defines the size of the table with respect to the table size. The parameters *x* and *y* are either absolute values in pixels for the height and width of the table or are percentages of the table size.
- `COLSPAN=`*n* – defines the number of columns the cell should span.
- `ROWSPAN=`*n* – defines the number of rows the cell should span.

- ALIGN=*direction* – defines how the cell contents are aligned horizontally. Valid options are *left*, *center* or *right*.
- VALIGN=*direction* – defines how the cells contents are aligned vertically. Valid options are *top*, *middle* or *baseline*.
- NOWRAP – informs the browser to keep the text on a single line (i.e. with no line breaks).

HTML script 22.3 produces some of the <TABLE> and <TD> options. In this case the text within each row is center aligned. On the second row, the second and third cells are merged using the COLSPAN=2 option. The first cell of the second and third rows has also been merged using the ROWSPAN=2 option. Figure 22.4 show the corresponding output. The table width has been increased to 90% of the full window, with a width of 50%.

📖 HTML script 22.3

```
<HTML><HEAD><TITLE> Fred Bloggs</TITLE></HEAD>
<BODY TEXT="#000000" BGCOLOR="#FFFFFF">
<H1>Fred Bloggs Home Page</H1>
I'm Fred Bloggs. Below is a tables of links.
<HR><P>
<TABLE BORDER=10 WIDTH=90% LENGTH=50%>
<TR>
   <TD><B>General</B></TD>
   <TD><A HREF="res.html">Research</TD>
   <TD><A HREF="cv.html">CV</TD>
   <TD><A HREF="paper.html">Papers Published</TD>
   <TD></TD>
</TR>
<TR>
   <TD ROWSPAN=2><B>HTML/Java Tutorials</B></TD>
   <TD><A HREF="intro.html">Tutorial 1</TD>
   <TD COLSPAN=2><A HREF="inter.html">Tutorial 2</TD>
</TR>
<TR>
   <TD><A HREF="java1.html">Tutorial 1</TD>
   <TD><A HREF="java2.html">Tutorial 2</TD>
   <TD><A HREF="java3.html">Tutorial 3</TD>
</TR>
</TABLE>
</BODY></HTML>
```

22.4 CGI scripts

CGI (Common Gateway Interface) scripts are normally written in either C or Perl and are compiled to produce an executable program. They can also come precompiled or in the form of a batch file. Perl has the advantage of being easily run on any computer, whereas a C program requires needs to be compiled for the server computer.

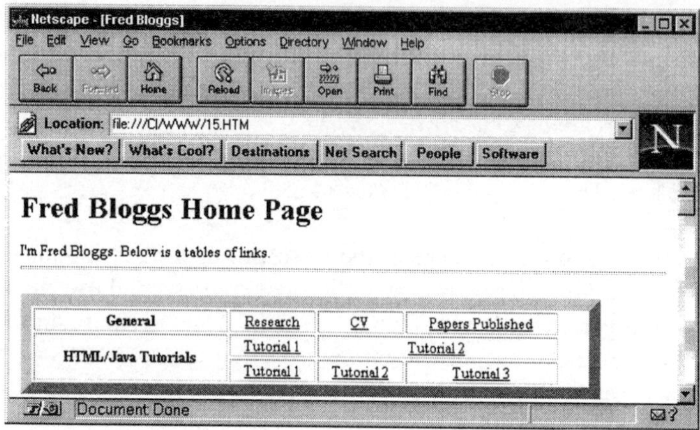

Figure 22.4 Alignments within a table created using HTML script 22.3

CGI scripts allow the user to interact with the server and to store and request data. They are often used in conjunction with forms and allow an HTML document to analyze, parse and store information received from a form. On most UNIX-type systems the default directory for CGI scripts is `cgi-bin`.

22.5 Forms

Forms are excellent methods of gathering data and can be used in conjunction with CGI scripts to collect data for future use.

A form is identified between the `<FORM>` and `</FORM>` tags. The method used to get the data from the form is defined using `METHOD="POST"`. The `ACTION` option defines the URL script to be run when the form is submitted. Data input is specified by the `<INPUT TYPE>` tag. HTML script 22.4 has the following parts:

- `<form action="/cgi-bin/AnyForm2" method="POST">` – which defines the start of a form, and when the `"submit"` option is selected the cgi script `/cgi-bin/AnyForm2` will be automatically run.
- `<input type="submit" value="Send Feedback">` – which causes the program defined in the action option in the `<form>` tag to be run. The button on the form will contain the text `"Send Feedback"`; Figure 22.5 shows the corresponding output screen.
- `<input type="reset" value="Reset Form">` – which resets the data in the form. The button on the form will contain the text `"Reset Form"`; Figure 22.5 shows the corresponding output screen.

- <input type="hidden" name="AnyFormTo" value= "f.bloggs @toytown.ac.uk"> – which passes a value of f.bloggs@toytown.ac.uk, having the parameter name of "AnyFormTo". The program AnyForm2 takes this parameter and automatically sends it to the email address defined in the value (i.e. f.bloggs @toytown.ac.uk).
- <input type="hidden" name="AnyFormSubject" value= "Feedback form"> – which passes a value of "Feedback form" having the parameter name "AnyFormSubject". The program AnyForm2 takes this parameter and adds the text "Feedback form" in the text sent to the email recipient (in this case, f.bloggs @toytown.ac.uk).
- Surname <input name="Surname"> – which defines a text input and assigns this input to the parameter name "Surname".
- <textarea name="Address" rows=2 cols=40> </textarea> – which defines a text input area having two rows and a width of 40 characters. The thumb bars appear at the right-hand side of the form if the text area exceeds more than two rows see Figure 22.5.

📖 HTML script 22.4

```
<HTML>
<HEAD>
<TITLE>Example form</TITLE>
</HEAD>
<H1><CENTER>Example form</CENTER></H1><P>
<form action="/cgi-bin/AnyForm2" method="POST">

<input type="hidden" name="AnyFormTo"
value="f.bloggs@toytown.ac.uk">
<input type="hidden" name="AnyFormSubject"
value="Feedback form">

Surname <input name="Surname"> First Name/Names <input
name="First Name"><P>

Address (including country)<P>
<textarea name="Address" rows=2 cols=40></textarea><P>

Business Phone <input name="Business Phone">Place of study (or
company) <input name="Study"><P>
E-mail    <input name="E-mail">  Fax Number <input name="Fax
Number"><P>

<input type="submit" value="Send Feedback"> <input type="reset"
value="Reset Form">
</Form>
</HTML>
```

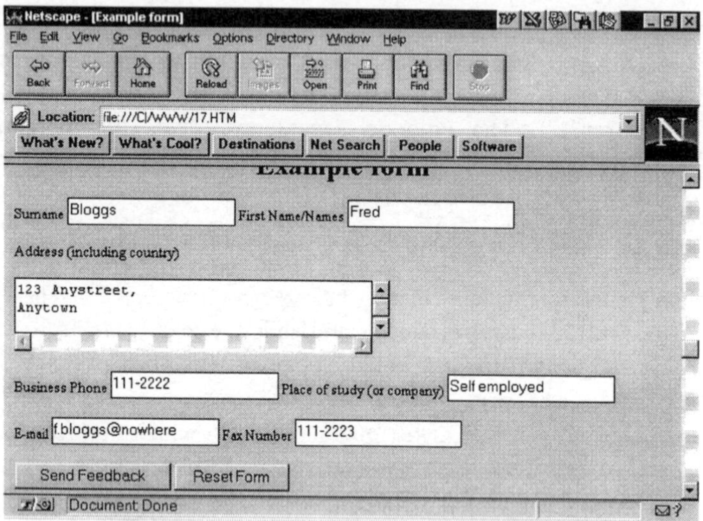

Figure 22.5 A form created using HTML script 22.4

In this case the recipient (`f.bloggs@toytown.ac.uk`) will receive an email with the contents:

```
Anyform Subject=Example form
Surname=Bloggs
First name=Fred
Address=123 Anystreet, Anytown
Business Phone=111-222
Place of study (or company)=Self employed
Email= f.bloggs@nowhere
Fax Number=111-2223
```

The extra options to the `<input>` tag are `size="`*n*`"`, where *n* is the width of the input box in characters, and `maxlength="`*m*`"`, where *m* is the maximum number of characters that can be entered, in characters. For example:

```
<input type="text size="15" maxlength="10">
```

indicates that the input type is text, the width of the box is 15 characters and the maximum length of input is 10 characters.

22.5.1 Input types

The type options to the `<input>` tag are defined in Table 22.1. HTML script 22.5 uses a few examples of input types and Figure 22.6 shows the corresponding output.

Table 22.1 Input type options

TYPE=	*Description*	*Options*
"text"	The input is normal text.	NAME="*nm*" where *nm* is the name that will be sent to the server when the text is entered. SIZE="*n*" where *n* is the desired box width in characters. SIZE="*m*" where *m* is the maximum number of input characters.
"password"	The input is a password which will be displayed with *'s. For example if the user inputs a 4-letter password then only **** will be displayed.	SIZE="*n*" where *n* is the desired box width in characters. SIZE="*m*" where *m* is the maximum number of input characters.
"radio"	The input takes the form of a radio button (such as ⊙ or O). It used to allow the user to select a single option from a list of options.	NAME="*radname*" where *radname* defines the name of the button. VALUE="*val*" where *val* is the data that will be sent to the server when the button is selected. CHECKED is used to specify that the button is initially set.
"checkbox"	The input takes the form of a checkbox (such as ☒ or ☐). It is used to allow the user to select several options from a list of options.	NAME="*chkname*" where *chkname* defines the common name for all the checkbox options. VALUE="*defval*" where *defval* defines the name of the option. CHECKED is used to specify that the button is initially set.

🕮 HTML script 22.5

```
<HTML><HEAD><TITLE>Example form</TITLE></HEAD>
<FORM METHOD="Post" >
<H2>Enter type of network:</H2><P>
<INPUT TYPE="radio" NAME="network" VALUE="ethernet"CHECKED>Ethernet
<INPUT TYPE="radio" NAME="network" VALUE="token">Token Ring
<INPUT TYPE="radio" NAME="network" VALUE="fddi">FDDI
<INPUT TYPE="radio" NAME="network" VALUE="atm">ATM
<H2>Enter usage:</H2><P>
<INPUT TYPE="checkbox" NAME="usage" VALUE="multi">Multimedia
<INPUT TYPE="checkbox" NAME="usage" VALUE="word">Word Processing
<INPUT TYPE="checkbox" NAME="usage" VALUE="spread">Spread Sheets
<P>Enter Password<INPUT TYPE="password" NAME="passwd" SIZE="10">
</FORM></HTML>
```

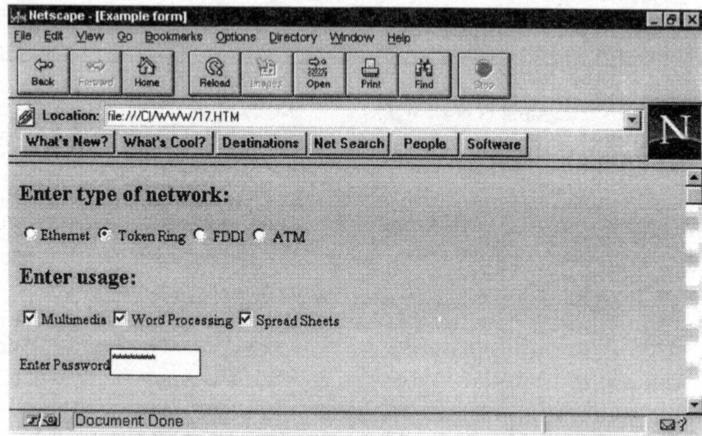

Figure 22.6 A window with different input options created using HTML script 22.5

22.5.2 Menus

Menus are a convenient method of selecting from multiple options. The `<SELECT>` tag is used to define the start of the menu options and the `</SELECT>` tag defines the end. Menu elements are then defined with the `<OPTION>` tag. The options defined within `<SELECT>` are:

- `NAME="`*name*`"` – which defines that *name* is the variable name of the menu. This is used when the data is collected by the server.
- `SIZE="`*n*`"` – which defines the number of options displayed in the menu.

HTML script 22.6 produces a menu. The additional options for the `<OPTION>` tag are:

- `SELECTED` – which defines the default selected option.
- `VALUE="`*val*`"` – where *val* defines the name of the data when it is collected by the server.

📖 HTML script 22.6

```
<HTML><HEAD><TITLE>Example form</TITLE> </HEAD>
<FORM METHOD="Post" >
Enter type of network:
<select Name="network" size="1">
<option>Ethernet
<option SELECTED>Token Ring
<option>FDDI
<option>ATM
</select></FORM></HTML>
```

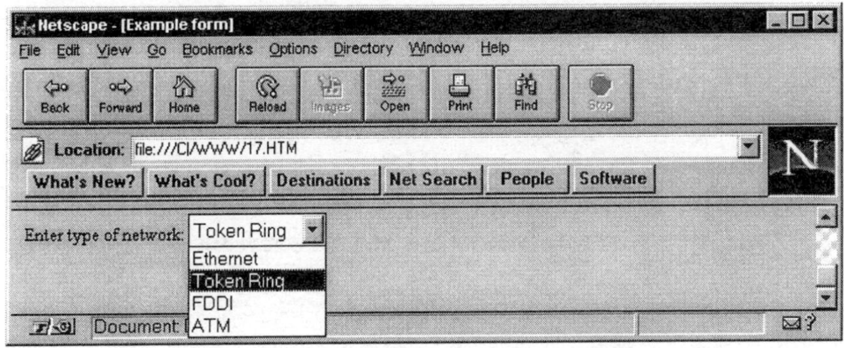

Figure 22.7 A menu created using HTML script 22.6

22.6 Multimedia

If the browser cannot handle all of the file types it may call on other application helpers to process the file. This allows other third-party programs to integrate into the browser. Figure 22.8 shows one configuration of the helper programs. The options in this case are:

- View in browser.
- Save to disk.
- Unknown: prompt user.
- Launch an application (e.g. audio playback program or MPEG viewer).

For certain applications the user can select whether the file is processed by the browser or by another application program. Helper programs make upgrades in helper applications relatively simple and they also allow new file types to be added with an application helper. Typically, when a program is installed that can be used with a browser, it will prompt the user to indicate whether it should automatically update the helper application list so it can handle the given file type(s).

Each file type is defined by the file extension, such as .ps for postscript files, .exe for a binary executable file, and so on. These file extensions have been standardized to meet the MIME (Multipurpose Internet Mail Extensions) specification. Table 22.2 shows some typical file extensions.

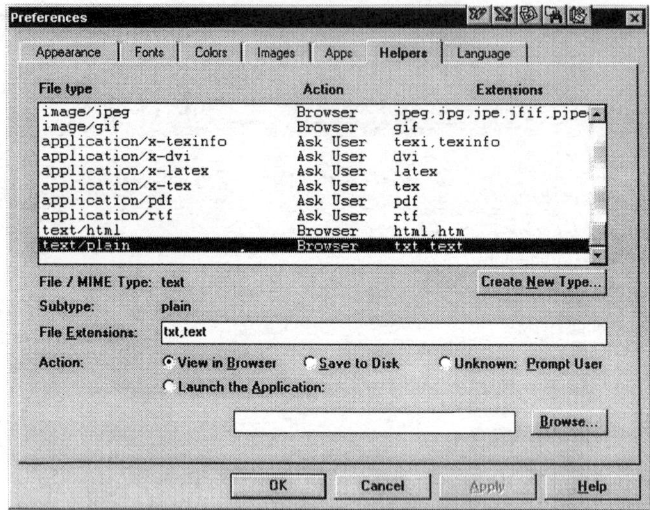

Figure 22.8 One configuration of help programs

Table 22.2 Input type options

Mime type	*Extension*	*Typical action*
application/octet-stream	exe, bin	Save
application/postscript	ps, ai, eps	Ask user
application/x-compress	Z	Compress program
application/x-gzip	gz	GZIP compress program
application/x-javascript	js, mocha	Ask user
application/x-msvideo	avi	Audio player
application/x-perl	pl	Save
application/x-tar	tar	Save
application/x-zip-compressed	zip	ZIP program
audio/basic	au, snd	Audio player
image/gif	gif	Browser
image/jpeg	jpeg, jpg, jpe	Browser
image/tiff	tif, tiff	Graphics viewer
image/x-MS-bmp	bmp	Graphics viewer
text/html	htm, html	Browser
text/plain	text, txt	Browser
video/mpeg	mpeg, mpg, mpe, mpv, vbs, mpegv	Video player
video/quicktime	qt, mov, moov	Video player

22.7 Exercises

22.1 Construct a WWW page with anchor points for the following text:

Select the network you wish to find out about:

Ethernet
Token ring
FDDI

Ethernet

Ethernet is the most widely used networking technology used in LAN (Local Area Network). In itself it cannot make a network and needs some other protocol such as TCP/IP or SPX/IPX to allow nodes to communicate. Unfortunately, Ethernet in its standard form does not cope well with heavy traffic. It has many advantages though, including:

- Networks are easy to plan and cheap to install.
- Network components are cheap and well supported.
- It is well-proven technology which is fairly robust and reliable.
- Simple to add and delete computers on the network.
- Supported by most software and hardware systems.

Token Ring

Token Ring networks were developed by several manufacturers, the most prevalent being the IBM Token Ring. Token Ring networks cope well with high network traffic loadings. They were at one time extremely popular but their popularity has since been overtaken by Ethernet. Token Ring networks have, in the past, suffered from network management problems and poor network fault tolerance. Token Ring networks are well suited to situations which have large amounts of traffic and work well with most traffic loadings. They are not suited to large networks or networks with physically remote stations. Their main advantage is that they cope better with high traffic rates than Ethernet, but they require a great deal of maintenance especially when faults occur or when new equipment is added or removed around the network. Many of these problems have now been overcome by MAUs (multistation access units), which are similar to the hubs used in Ethernet.

FDDI

A token-passing mechanism allows orderly access to a network. Apart from Token Ring the most commonly used token-passing network is the Fiber Distributed Data Interchange (FDDI) standard. This operates at 100 Mbps and, to overcome the problems of line breaks, has two concentric Token Rings. Fiber-optic cables have a much high specification than copper cables and allow extremely long connection. The maximum circumference of the ring is 100 km (62 miles), with a maximum 2 km between stations (in FDDI, nodes are also known as stations). It is thus an excellent mechanism for connecting networks across a city or over a campus. Up to 500 stations can connect to each ring with a maximum of 1000 stations for the complete network. Each station connected to the FDDI highway can be a normal station or a bridge to a conventional local area network, such as Ethernet or Token Ring.

22.2 Construct a WWW glossary page with the following terms:

Address	A unique label for the location of data or the identity of a communications device.
Address Resolution Protocol (ARP)	A TCP/IP process which maps an IP address to an Ethernet address.
American National Standards Institute (ANSI)	ANSI is a non-profit organization which is made up of expert committees that publish standards for national industries.
American Standard Code for Information Interchange (ASCII)	An ANSI-defined character alphabet which has since been adopted as a standard international alphabet for the interchange of characters.
Amplitude modulation (AM)	Information is contained in the amplitude of a carrier.
Amplitude-shift keying (ASK)	Uses two or more amplitudes to represent binary digits. Typically used to transmit binary data over speech-limited channels.
Application layer	The highest layer of the OSI model.
Asynchronous transmission	Transmission where individual characters are sent one-by-one. Normally each character is delimited by a start and stop bit. With asynchronous communication the transmitter and receiver only have to be roughly synchronized.

22.3 Construct a WWW page which can be used to enter a person's CV (use a form). The basic fields should be:

Name:
Address:
Email address:
Telephone number:
Experience:
Interests:
Any other information:

22.4 Write an HTML script to produce the following timetable.

<table>
<tr><td></td><td>9–11</td><td>11–1</td><td>1–3</td><td>3–5</td></tr>
<tr><td>Monday</td><td>Data Comms</td><td colspan="3">Networking</td></tr>
<tr><td>Tuesday</td><td colspan="2">Software Systems</td><td>Networking</td><td>Data Comms</td></tr>
<tr><td>Wednesday</td><td>Networking</td><td>FREE</td><td>Java</td><td>FREE</td></tr>
<tr><td>Thursday</td><td>Software Systems</td><td>C++</td><td>Networking</td><td>FREE</td></tr>
<tr><td>Friday</td><td>FREE</td><td colspan="3">Networking</td></tr>
</table>

22.5 Design your own home page with a basic user home page (`index.html`) which contains links to a basic CV page (perhaps it could be named `cv.html`) and a page which lists your main interests (`myinter.html`). Design one of the home pages with a list of links and another with a table of links. If possible incorporate graphics files into the pages.

23 Java/JavaScript

23.1 Introduction

Computer systems contain a microprocessor which controls the operation of the computer. The microprocessor only understands binary information and operates on a series of binary commands known as machine code. It is extremely difficult to write large programs in machine code, so that high-level languages are used instead. A low-level language is one which is similar to machine code and normally involves the usage of keyword macros to replace machine code instructions. High-level languages have a syntax that is almost like written English and thus makes programs easy to read and to modify. In most programs the actual operation of the hardware is invisible to the programmer. A compiler changes the high-level language into machine code. Typical high-level languages include C/C++, BASIC, COBOL, FORTRAN and Pascal; an example of a low-level language is 80486 Assembly Language.

Java is a high-level language which has been developed specifically for the WWW and is well suited to networked applications. It was originally developed by Sun Microsystems and is based on C++ (but with less of the difficulties of C++). Most new versions of Web browsers now support its usage. Java's main attributes are:

- It runs either as a standalone program or it can run within the Web browser. When run within the browser, the Java program is known as an applet.
- Java is a portable language and applets can run on any type of microprocessor (such as a PC based on the Intel 80486 or Pentium, or a Motorola-based computer).
- Java applets are independent of hardware and operating system. For example the program itself does not have to interface directly to the hardware such as a video adapter, or mouse. Typical high-level languages, such as C/C++ and Pascal, produce machine-dependent machine code, and can thus only be executed on a specific computer or operating system.
- Java allows for a client/server approach, where the applet can run on the remote computer, which thus reduces the loading on the local computer (typically the remote computer will be a powerful multitasking computer with enhanced computer architecture).

* A Java compiler creates standalone programs or applets. Many new versions of browsers have an integrated Java compiler.

Figure 23.1 shows the main functional differences between a high-level language, a Java applet and JavaScript. JavaScript is interpreted by the browser, whereas a Java applet is compiled to a virtual machine code which can be run on any computer system. The high-level language produces machine-specific code.

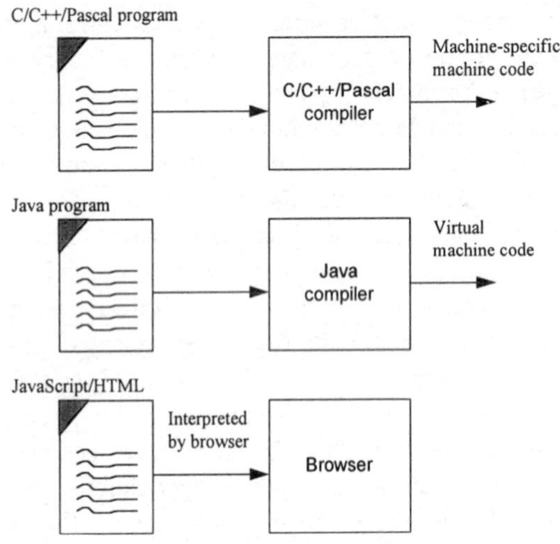

Figure 23.1 Differences between C++/Java and JavaScript

A normal C++ program allows access to hard-disk drives. This would be a problem on the Web as unsolicited users (hackers) or novice users could cause damage to the Web server. To overcome this, Java does not have any mechanism for file input/output (I/O). It can read standard file types (such as GIF and JPG) but cannot store changes to the Web server. A Java developers kit is available, free of charge, from `http://java.sun.com`.

The following HTML script contains JavaScript in boldface. Figure 23.2 gives the browser output.

JavaScript 23.1

```
<HTML> <HEAD><TITLE>My Java</TITLE></HEAD>
<BODY>
<SCRIPT language="javascript">
document.writeln("This is my first JavaScript");
for (i=0;i<10;i++)
 document.write("<center><font size=+1><b>Loop</b> ",i);
</SCRIPT></BODY></HTML>
```

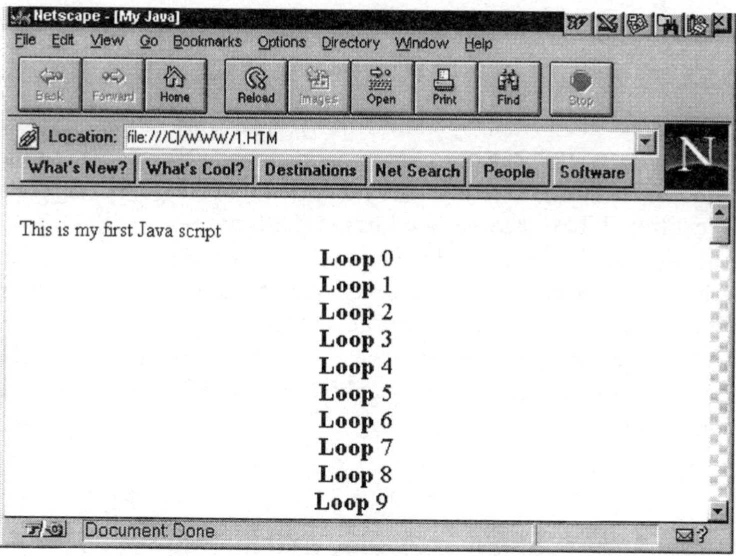

Figure 23.2 Browser output produced by JavaScript 23.1

23.2 JavaScript

Programming languages can either be compiled to produce an executable program or they can be interpreted while the user runs the program. Java is a program language which needs to be compiled before it is used. It thus cannot be used unless the user has the required Java compiler. JavaScript, on the other hand, is a language which is interpreted by the browser. It is similar in many ways to Java but allows the user to embed Java-like code into an HTML page. JavaScript supports a small number of data types representing numeric, Boolean and string values, and is supported by most modern WWW browsers, such as Microsoft Internet Explorer and Netscape Navigator.

HTML is useful when pages are short and do not contain expressions, loops or decisions. JavaScript allows most of the functionality of a high-level language for developing client and server Internet applications. It can be used to respond to user events such as mouse clicks, form input and page navigation.

A major advantage of JavaScript over HTML is that it supports the use of functions without any special declarative requirements. It is also simpler to use than Java because it has easier syntax, specialized built-in functionality and minimal requirements for object creation.

Important concepts in Java and JavaScript are objects. Objects are basically containers for values. The main differences between JavaScript and Java are:

- JavaScript is interpreted by the client, whereas Java is compiled on the server before it is executed.
- JavaScript is embedded into HTML pages, whereas Java applets are distinct from HTML and accessed from HTML pages.
- JavaScript has loose typing for variables (i.e. a variable's data type does not have to be declared), whereas Java has strong typing (i.e. a variable's data type must always declared before it is used).
- JavaScript has dynamic binding where object references are checked at run-time. Java has static binding where object references must exist at compile-time.

23.3 JavaScript values, variables and literals

JavaScript values, variable and literals are similar to the C programming language. Their syntax is discussed in this section.

23.3.1 Values

The four different types of values in JavaScript are:

- Numeric value, such as 12 or 91.5432.
- Boolean values which are either TRUE or FALSE.
- Strings types, such as "Fred Bloggs".
- A special keyword for a NULL value

Numeric values differ from most programming languages in that there is no explicit distinction between a real value (such as 91.5432) and an integer (such as, 12).

23.3.2 Data type conversion

JavaScript differs from Java in that variables do not need to have their data type defined when they are declared (loosely typed). Data types are then automatically converted during the execution of the program. Thus a variable could be declared with a numeric value as:

```
var value

  value = 19
```

and then in the same script it could be assigned a string value, such as:

```
value = "Enter your name >>"
```

The conversion between numeric values and strings in JavaScript is easy, as numeric values are automatically converted to an equivalent string. For example:

```
<HTML><HEAD><TITLE>My Java</TITLE></HEAD>
<BODY BGCOLOR="#ffffff">
<SCRIPT language="javascript">
var x,y,str

        x=13
        y=10
        str= x + " added to " + y + " is " + x+y
        document.writeln(str)

        z=x+y
        str= x + " added to " + y + " is " + z
        document.writeln(str)
</SCRIPT>
</BODY></HTML>
```

Sample run 23.1 gives the output from this script. It can be seen that `x` and `y` have been converted to a string value (in this case, `"13"` and `"10"`) and that `x+y` in the string conversion statement has been converted to `"1310"`. If a mathematical operation is carried out (`z=x+y`) then `z` will contain `23` after the statement is executed.

Sample run 23.1

```
13 added to 10 is 1310  13 added to 10 is 23
```

JavaScript provides several special functions for manipulating string and numeric values:

- The `eval` (*string*) function which converts a string to a numerical value.
- The `parseInt` (*string* [,*radix*]) function which converts a string into an integer of the specified radix (number base). The default radix is base-10.
- The `parseFloat` (*string*) function which converts a string into a floating-point value.

23.3.3 Variables

Variables are symbolic names for values within the script. A JavaScript identifier must either start with a letter or an underscore ("_"), followed by letters, an underscore or any digit (0–9). Like C, JavaScript is case sensitive so that variables with the same character sequence but with different cases for one or more characters are different. The following are different variable names:

```
i=5
I=10

valueA=3.543
VALUEA=10.543
```

23.3.4 Variable scope

A variable can be declared by either simply assigning it a value or by using the `var` keyword. For example the following declares to variables `Value1` and `Value2`:

```
var  Value1;
     Value2=23;
```

A variable declared within a function is taken as a local variable and can only be used within that function. A variable declared outside a function is a global variable and can be used anywhere in the script. A variable which is declared locally which is already declared as a global variable needs to be declared with the `var` keyword, otherwise the use of the keyword is optional.

23.3.5 Literals

Literal values have fixed values with the script. Various reserved forms can be used to identify special types, such as hexadecimal values, exponent format, and so on. With an integer the following are used:

- If the value is preceded by 0x then the value is a hexadecimal value (i.e. base 16). Examples of hexadecimal values are 0x1FFF, 0xCB.
- If the value is preceded by 0 then the value is an octal value (i.e. base 8). Examples of octal values are 0777, 010.
- If it is not preceded by either 0x or 0 then it is a decimal integer.

Floating-point values

Floating-point values are typically represented as a real value (such as 1.342) or in exponent format. Some exponent format value are:

Value	*Exponent format*
0.000001	1e-6
1342000000	1.342e9

Boolean

The true and false literals are used with Boolean operations.

Strings

In C a string is represented with double quotes ("*str*") whereas JavaScript accepts a string within double (") or single (') quotation marks. Examples of strings are:

```
"A string"
'Another string'
```

C uses an escape character sequence to represent special characters within a string. This character sequence always begins with a '\' character. For example, if the escape sequence '\n' appears in the string then this sequence is interpreted as a new-line sequence. Valid escape sequences are:

Character	*Meaning*	*Character*	*Meaning*
\b	backspace	\f	form feed
\n	new line	\r	carriage return
\t	tab	\\	backslash character
\"	prints a " character		

23.4 Expressions and operators

The expressions and operators used in Java and JavaScript are based on C and C++. This section outlines the main expressions and operators used in JavaScript.

23.4.1 Expressions

As with C, expressions are any valid set of literals, variables, operators, and expressions that evaluate to a single value. There are basically two types of expression, one which assign a value to a variable and the other which simply gives a single value. A simple assignment is:

```
value = 21
```

which assigns the value of 21 to `value` (note that the result of the expression is 21).

The result from a JavaScript expression can be:

- A numeric value.
- A string.
- A logical value (true or false).

23.5 JavaScript operators

Java and JavaScript have a rich set of operators; there are four main types:

- Arithmetic.
- Logical.
- Bitwise.
- Relational.

23.5.1 Arithmetic

Arithmetic operators operate on numerical values. The basic arithmetic operations are add (+), subtract (-), multiply (*), divide (/) and modulus division (%). Modulus division gives the remainder of an integer division. The following gives the basic syntax of two operands with an arithmetic operator.

```
operand operator operand
```

The assignment operator (=) is used when a variable 'takes on the value' of an operation. Other shorthand operators are used with it, including add equals (+=), minus equals (-=), multiply equals (*=), divide equals (/=) and modulus equals (%=). The following examples illustrate their uses.

Statement	*Equivalent*
`x+=3.0`	`x=x+3.0`
`voltage/=sqrt(2)`	`voltage=voltage/sqrt(2)`
`bit_mask *=2`	`bit_mask=bit_mask*2`

In many applications it is necessary to increment or decrement a variable by 1. For this purpose Java has two special operators; ++ for increment and -- for decrement. They can either precede or follow the variable. If they precede the variable, then a pre-increment/decrement is conducted, whereas if they follow it, a post-increment/decrement is conducted. Here are some .

Statement	Equivalent
`no_values++`	`no_values=no_values+1`
`i--`	`i=i-1`

Table 23.1 summarizes the arithmetic operators.

Table 23.1 Arithmetic operators

Operator	*Operation*	*Example*
-	subtraction, minus	5-4→1
+	addition	4+2→6
*	multiplication	4*3→12
/	division	4/2→2
%	modulus	13%3→1
+=	add equals	x += 2 is equivalent to x=x+2
-=	minus equals	x -= 2 is equivalent to x=x-2
/=	divide equals	x /= y is equivalent to x=x/y
*=	multiplied equals	x *= 32 is equivalent to x=x*32
=	assignment	x = 1
++	increment	Count++ is equivalent to Count=Count+1
--	decrement	Sec-- is equivalent to Sec=Sec-1

23.5.2 Relationship

The relationship operators determine whether the result of a comparison is TRUE or FALSE. These operators are greater than (>), greater than or equal to (>=), less than (<), less than or equal to (<=), equal to (==) and not equal to (!=). Table 23.2 lists the relationship operators.

Table 23.2 Relationship operators

Operator	*Function*	*Example*	*TRUE Condition*
>	greater than	(b>a)	when b is greater than a
>=	greater than or equal	(a>=4)	when a is greater than or equal to 4
<	less than	(c<f)	when c is less than f
<=	less than or equal	(x<=4)	when x is less than or equal to 4
==	equal to	(x==2)	when x is equal to 2
!=	not equal to	(y!=x)	when y is not equal to x

23.5.3 Logical (TRUE or FALSE)

A logical operation is one in which a decision is made as to whether the operation performed is TRUE or FALSE. If required, several relationship operations can be grouped together to give the required functionality. C assumes that a numerical value of 0 (zero) is FALSE and that any other value is TRUE. Table 23.2 lists the logical operators.

The logical AND operation will yields TRUE only if all the operands are TRUE. Table 23.4 gives the result of the AND (`&&`) operator for the operation `A && B`. The logical OR operation yields a `TRUE` if any one of the operands is `TRUE`. Table 23.4 gives the logical results of the `OR` (`||`) operator for the statement `A|| B`. Table 23.4 also gives the logical result of the NOT (!) operator for the statement `!A`.

Table 23.3 Logical operators

Operator	*Function*	*Example*	*TRUE condition*
`&&`	AND	`((x==1) && (y<2))`	when `x` equal 1 *and* `y` is less than 2
`||`	OR	`((a!=b) || (a>0))`	when `a` is not equal to `b` *or* `a` is greater than 0
`!`	NOT	`(!(a>0))`	when `a` is *not* greater than 0

Table 23.4 Logical operations

A	B	AND (`&&`)	OR (`||`)	NOT (`!A`)
FALSE	FALSE	FALSE	FALSE	TRUE
FALSE	TRUE	FALSE	TRUE	TRUE
TRUE	FALSE	FALSE	TRUE	FALSE
TRUE	TRUE	TRUE	TRUE	FALSE

23.5.4 Bitwise

The bitwise logical operators work conceptually as follows:

- The operands are converted to 32-bit integers and expressed as a series of bits (zeros and ones).
- Each bit in the first operand is paired with the corresponding bit in the second operand: first bit to first bit, second bit to second bit, and so on.
- The operator is applied to each pair of bits, and the result is constructed bitwise.

The bitwise operators are similar to the logical operators but they should not be confused as their operation differs. Bitwise operators operate directly on the individual bits of any operands, whereas logical operators determine whether a condition is TRUE or FALSE.

Numerical values are stored as bit patterns in either an unsigned integer format, signed integer (two's complement) or floating-point notation (an exponent and mantissa). Characters are normally stored as ASCII characters.

The basic bitwise operations are AND (`&`), OR (`|`), one's complement or bitwise inversion (~), XOR (^), shift left (<<) and shift right (>>). Table 23.5 gives the results of the AND bitwise operation on two bits *A* and *B*.

The Boolean bitwise instructions operate logically on individual bits. The XOR function yields a 1 when the bits in a given bit position differ, the AND function yields a 1 only when the given bit positions are both 1s. The OR operation gives a 1 when any one of the given bit positions are a 1. For example:

```
     00110011          10101111           00011001
AND  11101110      OR  10111111      XOR  11011111
     --------          --------           --------
     00100010          10111111           11000110
     ========          ========           ========
```

Table 23.5 Bitwise operations

A	*B*	*AND*	*OR*	*EX-OR*
0	0	0	0	0
0	1	0	1	1
1	0	0	1	1
1	1	1	1	0

To perform bit shifts, the <<, >> and >>> operators are used. These operators shift the bits in the operand by a given number defined by a value given on the right-hand side of the operation. The left shift operator (<<) shifts the bits of the operand to the left and zeros fill the result on the right. The sign-propagating right shift operator (>>) shifts the bits of the operand to the right and zeros fill the result if the integer is positive; otherwise it will fill with 1s. The zero-filled right shift operator (>>>) shifts the bits of the operand to the right and fills the result with zeros. The standard format is:

```
operand >>  no_of_bit_shift_positions
operand >>> no_of_bit_shift_positions
operand <<  no_of_bit_shift_positions
```

23.5.5 Precedence

There are several rules for dealing with operators:

- Two operators, apart from the assignment operator, should never be placed side by side. For example, `x * % 3` is invalid.

- Groupings are formed with parentheses; anything within parentheses will be evaluated first. Nested parentheses can also be used to set priorities.
- A priority level or precedence exists for operators. Operators with a higher precedence are evaluated first; if two operators have the same precedence, then the operator on the left-hand side is evaluated first. The priority levels for operators are as follows:

HIGHEST PRIORITY

`() [] .`	primary
`! ~ ++ -- -`	unary
`* / %`	multiplicative
`+ -`	additive
`<< >> >>>`	shift
`< > <= >=`	relational
`== !=`	equality
`&`	bitwise
`^`	
`|`	
`&&`	logical
`||`	
`= += -=`	assignment

LOWEST PRIORITY

The assignment operator has the lowest precedence. The following example shows how operators are prioritized in a statement (=> shows the steps in determining the result):

```
23 + 5 % 3 / 2 << 1   =>
23 + 2 / 2 << 1       =>
23 + 1 << 1           =>
23 + 2                => 25
```

23.5.6 Conditional expressions

Conditional expressions can produce one of two values depending on a condition. The syntax is:

```
(expression) ? value1 : value2
```

If the expression is true then `value1` is executed else `value2` is executed. For example:

```
(val >= 0) ? sign="postive" : sign="negative"
```

The expression will assign the string “positive” to `sign` if the value of `val` is greater than or equal to 0, else it will assign “negative”.

23.5.7 String operators

The normal comparison operators, such as <, >, >=, ==, and so on can be used with strings. In addition the concatenation operator (+) can be used to concatenate two string values together. For example,

```
str="This is " + "an example"
```

will result in the string

```
"This is an example"
```

23.6 JavaScript statements

JavaScript statements are similar to C and allow a great deal of control of the execution of a script. The basic categories are:

- Conditional statements, such as `if...else`.
- Repetitive statements, such as `for`, `while`, `break` and `continue`.
- Comments, using either the C++ style for single-line comments (`//`) or standard C multiline comments (`/*...*/`).
- Object manipulation statements and operators, such as `for...in`, `new`, `this`, and `with`.

23.7 Conditional statements

Conditional statements allow a program to make decisions on the route through a program.

23.7.1 `if...else`

A decision is made with the `if` statement. It logically determines whether a conditional expression is TRUE or FALSE. For TRUE, the program executes one block of code; FALSE causes the execution of another (if any). The keyword `else` identifies the FALSE block. Braces are used to define the start and end of the block.

Relationship operators (>,<,>=,<=,==,!=) yield TRUE or FALSE from their operation. Logical statements (`&&`, `||`, `!`) can then group them together to give the required functionality. If the operation is not a relationship, such as a bitwise or arithmetic operation, then any non-zero value is TRUE and a zero is FALSE.

The following is an example of the `if` statement syntax. If the statement block has only one statement the braces ({ }) can be excluded.

```
if (expression)
{
   statement block
}
```

The following is an example of an `else` extension.

```
if (expression)
{
   statement block1
}
else
{
   statement block2
}
```

It is possible to nest `if...else` statements to give a required functionality. In the next example, *statement block1* is executed if `expression1` is TRUE. If it is FALSE then the program checks the next expression. If this is TRUE the program executes *statement block2*, else it checks the next expression, and so on. If all expressions are FALSE then the program executes the final `else` statement block, in this case *statement block 3*:

```
if (expression1)
{
   statement block1
}
else if (expression2)
{
   statement block2
}
else
{
   statement block3
}
```

23.8 Loops

23.8.1 `for()`

Many tasks within a program are repetitive, such as prompting for data, counting values, and so on. The `for` loop allows the execution of a block of code for a given control function. The following is an example format; if there is only one statement in the block then the braces can be omitted.

```
for (starting condition;test condition;operation)
{
     statement block
}
```

where

`starting condition`	means the starting value for the loop
`test condition`	means if `test condition` is TRUE the loop will continue execution
`operation`	means the operation conducted at the end of the loop.

23.8.2 `while()`

The `while()` statement allows a block of code to be executed while a specified condition is TRUE. It checks the condition at the start of the block; if this is TRUE the block is executed, else it will exit the loop. The syntax is:

```
while (condition)
{
   :         :  statement block
   :         :
}
```

If the statement block contains a single statement then the braces may be omitted (although it does no harm to keep them).

23.9 Comments

Comments are author annotations that explain what a script does. Comments are ignored by the interpreter. JavaScript supports Java-style comments:

- Comments on a single line are preceded by a double-slash (`//`).
- Multiline comments can be preceded by `/*` and followed by `*/`.

The following example shows two comments:

```
// This is a single-line comment.
/* This is a multiple-line comment. It can be of any
length, and you can put whatever you want here. */
```

23.10 Functions

JavaScript supports modular design using functions. A function is defined in JavaScript with the `function` reserved word and the code within the function is defined within curly brackets. The standard format is:

```
function myfunct(param1, param2 ...)
{
   statements
   return(val)
}
```

where the parameters (`param1`, `param2`, and so on) are the values passed into the function. Note that the return value (`val`) from the function is only required when a value is returned from the function.

JavaScript 23.2 gives an example with two functions (`add()` and `mult()`). In this case the values, `value1` and `value2`, are passed into the variables, `a` and `b`, within the `add()` function; the result is then sent back from the function into `value3`.

📖 JavaScript 2

```
<HTML><TITLE>Example</TITLE>
<BODY BGCOLOR="#FFFFFF">

<SCRIPT>
var value1,value2,value3,value4;

value1=15;
value2=10;
value3=add(value1,value2)
value4=mult(value1,value2)
document.write("Added is ",value3)
document.write("<P>Multiplied is ",value4)

function add(a,b)
{
var   c
      c=a+b
      return(c)
}
function mult(a,b)
{
var   c
      c=a*b
      return(c)
}
</SCRIPT></FORM></HTML>
```

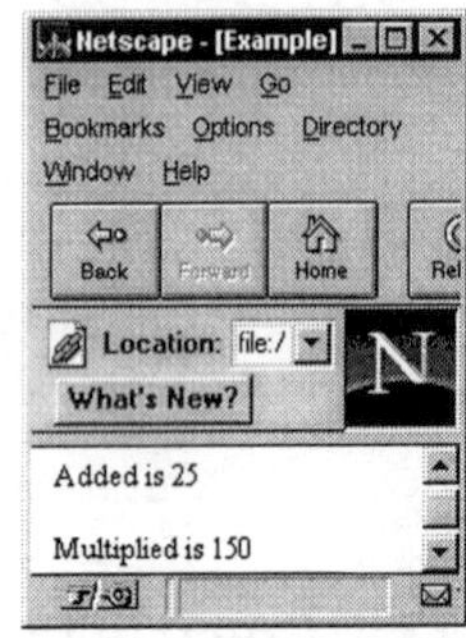

23.11 Objects and properties

JavaScript is based on a simple object-oriented paradigm, where objects are a construct with properties that are JavaScript variables. Each object has properties associated with it and can be accessed with the dot notation, such as:

objectName.propertyName

23.12 Document objects

The document object contains information on the currently opened document. HTML expressions can be displayed with the `document.write()` or `document.writeln()` functions. The standard format is:

```
document.write(exprA, [,exprB], ... [,exprN])
```

which displays one or more expressions to the specified window. To display to the current window, use `document.write()`. For a display to a specified window then the window reference is defined:

```
mywin=window.open("fred.html")
mywin.document.write("Hello")
```

Output is sent to the windows `mywin`.

The document object can also be used to display HTML properties. The standard HTML format is:

```
<BODY BACKGROUD="bgndimage" BGCOLOR="bcolor"
TEXT="fcolor" LINK="ufcolor" ALINK="actcolor"
VLINK="fcolor" </BODY>
```

These and other properties can be accessed within JavaScript using:

```
document.alinkColor        document.anchors
document.bgColor           document.fgColor
document.lastModified      document.linkColor
document.title             document.URL
document.vlinkColor
```

JavaScript 23.3 shows an example.

JavaScript 23.3

```
<HTML><TITLE>Example</TITLE>
<BODY BGCOLOR="#FFFFFF">

<SCRIPT>
document.write('ALINK color is ',
    document.alinkColor)
document.write('<P>BGCOLOR is ',
    document.bgColor)
document.write('<P>URL is ',document.URL)
document.write('<P>Title is',
    document.title)
</SCRIPT></HTML>
```

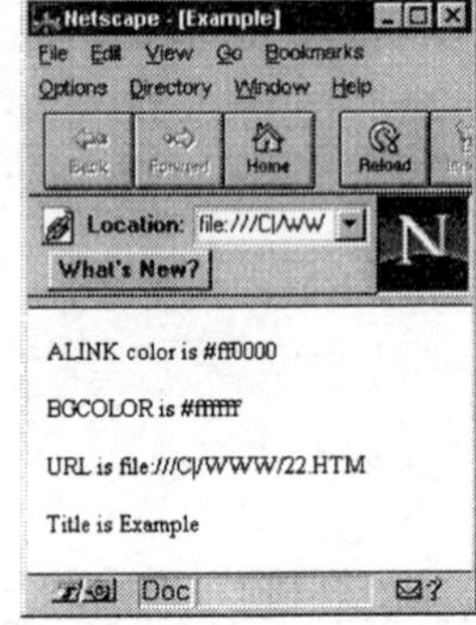

23.13 Event handling

JavaScript has event handlers which, on a certain event, cause other events to occur. For example, if a user clicks the mouse button on a certain menu option then the event handler can be made to carry-out a particular action, such as adding to numbers together. Table 23.8 outlines some event handlers.

Table 23.6 Event handers

Event	*Description*	*Example*	*Caused by*
onBlur	Blur events occur when the select or text field on a form loses focus.	`<INPUT TYPE="text" VALUE="" NAME="userName" onBlur="check(this.value)">` When the onBlur event occurs the JavaScript code required is executed (in this case the function `check()` is called.	select, text, textarea
onChange	The change event occurs when the select or text field loses focus and its value has been modified.	`<INPUT TYPE="button" VALUE="Compute" onClick="Calc(this.form)">` When the onChange event occurs the required JavaScript code is executed (in this case the function `Calc()` is called).	button, checkbox, radio, link, reset, submit
onClick	The click event occurs when an object on a form is clicked.	`<INPUT TYPE="button" VALUE="Calculate" onClick="go(this.form)">`	button, checkbox, radio, link, reset, submit

onFocus	The focus event occurs when a field receives input focus by tabbing with the keyboard or clicking with the mouse. Selecting within a field results in a select event and not a focus event.	`<INPUT TYPE="textarea" VALUE="" NAME="valueField" onFocus="valueCheck()">`	select, text, textarea
onLoad	The load event occurs when the browser finishes loading a window.	`<BODY onLoad="window.alert("Hello to my excellent page")>`	

onMouse Over	A mouseover event occurs each time the mouse pointer moves over an object.
	
	Note that a true value must be returned within the event handler.
		`<A HREF="http://home.html/"`
		
		`onMouseOver="window.status='Select this to go home'; return true">`
		`Go Home</A>`	

onSelect	A select event occurs when a user selects some of the text within a text or textarea field. The onSelect event handler executes JavaScript code when a select event occurs.	`<INPUT TYPE="text" VALUE="" NAME="valueField" onSelect="selectState()">`	
onUnload	An Unload event occurs when the browser quits from a window.	`<BODY onUnload="goodbye()">`	

JavaScript 23.4 shows an example.

📖 JavaScript 23.4

```
<HTML><HEAD><TITLE>Example form</TITLE></HEAD>

<BODY BGCOLOR="#ffffff">

<FORM>
Enter name<INPUT TYPE="text" NAME="myname"
   onBlur="testname(myname.value)">
</FORM>

<SCRIPT>
function testname(name)
{
  if (name!="fred") alert("You are not fred");
}
</SCRIPT></HTML>
```

23.13.1 Opening and closing windows

Windows are opened with `window.open()` and closed with `window.close()`. Examples are:

```
fredwin=window.open("fred.html");

fredwin.close(fredwin);
```

or to open a window it is possible to simply use `open()` and to close the current window the `close()` function is used. The standard format is:

[*winVar* =][`window`].open("*url*", "*winName*", ["*features*"])

where
winVar is the name of a new window which can refer to a given window.
winName is the window name given to the window.
features is a comma-separated list with any of the following:

toolbar[=yes|no]	location[=yes|no]	directories[=yes|no]
status[=yes|no]	menubar[=yes|no]	scrollbars[=yes|no]
resizable[=yes|no]	width=pixels	height=pixels

23.13.2 Window confirm

Window confirm is used to display a confirmatory dialogue box with a specified message and the OK and Cancel buttons. If the user selects the OK button then the function returns a TRUE, else it returns a FALSE. JavaScript 23.5 gives an example of window confirm. In this case when the Exit button is selected then the function `ConfirmExit()` is called. In this function the user is asked to confirm the exit using the confirmatory window. If the user selects OK then the window is closed (if it is the only window open then the browser quits).

📖 JavaScript 23.5

```
<HTML><TITLE>Example</TITLE>
<BODY BGCOLOR="#FFFFFF">

<form name="ExitForm">
<INPUT    TYPE="button"    VALUE="Exit"
onClick="ConfirmExit()">
</FORM>


<SCRIPT>
function ConfirmExit()
{
   if (confirm("Do you want to exit"))
   {
      prompt("Enter your password")
   }
}
</SCRIPT></HTML>
```

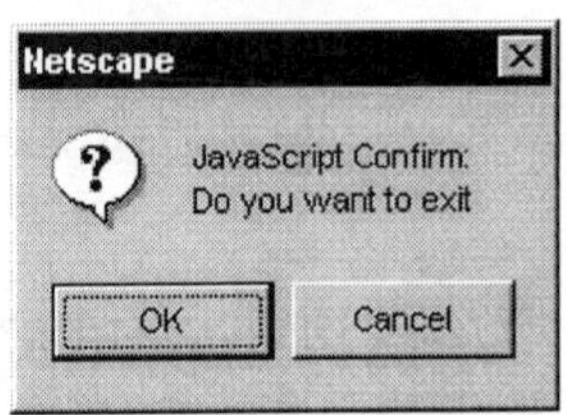

23.13.3 Window prompt

The window prompt displays a prompt dialogue box which contains a message and an input field. Its standard format is:

```
prompt("Message");
```

23.14 Object manipulation statements and operators

JavaScript has several methods in which objects can be manipulated; they include the `new` operator, the `this` keyword, the `for...in` statement, and the `with` statement.

23.14.1 this keyword

The `this` keyword is used to refer to the current object. The general format is:

```
this[.propertyName]
```

JavaScript 23.6 gives an example of the `this` keyword. Here `this` is used to pass the property values of the input form. They are then passed to the function `checkval()` when the `onBlur` event occurs.

📖 JavaScript 23.6

```
<HTML><TITLE>Example</TITLE>
<BODY BGCOLOR="#FFFFFF">

<FORM>
Enter a value<INPUT TYPE = "text" NAME = "inputvalue"
onBlur="checkval(this, 0,10)">

<SCRIPT>
function checkval(val, minval, maxval)
{
  if ((val.value < minval) || (val.value > maxval))
    alert("Invalid value (0-10)")
}

</SCRIPT></FORM></HTML>
```

23.14.2 new operator

The `new` operator is used to define a new user-defined object type or one of the predefined object types, such as array, Boolean, date, function and math. JavaScript 23.6 gives an example which creates an array object with six elements then assigns strings to each of the array elements. Note that in Java the first element of the array is indexed as 0.

📖 JavaScript 23.7

```
<HTML><HEAD><TITLE>Java Example</TITLE></HEAD>
<BODY BGCOLOR="#ffffff">
<SCRIPT language="javascript">

no_of_networks=6;
Networks = new Array(no_of_networks);

  Networks[0]="Ethernet"; Networks[1]="Token Ring"
  Networks[2]="FDDI";   Networks[3]="ISDN"
  Networks[4]="RS-232"; Networks[5]="ATM"

  for (i=0;i<no_of_networks;i++)
  {
    document.writeln("Network type "+Networks[i]);
    document.writeln("<P>");
  }

</SCRIPT></BODY></HTML>
```

```
Network type Ethernet
Network type Token Ring
Network type FDDI
Network type ISDN
Network type RS-232
Network type ATM
```

Typically the `new` operator is used to create new data objects. For example:

```
today = new Date()
Xmasday = new Date("December 25, 1997 00:00:00")
Xmasday = new Date(97,12,25)
```

23.14.3 for...in

The `for...in` statement is used to iterate a variable through all its properties. In general its format is:

```
for (variable in object)
{
    statements
}
```

23.14.4 with

The `with` statement defines a specified object for a set of statements. A `with` statement appears as follows:

```
with (object)
{
   statements
}
```

For example JavaScript 23.8 contains calls to the `Math` object for the PI property and cos and sin methods. JavaScript 23.9 then uses the `with` statement to define the Math, which object is the default object.

📖 JavaScript 23.8

```
<HTML><TITLE>Example</TITLE>
<BODY BGCOLOR="#FFFFFF">
<SCRIPT>
var area,x,y,radius
  radius=20

  area=Math.PI*radius*radius
  x=radius*Math.cos(Math.PI/4)
  y=radius*Math.sin(Math.PI/4)
  document.write("Area is ",area)
  document.write("<P>x is ",x, "<P>y is ",y)
</SCRIPT></FORM></HTML>
```

📖 JavaScript 23.9

```
<HTML><TITLE>Example</TITLE>
<BODY BGCOLOR="#FFFFFF">

<SCRIPT>
var area,x,y,radius
  radius=10
  with (Math)
  {
    area=PI*radius*radius
    x=radius*cos(PI/4)
    y=radius*sin(PI/4)
    document.write("Area is ",area)
    document.write("<P>x is ",x, "<P>y is ",y)
  }
</SCRIPT></FORM></HTML>
```

23.15 Exercises

23.1 Explain how Java differs from JavaScript.

23.2 Explain the main advantages of using Java compared with a high-level language, such as C++ or Pascal.

23.3 Implement the JavaScripts in this chapter and test them out.

23.4 Write a program in JavaScript that returns the square of a value entered by the user.

23.5 Write a program in JavaScript that enables the user to enter their name and have it tested for a match with "FRED", "BERT" or "FREDDY". If there is no match, then the browser exits. then the browser exits.

24 Windows NT

24.1 Introduction

Windows NT has provided an excellent network operating system. It communicates directly with many different types of networks, protocols and computer architectures. Windows NT and Windows 95 have the great advantage of other operating systems in that they have integrated network support. Operating systems now use networks to make peer-to-peer connections and also connections to servers for access to file systems and print servers. The three most widely used operating systems are MS-DOS, Microsoft Windows and UNIX. Microsoft Windows comes in many flavors; the main versions are outlined below and Table 24.1 lists some of their attributes.

- Microsoft Windows 3.*xx* – 16-bit PC-based operating system with limited multitasking. It runs from MS-DOS and thus still uses MS-DOS functionality and file system structure.
- Microsoft Windows 95 – robust 32-bit multitasking operating system (although there are some 16-bit parts in it) which can run MS-DOS applications, Microsoft Windows 3.*xx* applications and 32-bit applications.
- Microsoft Windows NT – robust 32-bit multitasking operating system with integrated networking. Networks are built with NT servers and clients. As with Microsoft Windows 95 it can run MS-DOS, Microsoft Windows 3.*x* applications and 32-bit applications.

Table 24.1 Windows comparisons

	Windows 3.1	*Windows 95*	*Windows NT*
Preemptive multitasking		✓	✓
32-bit operating system		✓	✓
Long file names		✓	✓
TCP/IP	✓	✓	✓
32-bit applications		✓	✓
Flat memory model		✓	✓
32-bit disk access	✓	✓	✓
32-bit file access	✓	✓	✓
Centralized configuration storage		✓	✓
OpenGL 3D graphics			✓

24.2 Novell NetWare networking

Novell NetWare is one of the most popular systems for PC LANs and provides file and print server facilities. The protocol used is SPX/IPX. This is also used by Windows NT to communicate with other Windows NT nodes and with NetWare networks. The Internet Packet Exchange (IPX) protocol is a network layer protocol for transportation of data between computers on a Novell network. IPX is very fast and has a small connectionless datagram protocol. Sequenced Packet Interchange (SPX) provides a communications protocol which supervises the transmission of the packet and ensures its successful delivery.

Novell uses the Open Data-Link Interface (ODI) standard to simplify network driver development and to provide support for multiple protocols on a single network adapter. It allows Novell NetWare drivers to be written to without concern for the protocol that will be used on top of them (similar to NDIS in Windows NT). The link support layer (LSL or LSL.COM) provides a foundation for the MAC layer to communicate with multiple protocols (similar to NDIS in Windows NT). The IPX.COM (or IPXODI.COM) program normally communicates with the LSL and the applications. The MAC driver is a device driver or NIC driver. It provides low-level access to the network adapter by supporting data transmission and some basic adapter management functions. These drivers also pass data from the physical layer to the transport protocols at the network and transport layers. NetWare and IPX/SPX are covered in more detail in the next chapter.

24.3 Servers, workstations and clients

Microsoft Windows NT is a 32-bit, preemptive, multitasking operating system. One of the major advantages it has over UNIX is that it can run PC-based software. A Windows NT network normally consists of a server and a number of clients. The server provides file and print servers as well as powerful networking applications, such as electronic mail applications, access to local and remote peripherals, and so on.

The Windows NT client can either:

- Operate as a standalone operating system.
- Connect with a peer-to-peer connection.
- Connect to a Windows NT server.

A peer-to-peer connection is when one computer logs into another computer. Windows NT provides unlimited outbound peer-to-peer connections and typically up to 10 simultaneous inbound connections.

24.4 Workgroups and domains

Windows NT assigns users to workgroups which are collection of users who are grouped together with a common purpose. This purpose might be to share resources such as file systems or printers, and each workgroup has its own unique name. With workgroups each Windows NT workstation interacts with a common group of computers on a peer-to-peer level. Each workstation then manages its own resources and user accounts. Workgroups are useful for small groups where a small number of users require to access resources on other computers

A domain in Windows NT is a logical collection of computers sharing a common user accounts database and security policy. Thus each domain must have at least one Windows NT server.

Windows NT is designed to operate with either workgroups or domains. Figure 24.1 illustrates the difference between domains and workgroups.

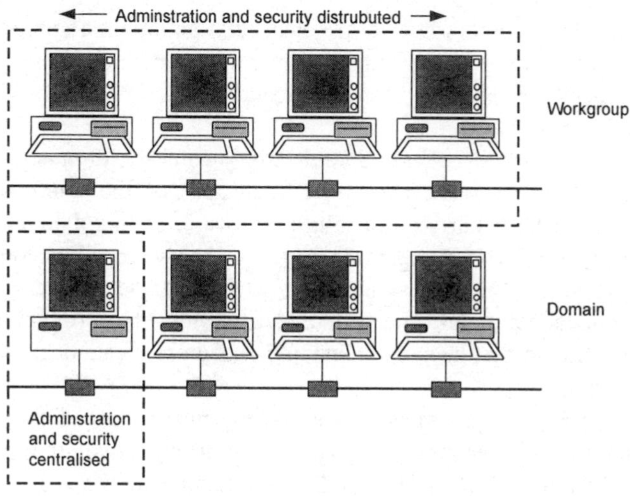

Figure 24.1 Workgroups and domains

Domains have the advantages that:

- Each domain forms a single administrative unit with shared security and user account information. This domain has one database containing user and group information and security policy settings.
- They segment the resources of the network so that users, by default, can view all networks for a particular domain.
- User accounts are automatically validated by the domain controller. This stops invalid users from gaining access to network resources.

24.5 User and group accounts

Each user within a domain has a user account and is assigned to one or more groups. Each group is granted permissions for the file system, accessing printers, and so on. Group accounts are useful because they simplify an organization into a single administrative unit. They also provide a convenient method of controlling access for several users who will be using Windows NT to perform similar tasks. By placing multiple users in a group, the administrator can assign rights and/or permissions to the group.

Each user on a Windows NT system has the following:

- A user name (such as `fred_bloggs`).
- A password (assigned by the administrator then changed by the user).
- The groups in which the user account is a member (e.g. `staff`).
- Any user rights for using the assigned computer.

Each time a user attempts to perform a particular action on a computer, Windows NT checks the user account to determine whether the user has the authority to perform that action (such as read the file, write to the file, delete the file, and so on).

Normally there are three main default user accounts: Administrator, Guest and an 'Initial User' account. The system manager uses the Administrator account to perform such tasks as installing software, adding/deleting user accounts, setting up network peripherals, installing hardware, and so on.

Guest accounts allow occasional users to log on and be granted limited rights on the local computer. The system manager must be sure that the access rights are limited so that hackers or inexperienced users cannot do damage to the local system.

The 'Initial User' account is created during installation of the Windows NT workstation. This account, assigned a name during installation, is a member of the Administrator's group and therefore has all the Administrator's rights and privileges.

After the system has been installed the Administrator can allocate new user accounts, either by creating new user accounts, or by copying existing accounts.

24.6 New user accounts

Typically the system manager will create new accounts by copying existing users accounts. The items copied directly from an existing user account to a new user account are as follows:

- The description of the user (such as `Fred Bloggs, Ext 4444`).
- Group account membership (such as `Production`).
- Profile settings (such as home directory).
- If set, the attribute to stop the user from changing their password (sometimes the manager does not want the user to change the default password).
- If set, the attribute that causes the password to remain unexpired (sometimes the system manager forces users to change their passwords from time to time).

The items which are cleared and completed by the system manager are:

- The username and full name.
- The attribute that prompts users to change their passwords when they next login (normally a default password is initially set up and is changed when a user initially logs in).
- The attribute which disables the account (the manager must reset this before a user can log in).

24.7 File systems

Windows NT supports three different types of file system:

- FAT (file allocation table) – as used by MS-DOS, OS/2 and Windows NT. A single volume can be up to 2 GB.
- HPFS (high performance file system) –a UNIX-style file system which is used by OS/2 and Windows NT. A single volume can be up to 8 GB. MS-DOS applications cannot access files.
- NTFS (NT file system) – as used by Windows NT. A single volume can be up to 8 GB. MS-DOS applications, themselves, cannot access the file system but they can when run with Windows NT.

The FAT file system is widely used and supported by a variety of operating systems, such as MS-DOS, Windows NT and OS/2. If a system is to use MS-DOS it must be installed with a FAT file system.

24.7.1 FAT

The standard MS-DOS FAT file and directory-naming structure allows an 8-character file name and a 3-character file extension with a dot separator (`.`) between them (the 8.3 file name). It is not case sensitive and the file name and extension cannot contain spaces and other reserved characters, such as:

`" / \ : ; | = , ^ * ? .`

With Windows NT and Windows 95 the FAT file system supports long file names which can be up to 255 characters. The name can also contain multiple spaces and dot separators. File names are not case sensitive, but the case of file names is preserved (a file named `FredDocument.XYz` will be displayed as `FredDocument.XYz` but can be accessed with any of the characters in upper or lower case.

Each file in the FAT table has four attributes (or properties): read-only, archive, system and hidden (as shown in Figure 24.2). The FAT uses a linked list where the file's directory entry contains its beginning FAT entry number. This FAT entry in turn contains the location of the next cluster if the file is larger than one cluster, or a marker that designates this is included in the last cluster. A file which occupies 12 clusters will have 11 FAT entries and 10 FAT links.

The main disadvantage with FAT is that the disk is segmented into allocated units (or clusters). On large-capacity disks these sectors can be relatively large (typically 512 bytes/sector). Disks with a capacity of between 256 MB and 512 MB use 16 sectors per cluster (8 kB) and disks from 512 MB to 1 GB use 32 sectors per cluster (16 kB). Drives up to 2 GB use 64 sectors per cluster (32 kB). Thus if the disk has a capacity of 512 MB then each cluster will be 8 kB. A file which is only 1kB will thus take up 8 kB of disk space (a wastage of 7 kB), and a 9 kB file will take up 16 kB (a wastage of 7 kB). Thus a file system which has many small files will be inefficient on a cluster-based system. A floppy disk normally use 1 cluster per sector (512 bytes).

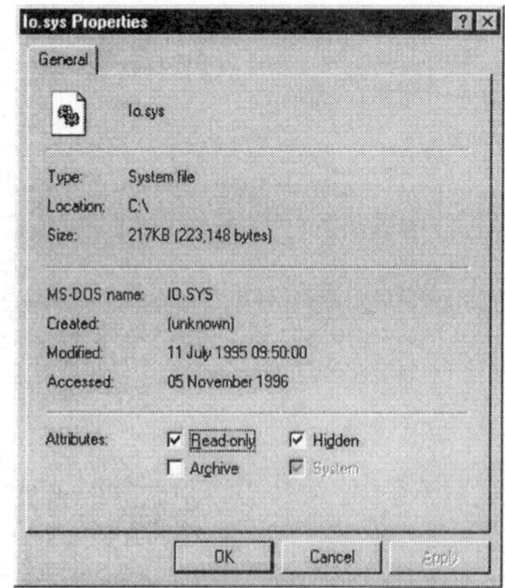

Figure 24.2 File attributes

Windows 95 and Windows NT support up to 255 characters in file names, unfortunately, MS-DOS and Windows 3.*xx* applications cannot read them. To accommodate this, every long file name has an autogenerated short file name (in the form `XXXXXXXX.YYY`). Table 24.2 shows three examples. The conversion takes the first six characters of the long name then adds a *~number* to the name to give it a unique name. File names with the same initial six characters are identified with different *numbers*. For example `Program Files` and `Program Directory` would be stored as `PROGRA~1` and `PROGRA~2`, respectively. Sample listing 24.1 shows a listing from Windows NT. The left hand column shows the short file name and the far right-hand column shows the long file name.

Table 24.2 File name conversions

Long file name	*Short file name*
`Program Files`	`PROGRA~1`
`Triangular.bmp`	`TRAING~1.BMP`
`Fredte~1.1`	`FRED.TEXT.1`

Sample listing 24.1

```
EXAMPL~1 DOC   4,608  05/11/96  23:36 Example Document 1.doc
EXAMPL~2 DOC   4,608  05/11/96  23:36 Example Document 2.doc
EXAMPL~3 DOC   4,608  05/11/96  23:36 Example Document 3.doc
EXAMPL~4 DOC   4,608  05/11/96  23:36 Example Document 4.doc
EXAMPL~5 DOC   4,608  05/11/96  23:36 Example Document 5.doc
EXAMPL~6 DOC   4,608  05/11/96  23:36 Example Document 6.doc
EXAMPL~7 DOC   4,608  05/11/96  23:36 Example Document 7.doc
EXAMPL~8 DOC   4,608  05/11/96  23:39 Example assignment A.doc
EXAMPL~9 DOC   4,608  05/11/96  23:40 Example assignment B.doc
EXAMP~10 DOC   4,608  05/11/96  23:40 Example assignment C.doc
```

24.7.2 HPFS (high-performance file system)

HPFS is supported by OS/2 and is typically used to migrate from OS/2 to Windows NT. It allows long file names of up to 254 characters with multiple extensions. As with the Windows 95 and Windows NT FAT system the file names are not case sensitive but preserve the case. HPFS uses B-tree format to store the file system directory structure. The B-tree format stores directory entries in an alphabetic tree, and binary searches are used to search for the target file in the directory list. The reserved characters for file names are:

```
" / \ : < > | * ?
```

24.7.3 NTFS (NT file system)

NTFS is the preferred file system for Windows NT as it makes more efficient usage of the disk and it offers increased security. It allows for file systems up to 16 EB (16 Exabytes, or 1 billion gigabytes, or 2^{64} bytes). As with HPFS it uses B-tree format for storing the file systems directory structure. Its main objectives are:

- To increase reliability. NTFS automatically logs all directory and file updates which can be used to redo or undo failed operations resulting from system failures such as power losses, hardware faults, and so on.
- To provide sector sparing (or hot fixing). When NTFS finds errors in a bad sector it causes the data in that sector to be moved to a different section and the bad sector to be marked as bad. No other data is then written to that sector. Thus the disk fixes itself as it is working and there is no need for disk repair programs (FAT only marks bad areas when formatting the disk).
- Increases file system size (up to 16 EB).
- To enhance security permissions.
- To support for POSIX requirements, such as case-sensitive naming, addition of a time stamp to show the time the file was last accessed and hard links from one file (or directory) to another.

The reserved characters for file names are:

" / \ : < > | * ?

24.8 Windows NT networking

Networks must use a protocol to transmit data. Typical protocols are:

- IPX/SPX – used with Novell NetWare, it accesses file and printer services.
- TCP/IP – used for Internet access and client/server applications.
- SNA DLC – used mainly by IBM mainframes and minicomputers.
- AppleTalk – used by Macintosh computers.
- NetBEUI – used in some small LANs (stands for NetBIOS Extended User Interface).

Novell NetWare is installed in many organization to create local area networks of PCs. It uses IPX/SPX for transmitting data and allows access to file servers and network printing services. TCP/IP is the standard protocol used when accessing the Internet and also for client/server applications (such as remote file transfer and remote login).

A major advantage of Windows NT is that networking is built into the operating system. Figure 24.3 shows how it is organized in relation to the OSI model. Windows NT has the great advantage of being protocol-independent and will work with most standard protocols, such as TCP/IP, IPX/SPX, NetBEUI, DLC and AppleTalk. The default protocol is NetBEUI.

Figure 24.4 shows a sample configuration for the Windows NT Client; the display shows the default network configurations.

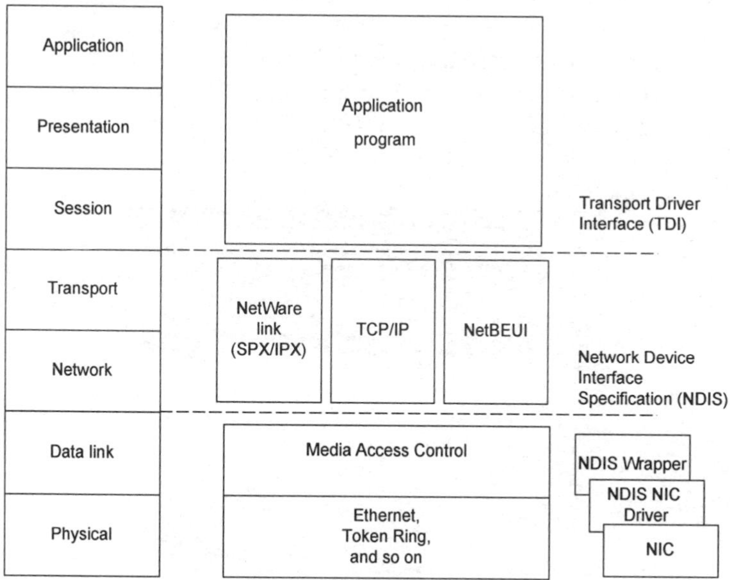

Figure 24.3 Windows NT network interfaces

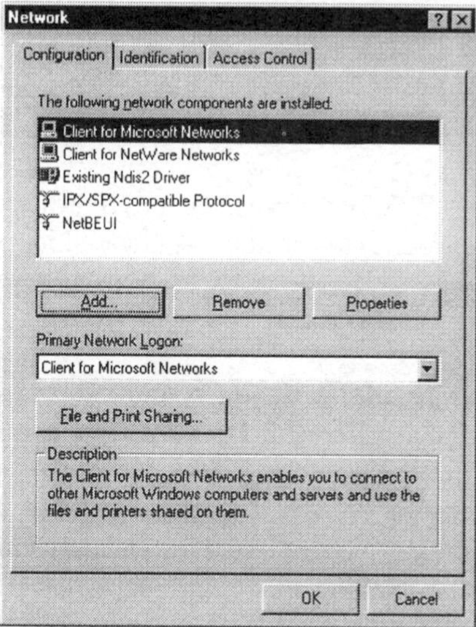

Figure 24.4 Windows NT network interfaces

There are two main boundaries in Windows NT and NDIS and TDI. The Network Device Interface Standard (NDIS) boundary layer interfaces to several network interface adapters (such as Ethernet, Token Ring, RS-232, modems, and so on) with different protocols. It allows for an unlimited number of network interface cards (NICs) and protocols to be connected to be used with the operating system. In Windows NT, a single software module, `NDIS.SYS`, (the NDIS wrapper) interfaces with the manufacturer-supplied NDIS NIC device driver. The wrapper provides a uniform interface between the protocol drivers (such as TCP/IP or IPX/SPX) and the NDIS device driver. Figure 24.5 shows the setup of the network adapter.

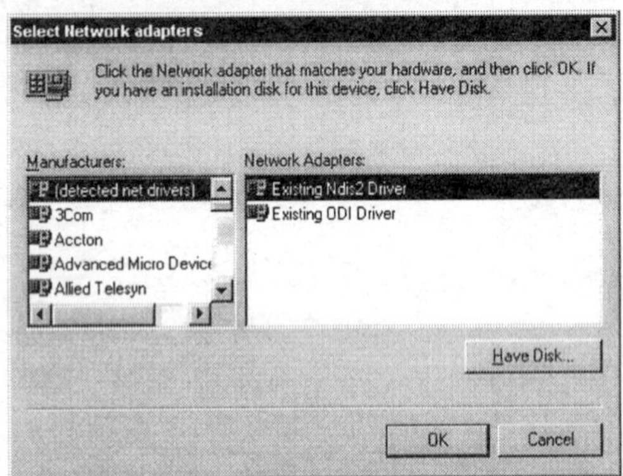

Figure 24.5 Windows NT network adapter drivers

24.8.1 TCP/IP

This will be covered in Section 24.9.

24.8.2 IPX/SPX

NetWare networks use SPX/IPX and is supported through Windows NT using the NWLink protocol stack. This protocol is covered in more detail in the next chapter.

24.8.3 NetBEUI

NetBEUI (NetBIOS Extended User Interface) has been used with network operating systems, such as Microsoft LAN manager and OS/2 LAN server. In Windows NT, the NetBEUI frame (NBF) protocol stack gives backward compatibility with existing NetBEUI implementations and also provides for enhanced implementations. NetBEUI is the standard technique that NT clients and servers use to intercommunicate.

NBF is similar to TCP/IP and SPX/IPX, it is used to establish a session between a client and a server, and also to provide the reliable transport of the data across the connection-oriented session. Thus NetBEUI tries to provide reliable data transfer through error checking and acknowledgment of each successfully received data packet. In the standard form of NetBEUI each packet must be acknowledged after its delivery. This is wasteful in time. Windows NT uses NBF which improves NetBEUI as it allows several packets to be sent before requiring an acknowledgment (called an adaptive sliding window protocol).

Each NetBEUI is assigned a 1-byte session number and thus allows a maximum of 254 simultaneously active session (as two of the connection numbers are reserved). NBF enhances this by allowing 254 connections to computers with 254 sessions for each connection (thus there is a maximum of 254×254 sessions).

24.8.4 AppleTalk

The AppleTalk protocol allows Windows NT to share a network with Macintosh clients. It can also act as an AppleShare server.

24.8.5 DLC

Data link control (DLC) is a communications protocol which is used with IBM mainframes. Windows NT interfaces to a DLC network.

24.9 Setting up TCP/IP networking on Windows NT

The default internetworking protocol for Windows NT/95 is NetBEUI. To use any Internet applications (such as FTP, TELNET, WWW browsers, and so on) the TCP/IP protocol must be installed. To achieve this the network icon in the control panel is selected, as shown in Figure 24.6. Next the network configuration screen is shown.

Windows NT has a DHCP (Dynamic Host Configuration Program) which assigns IP addresses from a pool of addresses. It relieves the system manager from assigning IP addresses to individual workstations and maintaining those addresses. Windows NT also has a name resolution service called WINS (Windows Internet Name Service). This program maps a computer name to an IP address; for example, `www.napier.ac.uk` is mapped to the IP address `146.176.131.10`. This facility is similar to the DNS server which is used on many TCP/IP networks. Note that a Windows NT server can support both WINS and DNS.

The settings for TCP/IP are set up by selecting Control Panel→Network and then, if they are not already set up, select Add→ Protocol→ Microsoft→ TCP/IP (as shown in Figure 24.7). After the network adapter is selected (such as NDIS) then select Properties from the TCP/IP option. This then gives the settings for:

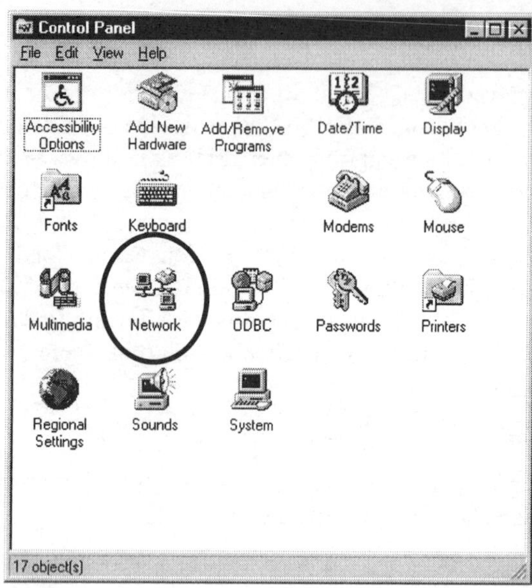

Figure 24.6 Windows NT selection of networking configuration

- TCP/IP properties – which is used to set the IP address of the host node. In the example in Figure 24.8 the node has an IP address of 146.176.151.130 and the subnet mask is 255.255.255.0.
- DNS configuration – which sets the IP address of the DNS server. In the case in Figure 24.9 there are two DNS servers, 146.176.150.62 and 146.176.151.99. The host is named `pc419` in the domain of `eece.napier.ac.uk`.
- Gateway – which is used to define the gateway node. In Figure 24.10 the gateway node is defined as 146.176.151.254.
- WINS server – which is used to define a WINS node (this functions as a DNS server). In Figure 24.11 the WINS node is defined as 146.176.151.50.

24.10 Windows sockets

A Windows socket (WinSock) are a standard method that allows nodes over a network to communicate with each other using a standard interface. It supports internetworking protocols such as TCP/IP, IPX/SPX, AppleTalk and NetBEUI. WinSock communicates through the TDI interface and uses the file `WINSOCK.DLL` or `WINSOCK32.DLL`. These DLLs (dynamic link libraries) contain a number of networking functions which are called in order to communicate with the transport and network layers (such as TCP/IP or SPX/IPX). As it communicates with these layers it is independent of the networking interface (such as Ethernet or FDDI).

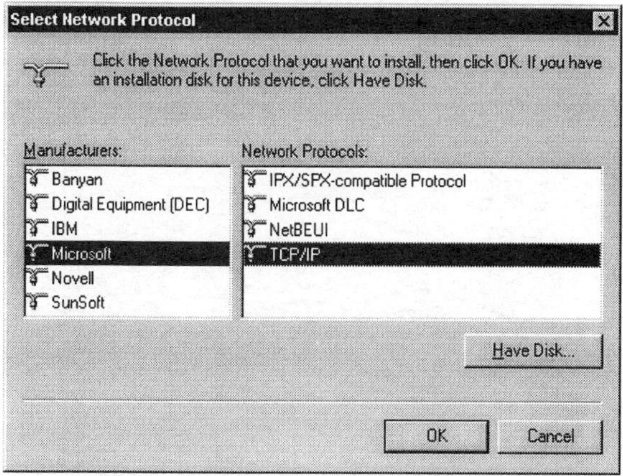

Figure 24.7 Setting-up for TCP/IP

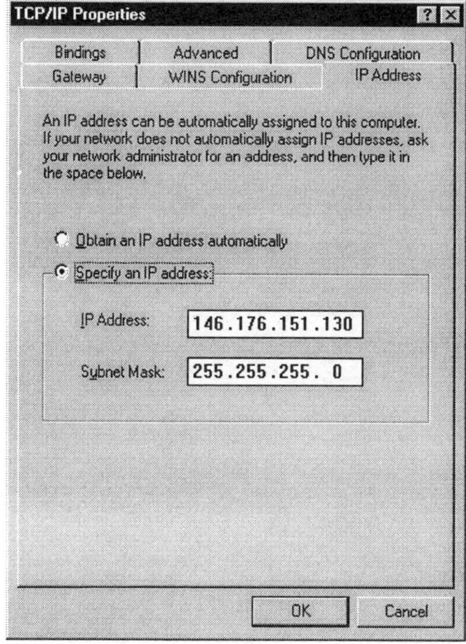

Figure 24.8 Setting-up an IP address

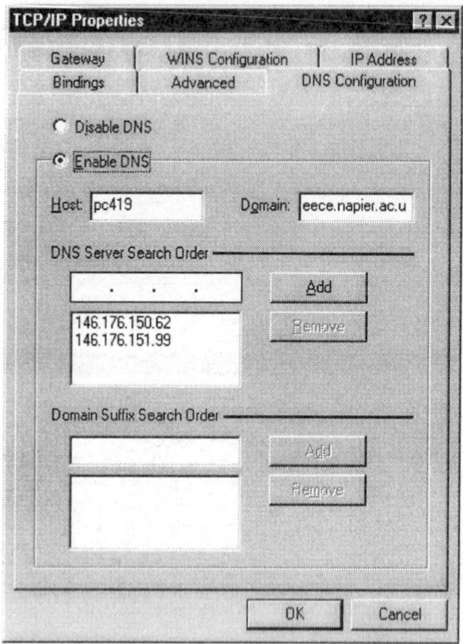

Figure 24.9 Windows NT DNS setup

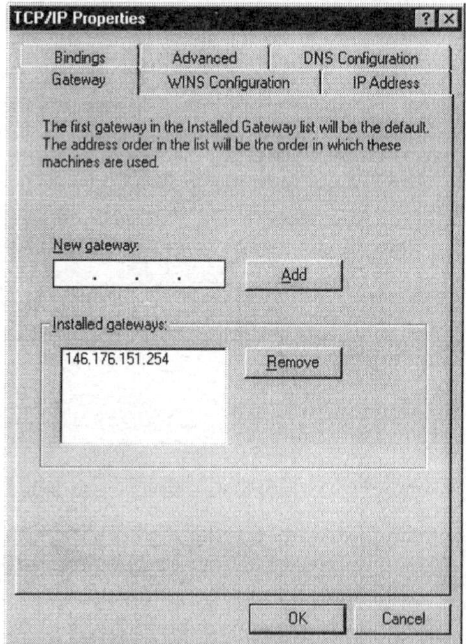

Figure 24.10 Setting up a gateway

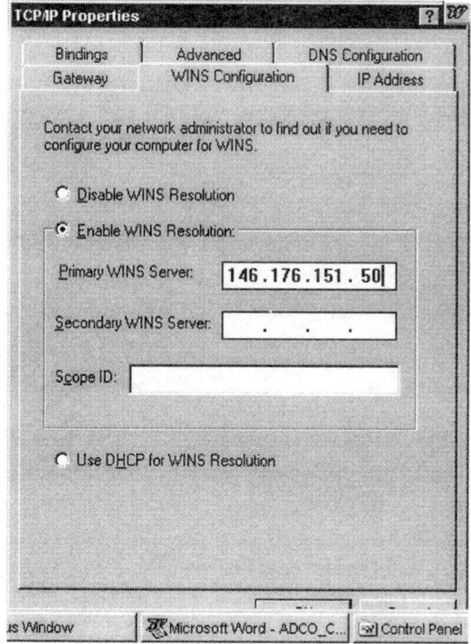

Figure 24.11 Windows NT WINS server IP address setup

24.11 Network dynamic data exchange (Net DDE)

Net DDE allows programs running on different computers on a network to exchange data. This is an extension of dynamic data exchange (DDE) which is used by programs running on the same computer to exchange data.

24.12 Robust networking

Windows NT provides fault tolerance in a number of ways. These are outlined in the following sections.

24.12.1 Disk mirroring

Windows NT servers support disk mirroring which protects against hard disk failure. It uses two partitions on different disk drives which are connected to the same controller. Data written to the first (primary) partition is mirrored automatically to the secondary partition. If the primary disk fails then the system uses the partition on the secondary disk. Mirroring also allows unallocated space on the primary drive to be allocated to the secondary drive. On a disk mirroring system the primary and secondary partitions have the same drive letter (such as C: or D:) and users are unaware that disks are being mirrored.

24.12.2 Disk duplexing

Disk duplexing means that mirrored pairs with different controllers. This have provides for fault tolerance on both disk and controller. Unfortunately it does not support multiple controllers connected to a single disk drive.

24.12.3 Striping with parity

Windows NT servers support disk striping with parity. This technique is based on RAID 5 (Redundant Array of Inexpensive Disks), where a number of partitions on different disks are combined to make one large logical drive. Data is written in stripes across all of the disk drives and additional parity bits. For example if a system has four disk drives then data is written to the first three disks and the parity is written to the fourth drive. Typically the stripe is 64 kB, thus 64 kBs will be written to drive 1, the same to drive 2 and drive 3, then the parity of the other three to the fourth. The following example illustrates the concept of RAID where a system writes the data `110`, `000`, `111`, `100` to the first three drives, this gives parity bits of `1`, `1`, `0` and `0`.

Disk 1	Disk 2	Disk 3	Disk 4 (Odd parity)
1	1	0	1
0	0	0	1
1	1	1	0
1	0	0	0

If one of the disk drives fails then the addition of the parity bit allows the bits on the failed disk to be recovered. For example if disk 3 fails then the bits from the other disk are simply XOR'ed together to generate the bits from the failed drive. If the data on the other disk drives is `111` then the recovered data gives `0`, `001` gives `0`, and so on.

The 64 kB stripes of data are also interleaved across the disks. The parity block is written to the first disk drive, then in the next block to the second, and so on. A system with four disk drives would store the following data:

Disk 1	Disk 2	Disk 3	Disk 4
Parity block 1	Data block A	Data block B	Data block C
Data block D	Parity block 2	Data block E	Data block F
Data block G	Data block H	Parity block 3	Data block I

Each of the data blocks will be 64 kB, which is also equal to the parity block.

The interlacing of the data ensures that the parity stripes are not all on the same disk, thus there is no single point of failure for the set.

Striping of data improves reading performance when each of the disk drives has a separate controller, because the data is simultaneously read by each of the controllers and simultaneously passed to the systems. It thus provides fast reading of data but only moderate writing performance (because the system must calculate the parity block).

The main advantages of RAID 5 can be summarized as:

- It recovers data when a single disk drive or controller fails (RAID level 0 does not use a parity block thus it cannot regenerate lost data).
- It allows a number of small partitions to be built into a large partition;
- Several disks can be mounted as a single drive.
- Performance can be improved with multiple disk controllers.

The main disadvantages of RAID 5 are:

- It requires increased memory because it generates of the parity block.
- Performance is reduced when one of the disks fails because of the need to regenerate the failed data.
- It increases the amount of disk space as it has an overhead due to the parity block (although the overhead is normally less than disk mirroring, which has a 50 % overhead).
- It requires at least three disk drives.

24.12.4 Tape backup

Windows NT provides for automatic tape based on the Maynard tape system. Tape backup allows data to be recovered when faults occur. Normally the system manager backs up the network at the start of the week and then does an incremental backup each day.

24.12.5 UPS services

Windows NT provides services to uninterruptable power supplies (UPSs). UPS systems provide power, from batteries, to a computer system when there is a glitch in the supply, power sags or power failure. Windows NT detects signals from a UPS unit and performs an orderly shutdown of applications, services and file systems as the stored energy in the UPS is depleted.

24.13 Security model

Windows NT treats all its resources as objects that can only be accessed by authorized users and services. Examples of objects are directories, printers, processes, ports, devices and files. On an NTFS partition the access to an object is controlled by the security descriptor (DS) structure which contains an access control list (ACL) and security identifier (SI). The SD contains the user (and group) accounts that have access and permissions to the object. The system always checks the ACL of an object to determine whether the user is allowed to access it.

The main parts of the SI are:

OWNER	Indicates the user account for the object.
GROUP	Indicates the group the object belongs to.
User ACL	The user-controller ACL.
System ACL	System manager controlled ACL.

The ACL file access rights are:

```
Full control          (All)   (All)
Read                  (RX)    (RX)
Change                (RWXD)  (RWXD)
Add                   (WX)
Change                (RWXD)
List                  (RX)
Change Permissions    (P)
```

24.14 TCP/IP applications

After the TCP/IP protocol has been installed (and all other TCP/IP drivers have been removed) then the system will be ready to run any TCP/IP applications. For example Windows NT and 95 are installed with Telnet and ftp. In Windows NT they are run by entering either Telnet or ftp from the run command, as illustrated in Figures 24.12 and 24.13.

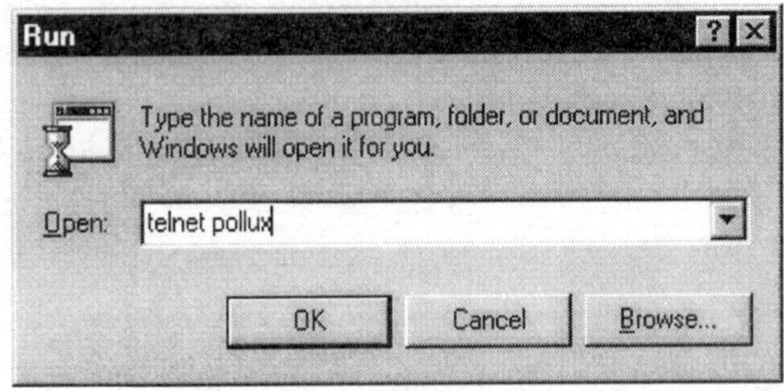

Figure 24.12 Running telnet

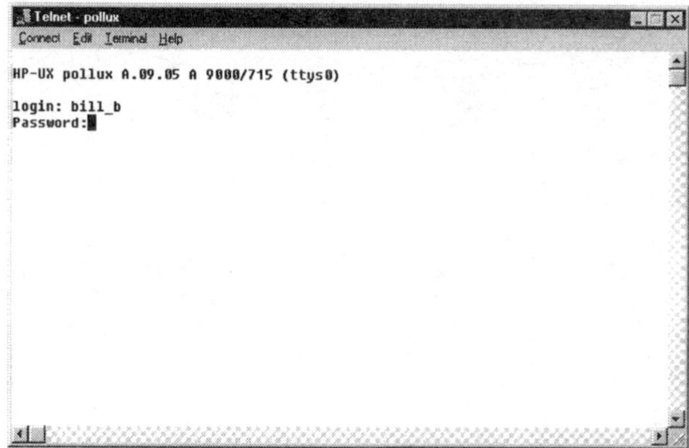

Figure 24.13 Example Telnet program

24.15 Windows NT network drives

Windows NT and 95 displays the currently mounted network drives within the group My Computer. Figure 24.14 shows drives which are either local (C: D: and E:) or mounted using NetWare (F: G: and so on). Windows NT and 95 also automatically scan the neighing networks to find network servers. An example is shown in Figures 24.15 and 21.15. Figure 24.15 shows the currently mounted servers (e.g. EECE_1) and by selecting the Global Network icon all the other connected local servers can be shown. Figure 24.16 shows an example network.

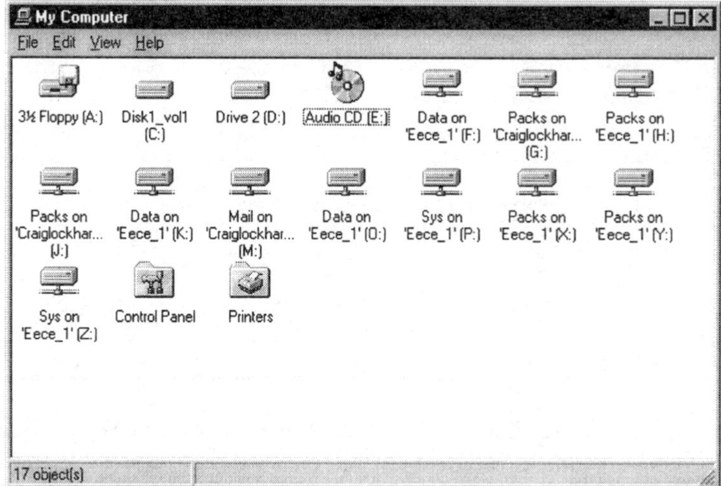

Figure 24.14 Mounted network and local drives

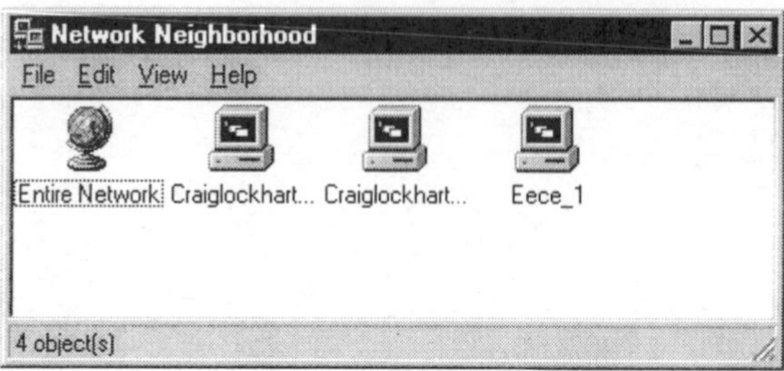

Figure 24.15 Network neighbourhood

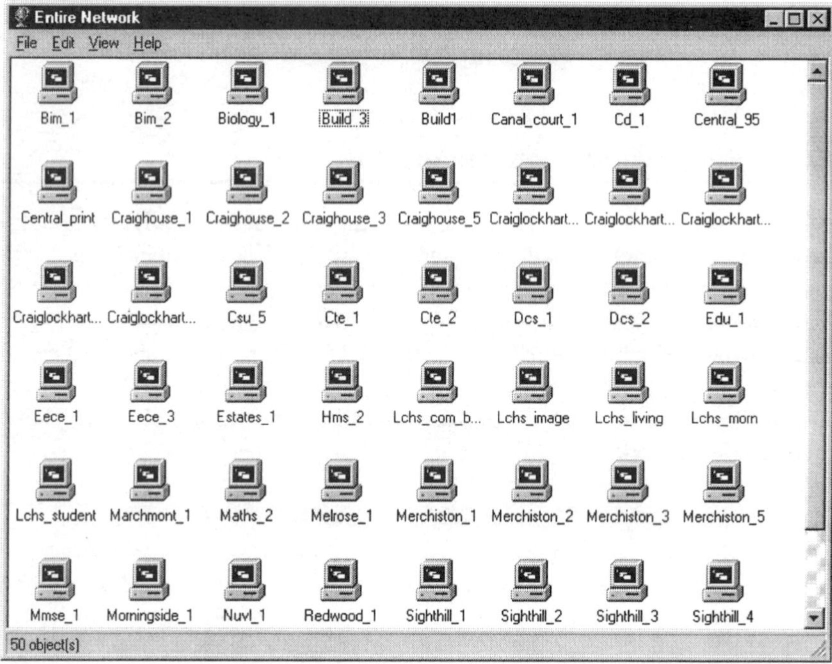

Figure 24.16 Local neighbourhood servers

24.16 Exercises

24.1 State the standard protocol that Windows NT and Windows 95 use to communicate between their clients and servers.

24.2 Explain how Windows NT deals with long file names and the DOS 8.3 format.

24.3 State the different file system formats that are used with Windows NT. Which file system is best when files are required to be undeleted and which is the best for determining bad sectors.

24.4 Explain how Windows NT uses NetBEUI.

24.5 If possible, investigate the TCP/IP settings for a Windows NT or Windows 95 node. Determine the node's IP address, the gateway it uses and its DNS server.

24.6 Explain the uses of WINS and DHCP. State the main advantage of DHCP.

24.7 Explain how application programs running on Windows NT use the file `WINSOCK.DLL`.

24.8 Explain how RAID 5 allows for robust networking.

24.9 Use the TELNET program in Windows NT or Windows 95 to log in to a remote computer.

24.10 Use the FTP program in Windows NT or Windows 95 to copy a file from a remote computer.

25 NetWare

25.1 Novell NetWare networking

Novell NetWare is one of the most popular network operating systems for PC LANs and provides file and print server facilities. Its default network protocol is normally SPX/IPX. This can also be used with Windows NT to communicate with other Windows NT nodes and with NetWare networks. The Internet Packet Exchange (IPX) protocol is a network layer protocol for transportation of data between computers on a NetWare network. IPX is very fast and has a small connectionless datagram protocol. The Sequenced Packet Interchange (SPX) provides a communications protocol which supervises the transmission of the packet and ensures its successful delivery.

25.2 NetWare and TCP/IP integration

NetWare is typically used in organizations and works well on a local network. Network traffic which travels out on the Internet or that communicates with UNIX networks must be in TCP/IP form. This section outlines possible methods used to integrate NetWare with TCP/IP traffic.

25.2.1 IP tunneling

IP tunneling encapsulates the IPX packet within the IP packet. This can then be transmitted into the Internet network. When the IP packet is received by the destination NetWare gateway; the IP encapsulation is stripped off. IP tunneling thus relies on a gateway into each IPX-based network that also runs IP. The NetWare gateway is often called an IP tunnel peer.

25.3 NetWare architecture

NetWare provides many services, such as file sharing, printer sharing, security, user administration and network management. The interface between the network interface card (NIC) and the SPX/IPX stack is ODI (Open Data-link Interface). NetWare clients run software which connects them to the server, the supported client operating systems are DOS, Windows, Windows NT, UNIX, OS/2 and Macintosh.

With NetWare Version 3, DOS and Windows 3 clients use a NetWare shell called NETx.COM. This shell is executed when the user wants to log into the network and stay resident. It acts as a commands redirector and processes requests which are either generated by application programs or from the keyboard. It then decides whether they should be handled by the NetWare network operating system or passed to the client's local DOS operating system. NETx builds its own tables to keep track of the location of network-attached resources rather than using DOS tables. Figure 25.1 illustrates the relationship between the NetWare shell and DOS, in a DOS-based client. Note that Windows 3 uses the DOS operating system, but Windows NT and 95 have their own operating systems and only emulate DOS. Thus, Windows NT and 95 do not need to use the NETx program.

The ODI allows NICs to support multiple transport protocols, such as TCP/IP and IPX/SPX, simultaneously. Also, in an Ethernet interface card, the ODI allows simultaneous support of multiple Ethernet frame types such as Ethernet 802.3, Ethernet 802.2, Ethernet II, and Ethernet SNAP. Figure 25.2 shows a configuration of the frame type for IPX/SPX protocol.

To install NetWare, the server must have a native operating system, such as DOS or Windows NT, and it must be installed on its own disk partition. NetWare then adds a partition in which the NetWare partition is added. This partition is the only area of the disk the NetWare kernel can access.

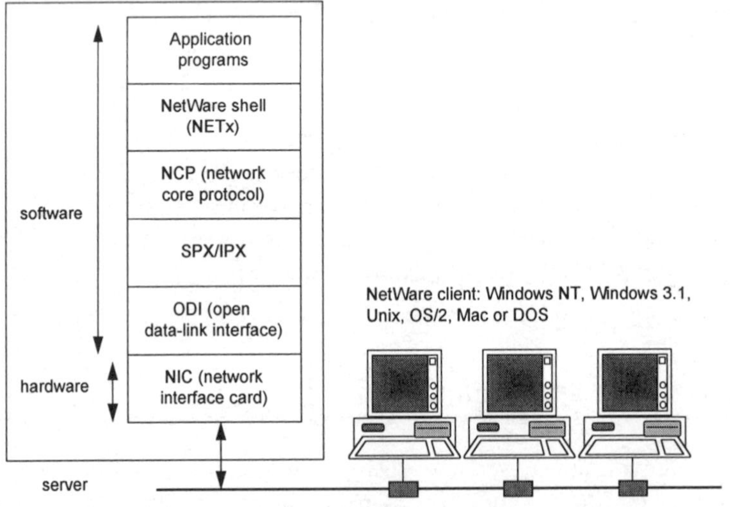

Figure 25.1 NetWare architecture

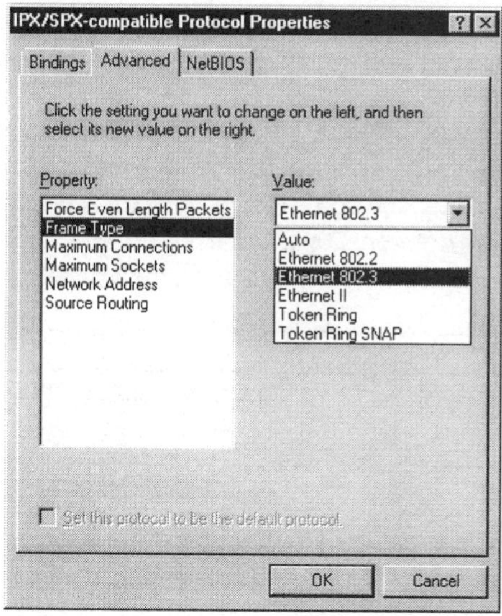

Figure 25.2 IPX/SPX frame types

25.3.1 NetWare loadable modules (NLMs)

NetWare allows enhancements from third-party suppliers using NLMs. The two main categories are:

- Operating systems enhancements – these allow extra operating system functions, such as a virus checker and also client hardware specific modules, such as a network interface drivers.
- Application programs – these programs actually run on the NetWare server rather than on the client machine.

25.3.2 Bindery services

NetWare must keep track of users and their details. Typically, NetWare must keep track of:

- User names and passwords.
- Groups and group rights.
- File and directory rights.
- Print queues and printers.
- User restrictions (such as allowable login times, the number of times a user can simultaneously log in to the network).
- User/group administration and charging (such as charging for user login).
- Connection to networked peripherals.

This information is kept in the bindery files. Whenever a user logs in to the network their login details are verified against the information in the bindery files.

The bindery is organized with objects, properties and values. Objects are entities that are controlled or managed, such as users, groups, printers (servers and queues), disk drives, and so on. Each object has a set of properties, such as file rights, login restrictions, restrictions to printers, and so on. Each property has a value associated with it. Here are some examples:

Object	*Property*	*Value*
User	Login restriction	Wednesday 9 am till 5pm
User	Simultaneous login	2
Group	Access to printer	No

Objects, properties and values are stored in three separate files which are linked by pointers on every NetWare server:

1. `NET$OBJ.SYS` (contains object information).
2. `NET$PROP.SYS` (contains property information).
3. `NET$VAL.SYS` (contains value information).

If multiple NetWare servers exist on a network then bindery information must be exchanged manually between the servers so that the information is the same on each server. In a multiserver NetWare 3.x environment, the servers send SAP (service advertising protocol) information between themselves to advertise available services. Then the bindery services on a particular server update their bindery files with the latest information regarding available services on other reachable servers. This synchronization is difficult when just a few servers exist but is extremely difficult when there are many servers. Luckily, NetWare 4.1 has addressed this problem with NetWare directory services; this will be discussed later.

25.4 NetWare protocols

NetWare uses IPX (Internet Packet Exchange) for the network layer and either SPX (Sequenced Packet Exchange) or NCP (NetWare Core Protocols) for the transport layer. The routing information protocol (RIP) is also used to transmit information between NetWare gateways. These protocols are illustrated in Figure 25.3.

OSI	NetWare	
Application	SAP	File server/ Application program
Transport	NCP	SPX
Network	IPX	
Data link	Ethernet/ Token Ring	
Physical		

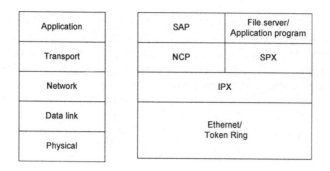

Figure 25.3 NetWare reference model

25.5 IPX

IPX performs a network function that is similar to IP. The higher information is passed to the IPX layer which then encapsulates it into IPX envelopes. It is characterized by:

- A Connectionless connection – each packet is sent into the network and must find its own way through the network to the final destination (connections are established with SPX)
- It is unreliable – as there is only basic error checking and no acknowledgment (acknowledgments are achieved with SPX).

IPX uses a 12-byte station address (whereas IP uses a 4-byte address). The IPX fields are:

- Checksum (2 bytes) – this field is rarely used in IPX, as error checking is achieved in the SPX layer. The lower-level data link layer also provides an error detection scheme (both Ethernet and Token Ring support a frame check sequence).
- Length (2 bytes) – this gives the total length of the packet in bytes (i.e. header+DATA). The maximum number of bytes in the DATA field is 546, thus the maximum length will be 576 bytes (2+2+1+1+12+12+546).
- Transport control (1 byte) –this field is incremented every time the frame is processed by a router. When it reaches a value of 16 it is deleted. This stops packets from traversing the network for an infinite time. It is also typically known as the time-to-live field or hop counter.
- Packet type (1 byte) –this field identifies the upper layer protocol so that the DATA field can be properly processed.
- Addressing (12 bytes) –this field identifies the address of the source and destination station. It is made up of three fields: a network address (4

bytes), a host address (6 bytes) and a socket address (2 bytes). The 48-bit host address is the 802 MAC LAN address. NetWare supports a hierarchical addressing structure where the network and host addresses identify the host station and the socket address identifies a process or application and thus supports multiple connections (up to 50 per node).

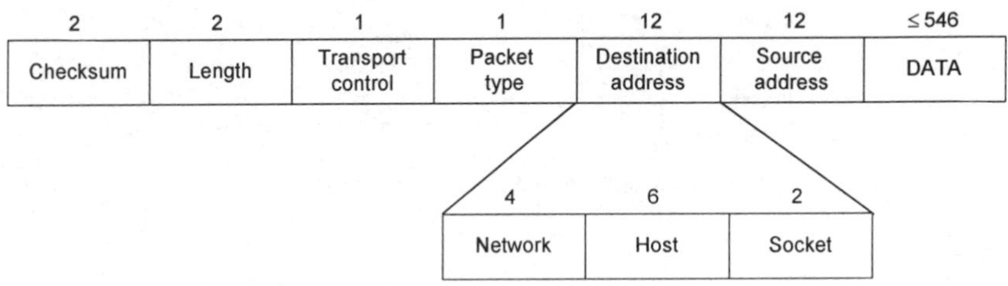

Figure 25.4 IPX packet format

25.5.1 SPX

On a NetWare network the level above IPX is either NCP or SPX. The SPX protocol sets up a virtual circuit between the source and the destination. Then all SPX packets follow the same path and will thus always arrive in the correct order. This type of connection is described as connection-oriented.

SPX also allows for error checking and an acknowledgment to ensure that packets are received correctly. Each SPX has flow control and also sequence numbers. Figure 25.5 illustrates an SPX packet and shows how the IPX header is added to the SPX packet.

The fields in the SPX header are:

- Connection control (1 byte) – this is a set of flags which assist the flow of data. These flags include an acknowledgment flag and an end-of-message flag.
- Datastream type (1 byte) – this byte contains information which can be used to determine the protocol or information contained within the SPX data field.

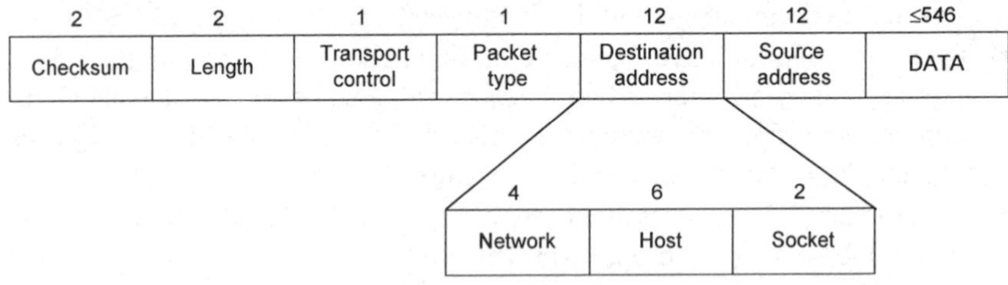

Figure 25.5 SPX packet format

- Destination connection ID (2 bytes) – the destination connection ID allows the routing of the packet through the virtual circuit.
- Source connection ID (2 bytes) – the source connection ID identifies the source station when it is transmitted through the virtual circuit.
- Sequence number (2 bytes) – this field contains the sequence number of the packet sent. When the receiver receives the packet, the destination error checks the packet and sends back an acknowledgment with the previously received packet number in it.
- Acknowledgment number (2 bytes) – this acknowledgment number is incremented by the destination when it receives a packet. It is in this field that the destination station puts the last correctly received packet sequence number.
- Allocation number (2 bytes) – this field informs the source station of the number of buffers the destination station can allocate to SPX connections.
- DATA (up to 534 bytes)

25.5.2 RIP

The NetWare Routing Information Protocol (RIP) is used to keep routers updated on the best routes through the network. RIP information is delivered to routers via IPX packets. Figure 25.6 illustrates the information fields in an RIP packet. The RIP packet is contained in the field which would normally be occupied by the SPX packet.

Routers are used within networks to pass packets from one network to another in an optimal way (and error-free with a minimumal time delay). A router reads IPX packets and examines the destination address of the node. If the node is on another network then it routes the packet in the required direction. This routing tends not to be fixed as the best route will depend on network traffic at given times. Thus the router needs to keep the routing tables up to date; RIP allows routers to exchange their current routing tables with other routers.

The RIP packet allows routers to request or report on multiple reachable networks within a single RIP packet. These routes are listed one after another (Figure 25.6 shows two routing entries). Thus each RIP packet has only one operation field, but has multiple entries of the network number, the number of router hops, and the number of tick fields, up to the length limit of the IPX packet.

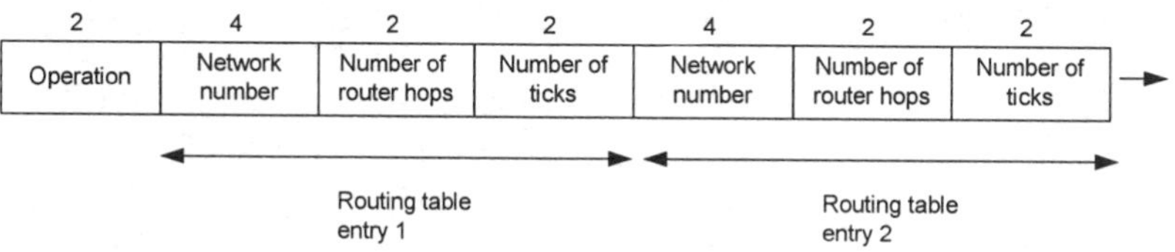

Figure 25.6 RIP packet format

The fields are:

- Operation (2 bytes) – this field indicates that the RIP packet is either a request or a response.
- Network number (4 bytes) – this field defines the assigned network address number to which the routing information applies.
- Number of router hops (2 bytes) – this field indicates the number of routes that a packet must go through in order to reach the required destination. Each router adds a single hop.
- Number of ticks (2 bytes) – this field indicates the amount of time (in 1/18 second) that it takes a packet to reach the given destination. Note that a route which has the fewest hops may not necessarily be the fastest.

RIP packets add to the general network traffic as each router broadcasts its entire routing table every 60 seconds. This shortcoming has been addressed by NetWare 4.1.

25.5.3 SAP

Every 60 seconds each server transmits a SAP (Service Advertising Protocol) packet which gives its address and tells other servers which services it offers. These packets are read by special agent processes running on the routers which then construct a database that defines which servers are operational and where they are located.

When the client node is first booted it transmits a request in the network asking for the location of the nearest server. The agent on the router then reads this request and matches it up to the best server. This choice is then sent back to the client. The client then establishes an NCP (NetWare Core Protocol) connection with the server, from which the client and server negotiate the maximum packet size. After this, the client can access the networked file system and other NetWare services.

Figure 25.7 illustrates the contents of a SAP packet. It can be seen that each SAP packet contains a single operation field and data on up to seven servers. The fields are:

- Operation type (2 bytes) – defines whether the SAP packet is server information request or a broadcast of server information.
- Server type (2 bytes) – defines the type of service offered by a server. These services are identified by a binary pattern, such as:

File server	0000 1000	Job server	0000 1001
Gateway	0000 1010	Print server	0000 0111
Archive server	0000 1001	SNA gateway	0010 0001
Remote bridge server	0010 0100	TCP/IP gateway	0010 0111
NetWare access server	1001 1000		

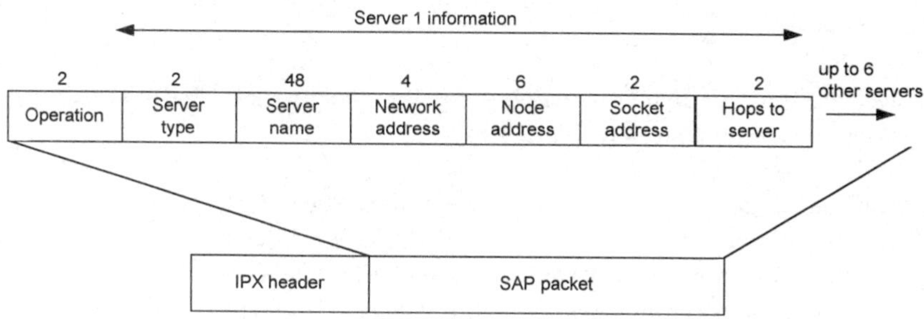

Figure 25.7 SAP packet format

- Server name (48 bytes) – which identifies the actual name of the server or host offering the service defined in the service type field.
- Network address (4 bytes) – which defines the address of the network to which the server is attached.
- Node Address (6 bytes) – which defines the actual MAC address of the server.
- Socket address (6 bytes) – which defines the socket address on the server assigned to this particular type of service.
- Hops to server (2 bytes) – which indicates the number of hops to reach the particular service.

25.5.4 NCP

The clients and servers communicate using the NetWare Core Protocols (NCPs). They have the following operation:

- The NETx shell reads the application program request and decides whether it should direct it to the server.
- If it does redirect, then it sends a message within an NCP packet, which is then encapsulated with an IPX packet and transmitted to the server.

Figure 25.8 illustrates the packet layout and encapsulation of an NCP packet.

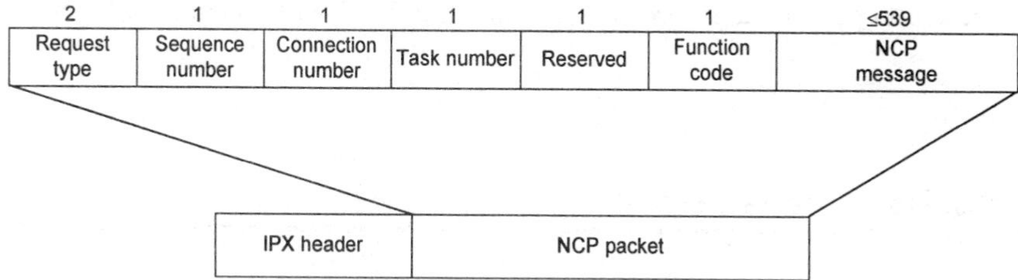

Figure 25.8 NCP packet format

The fields are:

- Request type (2 bytes) – which gives the category of NCP communications. Among the possible types are:

Busy message	1001 1001 1001 1001
Create a service	0001 0001 0001 0001
Service request from workstation	0010 0010 0010 0010
Service response from server	0011 0011 0011 0011
Terminate a service connection	0101 0101 0101 0101

For example the create-a-service request is initiated at login time and a terminate-a-connection request is sent at logout.

- Sequence number (1 byte) – which contains a request sequence number. The client reads the sequence number so that it knows the request to which the server is responding to.
- Connection number (1 byte) –a unique number which is assigned when the user logs into the server.
- Task number (1 byte) – which identifies the application program on the client issued the by service request.
- Function code (1 byte) – which defines the NCP message or commands. Example codes are:

Close a file	0100 0010	Create a file	0100 1101
Delete a file	0100 0100	Get a directory entry	0001 1111
Get file size	0100 0000	Open a file	0100 1100
Rename a file	0100 0101	Extended functions	0001 0110

  Extended functions can be defined after the 0001 0110 field.

- NCP message (up to 539 bytes) – the NCP message field contains additional information which is passed between the clients and servers. If the function code contains 0001 0110 then this field will contain subfunction codes.

25.6 Novel NetWare setup

NetWare 3.x and 4.1 use the Open Data-Link Interface (ODI) to interface NetWare to the NIC. Figure 25.9 shows how the NetWare 3.x fits into the OSI model. ODI is similar to NDIS in Windows NT and was developed jointly between Apple and Novell. It provides a standard vendor-independent method to interface the software and the hardware (Figure 25.10 shows that Windows NT can choose between NDIS2 and ODI).

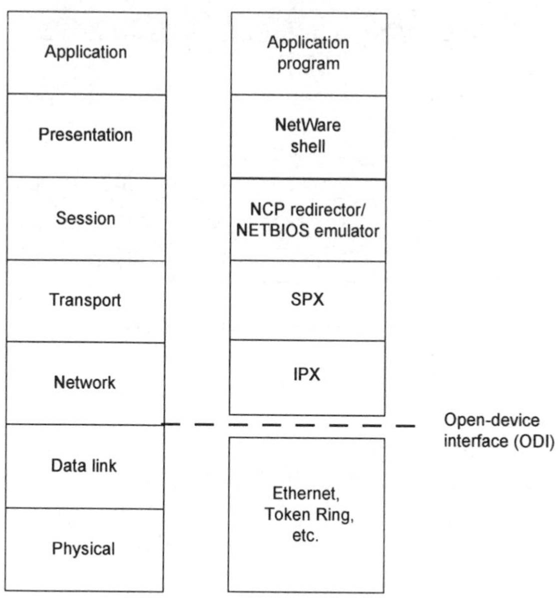

Figure 25.9 OSI model and NetWare 3.x

A typical login procedure for a NetWare 3.x network is:

```
LSL.COM
NE2000
IPXODI
NETx /PS=EECE_1
F:
LOGIN
```

The program `LSL` (link support layer) provides a foundation for the MAC layer to communicate with multiple protocols. An interface adapter driver (in this case `NE2000`) provides a MAC layer driver and is used to communicate with the interface card. This driver is known as a multilink interface driver (MLID). After this driver is installed, the program `IPXODI` is then installed. This program normally communicates with `LSL` and applications.

The `NETx` program communicates with the server and sets up a connection with the server `EECE_1`. This then sets up a local disk partition of `F:` (onto which the user's network directory will be mounted). Next the user logs in to the network with the command `LOGIN`.

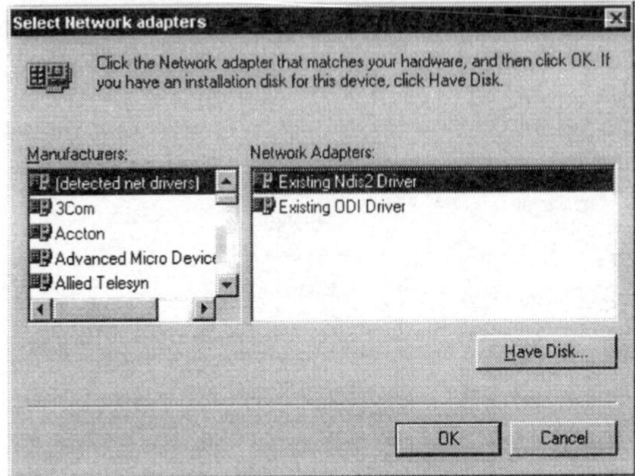

Figure 25.10 Network adapter driver

25.6.1 ODI

ODI allows users to load several protocol stacks (such as TCP/IP and SPX/IP) simultaneously for operation with a single NIC. It also allows support to link protocol drivers and to adapter drivers. Figure 25.11 shows the architecture of the ODI interface. The LSL layer supports multiple protocols and it reads from a file `NET.CFG`, which contains information on the network adapter and the protocol driver, such as the interface adapter, frame type and protocol.

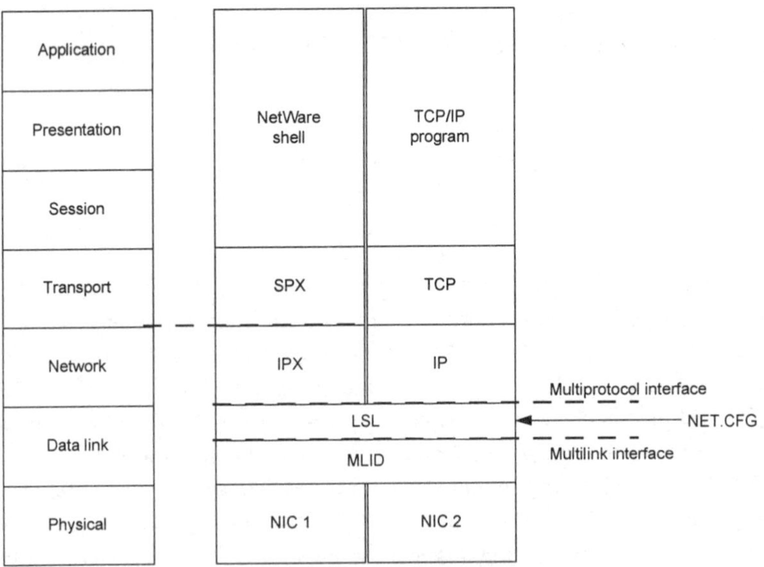

Figure 25.11 ODI architecture

Here is a sample NET.CFG file is:

```
Link Driver NE2000
   Int #1 11
   Port #1 320
   Frame Ethernet_II
   Frame Ethernet_802.3
   Protocol IPX 0 Ethernet_802.3
          Protocol Ethdev 0 Ethernet_II
```

This configuration file defines the interface adapter as using interrupt line 11, having a base address of 320h, operating IPX, Ethernet 802.3 frame type and following the Ethernet II protocol. Network interface card drivers (such as NE2000, from the previous setup) are referred to as a multilink interface driver (MLID).

25.7 NetWare 4.1

The main disadvantages of NetWare 3.x are:

- It uses SPX/IPX which is incompatible with TCP/IP traffic.
- It is difficult to synchronize servers with user information.
- The file structure is local to individual servers.
- Its server architecture is flat; it cannot be organized into a hierarchy structure.

These disadvantages have been addressed with NetWare 4.1 which has:

- A hierarchical server structure.
- etwork-wide users and groups.
- Global objects.
- System-wide login with a single password.
- Support for a distributed file system.

25.7.1 NetWare directory services (NDS)

One of the major changes between NetWare 3.x and NetWare 4.1 is NDS. A major drawback of the NetWare 3.x bindery files is that they were independently maintained on each server. NDS addresses this by setting up a single logical database which contains information on all network-attached resources. It is logically a single database but may be physically located on different servers over the network. As the database is global to the network, a user can log in to all authorized network-attached resources, rather than requiring to login into each separate server. Thus administation is focused on the single database.

As with NetWare 3.x bindery services, NDS organizes network resources by objects, properties, and values. NDS differs from the bindery services in that it defines two types of object:

- Leaf objects – which are network resources such as disk volumes, printers, printer queues, and so on.
- Container objects – which are cascadable organization units that contain leaf objects. A typical organizational unit might be company, department or group.

NDS organizes networked resourses in a hierachical or tree structure (as most organizations are structured). The top of the tree is the root object, to which there is only a single root for an entire global NDS database. Servers then use container objects to connect to branches coming off the root object. This structure is similar to the organization of a directory file structure and can be used to represent the hierarchical structure of an organization. Figure 25.12 illustrates a sample NDS database with root, container and leaf objects. In this case the organization splits into four main containers: electrical, mechanical, production and administation. Each of the containers has associated leaf objects, such as disk drivers, printer queues, and so on. This is a similar approach to Workgroups in Microsoft Windows.

To improve fault tolerance, NDS allows branches of the tree (or partitions) to be stored on multiple file servers. These copies are then synchronized to keep them up-to-date. Another advantage of replicating partions is that local copies of files can be stored so that network traffic is reduced.

25.7.2 Virtual loadable modules (VLMs)

The NETx redirector shell has been replaced with DOS client software known as the requester. Its main advantage is that it allows NetWare client to easily add or update their functionality by using VLMs. This is controlled through the DOS-based VLM management program (VLM.EXE). It differs from NETx in that the requester uses DOS tables of network-attached resources rather than creating and maintaining its own. The main difference between NETx and the requester is that it is the DOS system which controls whether the NetWare DOS request is called to handle network requests.

Various VLM modules can be added onto the client, such as:

- Bindery-based services.
- File management.
- IPX and NCP protocol stacks.

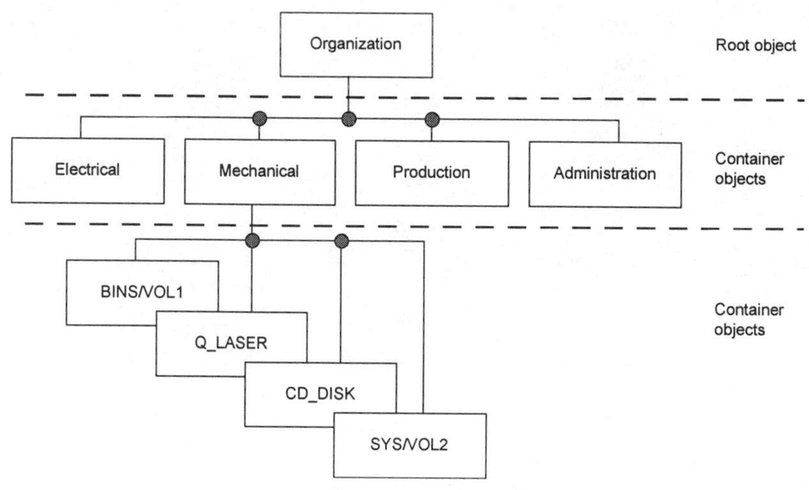

Figure 25.12 NDS structure

- NDS services.
- NetWare support for multiple protocol stacks (e.g. TCP/IP, SPX/IPX).
- NETx shell emulation.
- Printer redirector to network print queues.
- TCP/IP and NCP protocol stacks.

25.7.3 Fault tolerance

The previous chapter discussed how Windows NT servers use RAID (Redundant Array of Inexpensive Disks) with parity to allow the recovery of data when a disk drive fails. NetWare 4.1 allows disk mirroring of partitions when a disk drive fails.

Another major fault occurs when a server becomes inoperative. NetWare 4.1 uses a novel technique, known as SFT III, which allows server duplexing. In this technique the contents of the disk, memory and CPU are synchronized between primary and duplexed servers. When the primary server fails then the duplexed server takes over transparently. These servers are synchronized using the mirror server link (MSL), a dedicated link between the two servers, as illustrated in Figure 25.13. The MSL is a dedicated link because it prevents general network traffic from swamping the data.

It may seem expensive to have a backup server doing nothing apart from receiving data, but if it is costed with the loss of business or data when the primary server goes down then it is extermely cheap.

25.7.4 Communications protocols

NetWare 4.x has improved existing protocols and added the support for other standard network protocols, especially TCP/IP.

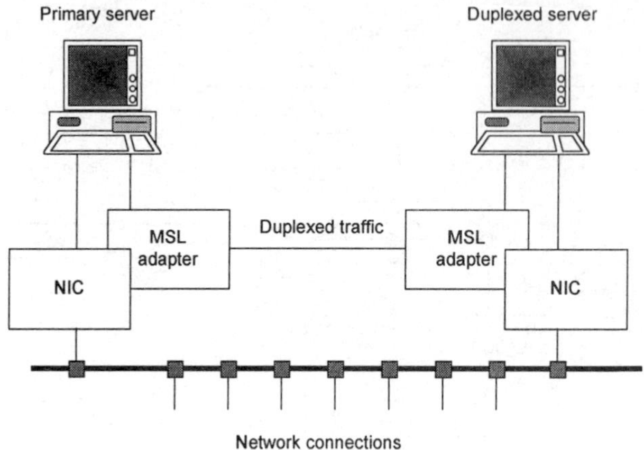

Figure 25.13 Mirror server link

TCP/IP

TCP/IP is supported with NetWare/IP which is included with NetWare 4.1; NetWare/IP servers can support IP, IPX or IP and IPX traffic.

Large IPX packets (LIPs)

Most networks have become less prone to error. Thus larger data packets can be transmitted with a low risk of errors occuring. LIP allows NetWare clients to increase the size of their DATA field by negotiating with routers as to the size of the IPX frame (normally its has a maximum of 576 bytes). Unfortunately an error in the packet causes the complete packet to be retransmitted (thus causing inefficencies). Also, the router must support the use of LIP. The software-based Novell router has a multiprotocol router which supports LIP. Unfortunately, other vendors may not support the LIP protocol.

NetWare Link-State Routing Protocol (NLSP)

RIP has several disadvantages, these include:

- File servers transmit their routing table every 60 seconds. This can have a great effect on the network loading, especially for interconnected networks.
- RIP only supports 16 hops before an RIP packet is discarded, thus limiting the physical size of the internetwork linking NetWare LAN

NLSP, which is included in NetWare 4.1, overcomes these problems. With the routing table, NLSP only broadcasts when a change occurs, with a mimum update of once every 2 hours. This can significantly reduce the router-to-router traffic. As with LIP, Novell routers support NLSP, but other vendors may not nescessarily support it.

NLSP supports an increased hop size. A great advantage with NLSP is that it can coexist with RIP and is thus backward compatible. This allows a gradual migration of network segments to NLSP.

25.7.5 NetWare 4.1 SMP

One of the great improvements in computer processing and power will be achievable through the use of parallel processing. This processing can either be realized using multiple local processors or network processors, called symmetrical multiprocessing (SMP). To maintain compatiblity with a previous release, NetWare 4.1 SMP loads the SMP kernel which works cooperatively with the operating system kernel. The main processor runs the main operating system while the SMP kernel runs the 2nd, 3rd and 4th processors.

25.7.6 Other enhancements

Other enhancements have been added, such as:

- File compression which is controllable on a file-by-file basis.
- Increased supervisor security.
- Increased support for printers (up to 255 can be connected).

25.8 Exercises

25.1 For a known NetWare network determine the following:

(a) Its version number.
(b) Its architecture.
(c) The connected perpherals (e.g. printers and tape backups).
(d) The number of user logins.
(e) File servers.
(f) The connections it makes with other NetWare servers.
(g) The connections it makes with the Internet.
(h) The location of bridges, routers or gateways.

25.2 Discuss the format of an IPX packet.

25.3 Discuss the format of an SPX packet.

25.4 Discuss the RIP, SAP and NCP packets.

25.5 Outline the main advantages of NetWare 4 over NetWare 3.

25.6 Explain how NetWare 4 uses NDS.

25.7 Explain how NetWare 4 allows server fault tolerance.

26 UNIX

26.1 Introduction

UNIX is an extremely popular operating system and dominates in the high-powered, multitasking workstation market. It is relatively simple to use and to administer, and also has a high degree of security. UNIX computers use TCP/IP communications which they uses to mount disk resources from one machine onto another. Its main characteristics are:

- Multiuser.
- Preemptive multitasking.
- Multiprocessing.
- Multithreaded applications.
- Memory management with paging (organizing programs so that the program is loading into pages of memory) and swapping (which involves swapping the contents of memory to disk storage).

The two main families of UNIX are UNIX System V and BSD (Berkeley Software Distribution) Version 4.4. System V is the operating system most often used and has descended from a system developed by the Bell Laboratories and was recently sold to SCO (Santa Cruz Operation). Popular UNIX systems are:

- AIX (on IBM workstations and mainframes).
- HP-UX (on HP workstations).
- Linux (on PC-based systems).
- OSF/1 (on DEC workstations).
- Solaris (on Sun workstations).

An initiative by several software vendors has resulted in a common standard for the user interface and the operation of UNIX. The user interface standard is defined by the common desktop environment (CDE). This allows software vendors to write calls to a standard CDE API (application-specific interface). The common UNIX standard has been defined as Spec 1170 APIs. Compliance with the CDE and Spec 1170 API are certified by X/Open, which is a UNIX standard organization.

26.2 Network setup

Modern UNIX-based systems tend to be based around four main components:

- UNIX operating system.
- TCP/IP communications.
- Network file system (NFS).
- X-Windows interface.

X-Windows presents a machine-independent user interface for client/server applications. This will be discussed in more detail in the next chapter. TCP/IP allows for network communications and NFS allows disk drives to be linked together to make a global file system.

26.3 TCP/IP protocols

UNIX uses the normal range of TCP/IP protocols, grouped into transport, routing, network addresses, users services, gateway and other protocols.

Routing

Routing protocols manage the addressing of the packets and provide a route from the source to the destination. Packets may also be split up into smaller fragments and reassembled at the destination. The main routing protocols are:

- ICMP (Internet Control Message Protocol) which support status messages for the IP protocol these may be errors or network changes that can affect routing.
- IP (Internet Protocol) which defines the actual format of the IP packet.
- RIP (Routing Information Protocol) which is a route determining protocol

Transport

The transport protocols are used by the transport layer to transport a packet around a network. The protocols used are:

- TCP (Transport Control Protocol) which is a connection-based protocol where the source and the destination make a connection and maintain the connection for the length of the communications.
- UDP (User Datagram Protocol) which is a connectionless server where there is no communication between the source and the destination.

Network addresses

The network address protocols resolve IP addresses with their symbolic names, and vice versa. These are:

- ARP (Address Resolution Protocol) determines the IP address of nodes on a network.
- DNS (Domain Name System) which determines IP addresses from symbolic names (such as `anytown.ac.uk` might be resolved to 112.123.33.22).

User services

These are applications to which users have direct access.

- BOOTP (Boot Protocol) which is typically used to start up a diskless networked node. Thus rather than reading boot information from its local disk it reads the data from a server. Typically it is used by X-Windows terminals.
- FTP (File Transfer Protocol) which is used to transfer files from one node to another.
- Telnet which is used to remotely log in to another node.

Gateway protocols

The gateway protocols provide help for the routing process, these protocols include:

- EGP (Exterior Gateway Protocol) which transfers routing information for external network.
- GGP (Gateway-to-Gateway Protocol) which transfers routing information between Internet gateways.
- IGP (Interior Gateway Protocol) which transfers routing information for internal networks.

Others services

Other important services provide support for networked files systems, electronic mail and time synchronization as well as helping maintain a global network database. The main services are:

- NFS (Network File System) which allows disk drives on remote nodes to be mounted on a local node and thus create a global file system. This will be discussed in more detail in Section 26.4.
- NIS (Network Information Systems) which maintain a network-wide data-

base for user accounts and thus allows users to log in to any computer on the network. Any changes to a user's accounts are made over the whole network.

- NTP (Network Time Protocol) which is used synchronize clocks of nodes on the network.
- RPC (Remote Procedure Call) which enables programs running on different nodes on a network to communicate with each other using standard function calls.
- SMTP (Simple Mail Transfer Protocol) which is a standard protocol for transferring electronic mail messages.
- SNMP (Simple Network Management Protocol) which maintains a log of status messages about the network.

26.4 NFS

The Network File System (NFS) allows computers to share the same files over a network and was originally developed by Sun Microsystems. It has the great advantage that it is independent of the host operating system and can provide data sharing among different types of systems (heterogeneous systems).

NFS uses a client/server architecture where a computer can act as an NFS client, an NFS server or both. An NFS client makes requests to access data and files on servers; the server then makes that specific resource available to the client.

NFS servers are passive and stateless. They wait for requests from clients and they maintain no information on the client. One advantage of servers being stateless is that it is possible to reboot servers without adverse consequences to the client.

The components of NFS are as follows:

- NFS remote file access may be accompanied by network information service (NIS).
- External data representation (XDR) is a universal data representation used by all nodes. It provides a common data representation if applications are to run transparently on a heterogeneous network or if data is to be shared among heterogeneous systems. Each node translates machine-dependent data formats to XDR format when sending and translating data. It is XDR that enables heterogeneous nodes and operating systems to communicate with each other over the network.
- Remote Procedure Call (RPC) provides the ability for clients to transparently execute procedures on remote systems of the network. NFS services run on top of the RPC, which corresponds to the session layer of the OSI model.

- Network lock manager (`rpc.lockd`) allows users to coordinate and control access to information on the network. It supports file locking and synchronizes access to shared files.

Figure 26.1 shows how the protocols fit into the OSI model.

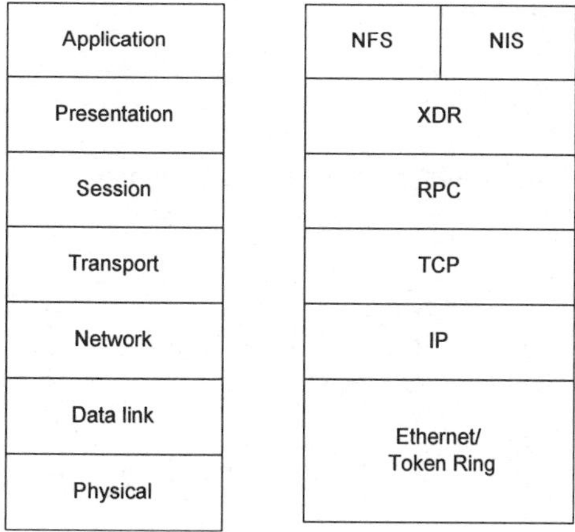

Figure 26.1 NFS services protocol stack

26.4.1 Network Information Service (NIS)

NIS is an optional network control program which maintains the network configuration files over a network. Previously it was named *Yellow Pages* (YP), but has had a name change as this is a registered trademark of the company British Telecommunications. It normally administers the network configuration files such as `/etc/group` (which defines the user groups), `/etc/hosts` (which defines the IP address and symbolic names of nodes on a network), `/etc/passwd` (which contains information, such as user names, encrypted passwords, home directories, and so on). Here is an except from a `passwd` file:

```
root:FDEc6.32:1:0:Super unser:/user:/bin/csh
fred:jt.06hLdiSDaA:2:4:Fred Blogs:/user/fred:/bin/csh
fred2:jtY067SdiSFaA:3:4:Fred Smith:/user/fred2:/bin/csh
```

This `passwd` file has three defined users, these are `root`, `fred` and `fred2`. The encrypted password is given in the second field (between the first and second colon). The third field is a unique number that defines the user (in this case `fred` is 2 and `fred2` is 3). The fourth field in this case defines the group number (which ties up with the `/etc/groups` file. An example of a `groups` file is given next. It can be seen from this file that group 4 is defined as `freds_group`, and contains three users: `fred`, `fred2` and `fred3`. The fifth field is simply a comment field and in this case it contains the user's names. In the next field each user's home directory is defined and the final field contains the initial UNIX shell (in this case it is the C-shell).

```
root::0:root
other::1:root,hpdb
bin::2:root,bin
sys::3:root,uucp
freds_grp::4:fred,fred2,fred3
```

A sample listing of a directory shows that a file owned by `fred` has the group name `freds_grp`.

```
> ls -l
-r-sr-xr-x   1 fred      freds_grp 24576   Apr 22   1997 file1
-r-xr-xr-x  13 fred      freds_grp 40      Apr 22   1997 file2
dr-xr-xr-x   2 fred      freds_grp 1024    Aug  5   14:01 myfile
-r-xr-xr-x   1 fred      freds_grp 32768   Apr 22   1997 text1.ps
-r-xr-sr-x   1 fred      freds_grp 24576   Apr 22   1997 text2.ps
-r-xr-xr-x   2 fred      freds_grp 16384   Apr 22   1997 temp1.txt
-r-xr-xr-x   1 fred      freds_grp 16384   Apr 22   1997 test.doc
```

An excerpt from the `/etc/hosts` file is shown next.

```
138.38.32.45     bath
198.4.6.3        compuserve
193.63.76.2      niss
148.88.8.84      hensa
146.176.2.3      janet
146.176.151.51   sun
```

The `/etc/protocols` file contains information with known protocols used on the Internet.

```
# The form for each entry is:
# <official protocol name><protocol number> <aliases>
#
# Internet (IP) protocols
#

ip       0  IP       # internet protocol, pseudo protocol number
icmp     1  ICMP     # internet control message protocol
ggp      3  GGP      # gateway-gateway protocol
tcp      6  TCP      # transmission control protocol
egp      8  EGP      # exterior gateway protocol
pup      12 PUP      # PARC universal packet protocol
udp      17 UDP      # user datagram protocol
hmp      20 HMP      # host monitoring protocol
xns-idp 22 XNS-IDP # Xerox NS IDP
rdp      27 RDP      # "reliable datagram" protocol
```

The /etc/netgroup file defines network-wide groups used for permission checking when doing remote mounts, remote logins, and remote shells. Here is a sample file:

```
# The format for each entry is: groupname  member1  member2 ...
#     (hostname, username, domainname)
engineering  hardware software (host3, mikey, hp)
hardware   (hardwhost1, chm, hp)   (hardwhost2, dae, hp)
software   (softwhost1, jad, hp)   (softwhost2, dds, hp)
```

NIS master server and slave server

With NIS, a single node on a network acts as the NIS master server, and there may be a number of NIS slave servers. The slave servers receive their NIS information from the master server. When a client first starts up it sends out a broadcast to all NIS servers (master or slaves) on the network and waits for the first to respond. The client then binds to the first that responds and addresses all NIS requests to that server. If this server becomes inoperative then an NIS client will automatically rebind to the first NIS server which responds to another broadcast.

Table 26.1 outlines the records which are used in the NIS database (or NIS map). This file consists of logical records with a search key and a related value for each record. For example, in the passwd.byname map, the users' login names are the keys and the matching lines from /etc/passwd are the values.

NIS domain

An NIS domain is a logical grouping of the set of maps contained on NIS servers. The rules for NIS domains are:

- All nodes in an NIS domain have the same domain name.
- Only one master server exists on an NIS domain.
- Each NIS domain can have zero or more slave servers.

An NIS domain is a subdirectory of /usr/etc/yp on each NIS server, where the name of the subdirectory is the name of the NIS domain. All directories that appear under /usr/etc/yp are assumed to be domains that are served by an NIS server. Thus to remove a domain being served, the user deletes the domain's subdirectory name from /etc/etc/yp on all of its servers.

The start up file on most UNIX systems is the /etc/rc file. This automatically calls the /etc/netnfsrc file which contains the default NIS domain name, which uses the program domainname. Appendix X gives an exert from the netnfsrc file.

Table 26.1 NIS database components

NIS map	*File maintained*	*Description*
group.bygid group.byname	`/etc/group`	Maintains user groups.
hosts.byaddr hosts.byname	`/etc/hosts`	Maintains a list of IP addresses and symbolic names.
netgroup.byhost netgroup.byuser	`/etc/netgroup`	Contains a mapping of network group names to a set of node, user and NIS domain names.
networks.byaddr networks.byname	`/etc/network`	Defines network-wide groups used for permission checking when doing remote mounts, remote logins, and remote shells.
passwd.byname passwd.byuid	`/etc/passwd`	Contains details, such as user names and encrypted passwords.
protocols.byname protocols.bynumber	`/etc/protocols`	Contains information with known protocols used on the Internet.
rpc.bynumber rpc.byname	`/etc/rpc`	Maps the RPC program names to the RPC program numbers and vice versa. This file is static; it is already correctly configured.
services.byname servi.bynp	`/etc/services`	
mail.byaddr mail.aliases	`/etc/aliases`	

26.4.2 NFS remote file access

To initially mount a remote directory (or file system) onto a local the superuser must do the following:

- On the server, exports the directory to the client.
- On the client, mounts (or imports) the directory.

For example if the remote directory /user is to be mounted on to be host miranda as the directory `/win`. To achieve these operation the following are setup:

1. The superuser logs on to the remote server and edits the file `/etc/exports` adding the `/user` directory.
2. The superuser then runs the program `exportfs` to make the `/user` directory available to the client.

```
% exportfs -a
```

3. The superuser then logs into the client and creates a mount point `/win` (empty directory).

```
% mkdir /mnt
```

4. The remote directory can then be mounted with:

```
% mount miranda:/user /win
```

NFS maintains the file `/etc/mnttab` which contains a record of the mounted file systems. The general format is:

```
 special_file_name  dir  type  opts  freq  passno  mount_time
cnode_id
```

where *mount_time* contains the time the file system was mounted using mount. Sample contents of `/etc/mnttab` could be:

```
/dev/dsk/c201d6s0 / hfs defaults 0 1 850144122 1
/dev/dsk/c201d5s0 /win hfs defaults 1 2 850144127 1
castor:/win /net/castor_win nfs rw,suid 0 0 850144231 0
miranda:/win /net/miranda_win nfs rw,suid 0 0 850144291 0
spica:/usr/opt /opt nfs rw,suid 0 0 850305936 0
triton:/win /net/triton_win nfs rw,suid 0 0 850305936 0
```

In this case there are two local drivers (`/dev/dsk/c201d6s0` is mounted as the root directory and `/dev/dsk/c201d5s0` is mounted locally as `/win`). There are also four remote directories which are mounted from remote servers (`castor`, `miranda`, `spica` and `triton`). The directory mounted from castor is the `/win` directory and it is mounted locally as `/net/castor_win`. `hfs` defines a UNIX format disk and `nfs` defines that the disk is mounted over NFS.

A disk can be unmounted from a system using the umount command, e.g.

```
% umount miranda:/win
```

26.4.3 NIS commands

NIS commands allow the maintainance of network information. The main commands are as follows:

- `domainname` which displays or changes the current NIS domain name.
- `ypcat` which lists the specified NIS map contents.
- `ypinit` which, on a master server, builds a map using the networking files in `/etc`. On a slave server the map is built using the master server.
- `ypmake` which is a script that builds standard NIS maps from files such as `/etc/passwd`, `/etc/groups`, and so on.
- `ypmatch` which prints the specified NIS map data (values) associated with one or more keys.
- `yppasswd` which can be used to change (or install) a user's password in the NIS `passwd` map.
- `ypwhich` which is used to print the host name of the NIS server supplying NIS services to an NIS client.
- `ypxfr` which transfers the NIS map from one slave server to another.

For example the command:

```
ypcat group.byname
```

lists the group name, the group ID and the members of the group. Here is an example of changing a user's password for the NIS domain. In this case, the user `bill_b` changes the network-wide password on the master server `pollux`.

```
% yppasswd
Changing NIS password for bill_b...
Old NIS password: ********
New password: *******
Retype new password: *******
The NIS passwd has been changed on pollux, the master NIS passwd
server.
```

The next example uses the `ypcat` program.

```
% ypcat group
students:*:200:msc01,msc02,msc03,msc04
nogroup:*:-2:
daemon::5:root,daemon
users::20:root,msc08
other::1:root,hpdb
root::0:root
```

The next example shows the `ypmake` command file which rebuild the NIS database.

```
# ypmake
For NIS domain eece:
The passwd map(s) are up-to-date.
```

```
Building the group map(s)... group build complete.
   Pushing the group map(s):  group.bygid  group.byname
The hosts map(s) are up-to-date.
The networks map(s) are up-to-date.
The rpc map(s) are up-to-date.
The services map(s) are up-to-date.
The protocols map(s) are up-to-date.
The netgroup map(s) are up-to-date.
The vhe_list map(s) are up-to-date.
ypmake complete:
```

26.5 Network configuration files

The main files used to set up networking are as follows:

`/etc/checklist` is a list of directories or files are automatically mounted at boot time.

`/etc/exports` contains a list of directories or files that clients may import.

`/etc/inded.conf` contains information about servers started by `inetd`.

`/etc/netgroup` contains a mapping of network group names to a set of node, user, and NIS domain names.

`/etc/netnfsrc` is automatically started at run time and initiates the required daemons and servers, and defines the node as a client or server.

`/etc/rpc` maps the RPC program names to the RPC program numbers and vice versa.

`/usr/adm/inetd.sec` checks the Internet address of the host requesting a service against the list of hosts allowed to use the service.

26.5.1 Daemons

Networking programs normally initiate networking daemons which are background processes and are always running. Their main function is to wait for a request to perform a task. Typical daemons are:

`biod` which is asynchronous block I/O daemons for NFS clients.

`inetd` which is an Internet daemon that listens to service ports. It listens for service requests and calls the appropriate server. The server it calls depends on the contents of the `/etc/inetd.conf` file.

`nfsd` which is the NFS server daemon. It is used by the client for reading and writing to a remote directory and it sends a request to the remote server `nfsd` process.

`pcnfsd` which is a PC user authentication daemon.

`portmap` which is an RPC program to port number conversion daemon. When a client makes an RPC call to a given program number, it first contacts `portmap` on the server node to determine the port number where RPC requests should be sent.

Here is an extract from the processes that run a networked UNIX workstation:

```
 UID    PID  PPID  C     STIME TTY       TIME COMMAND
root    100     1  0  Dec   9  ?         0:00 /etc/portmap
root    138     1  0  Dec   9  ?         0:00 /etc/inetd
root     93     1  0  Dec   9  ?         0:00 /etc/rlbdaemon
root    104     1  0  Dec   9  ?         9:20 /usr/etc/ypserv
root    106     1  0  Dec   9  ?         0:00 /etc/ypbind
root    122   120  0  Dec   9  ?         0:00 /etc/nfsd 4
root    120     1  0  Dec   9  ?         0:00 /etc/nfsd 4
root    116     1  0  Dec   9  ?         0:00 /usr/etc/rpc.yppasswdd
root    123   120  0  Dec   9  ?         0:00 /etc/nfsd 4
root    124   120  0  Dec   9  ?         0:00 /etc/nfsd 4
root    125     1  0  Dec   9  ?         0:02 /etc/biod 4
root    126     1  0  Dec   9  ?         0:02 /etc/biod 4
root    127     1  0  Dec   9  ?         0:02 /etc/biod 4
root    128     1  0  Dec   9  ?         0:02 /etc/biod 4
root    131     1  0  Dec   9  ?         0:00 /etc/pcnfsd
root    133     1  0  Dec   9  ?         0:00 /usr/etc/rpc.statd
root    135     1  0  Dec   9  ?         0:00 /usr/etc/rpc.lockd
root   4652     1  0 14:33:15 ?          0:00 /etc/pcnfsd
root   4649     1  0 14:33:15 ?          0:00 /usr/etc/rpc.mountd
```

26.6 Sample startup file

The following is an exert from an `/etc/rc` file from a UNIX workstation. This file is automatically called when the workstation is booted. A sample `rc` file is given in Appendix F.

26.7 Exercises

26.1 Outline the main protocols used on networked UNIX systems.

26.2 Explain how NFS is set up on a UNIX system and describe the protocols it uses.

26.3 Explain how NIS is set up on a UNIX system and the files it uses.

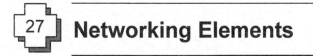

27 Networking Elements

27.1 LANs, WANs, and MANs

Computer systems operate on digital data. They can communicate with other digital equipment over a local area network (LAN). A LAN is defined as a collection of computers within a single office or building that connect to a common electronic connection – commonly known as a network backbone. A LAN can be connected to other networks either directly or through a wide area network (WAN), as illustrated in Figure 27.1.

A WAN normally connects networks over a large physical area, such as in different buildings, towns or even countries. Figure 27.1 shows three local area networks, LAN A, LAN B and LAN C, some of which are connected by the WAN. A modem connects a LAN to a WAN when the WAN connection is an analogue line. For a digital connection a gateway connects one type of LAN to another LAN, or WAN, and a bridge connects a LAN to similar types of LAN.

The public switched telecommunications network (PSTN) provides long-distance analogue lines. These public telephone lines can connect one network line to another using circuit switching. Unfortunately, they have a limited bandwidth and can normally only transmit frequencies from 400 to 3 400 Hz. For a telephone line connection a modem is used to convert the digital data into a transmittable form. Figure 27.2 illustrates the connection of computers to a PSTN. These computers can connect to the WAN through a service provider (such as CompuServe) or through another network which is connected by modem. The service provider has the required hardware to connect to the WAN.

A public switched data network (PSDN) allows the direct connection of digital equipment to a digital network. This has the advantage of not requiring the conversion of digital data into an analogue form. The integrated services digital network (ISDN) allows the transmission of many types of digital data into a truly global digital network. Transmittable data types include digitized video, digitized speech and computer data. Since the switching and transmission are digital, fast access times and relatively high bit rates are possible. Typical base bit rates may be 64 kbps. All connections to the ISDN require network termination equipment (NTE).

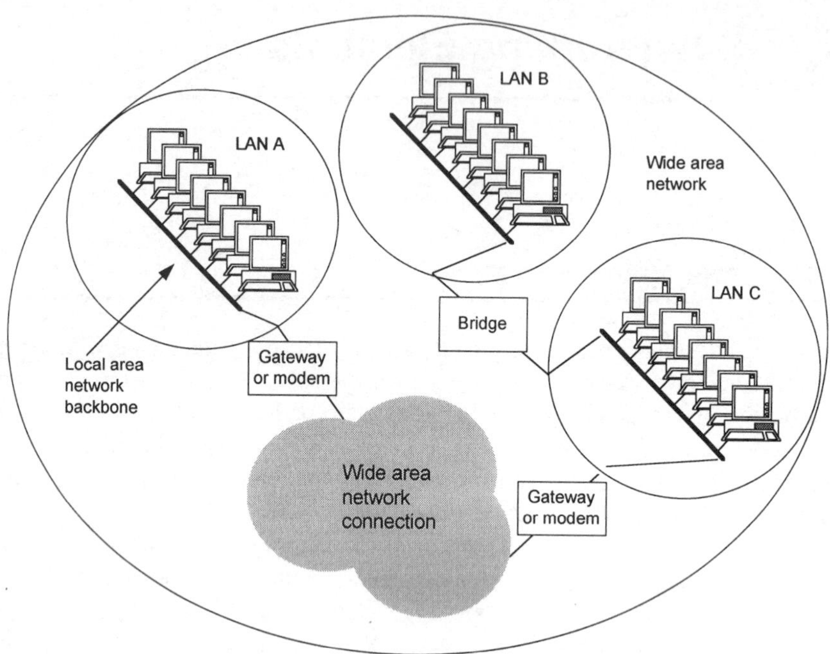

Figure 27.1 Interconnection of LANs to make a WAN

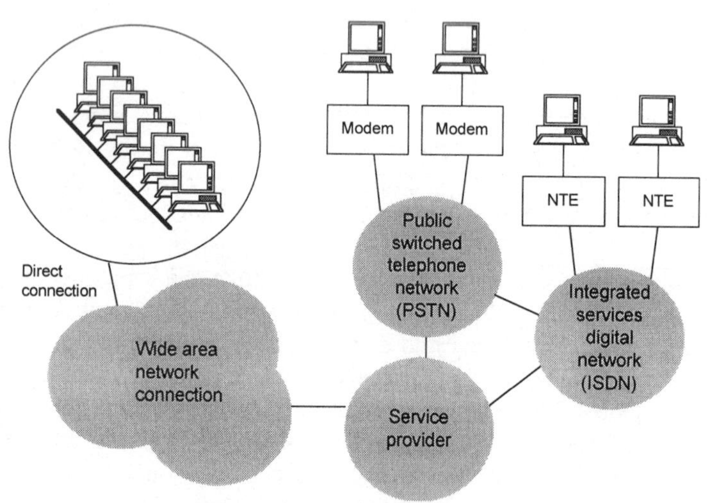

Figure 27.2 Connection of nodes to a PSTN

27.2 OSI model

An important concept in understanding data communications is the OSI (open systems interconnection) model. It allows manufacturers of different systems to interconnect their equipment through standard interfaces. It also allows software and hardware to integrate well and be portable on differing systems. International Standards Organization (ISO) developed the model and it is shown in Figure 27.3.

Data is passed from the top layer of the transmitter to the bottom then up from the bottom layer to the top on the recipient. But each layer on the transmitter communicates directly with the recipient's corresponding layer. This creates a virtual data flow between layers.

The top layer (the application layer) initially gets data from an application and appends it with data that the recipient's application layer will read. This appended data passes to the next layer (the presentation layer). Again it appends its own data, and so on, down to the physical layer. The physical layer is then responsible for transmitting the data to the recipient. The data sent can be termed a data packet or data frame.

Figure 27.4 shows the basic function of each of the layers. The physical link layer defines the electrical characteristics of the communications channel and the transmitted signals. This includes voltage levels, connector types, cabling, and so on.

The data link layer ensures that the transmitted bits are received in a reliable way. This includes adding bits to define the start and end of a data frame, adding extra error detection/correction bits and ensuring that multiple nodes do not try to access a common communications channel at the same time.

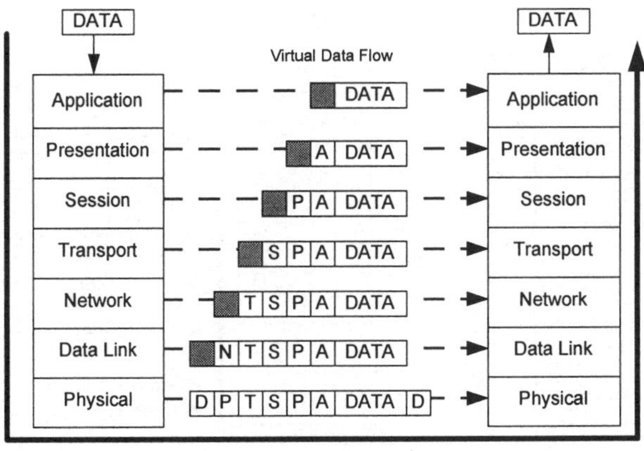

Figure 27.3 Seven-layer OSI model

The network layer routes data frames through a network. If data packets require to go out of a network then the transport layer routes them through interconnected networks. Its task may involve splitting up data for transmission and reassembling it upon reception.

The session layer provides an open communications path to the other system. It involves setting up, maintaining and closing down a session. The communications channel and the internetworking of the data should be transparent to the session layer.

The presentation layer uses a set of translations that allow the data to be interpreted properly. It may have to carry out translations between two systems if they use different presentation standards such as different character sets or different character codes. For example on a UNIX system a text file uses a single ASCII character for new line (the carriage return). Whereas, on a DOS-based system there are two, the line feed and the carriage return. The presentation layer would convert from one computer system to another so that the data could be displayed correctly, in this case by either adding or taking away a character. The presentation layer can also add data encryption for security purposes.

The application layer provides network services to application programs such as file transfer and electronic mail.

Figure 27.5 shows an example with two interconnected networks, Network A and Network B. Network A has four nodes N1, N2, N3 and N4, and Network B has nodes N5, N6, N7 and N8. If node N1 were tp communicate with node N7 then a possible path would be via N2, N5 and N6. The data link layer ensures that the bits transmitted between nodes N1 and N2, nodes N2 and N5, and so on, are transmitted in a reliable way.

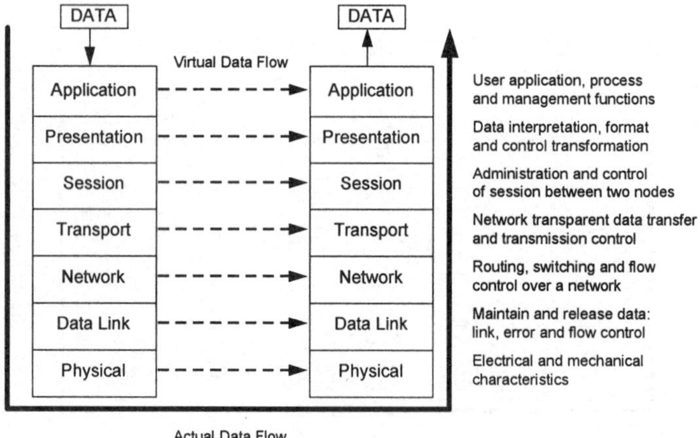

Figure 27.4 ISO open systems interconnection (OSI) model

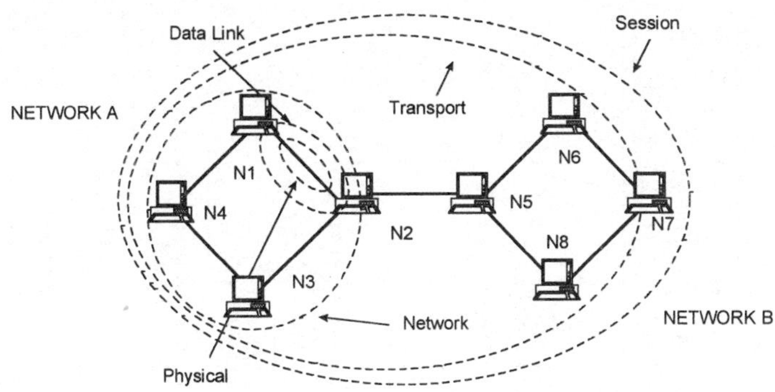

Figure 27.5 Scope of concern of OSI layers

The network layer would then be responsible for routing the data packets through Network A and through Network B. The transport layer routes the data through interconnections between the networks. In this case it would route data packets from N2 to N5. If other routes existed between N1 and N7 it might use another route.

27.3 Communications standards and the OSI model

The following sections look at practical examples of data communications and networks, and how they fit into the layers of the OSI model. Unfortunately, most currently available technologies do not precisely align with the layers of this model. For example, RS-232 provides a standard for a physical layer but it also includes some data link layer functions, such as adding error detection and framing bits for the start and end of a packet.

Figure 27.6 shows the main technologies covered in this book. They are split into three basic sections: asynchronous data communication, local area networks (LANs) and wide area networks (WANs).

The most popular types of LAN are Ethernet and Token Ring. Standards for Ethernet include Ethernet 2.0 and IEEE 802.3 (with IEEE 802.2). For Token Ring the standards are IBM Token Ring and IEEE 802.5 (with IEEE 802.2). Ethernet uses carrier sense multiple access/ collision detect (CMSA/CD) technology which is why the IEEE standard includes the name CSMA/CD.

One of the main standards for the interconnection of networks is the Transport Control Protocol/Internet Protocol (TCP/IP). IP routes data packets through a network and TCP routes data packets between interconnected network. An equivalent to the TCP/IP standard used in some PC networks; is called SPX/IPX.

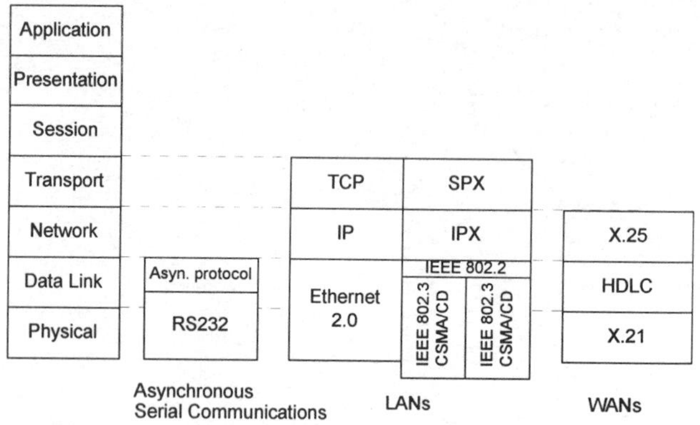

Figure 27.6 ISO open systems interconnection (OSI) model

For digital connections to WANs the main standards are CCITT X.21, HDLC and CCITT X.25.

27.4 Standards agencies

There are six main international standards agencies that define standards for data communications system. They are the International Standards Organization (ISO), the Comité Consultatif International Télégraphique et (CCITT), the Electrical Industries Association (EIA), the International Telecommunications Union (ITU), American National Standards Institute (ANSI) and the Institute of Electrical and Electronic Engineers (IEEE).

The ISO and the IEEE have defined standards for the connection of computers to local area networks and the CCITT (now know as the ITU) has defined standards for the interconnection of national and international networks. The CCITT standards covered in this book split into three main sections: these are asynchronous communications (V.xx standards), PSDN connections (X.xxx standards) and ISDN (I.4xx standards). The main standards are given in Table 1. The EIA has defined standards for the interconnection of computers using serial communications. The original standard was RS-232-C; this gives a maximum bit rate of 20 kbps over 20 m. It has since defined several other standards, including RS-422 and RS-423, which provide a data rate of 10 Mbps.

27.5 Network cable types

The cable type used on a network depends on several parameters, including:

- The data bit rate.
- The reliability of the cable.
- The maximum length between nodes.
- The possibility of electrical hazards.
- Power loss in the cables.
- Tolerance to harsh conditions.
- Expense and general availability of the cable.
- Ease of connection and maintenance.
- Ease of running cables, and so on.

Table 27.1 Typical standards

Standard	*Equivalent ISO/CCITT*	*Description*
EIA RS-232C	CCITT V.28	Serial transmission up to 20 kps/ 20 m
EIA RS-422	CCITT V.11	Serial transmission up to 10 Mbps/ 1200 m
EIA RS-423	CCITT V.10	Serial transmission up to 300 kbps/ 1200 m
ANSI X3T9.5		LAN: Fibre Optic FDDI standard
IEEE 802.2	ISO 8802.2	LAN: IEEE standard for logical link control
IEEE 802.3	ISO 8802.3	LAN: IEEE standard for CSMA/CD
IEEE 802.4	ISO 8802.4	LAN: Token passing in a Token Ring network
IEEE 802.5	ISO 8802.5	LAN: Token Ring topology
	CCITT X.21	WAN: Physical layer interface to a PSDN
HDLC	CCITT X.212/ 222	WAN: Data layer interfacing to a PSDN
	CCITT X.25	WAN: Network layer interfacing to a PSDN
	CCITT I430/1	ISDN: Physical layer interface to an ISDN
	CCITT I440/1	ISDN: Data layer interface to an ISDN
	CCITT I450/1	ISDN: Network layer interface to an ISDN

The main types of cables used in networks are twisted-pair, coaxial and fiber-optic, they are illustrated in Figure 27.7. Twisted-pair and coaxial cables transmit electric signals, whereas fiber-optic cables transmit light pulses. Twisted-pair cables are not shielded and thus interfere with nearby cables. Public telephone lines generally use twisted-pair cables. In LANs they are generally used up to bit rates of 10 Mbps and with maximum lengths of 100 m.

Coaxial cable has a grounded metal sheath around the signal conductor. This limits the amount of interference between cables and thus allows higher data rates. Typically they are used at bit rates of 100 Mbps for maximum lengths of 1 km.

The highest specification of the three cables is fiber-optic. This type of cable allows extremely high bit rates over long distances. Fiber-optic cables do not interfere with nearby cables and give greater security, more protection from electrical damage by external equipment and greater resistance to harsh environments; they are also safer in hazardous environments.

A typical bit rate for a LAN using fiber-optic cables is 100 Mbps, in other applications this reaches several gigabits per second. The maximum length of the fiber-optic cable depends on the electronics in the transmitter and receiver, but a single length of 20 km is possible.

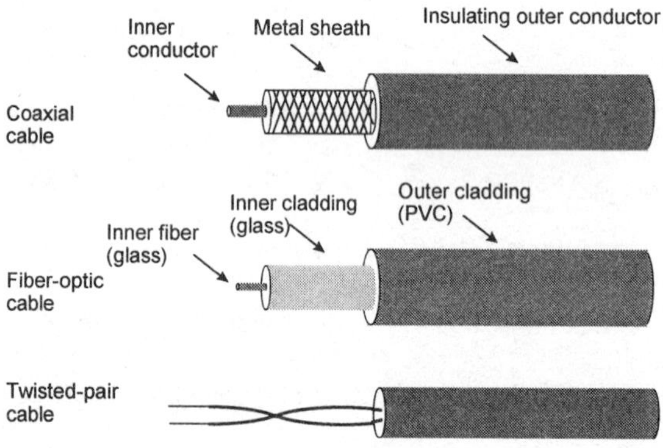

Figure 27.7 Type of network cable

27.6 LAN topology

Computer networks are ever expanding, and a badly planned network can be inefficient and error prone. Unfortunately networks tend to undergo evolutionary change instead of revolutionary change and they can become difficult to manage if not planned properly. Most modern network have a backbone which is a common link to all the networks within an organization. This backbone allows users on different network segments to communicate and also allows data into and out of the local network. Figure 27.8 shows that a local area network contains various segments: LAN A, LAN B, LAN C, LAN D, LAN E and LAN F. These are connected to the local network via the BACKBONE 1. Thus if LAN A talks to LAN E then the data must travel out of LAN A, onto BACKBONE1, then into LAN C and through onto LAN E.

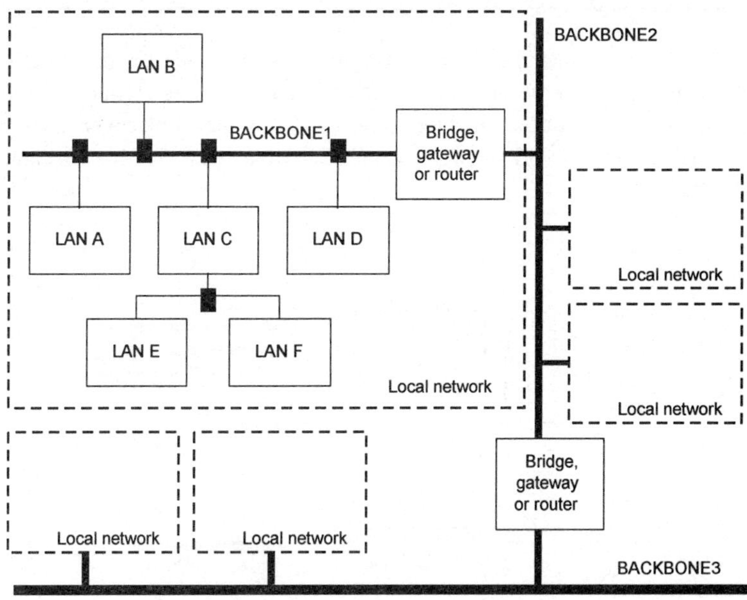

Figure 27.8 Interconnection of local networks

Networks are partitioned from other networks using a bridge, a gateway or a router. A bridge links two networks of the same type, such as Ethernet to Ethernet, or Token Ring to Token Ring. A gateway connects tow networks of dissimilar type. Routers operate rather like gateways and can either connect two similar networks or two dissimilar networks. The key operation of a gateway, bridge or router is that it only allows data traffic through itself when the data is intended for another network which is outside the connected network. This filters traffic and stops traffic not indented for the network from clogging up the backbone. Modern bridges, gateways and routers are intelligent and can determine the network topology.

A spanning-tree bridge allows multiple network segments to be interconnected. If more than one path exists between individual segments then the bridge finds alternative routes. This is useful in routing frames away from heavy traffic routes or around a faulty route. Conventional bridges can cause frames to loop around forever. Spanning-tree bridges have built-in intelligence and can communicate with other bridges. This allows them to build up a picture of the complete network and thus to make decisions on where frames are routed.

27.7 Internetworking connections

Networks connect to other networks through repeaters, bridges or routers. A repeater corresponds to the physical layer and always routes signals from one network segment to another. Bridges route using the data link layer and routers route using the network layer. Figure 27.9 illustrates the three interconnection types.

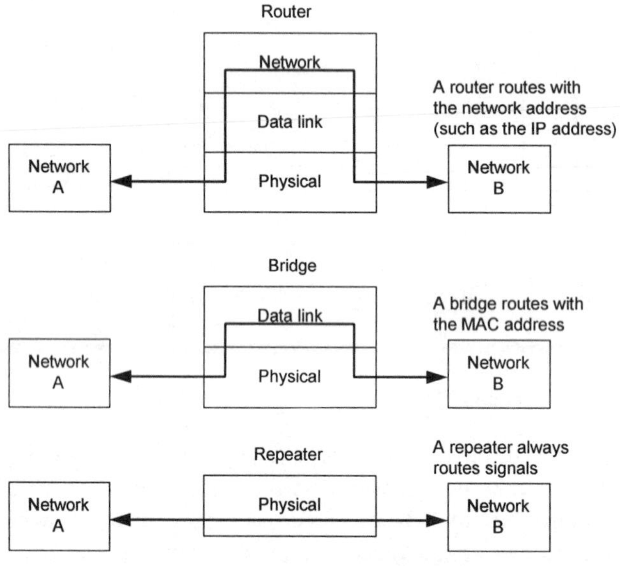

Figure 27.9 Repeaters, bridges and routers

27.7.1 Repeaters

All types of network connections suffer from attenuation and pulse distortion; for a given cable specification and bit rate , each has a maximum length of cable. Repeaters can be used to increase the maximum interconnection length and will do the following:

- Clean signal pulses.

- Pass all signals between attached segments.
- Boost signal power.
- Possibly translate between two different media types (e.g. fiber-optic to twisted-pair cable).

27.7.2 Bridges

Bridges filter input and output traffic so that only packets intended for a network are actually routed into the network and only packet intended for the outside are allowed out of the network.

The performance of a bridge is governed by two main factors (which are measured in packets per second or frames per second):

- The filtering rate – the bridge reads the MAC address of the Ethernet/Token Ring/FDDI node and then decides if it should forward the packet into the internetwork. When a bridge reads the destination address on an Ethernet frame or Token Ring packet and decides whether or not that packet should be allowed access to the internetwork Filter rates for bridges ranges from around 5000 to 70 000 pps (packets per second). A typical bridge has a filtering rate of 17 500 pps.
- The forward rate– once the bridge has decided to route the packet into the internetwork then the bridge must forward the packet onto the internetwork media. This is a forwarding operation and typical rates range from 500 to 140 000 pps. A typical forwarding rate is 90 000 pps.

A typical Ethernet bridge has the following specifications:

Bit rate:	10 Mbps
Filtering rate:	17 500 pps
Forwarding rate:	11 000 pps
Connectors:	2 DB15 AUI (female), 1 DB9 male console port, 2 BNC (for 10BASE2) or 2 RJ-45 (for 10BASE-T)
Algorithm:	Spanning tree protocol. It automatically learns the addresses of all devices on both interconnected networks and builds a separate table for each network

And for a Token Ring bridge:

Bit rate:	4/16 Mbps
Filtering rate:	120 000 pps
Forwarding rate:	3400 pps
Connectors:	1 DB9 male console port, 2 DB9 connectors
Algorithm:	Source routing transparent

Spanning-tree architecture (STA) bridges

The spanning-tree algorithm has been defined by the standard IEEE 802.1. It is normally implemented as software on STA-compliant bridges On power-up they automatically learn the addresses of all nodes on both interconnected networks and build a separate table for each network.

They can also have two connections between two LANs so that when the primary path becomes disabled, the spanning-tree algorithm can reenable the previously disabled redundant link. The path management is achieved by each bridge communicating using configuration bridge protocol data units (configuration BPDU).

Source route bridging

With source route bridging a source device, not the bridge, is used to send special explorer packets which are then used to determine the best path to the destination. Explorer packets are sent out from the source routing bridges until they reach their destination workstation. Then each source routing bridge along the route enters its address in the routing information field of the explorer packet. The destination node then sends back the completed RIF field to the source node. When the source device (normally a PC) has determined the best path to the destination, it sends the data message along with the path instructions to the local bridge. It then forwards the data message according to the received path instructions.

Although the source routing bridge receives the data, there is a 7-hop limit on the number of internetwork connections. This is because of the limited space in the router information field (RIF) of the explorer packet.

27.7.3 Routers

Routers examine the network address field (such as IP or IPX) and determine the best route for the packet. They have the great advantage that they normally support several different types of network layer protocol.

Routers which only read one type of protocol will normally have high filtering and forwarding rates. If they support multiple protocols then there is normally an overhead in that the router must detect the protocol and look in the correct place for the destination address.

Typical network layer protocols and their associated network operating systems or upper layer protocols are:

AFP	AppleTalk	IP	TCP/IP	IPX	NetWare
OSI	Open Systems	VIP	Vines	XNS	3Com

Routers can also be used to connect to datalink layer protocols without network layer addressing schemes. Typically the datalink frames are encapsulated within a routeable network layer protocol (such as IP).

Routing protocols

Routers need to communicate with other routers so they can exchange routing information. Most network operating systems have associated routing protocols which support the transfer of routing information. Typical routing protocols and their associated network operating systems are:

- BGP (Border Gateway Protocol) – TCP/IP.
- EGP (Exterior Gateway Protocol) – TCP/IP.
- IS-IS (Immediate System to Intermediate Systems) – DECnet, OSI.
- NLSP (NetWare Link State Protocol) – NetWare 4.1.
- OSPF (Open Shortest Path First) – TCP/IP.
- RIP (Routing Information Protocol) – XNS, NetWare, TCP/IP.
- RTMP (Routing Table Maintenance Protocol) – AppleTalk.

Most routers support IP, IPX/SPX and AppleTalk network protocol using RIP, and EGP. the main Internet-based protocols are discussed in the next section. In the past RIP was the most popular router protocol standard. Its widespread use is due in no small part to the fact that it was distributed along with the Berkeley Software Distribution (BSD) of UNIX (from which most commercial versions of UNIX are derived). It suffers from several disadvantages and has been largely replaced by OSFP and EGP. These protocols have an advantage over RIP in that they can handle large internetworks as well as reducing routing table update traffic.

RIP uses a distance vector algorithm which measures the number of hops, up to a maximum of 16, to the destination router. The OSPF and EGP protocols use a link state algorithm which can decide between multiple paths to the destination router. They are based, not only hops, but on other parameters such as delay capacity, reliability and throughput.

With distance vector routing each router maintains its table by communicating with neighboring routers. The number of hops in its own table is then computed as it knows the number of hops to local routers. Unfortunately the routing table can take some time to be updated when changes occur, because it takes time for all the routers to communicate with each other (known as slow convergence).

OSPF, EGP, BGP and NLSP use link state protocols and differ from distance vector routing in that they use network information received from all routers on a given network, rather than just from neighboring routes. This then overcomes slow convergence.

27.7.4 Example bridge/router

Here is an example of a typical bridge/router, which has a RISC-based processor, and can operate as a bridge and/or a router:

WAN port: DB25 connector
LAN port: 2 BNC, 2 AUI (for Ethernet connections)
WAN line speed: 256 kbps
Forwarding rate: 59 kpps
Filtering rate: 14.88 kpps
Routable protocol: IP, IPX/SPX, AppleTalk
Routing protocol: RIP, HELLO, EGP
WAN interface: RS-232/V.24, V.35, RS-530

27.8 Internet routing protocols

27.8.1 RIP

Figure 27.10 recalls from Chapter 25 the RIP packet format. The fields are:

- Operation (2 bytes) – this field gives an indication that the RIP packet is either a request or a response. The first 8 bits of the field give the command/request name and the next 8 bits give the version number.
- Network number (4 bytes of IP addresses) – this field defines the assigned network address number to which the routing information applies (note that, although 4 bytes are shown, there are in fact 14 bytes reserved for the address. In RIP version 1 (RIPv1), with IP traffic, 10 of the bytes were unused; RIPv2 uses the 14-byte address field for other purposes, such as subnet masks.
- Number of router hops (2 bytes) – this field indicates the number of routes that a packet must go through in order to reach the required destination. Each router adds a single hop, the minimum number is 1 and the maximum is 16.
- Number of ticks (2 bytes) – this field indicates the amount of time (in 1/18 second) it will take for a packet to reach a given destination. Note that a route which has the fewest hops may not necessarily be the fastest route.

RIP packets add to the general network traffic as each router broadcasts its entire routing table every 30–60 seconds.

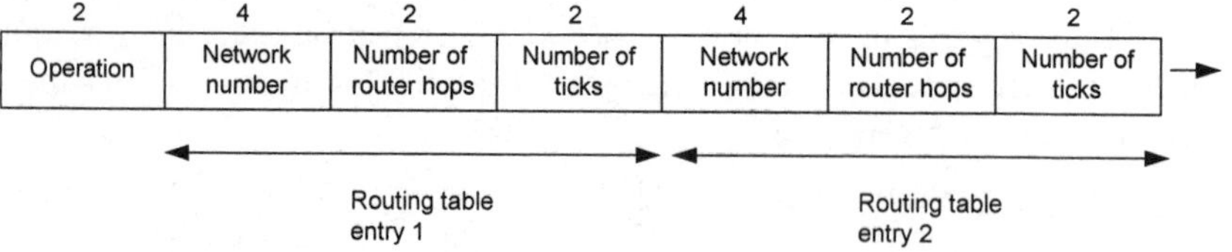

Figure 27.10 RIP packet format

OSPF

The OSPF is an open, non-proprietary standard which was created by the IEFF (Internet Engineering Task Force), a task force of the IAB. It is a link state routing protocol and is thus able to maintain a complete and more current view of the total internetwork than distance vector routing protocols. Link state routing protocols have the features:

- They use link state packets (LSPs) which are special datagrams that determine the names of and the cost or distance to any neighboring routers and associated networks.
- Any information learned about the network is then passed to all known routers, and not just neighboring routers, using LSPs. Thus all routers have a fuller knowledge of the entire internetwork than the view of only the immediate neighbors (as with distance vector routing).

OSPF adds to these features with:

- Additional hierarchy. OSPF allows the global network to be split into areas. Thus a router in a domain does not necessarily have to know how to reach all the networks with a domain, it simply has to send to the right area.
- Authentication of routing messages using an 8-byte password. This length is not long enough to stop unauthorized users from causing damage. Its main purpose is to reduce the traffic from misconfigured routers. Typically a misconfigured router will inform the network that it can reach all nodes with no overhead.
- Load balancing. OSPF allows multiple routes to the same place to be assigned the same cost and will cause traffic to be distributed evenly over those routes.

Figure 27.11 shows the OSPF header. The fields in the header are:

- A version number (1 byte) which, in current implementations, has the version number of 2.
- The type field (1 byte) which can range from 1 to 5. Type 1 is the HELLO message and the others are to request, send and acknowledge the receipt of link state messages. HELLO messages are used by nodes to convince their neighbors that they are alive and reachable. If a router fails to receive these messages from one of its neighbors for some period of time, it assumes that the node is no longer directly reachable and updates its link state information accordingly.
- SourceAddr (4 bytes) identifies the sender of the message.

- Areald (4 bytes) is an identifier to the area in which the node is located.
- An authentication field can either be set to 0 (none) or 1. If it is set to 1 then the authentication contains an 8-byte password.

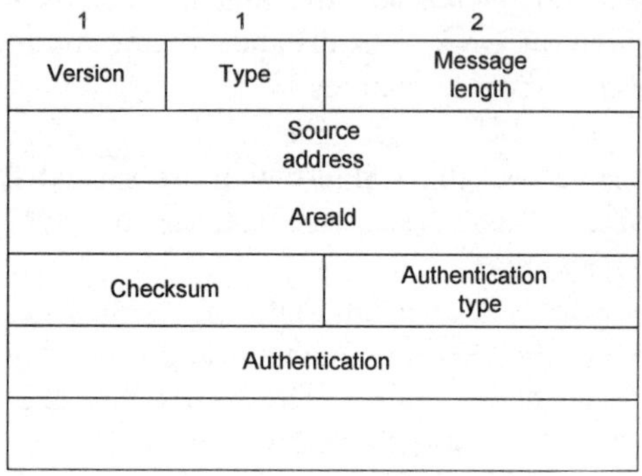

Figure 27.11 OSPF header format

EGP/BGP

The two main interdomain routing protocols in recent history are EGP and BGP. EGP suffers from several limitations. Its principle limitation is that it treats the Internet as a tree like structure. A treelike structure, as illustrated in Figure 27.12, is normally made up of parents and children, with a single backbone. A more typical topology for the Internet is illustrated in Figure 27.13.

BGP is an improvement on EGP (the fourth version of BGP is known as BGP-4). Unfortunately it is more complex than EGP, but not as complex as OSPF.

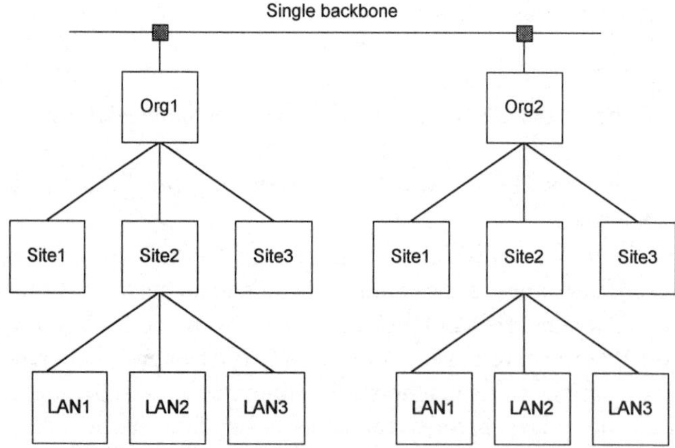

Figure 27.12 Tree-like topology

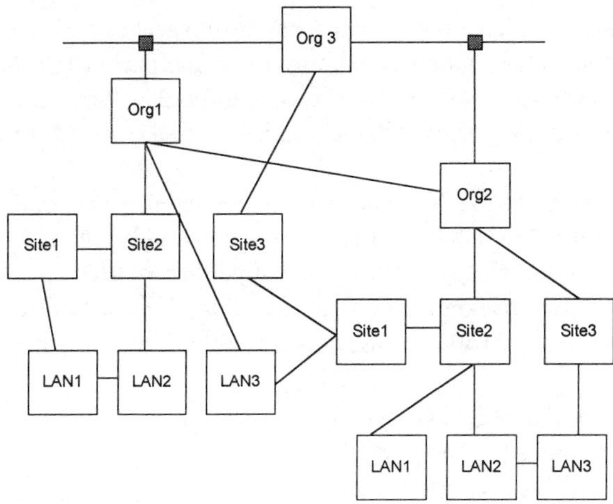

Figure 27.13 Network with multiple backbones

BGP assumes that the Internet is made up of an arbitrarily interconnected set of nodes. It then assumes the Internet connects to a number of AANs (autonomously attached networks), as illustrated in Figure 27.14. These may create boundaries around an organization, an Internet service provider, and so on. It then assumes that, once they are in the AAN, the packets the packets will be properly routed.

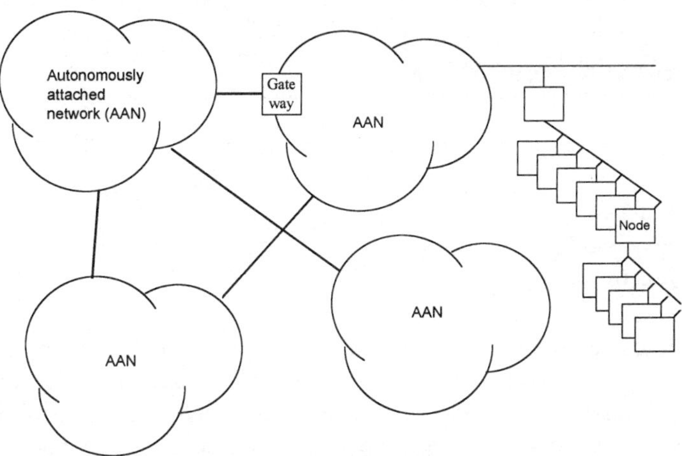

Figure 27.14 Autonomously attached networks

BGP differs in that it tries to find any paths through the network. Thus the main goal is reachability instead of the number of hops to the destination. So finding a path which is nearly optimal is a good achievement. The AAN administrator selects at least one node to be a BGP speaker and also one or more border gateways. These gateways simply route the packet into and out of the AAN. The border gateways are the routers through which packets reach the AAN. Most routing algorithms try to find the quickest way through the network.

The speaker on the AAN broadcasts its reachability information to all the networks within its AAN. This information states only whether a destination AAN can be reached; it does not describe any other metrics.

The BGP update packet also contains information on routes which cannot be reached (withdrawn routes). The content of the BGP-4 update packet is:

- Unfeasible routes length (2 bytes).
- Withdrawn routes (variable length).
- Total path attribute length (2 bytes).
- Path attributes (variable length).
- Network layer reachability information (variable length).

The network layer reachability information can contain extra information, such as 'use AAN 1 in preference to AAN 2'.

An important point is that BGP is not a distance vector or link state protocol because it transmits complete routing information instead of partial information.

27.9 Network topologies

The three basic topologies for LANs are shown in Figure 27.15:

- A star network.
- A ring network.
- A bus network.

Other topologies are either a combination of two or more of the basic topologies or they are derivatives of them. A typical topology is a tree topology. This is essentially a star and a bus network combined, as illustrated in Figure 27.16. A concentrator is used to connect the nodes onto the network.

27.9.1 Star network

In a star topology, a central server switches data around the network. Data traffic between nodes and the server will thus be relatively low. Its main advantages are:

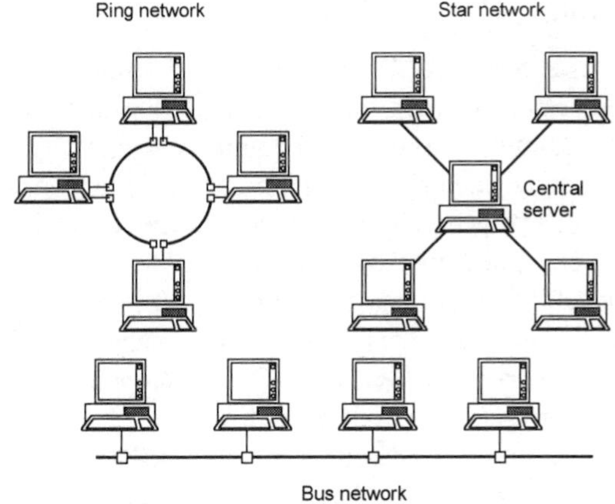

Figure 27.15 Network topologies

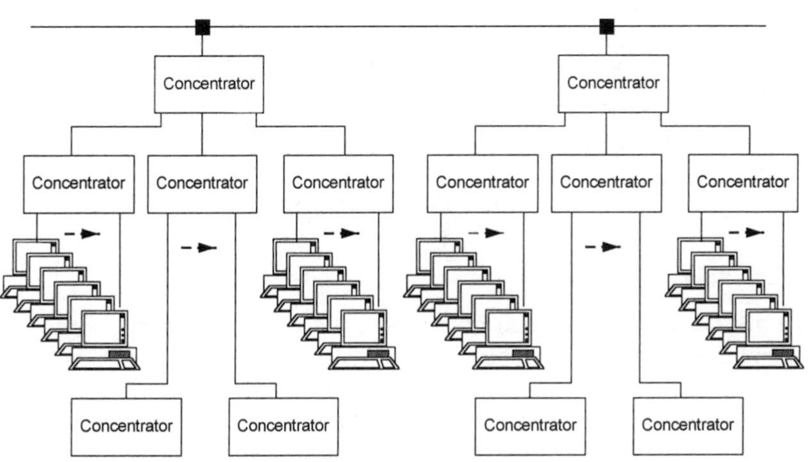

Figure 27.16 Tree topology

- Since the data rate is relatively low between the central server and the node, a low-specification twisted-pair cable can be used to connect the nodes to the server.
- A fault on one of the nodes will not affect the rest of the network. Typically, mainframe computers use a central server with terminals connected to it.

The main disadvantage of this topology is that the network is highly dependent upon the operation of the central server. If it were to slow down significantly then the network would become slow. And if it were to become inoperative then the complete network would shut down.

27.9.2 Bus network

A bus network uses a multidrop transmission medium, as shown in Figure 27.17. All nodes on the network share a common bus and all share communications. This allows only one device to communicate at a time. A distributed medium access protocol determines which station is to transmit. As with the ring network, data packets contain source and destination addresses. Each station monitors the bus and copies frames addressed to itself.

Twisted-pair cables gives data rates up to 100 Mbps. Coaxial and fiber-optic cables give higher bit rates and longer transmission distances. A bus network is a good compromise over the other two topologies as it allows relatively high data rates. Also, if a node goes down then it does not affect the rest of the network. The main disadvantage of this topology is that it requires a network protocol to detect when two nodes are transmitting at the same time. A typical bus network is Ethernet 2.0.

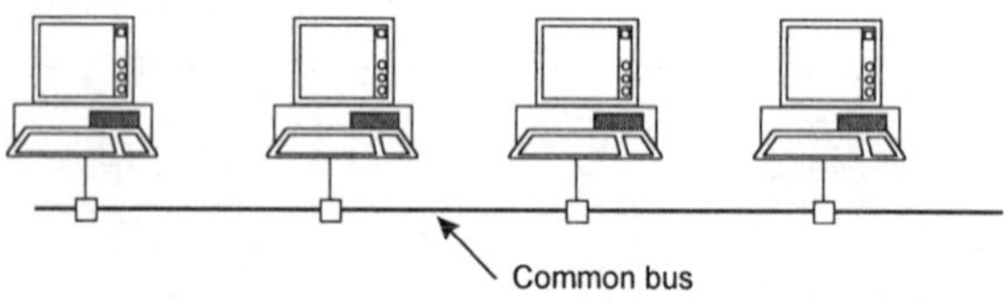

Figure 27.17 Bus topology

27.9.3 Ring network

In a ring network nodes, connect to make a closed electronic loop. Each node communicates with its neighbor and thus data is passed from one node to the next until it reaches its destination. Normally, to provide an orderly flow of data, an electronic token passes around the ring; this is known as a token passing ring network, as shown in Figure 27.18. A node must capture the token before it can transmit a data frame. A distributed control protocol determines the sequence in which nodes transmit.

In a manner similar to the star network each link between nodes is basically a point-to-point link. This allows almost any transmission medium to be used. Typically twisted-pair cables allow a bit rate of up to 16 Mbps, but coaxial and fiber-optic cables are normally used for extra reliability and higher data rates. A typical ring network is IBM Token Ring. The main advantage of a Token Ring network is that all nodes on the network have an equal chance of transmitting data. Unfortunately it suffers from several problems. The most severe is that if one of the nodes goes down then the whole network may go down.

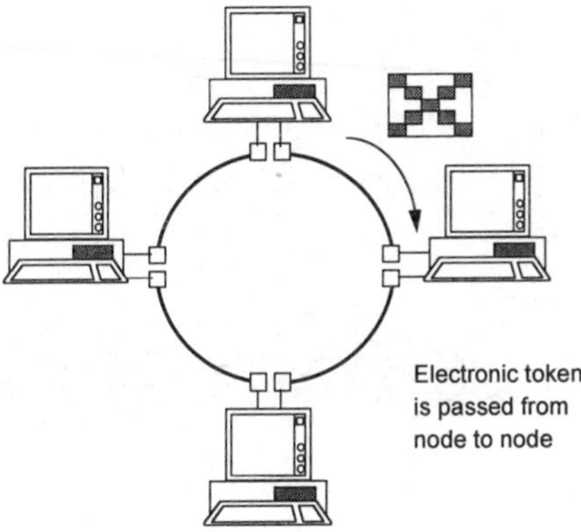

Figure 27.18 Token passing ring network

27.10 Network loading

The data traffic on a network will not be constant over time. It will vary each minute, each hour, each day, each month and so on. Loading will tend to be similar from day to day. An example loading is given in Figure 29.19. It shows that, in this case, the peak network traffic occurs at around 11 am and also between 3 pm and 5 pm. Network managers often have to perform network maintenance and file backups. In this case the network manager has tried to even out network traffic by performing network backups at times when the network loading was low (i.e. at night).

Much of the network traffic is due to disk transfers. By analyzing network statistics it is possible to determine when hot spots occur. From these statistics the network manager may ask some users to change the way they operate. For example, by staggering lunch breaks the network loading traffic could be evened out. The system manager may also even out the network traffic by allowing only certain applications (or users) to use the network at certain times of the day. It is usual for heavy processing or network-intensive tasks to run during periods when there is a light network loading, typically at night.

27.11 Exercises

27.1 Explain why bridges and routers help to reduce internetwork traffic.

27.2 Distinguish between a repeater, a bridge and a router.

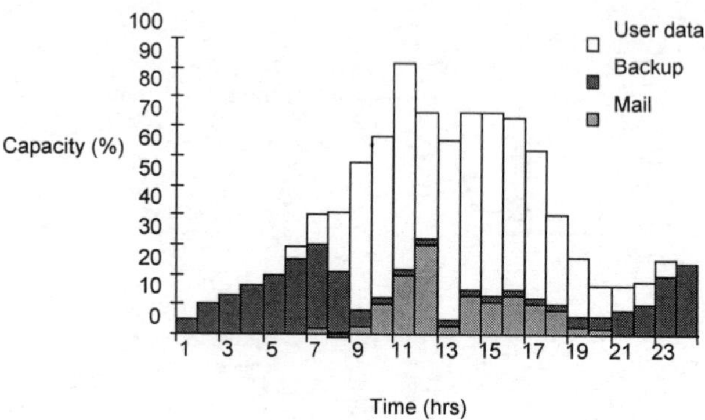

Figure 27.19 Example of network traffic over 24 hours

27.3 Distinguish between a spanning tree bridge and a source route bridge.

27.4 Outline pros and cons of the main routing protocols.

27.5 Discuss the pros and cons main network topologies.

28 Ethernet

28.1 Introduction

Ethernet is the most widely used networking technology used in LANs (local area networks). On its own, Ethernet cannot make a network; it needs some other protocol such as TCP/IP or SPX/IPX to allow nodes to communicate. Unfortunately, Ethernet in its standard form does not cope well with heavy traffic. But has many advantages, including:

- Its networks are easy to plan and cheap to install.
- Its network components are cheap and well supported.
- It is well-proven technology which is fairly robust and reliable.
- It is simple to add and delete computers on the network.
- It is supported by most software and hardware systems.

A major problem with Ethernet is that because computers compete for access to the network there is no guarantee that a particular computer will get access within a given time. And contention causes problems when two computers try to communicate at the same time; they must both back off and no data can be transmitted. In its standard form Ethernet allows a bit rate of 10 Mbps. New standards for fast Ethernet systems minimize the problems of contention and also increase the bit rate to 100 Mbps. Ethernet uses coaxial or twisted-pair cable.

DEC, Intel and the Xerox Corporation initially developed Ethernet, and the IEEE 802 committee have since defined standards for it. The most common standards for Ethernet are Ethernet 2.0 and IEEE 802.3. It uses a shared-media, bus-type network topology where all nodes share a common bus. It is a contention-type network where only one node communicates at a time. Data is transmitted in frames which contain the MAC (media access control) source and destination addresses of the sending and receiving node, respectively. The local shared-media is known as a segment. Each node on the network monitors the segment and copies any frames addressed to itself.

Ethernet uses carrier sense, multiple access with collision detection (CSMA/CD). On a CSMA/CD network, nodes monitor the bus (or Ether) to determine if it is busy. A node wishing to send data waits for an idle condition then transmits its message. Unfortunately collision can occur when two nodes transmit at the same time, thus nodes must monitor the cable when they transmit. When this happens both nodes stop transmitting frames and transmit a jamming signal. This informs all nodes on the network that a collision has occurred. Each of the nodes then waits a random period of time before attempting a retransmission. As each node has a random delay time, there can be a prioritization of the nodes on the network. Nodes thus contend for the network and are not guaranteed access to it. Collisions generally slow down the network. Each node on the network must be able to detect collisions and must be capable of transmitting and receiving simultaneously.

28.2 IEEE standards

The IEEE is the main standards organization for LANs; it calls the standard for Ethernet CSMA/CD (carrier sense multiple access/collision detect). Figure 28.1 shows how the IEEE standards for Token Ring and CSMA/CD fit into the OSI model. The two layers of the IEEE standards correspond to the physical and data link layers of the OSI model. A Token Ring network uses IEEE 802.5 (ISO 8802.5) and a CSMA/CD network uses IEEE 802.3 (ISO 8802.3). On Ethernet networks, most hardware will comply with IEEE 802.3 standard. The object of the MAC layer is to allow many nodes to share a single communication channel. It also adds start and end frame delimiters, error detection bits, access control information and source and destination addresses.

The IEEE 802.2 (ISO 8802.2) logical link control (LLC) layer conforms to the same specification for both types of network.

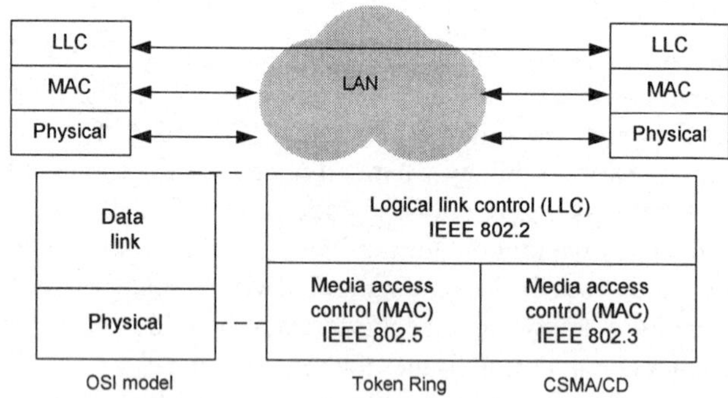

Figure 28.1 Standards for IEEE 802 LANs

28.3 Ethernet - media access control (MAC) layer

When sending data the MAC layer takes the information from the LLC link layer. Figure 28.2 shows the IEEE 802.3 frame format. It contains 2 or 6 bytes for the source and destination addresses (16 or 48 bits each), 4 bytes for the CRC (32 bits), 2 bytes for the LLC length (16 bits). The LLC part may be up to 1500 bytes long. The preamble and delay components define the start and end of the frame. The initial preamble and start delimiter are, in total, 8 bytes long and the delay component is a minimum of 96 bytes long.

A 7-byte preamble precedes the Ethernet 802.3 frame. Each byte has a fixed binary pattern of `10101010` and each node on the network uses it to synchronize their clocks and transmission timings. It also informs nodes that a frame is to be sent and for them to check the destination address in the frame.

The end of the frame is a 96-byte delay period which provides the minimum delay between two frames. This slot time delay allows for the worst-case network propagation delay.

The start delimiter field (SDF) is a single byte (or octet) of `10101011`. It follows the preamble and identifies that there is a valid frame being transmitted. Most Ethernet systems uses a 48-bit MAC address for the sending and receiving nodes. Each Ethernet node has unique MAC address, which is normally defined using hexadecimal digits, such as:

```
4C - 31 - 22 - 10  - F1 - 32
```

or `4C31 : 2210: F132`

A 48-bit address field allows 2^{48} different addresses (or approximately 281 474 976 710 000 different addresses).

The LLC length field defines whether the frame contains information or whether it can be used to define the number of bytes in the logical link field. The logical link field can contain up to 1500 bytes of information and has a minimum of 46 bytes; its format is given in Figure 28.3. If the information is greater than the upper limit then multiple frames are sent. Also, if the field is less than the lower limit then it is padded with extra redundant bits.

7 bytes	1 byte	6 bytes	6 bytes	2 bytes	46 to 1500 bytes	4 bytes	96 bytes
Preamble	Start delimiter	Destination address	Source address	Length	Logical link control	FCS	Delay
10101...0101010	10101011						

Figure 28.2 IEEE 802.3 frame format

The 32-bit frame check sequence (FCS) is an error detection scheme. It is used to determine transmission errors and is often called a cyclic redundancy check (CRC) or simply a checksum.

28.3.1 Ethernet II

The first standard for Ethernet was Ethernet I. Most currently available systems implement either Ethernet II or IEEE 802.3 (although most networks are now defined as being IEEE 802.3 compliant). An Ethernet II frame is similar to the IEEE 802.3 frame; it consists of 8 bytes of preamble, 6 bytes of destination address, 6 bytes of source address, 2 bytes of frame type, between 46 and 1500 bytes of data, and 4 bytes of the frame check sequence field.

When the protocol is IPX/SPX the type field contains the bit pattern 1000 0001 0011 0111, but when the protocol is TCP/IP the type field contains 0000 1000 0000 0000.

28.4 IEEE 802.2 and Ethernet SNAP

The LLC is embedded in the Ethernet frame and is defined by the IEEE 802.2 standard. Figure 28.3 illustrates how the LLC fields are inserted into the IEEE 802.3 frame. The DSAP and SSAP fields define the types of network protocol used. A SAP code of 1110 0000 identifies the network operating system layer as NetWare, whereas 0000 0110 identifies the TCP/IP protocol. These SAP numbers are issued by the IEEE. The control field is, among other things, for the sequencing of frames.

In some cases it was difficult to modify networks to be IEEE 802-compliant. Thus an alternative method was to identify the network protocol, known as Ethernet SNAP (Subnetwork Access Protocol). This was defined to ease the transition to the IEEE 802.2 standard and is illustrated in Figure 28.4. It simply adds an extra two fields to the LLC field to define an organization ID and a network layer identifier. NetWare allows for either Ethernet SNAP or Ethernet 802.2 (as Novell used Ethernet SNAP to translate to Ethernet 802.2).

Non-compliant protocols are identified with the DSAP and SSAP code of 1010 1010, and a control code of 0000 0011. After these fields:

- Organization ID which indicates where the company developed the embedded protocol belongs. If this field contains all zeros it indicates a non-company-specific generic Ethernet frame.
- EtherType field which defines the networking protocol. A TCP/IP protocol uses 0000 1000 0000 0000 for TCP/IP, while NetWare uses 1000 0001 0011 0111. NetWare frames adhering to this specification are known as NetWare 802.2 SNAP.

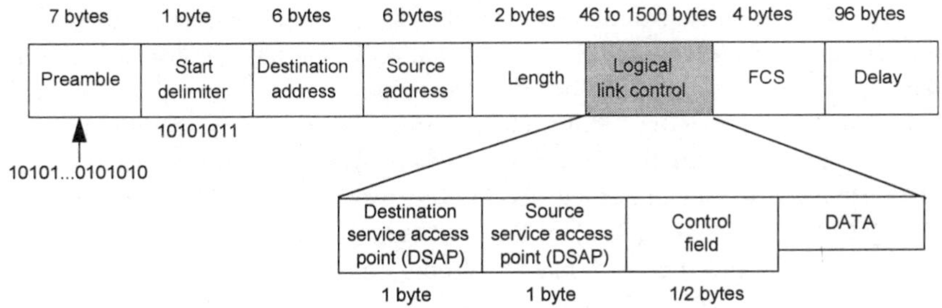

Figure 28.3 Ethernet IEEE 802.3 frame with LLC

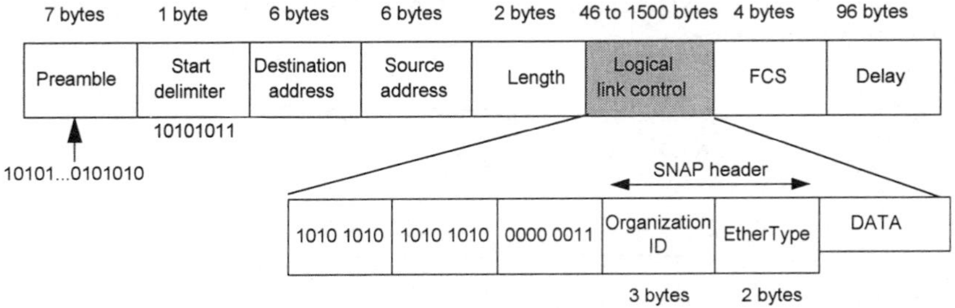

Figure 28.4 Ethernet IEEE 802.3 frame with LLC containing SNAP header

28.4.1 LLC protocol

The 802.3 frame provides some of the data link layer functions, such as node addressing (source and destination MAC addresses), the addition of framing bits (the preamble) and error control (the FCS). The rest of the functions of the data link layer are performed with the control field of the LLC field these functions are:

- Flow and error control. Each data frame sent has a frame number. A control frame is sent from the destination to a source node informing that it has or has not received the frames correctly.
- Sequencing of data. Large amounts of data are sliced and sent with frame numbers. The spliced data is then reassembled at the destination node.

Figure 28.5 shows the basic format of the LLC frame. There are three principal types of frame: information, supervisory and unnumbered. An information frame contains data, a supervisory frame is used for acknowledgment and flow control, and an unnumbered frame is used for control purposes. The first 2 bits of the control field determine which type of frame it is. If they are 0X (where X is a don't care) then it is an information frame, 10 specifies a supervisory frame and 11 specifies an unnumbered frame.

An information frame contains a send sequence number in the control field which ranges from 0 to 127. Each information frame has a consecutive number, N(S) (note that there is a rollover from frame 127 to frame 0). The destination node acknowledges that it has received the frames by sending a supervisory frame. The function of the supervisory frame is specified by the 2-bit S-bit field. This can either be set to Receiver Ready (RR), Receiver Not Ready (RNR) or Reject (REJ). If an RNR function is set then the destination node acknowledges that all frames up to the number stored in the receive sequence number N(R) field were received correctly. An RNR function also acknowledges the frames up to the number N(R), but informs the source node that the destination node wishes to stop communicating. The REJ function specifies that frame N(R) has been rejected and all other frames up to N(R) are acknowledged.

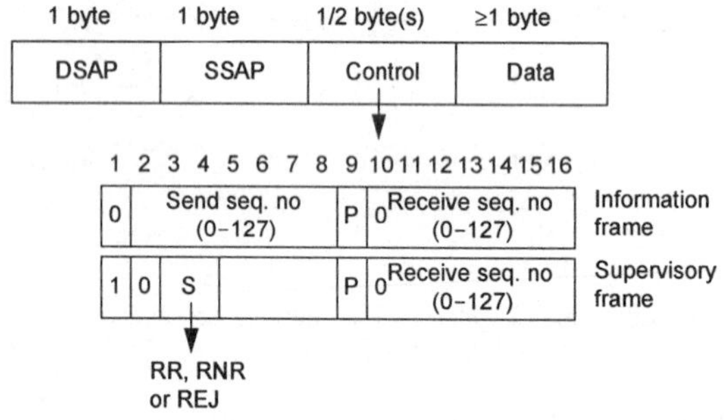

Figure 28.5 LLC frame format

28.5 OSI and the IEEE 802.3 standard

Ethernet fits into the data link and the physical layer of the OSI model. These two layers only deal with the hardware of the network. The data link layer splits into two parts: the LLC and the MAC layer.

The IEEE 802.3 standard splits into three sublayers:

- MAC (media access control).
- Physical signaling (PLS).
- Physical media attachment (PMA).

The interface between PLS and PMA is called the attachment unit interface (AUI) and the interface between PMA and the transmission media is called the media dependent interface (MDI). This grouping into modules allows

Ethernet to be very flexible and to support a number of bit rates, signaling methods and media types. Figure 28.5 illustrates how the layers interconnect.

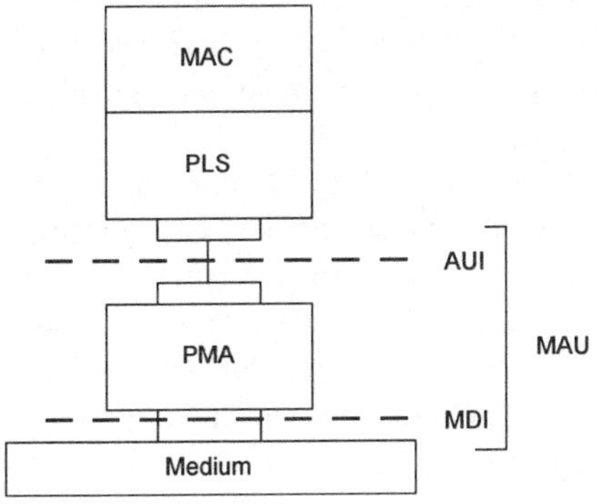

Figure 28.6 Organization of the IEEE 802.3 standard

28.5.1 Media access control (MAC)

CSMA/CD is implemented in the MAC layer. The functions of the MAC layers are:

- When sending frames: receive frames from LLC; control whether the data fills the LLC data field, if not add redundant bits; make the number of bytes an integer, and calculate the FCS; add the preamble, SFD and address fields to the frame; send the frame to the PLS in a serial bitstream.
- When receiving frames: receive one frame at a time from the PLS in a serial bitstream; check whether the destination address is the same as the local node; ensures the frame contains an integer number of bytes and the FCS is correct; remove the preamble, SFD, address fields, FCS and remove redundant bits from the LLC data field; send the data to the LLC.
- Avoid collisions when transmitting frames and keep the right distance between frames by not sending when another node is sending; when the medium gets free, wait a specified period of time before starting to transmit.
- Handle any collision that appears by sending a jam signal; generating a random number and backing off from sending during that random time.

28.5.2 Physical signaling (PLS) and physical medium attachment (PMA)

PLS defines transmission rates, types of encoding/decoding and signaling methods. In PMA a further definition of the transmission media is accomplished, such as coaxial, fiber or twisted-pair. PMA and MDI together form the media attachment unit (MAU), often known as the transceiver.

28.6 Ethernet transceivers

Ethernet requires a minimal amount of hardware. The cables used to connect it are either unshielded twisted-pair cable (UTP) or coaxial cables. These cables must be terminated with their characteristic impedance, which is 50 Ω for coaxial cables and 100 Ω for UTP cables.

Each node has transmission and reception hardware to control access to the cable and also to monitor network traffic. The transmission/reception hardware is called a transceiver (short for *trans*mitter/re*ceiver*) and a controller builds up and strips down the frame. The transceiver builds the transmits bits at a rate of 10 Mbps thus the time for one bit is $1/10 \times 10^6$ which is 0.1 μs.

The Ethernet transceiver transmits onto a single ether. When none of the nodes are transmitting then the voltage on the line is +0.7 V. This provides a carrier sense signal for all nodes on the network; it is also known as the heartbeat. If a node detects this voltage then it knows that the network is active and that no nodes are currently transmitting.

Thus when a node wishes to transmit a message it listens for a quiet period. Then if two or more transmitters transmit at the same time, a collision results. When they detect the signal, each node transmits a 'jam' signal. The nodes involved in the collision then wait for a random period of time (ranging from 10 to 90 ms) before attempting to transmit again. Each node on a network also awaits a retransmission. Thus collisions are inefficient in a network as they stop nodes from transmitting. Transceivers normally detect a collision by monitoring the DC (or average) voltage on the line.

When transmitting, a transceiver unit transmits the preamble of consecutive 1's and 0's. The coding used is a Manchester code which represents a 0 as a high-to-low voltage transition and a 1 as a low-to-high voltage transition. A low voltage is –0.7 V and a high is +0.7 V. Thus when the preamble is transmitted the voltage will change between +0.7 and –0.7 V; this is illustrated in Figure 28.7. If after the transmission of the preamble no collisions are detected then the rest of the frame is sent.

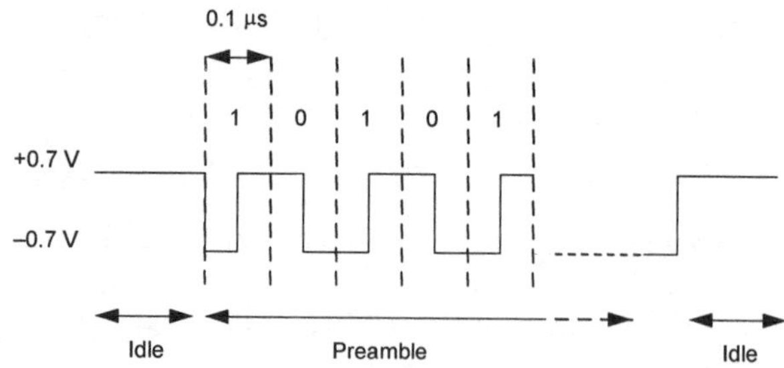

Figure 28.7 Ethernet digital signal

28.7 NIC

When receiving data, the function of the NIC is to copy all data transmitted on the network, decode it and transfer it to the computer. An Ethernet NIC basically contains three parts:

- Physical medium interface. The physical medium interface corresponds to the PLS and PMA in the standard and is responsible for the electrical transmission and reception of data. It consists of two parts: the transceiver, which receivers and transmits data from or onto the transmission media; and a code converter that encodes/decodes the data. It also recognizes a collision of the media.
- Data link controller. The controller corresponds to the MAC layer.
- Computer interface.

It can be split into four main functional blocks:

- Network interface.
- Manchester decoder.
- Memory buffer.
- Computer interface.

28.7.1 Network interface

The network interface function is to listen, recreate the waveform transmitted on the cable into a digital signal and transfer the digital signal to the Manchester decoder. The network interface consists of three parts:

- BNC/RJ-45 connector.
- Reception hardware. The reception hardware translates the waveforms transmitted on the cable to digital signals then copies them to the Manchester decoder.
- Isolator. The isolator is connected directly between the reception hardware and the rest of the Manchester decoder; it guarantees that no noise from the network affects the computer, and vice versa.

The reception hardware is called a receiver and is the main component in the network interface. Basically it has the function of an earphone, listening and copying the traffic on the cable. Unfortunately, the Ether and transceiver electronics are not perfect. The transmission line contains resistance and capacitance which distort the shape of the bitstream transmitted onto the Ether. Distortion in the system causes pulse spreading., which leads to intersymbol interference. These is also a possibility of noise affecting the digital pulse as it propagates through the cable. Therefore the receiver also need to recreate the digital signal and filter noise.

Figure 28.8 shows a block diagram of an Ethernet receiver. The received signal goes through a buffer with high input impedance and low capacitance to reduce the effects of loading on the coaxial cable. An equalizer passes high frequencies and attenuates low frequencies from the network, flattening the network passband. A 4-pole Bessel low-pass filter provides the average DC level from the received signal. The squelch circuit activates the line driver only when it detects a true signal. This prevents noise activating the receiver.

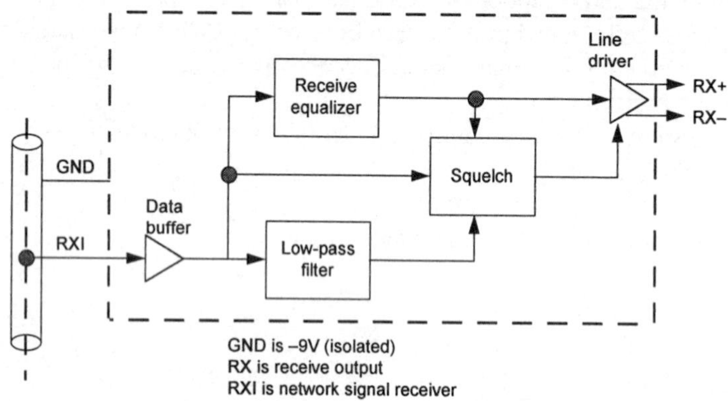

Figure 28.8 Ethernet receiver block diagram

28.7.2 Manchester decoder

Manchester coding has the advantage of embedding timing (clock) information within the transmitted bits. A positively edged pulse (low→high) represents a 1 and a negatively edged pulse (high→low) a 0, as shown in Figure 28.9. Another advantage of this coding method is that the average voltage is always zero when used with equal positive and negative voltage levels.

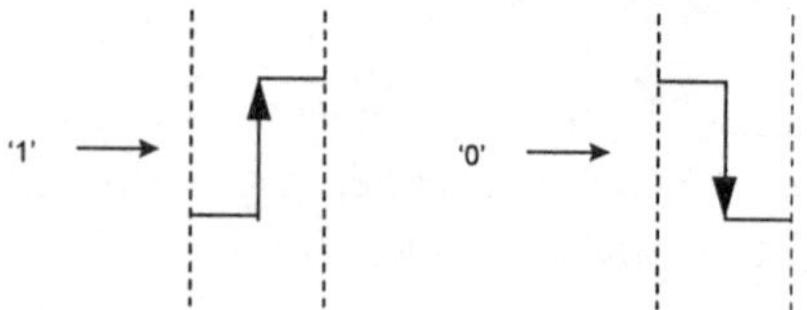

Figure 28.9 Manchester encoding

Figure 28.10 is an example of transmitted bits using Manchester encoding. The receiver passes the received Manchester-encoded bits through a low-pass filter. This extracts the lowest frequency in the received bitstream, i.e., the clock frequency. With this clock the receiver can then determine the transmitted bit pattern.

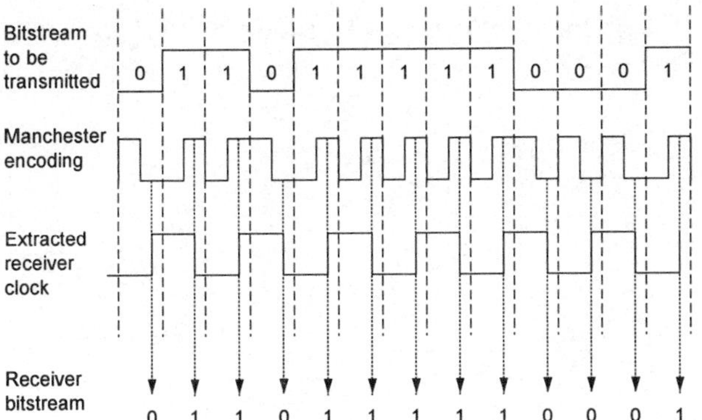

Figure 28.10 Example of Manchester coding

For Manchester decoding, the Manchester-encoded signal is first synchronized to the receiver (called bit synchronization). A transition in the middle of each bit cell is used by a clock recovery circuit to produce a clock pulse in the center of the second half of the bit cell. In Ethernet the bit synchronization is achieved by deriving the clock from the preamble field of the frame using a clock and data recovery circuit. Many Ethernet decoders used the SEEQ 8020 Manchester code converter, which uses a phase-locked loop (PLL) to recover the clock. The PLL is designed to lock onto the preamble of the incoming signal within 12-bit cells. Figure 28.11 shows a circuit schematic of bit synchronization using Manchester decoding and a PLL.

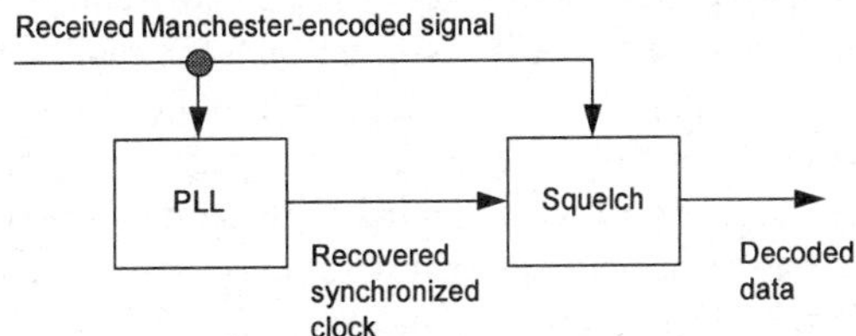

Figure 28.11 Manchester decoding with bit synchronization

The PLL is a feedback circuit which is commonly used for the synchronization of digital signals. It consists of a phase detector (such as an EX-OR gate) and a voltage-controlled oscillator (VCO) which uses a crystal oscillator as a clock source. The frequency of the crystal is twice the frequency of the received signal. It is so constant that it only needs irregular and small adjustments to be synchronized to the received signal. The function of the phase detector is to find irregularities between the two signals and adjusts the VCO to minimize the error. This is accomplished by comparing the received signals and the output from the VCO. When the signals have the same frequency and phase the PLL is locked. Figure 28.12 shows the PLL components and the function of the EX-OR.

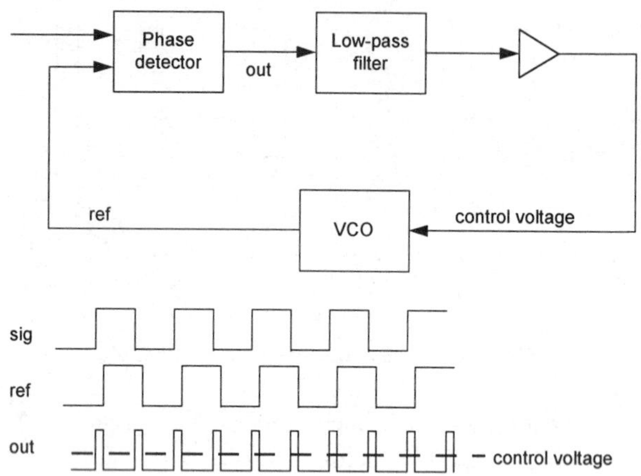

Figure 28.12 PLL and example waveform for the phase detector

28.7.3 Memory buffer

The rate at which data is transmitted on the cable differs from the data rate used by the receiving computer, and the data appears in bursts. To compensate for the difference between the data rate, a first-in first-out (FIFO) memory buffer is used to produce a constant data rate. An important condition is that the average data input rate should not exceed the frequency of the output clock; if this is not the case the buffer will be filled up regardless of its size.

A FIFO is a RAM that uses a queuing technique where the output data appears in the same order that it went in. The input and output are controlled by separate clocks, and the FIFO keeps track of the data that has been written and the data that has been read and can thus be overwritten. This is achieved with a pointer. Figure 28.13 shows a block diagram of the FIFO configuration. The FIFO status is indicated by flags, the empty flag (EF) and the full flag (FF), which show whether the FIFO is either empty or full.

28.7.4 Ethernet implementation

The completed circuit for the Ethernet receiver is given in Appendix G and is outlined in Figure 28.14. It uses the SEEQ Technologies 82C93A Ethernet transceiver as the receiver and the SEEQ 8020 Manchester code converter which decodes the Manchester code. A transformer and DC-to-DC converter isolates the SEEQ 82C92A and the network cable from the rest of the circuit (and the computer). The isolated DC-to-DC converter converts a 5 V supply to the –9 V needed by the transceiver.

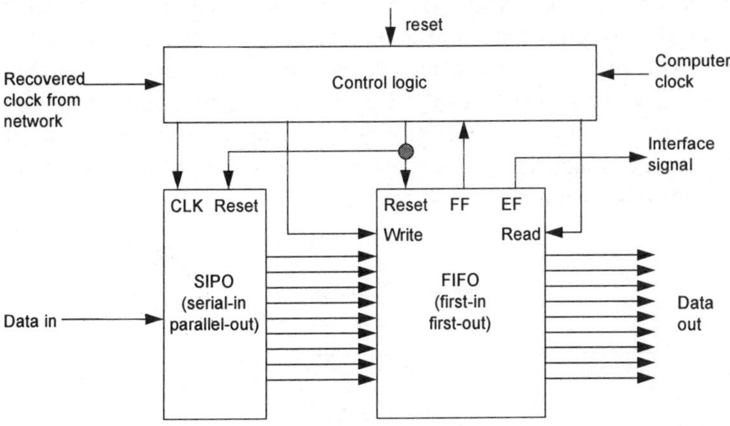

Figure 28.13 Memory buffering

The memory buffer used is the AMD Am7204 FIFO which has 4096 data words with 9-bit words (but only 8 bits are actually used). The output of the circuit is 8 data lines, the control lines $\overline{FF}$, $\overline{EF}$, $\overline{RS}$, $\overline{R}$ and $\overline{W}$, and the +5 V and GND supply rails.

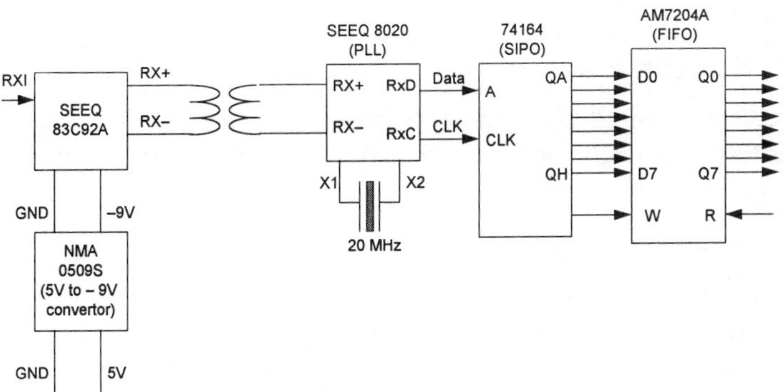

Figure 28.14 Ethernet receiver

28.8 Standard Ethernet limitations

The standard Ethernet CSMA/CD specification places various limitations on maximum cables lengths. This limitation is due to maximum signal propagation times and the clock period.

28.8.1 Length of segments

Twisted-pair and coaxial cables have a characteristic impedance. A cable must be terminated with the correct characteristic impedance so that there is no loss of power and no reflections at terminations. Twisted-pair cables must be terminated with its characteristic impedance (for twisted-pair cables it is normally 100 Ω and for coaxial cables it is 50 Ω). The Ethernet connection may consist of many spliced coaxial sections. One or many sections constitute a cable segment, which is a standalone network. A segment must not exceed 500 m. This is shown in Figure 28.15.

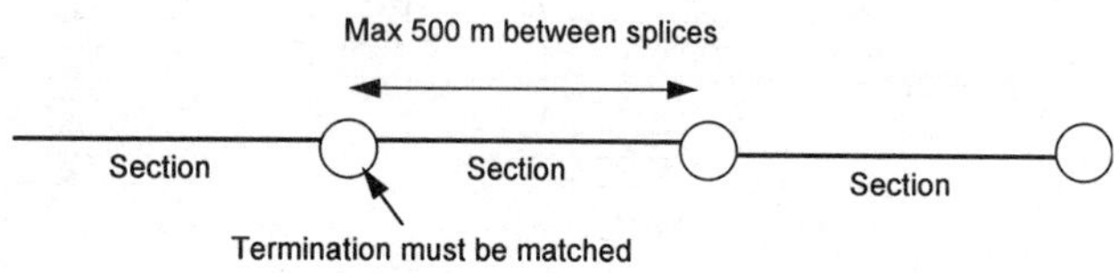

Figure 28.15 Connection of sections

28.8.2 Repeater lengths

A repeater is added between segments to boost the signal. A maximum of two repeaters can be inserted into the path between two nodes. The maximum distance between two nodes connected via repeaters is 1500 m, this is illustrated in Figure 28.16.

28.8.3 Maximum links

The maximum length of a point-to-point coaxial link is 1500 m. A long run such as this is typically used as a link between two remote sites within a single building.

28.8.4 Distance between transceivers

Transceivers should not be placed closer than 2.5 m. Additionally, each segment should not have more than 100 transceiver units, as illustrated in Figure 28.17. Transceivers which are placed too close to each other can cause transmission interference and also an increased risk of collision.

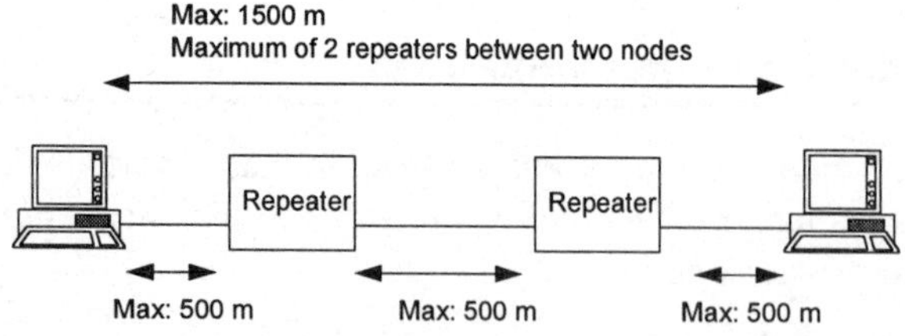

Figure 28.16 Maximum number of repeaters between two nodes

Each node transceiver lowers network resistance and dissipates the transmission signal. A sufficient number of transceivers reduces the electrical characteristic of the network below the specified operation threshold.

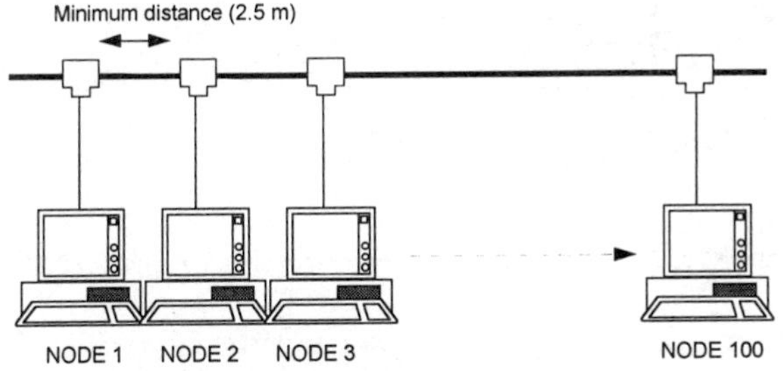

Figure 28.17 Connection of sections

28.9 Ethernet types

There are five main types of standard Ethernet:

- Standard, or thick-wire, Ethernet (10BASE5).
- Thinnet, or thin-wire Ethernet, or Cheapernet (10BASE2).
- Twisted-pair Ethernet (10BASE-T).
- Optical fiber Ethernet (10BASE-FL).
- Fast Ethernet (100BASE-TX or 100VG-Any LAN).

The thin- and thick-wire types connect directly to an Ethernet segment; they are shown in Figure 28.18 and Figure 28.19. Standard Ethernet, 10BASE5, uses a high-specification cable (RG-50) and N-type plugs to connect the transceiver to the Ethernet segment. A node connects to the transceiver using a 9-pin D-type connector. A vampire (or bee-sting) connector can be used to clamp the transceiver to the backbone cable.

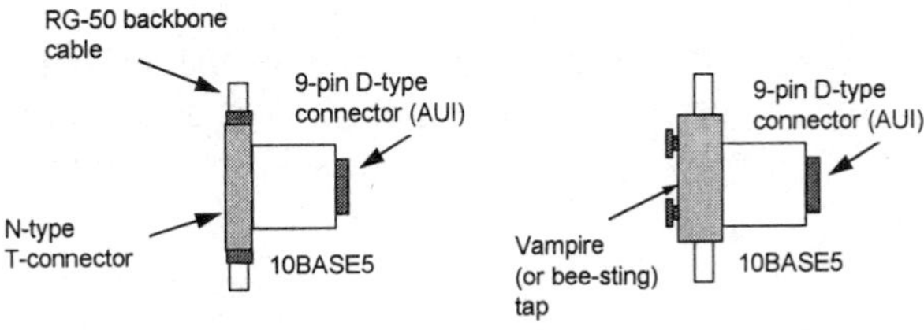

Figure 28.18 Ethernet connections for thick Ethernet

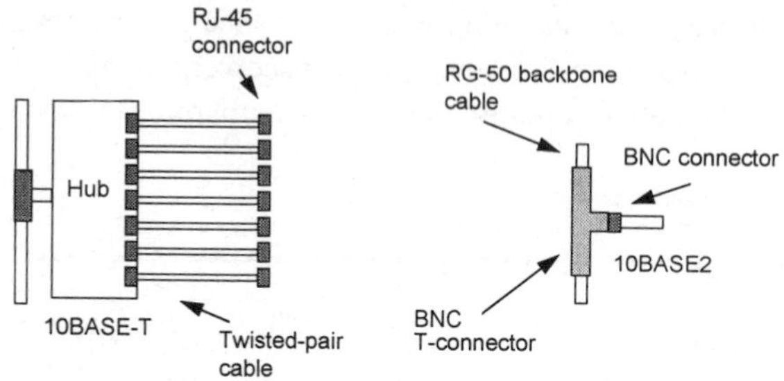

Figure 28.19 Ethernet connections for thin Ethernet and 10BASE-T

Thin-wire, or Cheapernet, uses a lower specification cable (it has a lower inner conductor diameter). The cable connector required is also of a lower specification, i.e. BNC rather than N-type connectors. In standard Ethernet the transceiver unit is connected directly onto the backbone tap. On a Cheapernet network the transceiver is integrated into the node.

Many modern Ethernet connections are to a 10BASE-T hub which connects UTP cables to the Ethernet segment. An RJ-45 connector is used for 10BASE-T. The fiber-optic type, 10BASE-FL, allows long lengths of interconnected lines, typically up to 2 km. They use either SMA connectors or an ST connector. SMA connectors are screw-on connectors whereas ST connectors are push-on. Table 28.1 shows the basic specifications for the different types.

28.10 Twisted-pair hubs

Twisted-pair Ethernet (10BASE-T) nodes normally connect to the backbone using a hub, as illustrated in Figure 28.20. Connection to the twisted-pair cable is via an RJ-45 connector. The connection to the backbone can either be to thin- or thick-Ethernet. Hubs can also stackable where one hub connects to another. This leads to concentrated area networks (CANs) and limits the amount of traffic on the backbone. Twisted-pair hubs normally improves network performance.

10BASE-T uses two twisted-pair cables, one for the transmit and one for the receive. A collision occurs when the node (or hub) detects that it is receiving data when it is currently transmitting data, as illustrated in Figure 28.21.

Table 28.1 Ethernet network parameters

Parameter	*10BASE5*	*10BASE2*	*10BASE-T*
Common name	Standard or thick-wire Ethernet	Thinnet, thin-wire Ethernet	Twisted-pair Ethernet
Data rate	10 Mbps	10 Mbps	10 Mbps
Maximum segment length	500 m	200 m	100 m
Maximum nodes on a segment	100	30	3
Maximum number of repeaters	2	4	4
Maximum nodes per network	1024	1024	
Minimum node spacing	2.5 m	0.5 m	No limit
Location of trans-ceiver electronics	Located at the cable connection	Integrated within the node	In a hub
Typical cable type	RG-50 (0.5 in diameter)	RG-6 (0.25 in diameter)	UTP cables
Connectors	N-type	BNC	RJ-45/ Telco
Transceiver cable connector	15-pin D-type	BNC	

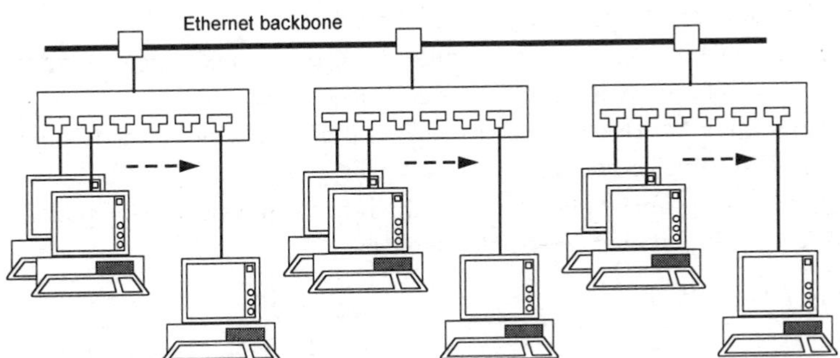

Figure 28.20 10BASE-T connection

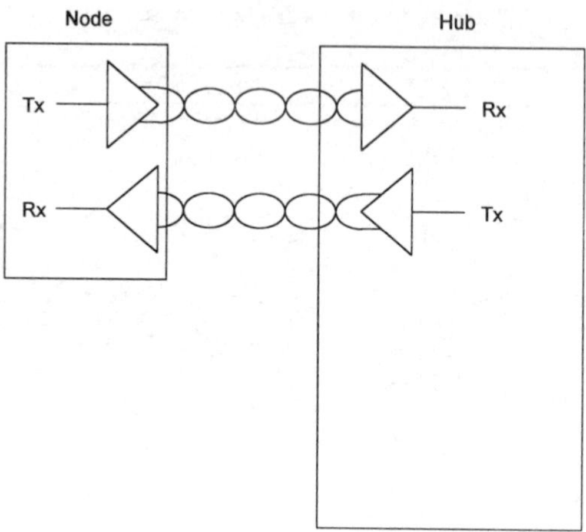

Figure 28.21 10BASE-T connections

28.11 100 Mbps Ethernet

Standard 10 Mbps Ethernet does not performance well when many users running multimedia applications. Two improvements to the standard are Fast Ethernet and 100VG-AnyLAN. The IEEE have defined standards for both of them, IEEE 802.3u for Fast Ethernet and 802.12 for 100VG-AnyLAN. They are supported by many manufacturers and use bit rates of 100 Mbps. This gives at least 10 times the performance of standard Ethernet.

New standards relating to 100 Mbps Ethernet are now becoming popular:

- 100BASE-TX (twisted-pair) – which uses 100 Mbps over two pairs of Cat-5 UTP cable or two pairs of Type 1 STP cable.
- 100BASE-T4 (twisted-pair) – which is the physical layer standard for 100 Mbps bit rate over Cat-3, Cat-4 or Cat-5 UTP.
- 100VG-AnyLAN (twisted-pair) – which uses 100 Mbps over two pairs of Cat-5 UTP cable or two pair of Type 1 STP cable.
- 100BASE-FX (fiber-optic cable) – which is the physical layer standard for 100 Mbps bit rate over fiber-optic cables.

28.11.1 100BASE-T

Fast Ethernet, or 100BASE-T, is simply 10BASE-T running at 10 times the bit rate. It is a natural progression from standard Ethernet and thus allows existing Ethernet networks to be easily upgraded. Unfortunately, as with standard Ethernet, nodes contend for the network, reducing the network efficient when there are high traffic rates. Also, as it uses collision detect, the maximum segment length is limited by the amount of time for the farthest nodes on a network to properly detect collisions. On a Fast Ethernet network with twisted-pair copper cables this distance is 100 m and for a fiber-optic link it is 400 m. Table 28.2 outlines the main network parameters for Fast Ethernet.

Since 100BASE-T standards are compatible with 10BASE-T networks then the network allows both 10 Mbps and 100 Mbps bit rates on the line. This makes upgrading simple as the only additions to the network are dual-speed interface adapters. Nodes with the 100 Mbps capabilities can communicate at 100 Mbps, but they can also communicate with slower nodes, at 10 Mbps.

28.11.2 100BASE-4T

100BASE-4T allows the use of standard Cat-3 cables. These contain eight wires made up of four twisted pairs. 100BASE-4T uses all of the pairs to transmit at 100 Mbps. This differs from 10BASE-T in that 10BASE uses only two pairs, one for the transmit and one for the receive. 100BASE-T allows compatibility with 10BASE-T in that the first two pairs (Pair 1 and Pair 2) are used in the same way as 10BASE-T connections (as was illustrated in Figure 28.21). 100BASE-T then uses the other two pairs (Pair 3 and Pair 4) with half-duplex links between the hub and the node. The connections are illustrated in Figure 28.22.

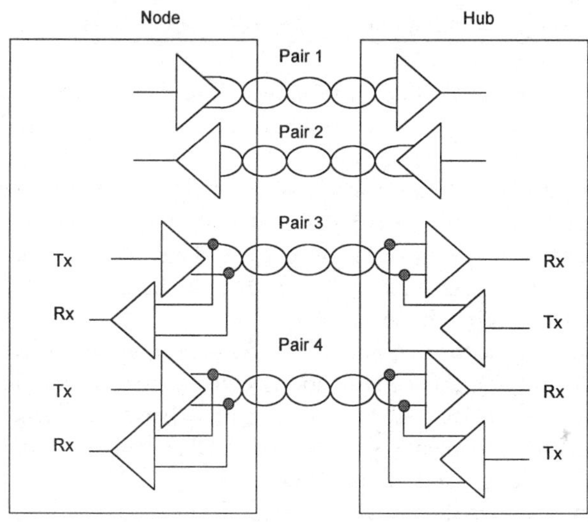

Figure 28.22 100BASE-T connections

Table 28.2 Fast Ethernet network parameters

	100BASE-TX	*100VG-AnyLAN*
Standard	IEEE 802.3u	IEEE 802.12
Bit rate	100 Mbps	100 Mbps
Actual throughput	Up to 50 Mbps	Up to 96 Mbps
Maximum distance (hub to node)	100 m (twisted-pair, CAT-5) 400 m (fiber)	100 m (twisted-pair, CAT-3) 200 m (twisted-pair, CAT-5) 2 km (fiber)
Scaleability	None	Up to 400 Mbps
Advantages	Easy migration from 10BASE-T	Greater throughput, greater distance

28.11.3 Line code

100BASE-4T uses four separate Cat-3 twisted-pair wires. The maximum clock rate that can be applied to Cat-3 cable is 30 Mbps. Thus some mechanism must be devised which can reduce the line bit rate to under 30 Mbps but give a symbol rate of 100 Mbps. This is achieved with a 3-level code (+, – and 0) and is known as 8B6T. This code converts 8 binary digits into 6 ternary symbols. Part of the table is given in Appendix I.7, the first 6 codes are:

Data byte	*Code*	*Data byte*	*Code*	*Data byte*	*Code*
00000000	–+00–+	00000001	0–+–+0	00000010	0–+0–+
00000011	0–++0–	00000100	–+0+0–	00001001	+0—+0

Thus the bit sequence 00000000 will be coded as a negative voltage, a positive voltage, a zero voltage, a zero voltage, a negative voltage and a positive voltage.

The maximum base frequency for a 100 Mbps signal will be produced when the input bitstream is 010101010....01010. As each bit lasts 10 ns then the period between consecutive levels is 20 ns. Thus the minimum frequency contained will be 50 MHz. This is greater than the bandwidth of Cat-3 cable, so it would not pass through the cable.

With a 3-level code the maximum base frequency will occur with the code +–+–+–+–+–...+–. If this code is not included and the maximum transition is made as +–0+–0+–0+–...0+– then period between changes becomes 40 ns. The minimum frequency will then be 25MHz, which is within the passband of the Cat-3 cable. This is illustrated in Figure 28.23.

Apart from reducing the frequencies with the digital signal, the 8B6T code has the advantage of reducing the DC content of the signal. Most of the codes contain the same number of positive and negative voltages. This is because

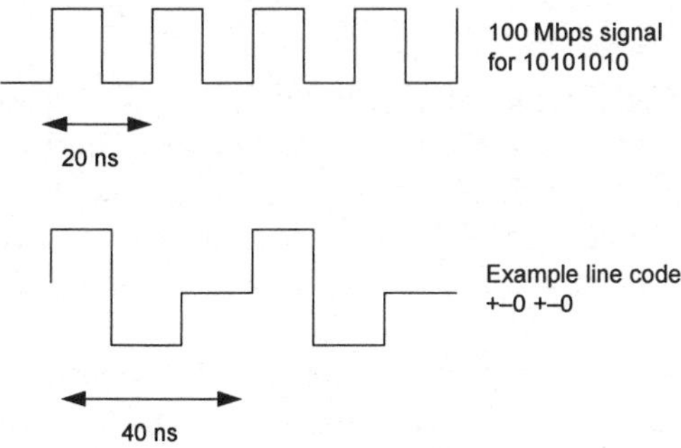

Figure 28.23 Example codes showing repetition period.

only 256 of the possible 729 (3^6) codes are actually used. The codes are also chosen to have at least two transitions in every code word, thus the clock information is embedded into signal.

Unfortunately it is not possible to have all codes with the same number of negative voltages as positive voltages. Thus there are some codes which have a different number of negatives and positives these include:

0100 0001 +0−00++
0111 1001 +++−0−

Most transceiver circuits use a transformer to isolate the external equipment from the computer equipment. These transformers do not allow the passage of DC current. Thus if the line code has a sequence which consistently has more positives than negatives, the DC current will move away from its zero value. As this does not pass across the transformer, the receive bitstream on the output of the transformer can reduce the amplitude of the received signal (and may thus cause errors). This phenomenon is known as DC wander. A code which has more positive levels than negative levels is defined as having a weighing of +1.

The technique used to overcome this is to invert consecutive codes which have a weighing of +1. For example suppose the line code were:

+0++−− ++0+−− +++−−0 +++−−0

it would actually be coded as:

+0++−− −−0−++ +++−−0 −−−++0

The receiver detects the –1 weighted codes as an inverted pattern.

28.11.4 100VG-AnyLAN

The 100VG-AnyLAN standard (IEEE 802.12) was developed mainly by Hewlett Packard and overcomes the contention problem by using a priority based round-robin arbitration method, known as demand priority access method (DPAM). Unlike Fast Ethernet, nodes always connect to a hub which regularly scans its input ports to determine whether any nodes have requests pending.

100VG-AnyLAN has the great advantage that it supports both IEEE 802.3 (Ethernet) and IEEE 802.5 (Token Ring) frames and can thus integrate well with existing 10BaseT and Token Ring networks.

100VG-Any has an in-built priority mechanism with two priority levels: a high priority request and a normal priority request. A normal priority request is used for non-real-time data, such as data files, and so on. High priority requests are used for real-time data, such as speech or video data. At present there is limited usage of this feature and there is no support mechanism for this facility after the data has left the hub.

100VG-AnyLAN allows up to seven levels of hubs (i.e. one root and six cascaded hubs) with a maximum distance of 150 m between nodes. Unlike, other forms of Ethernet, it allows any number of nodes to be connected to a segment. Appendix I.6 contains more information on the data transmission.

28.11.5 Connections

100BASE-TX, 100BASE-T4 and 100VG-AnyLAN use the RJ-45 connector, which has eight connections. 100BASE-TX uses pairs 2 and 3, whereas 100BASE-4 and 100VG-AnyLAN use pairs 1, 2, 3 and 4. The connections for the cables are defined in Table 28.3. The white/orange color identifies the cable which is white with an orange stripe, whereas orange/white identifies an orange cable with a white stripe.

Table 28.3 Cable connections for 100BASE-X

Pin	*Cable color*	*Cable color*	*Pair*
1	white/orange	white/orange	Pair 4
2	orange/white	orange/white	Pair 4
3	white/green	white/green	Pair 3
4	blue/white	blue/white	Pair 3
5	white/blue	white/blue	Pair 1
6	green/white	green/white	Pair 1
7	white/brown	white/brown	Pair 2
8	brown/white	brown/white	Pair 2

28.11.6 Migration to Fast Ethernet

If an existing network is based on standard Ethernet then, in most cases, the best network upgrade is either to Fast Ethernet or 100VG-AnyLAN. Since the protocols and access methods are the same there is no need to change any of the network management software or application programs. The upgrade path for Fast Ethernet is simple and could be:

- Upgrade high data rate nodes, such as servers or high powered workstations to Fast Ethernet.
- Gradually upgrade NICs (network interface cards) on Ethernet segments to cards which support both 10BASE-T and 100BASE-T. These cards automatically detect the transmission rate to give either 10 or 100 Mbps.

The upgrade path to 100VG-AnyLAN is less easy as it relies on hubs and, unlike Fast Ethernet, most NICs have different network connectors, one for 10BASE-T and the other for 100VG-AnyLAN (although it is likely that more NICs will have automatic detection). A possible path could be:

- Upgrade high data rate nodes, such as servers or high-powered workstations to 100VG-AnyLAN.
- Install 100VG-AnyLAN hubs.
- Connect nodes to 100VG-AnyLAN hubs and change over connectors.

It is difficult to assess the performance differences between Fast Ethernet and 100VG-AnyLAN. Fast Ethernet uses a well-proven technology but suffers from network contention. 100VG-AnyLAN is a relatively new technology and the handshaking with the hub increases delay time. The maximum data throughput of a 100BASE-TX network is limited to around 50 Mbps, whereas 100VG-AnyLAN allows rates up to 96 Mbps.

The 100BASE-TX standard does not allow future upgrading of the bit rate, whereas, 100VG-AnyLAN allows possible upgrades to 400 Mbps.

28.12 Ethernet security

Ethernet itself provides no security to be it is designed as a simple and open physical medium for data transmission. Furthermore it is not immune to snooping and spying. The vulnerabilities of the Ethernet are:

- It is an open architecture where any node can transmit and/or receive.
- It uses broadcast communications.
- It is easy to tap into.
- It has no hardware security.
- It is easy to jam a network.

28.13 Exercises

28.1 Discuss how a destination node uses the LLC layer to indicate that frames have been received in error.

28.2 Discuss the main reasons for the preamble in an Ethernet frame.

28.3 Discuss the limitations of the different types of Ethernet.

28.4 Discuss 100 Mbps Ethernet technologies with respect to the how they operate and their typical parameters.

28.5 A node has a binary network address of `0011 1111 0101 1111 1000 1000 0101 0000 0000 1000 0111 1111`. Determine its hexadecimal address; that is, the address in the form `XXXX:XXXX:XXXX`. Table 28.4 shows the conversion of binary digits into hexadecimal.

The binary number `0111010111000000` is converted into hexadecimal by the following:

Binary	0111	0101	1100	0000
Hex	7	5	C	0

28.6 Explain the usage of Ethernet SNAP.

28.7 State the main advantage of Manchester coding and show the bit pattern for the bit sequence:

01111010101101010001011010

28.8 Explain the main functional differences between 100BASE-T, 100BASE-4T and 100VG-AnyLAN.

Table 28.4 Decimal, binary and hexadecimal conversions

Decimal	*Binary*	*Hex*	*Decimal*	*Binary*	*Hex*
0	0000	0	8	1000	8
1	0001	1	9	1001	9
2	0010	2	10	1010	A
3	0011	3	11	1011	B
4	0100	4	12	1100	C
5	0101	5	13	1101	D
6	0110	6	14	1110	E
7	0111	7	15	1111	F

29 Token Ring

29.1 Introduction

Token Ring networks were developed by several manufacturers, the most prevalent being the IBM Token Ring. Token Ring networks cope well with high network traffic loadings. They were at one time extremely popular but their popularity has since been overtaken by Ethernet. Token Ring networks have, in the past, suffered from network management problems and poor network fault tolerance.

Token Ring networks are well suited to situations which have large amounts of traffic and also work well with most traffic loadings. They are not suited to large networks or networks with physically remote stations. The main advantage of Token Ring is that it copes better with high traffic rates than Ethernet, but requires a great deal of maintenance, especially when faults occur or when new equipment is added to or removed from the network. Many of these problems have now been overcome by MAUs (multistation access units), which are similar to the hubs using in Ethernet.

The IEEE 802.5 standard specifies the MAC layer for a Token Ring network with a bit rate of either 4 Mbps or 16 Mbps. There are two main types of Token Ring networks. Type 1 Token Ring uses Type 1 Token Ring cable (shielded twisted-pair) with IBM style universal connectors. Type 3 Token Ring use either Cat-3, Cat-4 or Cat-5 unshielded twisted-pair cables with modular connectors. Cat-3 has the advantage of being cheap to install and is typically used in telephone connections. Unfortunately the interconnection distance is much less than for Cat-4 and Cat-5 cables.

29.2 Operation

A Token Ring network circulates an electronic token (named a control token) around a closed electronic loop. Each node on the network reads the token and repeats it to the next node. The control token circulates around the ring even when there is no data being transmitted.

Nodes on a Token Ring network wishing to transmit must await a token.

When they get it, they fill a frame with data and add the source and destination addresses then send it to the next node. The data frame then circulates around the ring until it reaches the destination node. It then reads the data into its local memory area (or buffer) and marks an acknowledgment on the data frame. This then circulates back to the source (or originating) node. When it receives the frame it tests it to determine whether it contains an acknowledgment. If it does then the source nodes knows that the data frame was received correctly, else the node is not responding. If the source node has finished transmitting data then it transmits a new token, which can be used by other nodes on the ring.

Figure 29.1(a)–(d) shows a typical interchange between node B and node A. Initially, in (a), the control token circulates between all the nodes. This token does not contain any data and is only 3 bytes long. When node B finally receives the control token it then transmits a data frame, as illustrated in (b). This data frame is passed to node C, then to node D and finally onto A. Node A will then read the data in the data frame and return an acknowledgment to node B, as illustrated in (c). After node B receives the acknowledgment, it passes a control token onto node C and this then circulates until a node wishes to transmit a data frame. No nodes are allowed to transmit data unless they have received a valid control token. A distributed control protocol determines the sequence in which nodes transmit. This gives each node equal access to the ring.

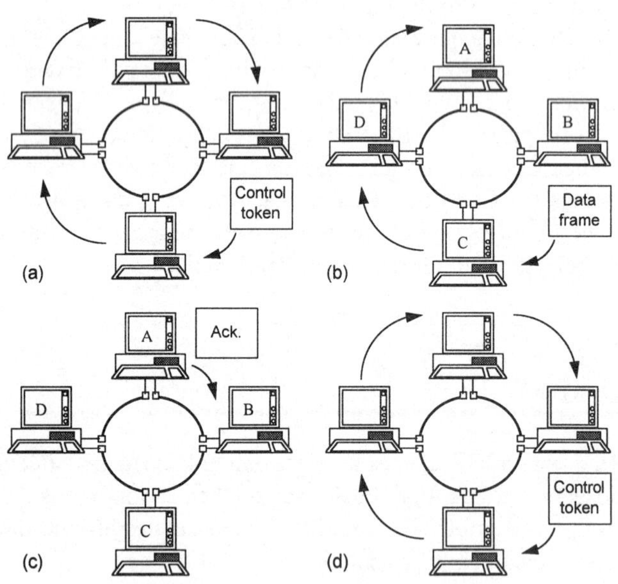

Figure 29.1 Example data exchange

29.3 Token Ring - media access control (MAC)

Token passing allows nodes controlled access to the ring. Figure 29.2 shows the token format for the IEEE 802.5 specification. There are two main types of frame: a control token and a data frame. A control token contains only a start and end delimiter, and an access control (AC) field. A data frame has start and end delimiters (SD/ED), an access control field, a frame control field (CF), a destination address (DA), a source address (SA), frame check sequence (FCS), data and a frame status field (FS).

The access control and frame control fields contain information necessary for managing access to the ring. This includes priority reservation, priority information and information on whether the data is user data or control information. It also contains an express indicator which informs networked nodes that an individual node requires immediate action from the network management node.

The destination and source addresses are 6 bytes in length. Logical link control information has variable length and is shown in Figure 29.2. It can either contain user data or network control information.

The frame check sequence (FCS) is a 32-bit cyclic redundancy check (CRC) and the frame control field is used to indicate whether a destination node has read the data in the token.

The start and end delimiters are special bit sequences which define the start and end of the frame and thus cannot occur anywhere within the frame. As with Ethernet the bits are sent using Manchester coding. The start and end delimiters violate the standard coding scheme. The standard Manchester coding codes a 1 as a low-to-high transition and a 0 as a high-to-low transition. In the start and end delimiters, two of the bits within the delimiters are set to either a high level (`H`) or a low level (`L`). These bits disobey the standard coding as there is no change in level, i.e. from a high to a low or a low to a high. When the receiver detects this violation and the other standard coded bits in the received bit pattern, it knows that the accompanying bits are a valid frame. The coding is as follows:

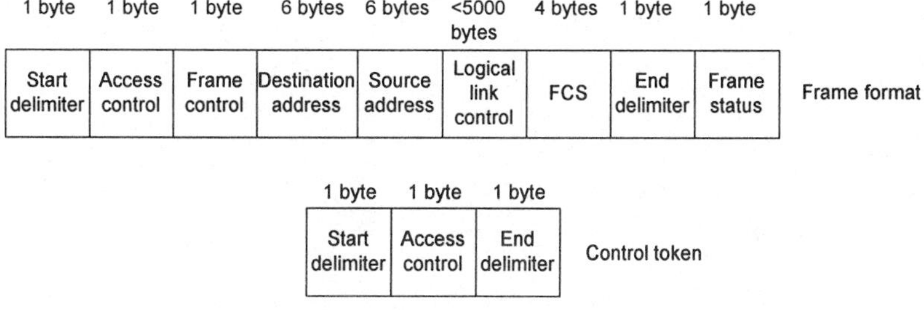

Figure 29.2 IEEE 802.5 frame format

- If the preceding bit is a 1 then the start delimiter is HLOHL000, else
- If the preceding bit is a 0 then the start delimiter is LHOLH000.

They are shown in Figure 29.3. The end delimiter is similar to the start delimiter, but 0's are replaced by 1's. An error detection bit (E) and a last packet indicator bit (I) are added.

If the bit preceding the end delimiter is a 1 then the end delimiter is HL1HL1IE. If it is a 0 then it is LH1LH1IE. The E bit is used for error detection and is initially set by the originator to a 0. If any of the nodes on the ring detects an error the E bit is set to a 1. This indicates to the originator that the frame has developed an error as it was sent. The I bit determines whether the data being sent in a frame is the last in a series of data frames. If the I bit is a 0 then it is the last, else it is an intermediate frame.

The access control field controls the access of nodes on the ring. It takes the form of PPPTMRRR, where:

PPP – indicates the priority of the token; this indicates which type of token the destination node can transmit.

T – is the token bit and is used to discriminate between a control token and a data token.

M – is the monitor bit and is used by an active ring monitor node to stop tokens from circulating around a network continuously.

RRR – are the reservation bits and allow nodes with a high priority to request the next token.

The frame control field contains control information for the MAC layer. It takes the form FFDDDDDD, where:

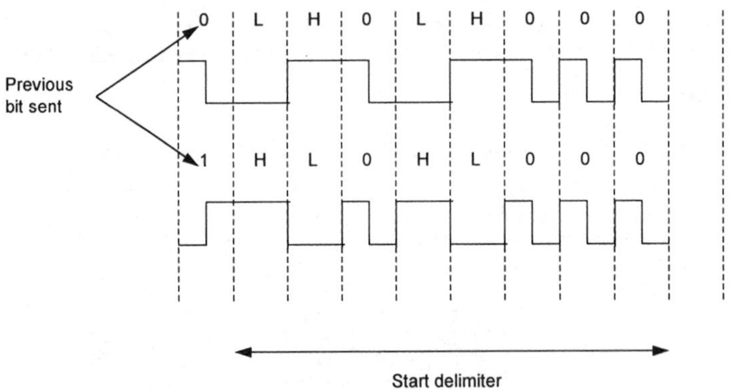

Figure 29.3 Start delimiter

FF – indicates whether the frame is a data frame; if it is not then the DDDDDD bits control the operation of the Token Ring MAC protocol.

DDDDDD – controls the operation of the Token Ring MAC protocol.

The source and destination addresses can either be 2 or 6 bytes (that is, 16 or 48 bits) in length. This size must be the same for all nodes on a ring. The first bit specifies the type of address. If it is a 0 then the address is an individual node address, else it is a group address. An individual node address is used to transmit to a single node, whereas a group address transmits to all nodes with the same group address. The source address will always by an individual address as it indicates the node which originated the token. A special destination address of all 1's is used to transmit to all nodes on a ring.

The frame status field contains information on how a frame has been operated upon as it circulates round the ring. It takes the form ACXXACXX, where:

A – indicates if the destination address has been recognized. It is initially set to a 0 by the source node and is set to a 1 when the destination reads the data. If the source node detects that this bit has not been set then it knows the destination is either not present on the network or is not responding.

C – indicates that a destination node has copied a frame into its memory. This bit is also initially set to a 0 by the source node. When the destination node reads the data from the frame it is set to a 1. By testing this bit and the A bit, the source node can determine whether the destination node is active but not reading data from the frame.

29.4 Token Ring maintenance

A Token Ring system requires considerable maintenance; it must perform the following function:

- Ring initialization – when the network is started, or after the ring has been broken, it must be reinitialized. A cooperative decentralized algorithm sorts out which nodes start a new token, which goes next, and so on.
- Adding to the ring – if a new node is to be physically connected to the ring then the network must be shut down and reinitialized.
- Deletion from the ring – a node can disconnect itself from the ring by joining together its predecessor and its successor. Again, the network may have to be shut down and reinitialized.

- Fault management – typical Token Ring errors occur when two nodes think it is their turn to transmit or when the ring is broken as no node thinks that it is their turn.

29.5 Token Ring multistation access units (MAUs)

The problems of connecting and deleting nodes to or from a ring network are significantly reduced with a multistation access unit (MAU). Figure 29.4 shows two 3-way MAUs connected to produce a 6-node network. Normally, an MAU allows nodes to be switched in and out of a network using a changeover switch or by automatic electronic switching (known as auto-loopback). This has the advantage of not shutting down the network when nodes are added and deleted or when they develop faults.

If the changeover switches in Figure 29.4 are in the down position then the node is bypassed; if they are in the up position then the node connects to the ring.

A single coaxial (or twisted-pair) cable connects one MAU to another and two coaxial (or twisted-pair) cables connect a node to the MAU (for the in ports and the out ports). Most modern application use STP cables.

The IBM 8228 is a typical passive MAU. It can operate at 4 Mbps or 16 Mbps and has 10 connection ports, i.e. 8 passive node ports along with ring in (RI) and ring out (RO) connections. The maximum distance between MAUs is typically 650 m (at 4 Mbps) and 325 m (at 16 Mbps).

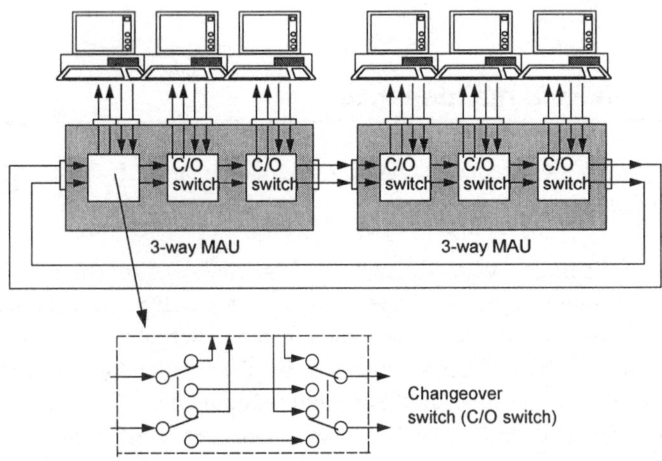

Figure 29.4 Six-node Token Ring network with two MAUs

Most MAUs either have 2, 4 or 8 ports and can automatically detect the speed of the node (i.e. either 4 or 16 Mbps). Figure 29.4 shows a 32-node Token Ring network using four 8-port MAUs. Typical connectors are RJ-45 and IBM Type A connectors. The ring cable is normally either twisted-pair (Type 3), fiber-optic or coaxial cable (Type 1). MAU are intelligent devices and can detect faults on the cables supplying nodes then isolate them from the rest of the ring. Most MAUs are passive devices in that they do not require a power supply. If there are large distances between nodes then an active unit is normally used.

Modern Token Ring networks normally use twisted-pair cables instead of coaxial cables. These twisted-pair cables can either be high-specification shielded twisted-pair (STP) or lower-specification unshielded twisted-pair (UTP). Cabling is discussed in the next section.

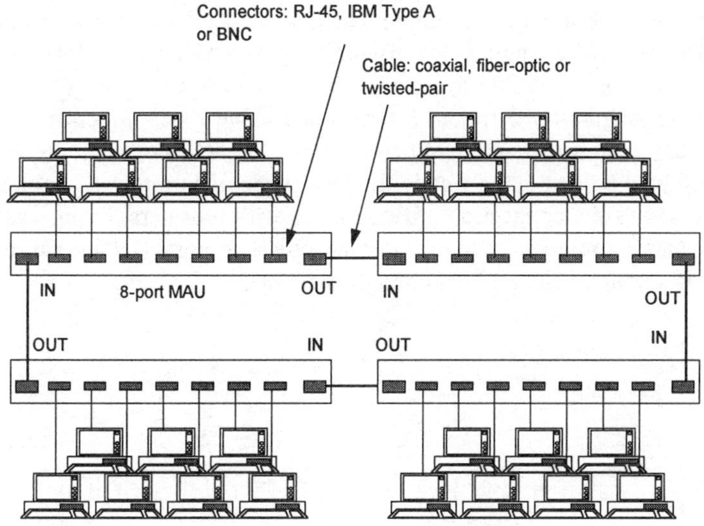

Figure 29.5 A 32-node Token Ring network with 4 MAUs

29.6 Cabling and connectors

There are two main types of cabling used in Token Ring networks: Type 1 and Type 3. Type 1 uses STP (shielded twisted-pair) cables with IBM style male-female connectors. Type 3 networks uses Cat-3 or Cat-5 UTP (unshielded twisted-pair) cables with RJ-45 connectors. Unfortunately, Cat-3 cables are unshielded thus the maximum length of the connection is reduced. Type 1 networks can connect up to 260 nodes, whereas Type 3 networks can only connect up to 72 nodes.

A further source of confusion comes from the two different types of modern STP cables used in Token Ring networks. IBM type 1 cable has four cores with a screen tinned copper braid around them. Each twisted-pair is screened from each other with aluminized polyester tape. The characteristic impedance of the twisted-pairs is 150 Ω. The IBM type 6 cable is a lightweight cable which is preferred in office environments. It has a similar construction but, because it has a thinner core, signal loss is higher.

29.7 Repeaters

A repeater is used to increase either main-ring or lobe lengths in a Token Ring LAN. The main-ring length is the distance between MAUs. The lobe length is the distance from an MAU to a node. Table 29.1 shows some typical maximum cable lengths for different bit rates and cable types. Fiber-optic cables provide the longest distances with a range of 1 km. The next best are STP cables, followed by Cat-5 and finally the lowest specification Cat-3 cables. Figure 29.6 shows the connection of two MAUs with repeaters. In this case four repeaters are required as each repeater has only two ports (IN and OUT). The token will circulate clockwise around the network.

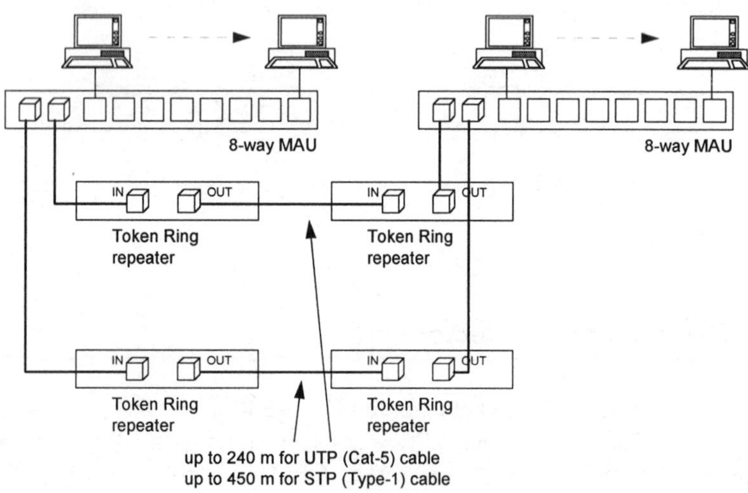

Figure 29.6 16-node Token Ring network with repeaters

Table 29.1 Typical maximum cable lengths for different cables and bit rates

Type	*Bit rate*	*Cable type*	*Maximum distance*
Type 1	4 Mbps	STP	730 m
Type 3	4 Mbps	UTP (Cat-3 cable)	275 m
Type 3	16 Mbps	UTP (Cat-5 cable)	240 m
Type 1	16 Mbps	STP	450 m
Type 1	16 Mbps	Fiber	1000 m

29.8 Jitter suppression

Jitter can be a major problem with Token Ring networks. It is caused when the nodes on the network operate with different clock rates. It can lead to network slowdown, data corruption and station loss. Jitter is the reason that the number of nodes on a Token Ring is limited to 72 at 16 Mbps. With a jitter suppressor the number of nodes can be increased to 256 nodes. It also allows Cat-3 cable to be used at 16 Mbps. Normally a Token Ring Jitter Suppresser is connected to a group of MAUs. Thus the network in Figure 29.6 could have one jitter suppresser unit connected to two of the MAUs (this would obviously limit to 7 the number of nodes connected to these MAUs).

29.9 Exercise

29.1 Discuss how Token Ring defines the start and end delimiters.

29.2 Discuss the main problems of a Token Ring network and describe some methods which can be used to overcome them.

29.3 Discuss how 4-core STP cables are screened to prevent crosstalk between the two pairs.

29.4 Discuss the main advantages of MAUs in a Token Ring network.

30 FDDI

30.1 Introduction

A token-passing mechanism allows orderly access to a network. Apart from Token Ring the most commonly used token-passing network is the Fiber Distributed Data Interchange (FDDI) standard. This operates at 100 Mbps and, to overcome the problems of line breaks, has two concentric Token Rings, as illustrated in Figure 30.1. Fiber optic cables have a much high-specification than copper cables and allow extremely long connections. The maximum circumference of the ring is 100 km (62 miles), with a maximum 2 km between stations (FDDI nodes are also known as stations). It is thus an excellent mechanism for connecting networks across a city or over a campus. Up to 500 stations can connect to each ring with a maximum of 1000 stations for the complete network. Each station connected to the FDDI highway can be a normal station or a bridge to a conventional local area network, such as Ethernet or Token Ring.

The two rings are useful for fault conditions but are also used for separate data streams. This effectively doubles the data-carrying capacity of FDDI (to 200 Mbps). However, if the normal traffic is more than the stated carrying capacity, or if one ring fails, then its performance degrades.

The main features of FDDI are:

- Point-to-point Token Ring topology.
- A secondary ring for redundancy.
- Dual counterrotating ring topology.
- Distributed clock for the support of large numbers of stations on the ring.
- Distributed FDDI management – equal rights and duties for all stations.
- Data integrity ensured through sophisticated encoding techniques.

30.2 Operation

As with Token Ring, FDDI uses a token passing medium access method. Unlike Token Ring there are two types of (control) token:

- A restricted token – which is the normal token. The restricted token circulates around the network. A station wishing to transmit data captures the unrestricted token. It then transmits frames for a period of time made up of a fixed part (T_f) and a variable part (T_v). The variable time depends on the traffic on the ring. When the traffic is light a station may keep the token much longer than when it is heavy.
- An unrestricted token – which is used for extended data interchange between two stations. To enter into an extended data interchange a station must capture the unrestricted token and change it to a restricted token. This token circulates round the network until the exchange is complete, or the extended time is over. Other stations on the network may use the ring only for the fixed period of time T_f. Once complete the token is changed back to an unrestricted type.

FDDI uses a timed token-passing protocol to transmit data because a station can hold the token no longer than a specified amount of time. Therefore, there is a limit to the amount of data that a station can transmit on any given opportunity.

The sending station must always generate a new token once it has transmitted its data frames. The station directly downstream from a sending station has the next opportunity to capture the token. This feature and the timed token ensures that the ring's capacity is divided almost equally among the stations on the ring.

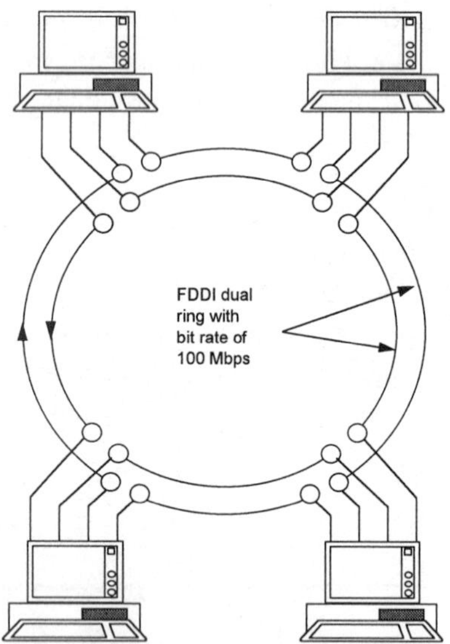

Figure 30.1 FDDI network

30.3 FDDI layers

The ANSI-defined FDDI standard defines four key layers:

- Media access control (MAC) layer.
- Physical layer (PHY).
- Physical media dependent (PMD).
- Station management (SMT) protocol.

FDDI covers the first two layers of the OSI model; Figure 30.2 shows how these layers fit into the model.

The MAC layer defines addressing, scheduling and data routing. Data is formed into data packets with a PHY layer. It encodes and decodes the packets into symbol streams for transmission. Each symbol has 4 bits and FDDI then uses the 4B/5B encoding to encode them into a group of 5 bits. The 5 bits are chosen to contain, at most, two successive zeros. Table 30.1 shows the coding for the bits. This type of coding ensures that there will never be more than four consecutive zeros. FDDI uses NRZI (non-return to zero with inversion) to transmit the bits. With NRZI a 1 is coded with an alternative light (or voltage) level transition for each 1, and a zero does not change the light (or voltage) level. Figure 30.3 shows an example.

In this case the input bitsteam is 1011, 0111 and 0000. This is encoded, as in Table 30.1, as 10111, 01111 and 11110. This is then transmitted in NRZI. The first bit sent is a 1 and is represented as a high-to-low transition. The next bit is a zero and thus has no transition. After this a 1 is encoded, this will be transmitted as a low-to-high transition, and so on. The main advantage of NRZI coding is that the timing information is inherent within the transmitted signal and can be easily filtered out.

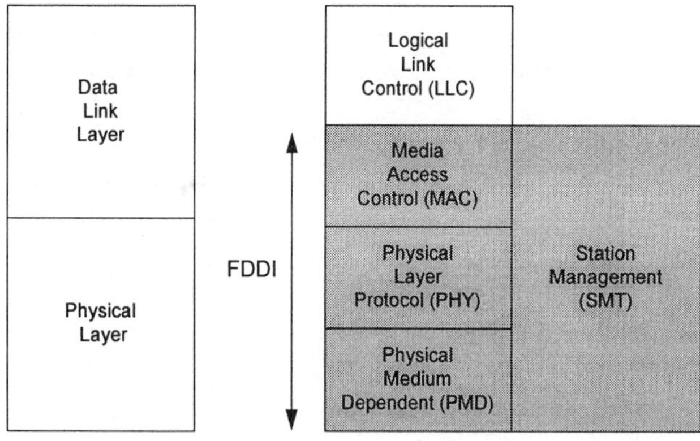

Figure 30.2 FDDI network

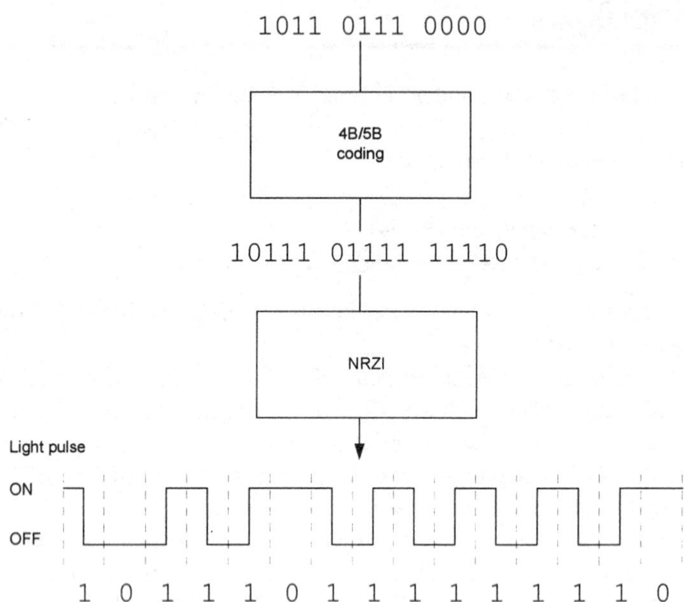

Figure 30.3 Example of bit encoding

Apart from the 16 encoded bit patterns given Table 30.1 there are also 8 control and 8 violation patterns. Table 30.2 shows the 8 other control symbols (QUIET, HALT, IDLE, J, K, T, R and S) and 8 other violation symbols (the encoded bitstream binary values are 00001, 00010, 00011, 00101, 00110, 01000, 01100 and 10000).

The coding of the data symbols is chosen so that there are no more than three consecutive zeros in a row. This is necessary to ensure that all the stations on the ring have their clocks synchronized with all the others. Each station on an FDDI has its own independent clock. The control and violation symbols allow the reception of four or more zero bits in a row.

Table 30.1 4B/5B coding

Symbol	*Binary*	*Bitstream*	*Symbol*	*Binary*	*Bitstream*
0	0000	11110	8	1000	10010
1	0001	01001	9	1001	10011
2	0010	10100	A	1010	10110
3	0011	10101	B	1011	10111
4	0100	01010	C	1100	11010
5	0101	01011	D	1101	11011
6	0110	01110	E	1110	11100
7	0111	01111	F	1111	11101

Table 30.2 4B/5B coding

Symbol	*Bitstream*	*Symbol*	*Bitstream*
QUIET	00000	K	10001
HALT	00100	T	01101
IDLE	11111	R	00111
J	11000	S	11001

30.4 SMT protocol

The SMT protocol handles the management of the FDDI ring, which includes:

- Adjacent neighbor indication.
- Fault detection and reconfiguration.
- Insertion and deletion from the ring.
- Traffic statistics monitoring.

30.5 Physical connection management

Within each FDDI station there are SMT entities called PCM (physical connection management). The number of PCM entities with a station is exactly equal to the number of ports the station has. This is because each PCM is responsible for one port. The PCM entities are the parts of the SMT which control the ports. In order to make a connection, two ports must be physically connected to each other by means of a fiber or copper cable.

30.6 Fault tolerance method

When a station on a ring malfunctions or there is a break in one of the rings then the rest of the stations can still use the other ring. When a station on the network malfunctions then both of the rings may become inoperative. FDDI allows other stations on the network to detect this and to implement a single rotating ring. Figure 30.4 shows an FDDI network with four connected stations. In this case the link between the upper stations have developed a fault. These stations will quickly determine that there is a fault in both cables and will inform the other stations on the network to implement a single rotating ring with the outer ring transmitting in the clockwise direction and the inner ring in the counterclockwise direction. This fault tolerance method also makes it easier to insert and delete stations from the ring.

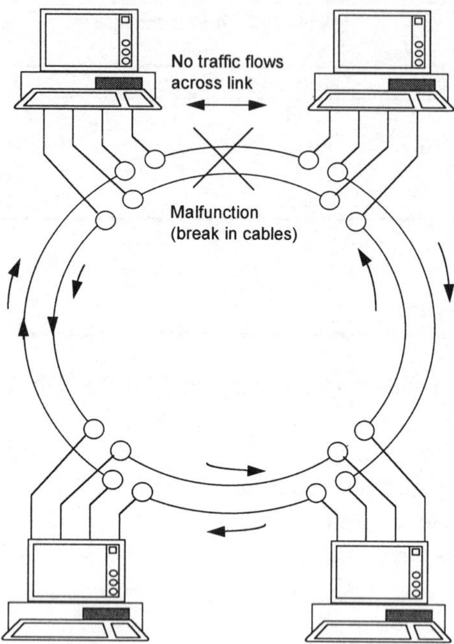

Figure 30.4 Fault tolerant network

30.7 FDDI token format

A token circulates around the ring until a station captures it and then transmits its data in a data frame (see next section). Figure 30.5 shows the basic token format. The preamble (PA) field has four or more symbols of idle (bitstream of 11111). This is followed by the start delimiter (SD) which has a fixed pattern of 'J' (bitstream of 11000) and 'K' (bitstream of 10001). The end delimiter (ED) is two 'T' symbols (01101). The start and end delimiter bit patterns cannot occur anywhere else in the frame as they violate the standard 4B/5B coding (i.e. they may contain three or more consecutive zeros).

30.8 FDDI Frame format

Figure 30.6 shows the FDDI data frame format, which is similar to the IEEE 802.6 frame. The PA, SD and ED fields are identical to the token fields. The frame control field (FC) contains information on the kind of frame that is to follow in the INFO field.

The fields are:

2+ bytes	1 byte	1 byte	1 byte
PA	SD	FC	ED

PA – preamble (4 or more symbols of idle)
SD – start delimiter ('J' and 'K')
FC – frame control (2 symbols)
ED – end delimiter (two 'T' symbols)

Figure 30.5 FDDI token format

PA	SD	FC	DA	SA	INFO	FCS	ED	FS

PA – preamble (2+ bytes)
SD – starting delimiter (1 byte)
FC – frame control (1 byte)
DA – destination address (6 bytes)
SA – source address (6 bytes)
INFO – information (N bytes)
FCS – frame check sequence (4 bytes)
ED – end delimiter (1/2 symbols)
FS – frame status (3 symbols)

Figure 30.6 FDDI data frame format.

- Preamble. The preamble field contains 16 idle symbols which allow stations to synchronize their clocks.
- Start delimiter. This contains a fixed field of the 'J' and 'K' symbols.
- Control field. The format of the control field is SAFFxxxx where S indicates whether it is synchronous or asynchronous. A indicates whether it is a 16-bit or 48-bit address; FF indicates whether this is an LLC (01), MAC control (00) or reserved frame. For a control frame, the remaining 4 bits (xxxx) are reserved for control types. Typical (decoded) codes are:

  0100 0000 – void frame
  1100 0010 – MAC frame
  0101 0101 – station management frame
  0101 0000 – LLC frame

  When the frame is a token the control field contains either 10000000 or 11000000.
- The destination address (DA) and source address (SA) are 12-symbol (6-byte) codes which identify the address of the station from where the frame has come or to where the frame is heading. Each station has a unique address and each station on the ring compares it with its own address. If the frame is destined for a particular station, that station copies the frame's contents into its buffer.

  A frame may also be destined for a group of stations. Group addresses are identified by the start bit 1. If the start bit is 0 then the frame is destined for an individual station. A broadcast address of all 1's is used to send information to all the stations on the ring.

Station addresses can either be locally or globally administered. For global addresses the first six symbols are the manufacturer's OUI; each manufacturer has a unique OUI for all its products. The last six symbols of the address differentiate between stations of the same manufacturer.

The second bit in the address field identifies whether the address is local or global. If it is set (a 1) then it is a locally administered address, if it is unset then it is a globally administered address. In a locally administered network the system manager sets the addresses of all network stations.

- The information field (INFO) can contain from 0 to 4478 bytes of data. Thus the maximum frame size will be as follows:

Field	*Number of bytes*
Start delimiter	1
Frame control	1
Destination address	6
Source address	6
DATA (maximum)	4478
Frame check sequence	4
End delimiter	2
End of frame sequence	2
TOTAL	4500

- The frame check sequence contains a 32-bit CRC which is calculated from the FC, DA, SA and information fields.
- The ending delimiter contains two 'T' symbols.
- The frame status contains extra bits which identify the current status, such as frame copied indicators (F), errors detected (E) and address recognized (A).

FDDI supports either synchronous or asynchronous traffic, but the terms are actually confusing. Frames that are transmitted during their capacity allocation are known as synchronous.

30.9 MAC protocol

The MAC protocol of FDDI MAC is similar to IEEE 802.5. It can be thought of as train, filled with passengers, traveling around a track. The train travels around the ring continuously. When a passenger wishes to get on the train they get in front of the train, which then pushes them round the ring. The passenger then travels around the ring, and delivers their message to the destination station. The passenger stays on the train until the train reaches the source again, where they will get off. Other passengers can get on the train and deliver their own messages while there are others on the train. These passenger go in between the train and the existing passengers.

The actual operational parts work like this:

- A node cannot transmit until it captures a token (the train).
- When a node captures the token, it transmits its frame then the token.
- The frame travels around the ring from source back to source station.
- If another station wishes to transmit a frame it waits for the end of the frames currently on the ring, adds its frame after these frames then appends the token onto the end.
- The station which initiates a frame is responsible for taking it off the ring.
- Each station reads the frame as it circulates around the ring.
- Each station can modify the status bits as the frame passes. If a station detects an error then it sets the E bit; if it has copied the frame into its buffer then it sets the C bit; if it detects its own address it sets the A bit.
- Thus a correctly received frame will be received back at the source node with the C and A bits set.
- A frame which is recieved back at the source with the E bit is not automatically retransmitted. A message is sent to higher layer in the protocol (such as the LLC layer).

30.10 Applications of FDDI networks

As was seen in Chapter 28, Ethernet is an excellent method of attaching stations to a network cheaply but is not a good transport mechanism for a backbone network or with high traffic levels. It also suffers, in its standard form, from a lack of speed. FDDI networks overcome these problems as they offer a much higher bit rate, higher reliability and longer interconnections. Thus typical applications of FDDI networks are:

- As a backbone network in an internetwork connection.
- Any applications which requires high security and/or a high degree of fault tolerance. Fiber-optic cables are generally more reliable and are difficult to tap into without it being detected.
- As a subnetwork connecting high-speed computers and their peripheral devices (such as storage units).
- As a network connecting stations where an application program requires high-speed transfers of large amounts of data (such as computer-aided design – CAD). Maximum data traffic for an FDDI network is at least 10 times greater than for standard Ethernet and Token Ring networks. As it is a token-passing network it is less susceptible to heavy traffic loads than Ethernet.

30.11 FDDI backbone network

The performance of the network backbone is extremely important as many users on the network depend on it. If the traffic is too heavy, or if it develops a fault, then it affects the performance of the whole network. An FDDI backbone helps with these problems because it has a high bit rate and normally increases the reliability of the backbone.

Figure 30.7 shows an FDDI backbone between four campuses. In this case, the FDDI backbone only carries traffic which is transmitted between campuses. This is because the router only routes traffic out of the campus network when it is intended for another campus. As tokens circulate round both rings, two data frames can be transmitted round the rings at the same time.

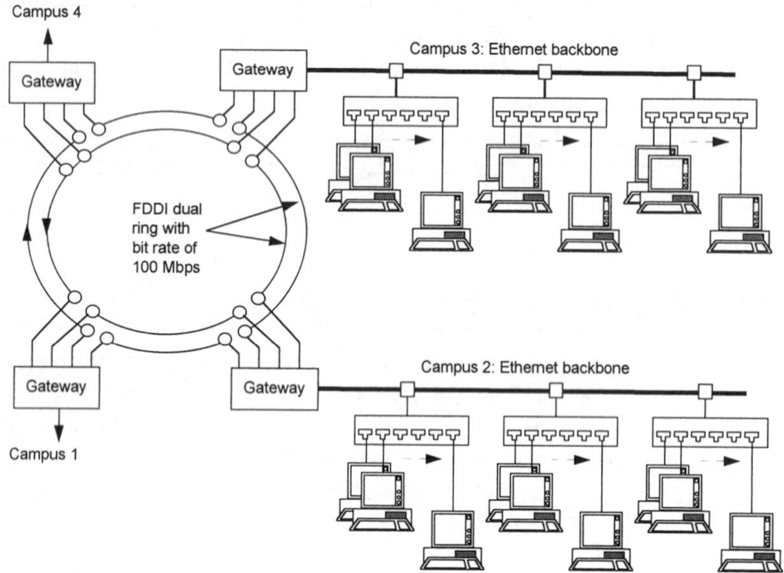

Figure 30.7 FDDI backbone network

30.12 FDDI attachments

There are four types of station which can attach to an FDDI network:

- Dual attachment stations (DAS).
- Single attachment stations (SAS).
- Dual attachment concentrators (DAC).
- Single attachment concentrators (SAC).

Figure 30.8 shows an FDDI network configuration that includes all these types of station. An SAS connects to the FDDI rings through a concentrator, so it is easy to add, delete or change its location. The concentrator automatically bypasses disconnected stations.

Each DAS and DAC requires four fibers to connect it to the network: Primary In, Primary Out, Secondary In and Secondary Out. The connection of an SAS only requires two fibers. Normally Slave In and Slave Out on the SAS are connected to the Master Out and Master In on the concentrator unit, as shown in Figure 30.9.

FDDI stations attach to the ring using a media interface connector (MIC). An MIC receptacle connects to the stations and an MIC plug on the network end of the connection. A dual attachment station has two MIC receptacles. One provides Primary Ring In and Secondary Ring Out, the other has Primary Ring Out and Secondary Ring In, as illustrated in Figure 30.10.

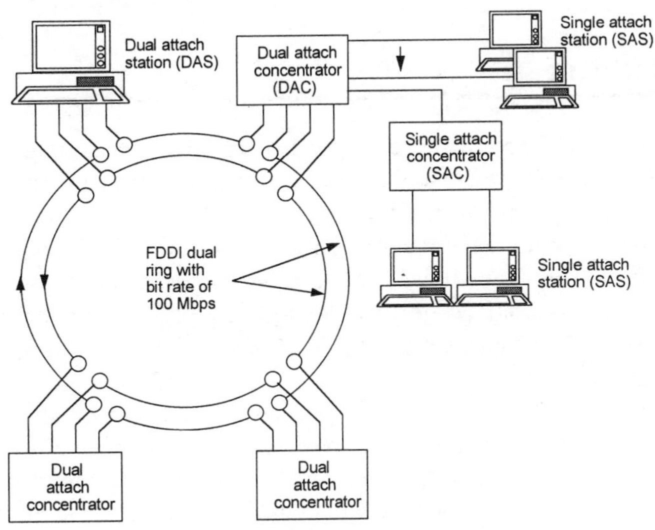

Figure 30.8 FDDI network configuration

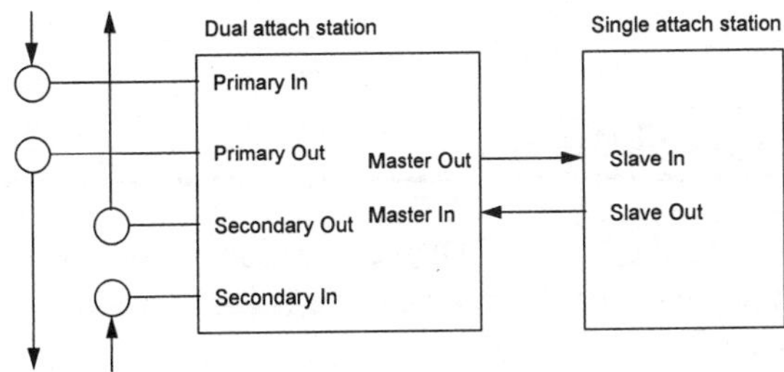

Figure 30.9 Connection of a DAS and a SAS

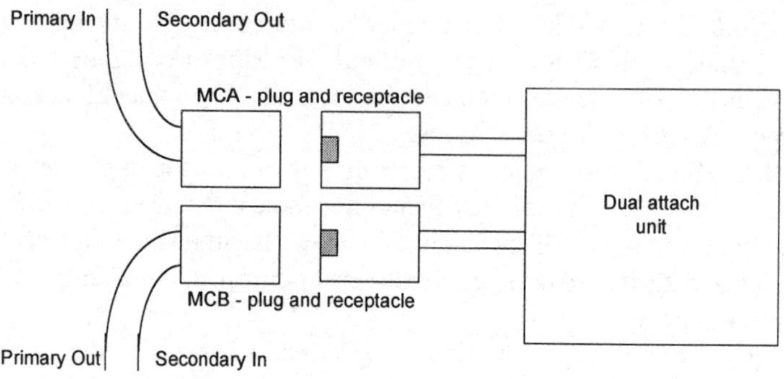

Figure 30.10 Connection of dual attach units

30.13 FDDI media

FDDI networks can use two types of fiber-optic cable, either single-mode or multimode. The mode refers to the angle at which light rays are reflected and propagated through the fiber core. Single-mode fibers have a narrow core, such as 10 μm for the core and 125 μm for the cladding (known as 10/125 micron cable). This type allows light to enter only at a single angle. Multi-mode fiber has a relatively thick core, such as 62.5 μm for the core and 125 μm for the cladding (known as 62.5/125 micron cable). Multi-mode cable reflects light rays at many angles. The disadvantage of these multiple propagation paths is that it can cause the light pulses to spread out and thus limit the rate at which data is accurately received. Thus, single-mode fibers have a higher bandwidth than multimode fibers and also allow longer interconnection distances. The fibers most commonly used in FDDI are 62.5/125, and this type of cable is defined in the ANSI X3T9.5 standard.

30.14 FDDI specification

Table 30.3 describes the basic FDDI specification. Notice that the maximum interconnection distance for multimode cable is 2 km as compared with 20 km for single-mode cable.

Table 30.3 Basic FDDI specification

Parameter	*Description*
Topology	Token Ring
Access method	Time token passing
Transport media	Optical fiber, shield twisted pair, unshielded twisted pair
Maximum number of stations	500 each ring (1000 total)
Data rate	100 Mbps
Maximum data packet size	4 500 bytes
Maximum total ring length	100 km
Maximum distance between stations	2 km (for multimode fiber cable) 20 km (for single-mode fiber cable)
Attenuation budget	11 dB (between stations) 1.5 dB/km at 1300 nm for 62.5/125 fiber
Link budget	< 11 dB

30.15 FDDI-II

FDDI-II is an upward-compatible extension to FDDI that adds the ability to support circuit-switched traffic in addition to the data frames supported by the original FDDI. With FDDI-II, it is possible to set up and maintain a constant data rate connection between two stations.

The circuit-switched connection consists of regularly repeating time slots in the frame, often called an isochronous frame. This type of data is common when real-time signals, such as speech and video, are sampled. For example, speech is sampled 8000 times per second, whereas high-quality audio is sampled at 44 000 times per second.

Figure 30.11 shows a layer diagram of an FDDI-II station. The physical layer and the station management are the same as the original FDDI. Two new layers have been added to the MAC layer; known as hybrid ring control, they consist of:

- Hybrid multiplexer (HMUX).
- Isochronous MAC (IMAC).

The IMAC module provides an interface between FDDI and the isochronous service, represented by the circuit-switched multiplexer (CS-MUX). The HMUX multiplexes the packet data from the MAC and the isochronous data from the IMAC.

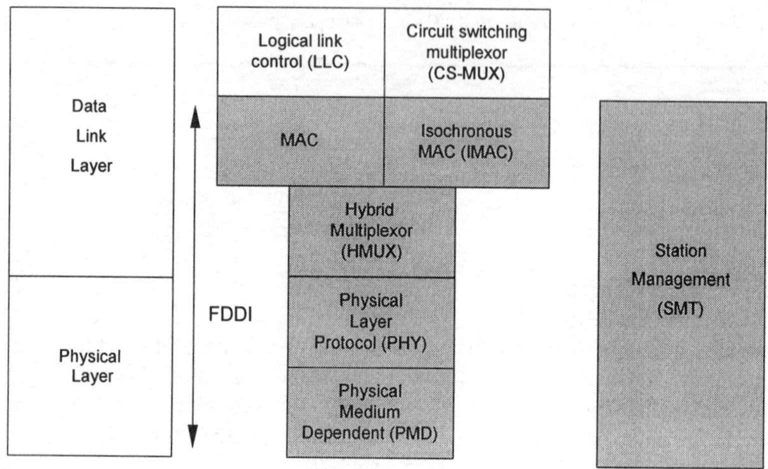

Figure 30.11 FDDI-II layered model

An FDDI-II network operates in a basic mode or a hybrid mode. In the basic mode the FDDI ring operates as the original FDDI specification where tokens rotate around the network. In the hybrid mode a connection can either be circuit-switched or packet-switched. It uses a continuously repeating protocol data unit known as a cycle. A cycle is a data frame that is similar to synchronous transmission systems. The content of the cycle is visible to all stations as it circulates around the ring. A station called the cycle master generates a new cycle 8 000 times a second. At 100 Mbps, this gives a cycle size of 12 500 bits. As each cycle completes its circuit of the ring, its is stripped by the cycle master. Figure 30.12 shows the two different types of transmission.

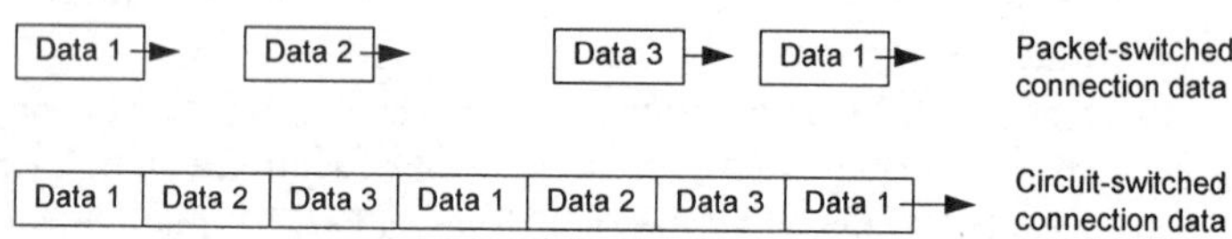

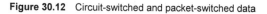

Figure 30.12 Circuit-switched and packet-switched data

30.16 Standards

The FDDI standard was defined by the ANSI committee X3T9.5. This has since been adopted by the ISO as ISO 9314 which defines FDDI using five main layers:

- ISO 9314-1: Physical Layer Protocol (PHY).
- ISO 9314-2: Media Access Control (MAC).
- ISO 9314-3: Physical Media Dependent (PMD).
- ISO 9314-4: Station Management (SMT).
- ISO 9314-5: Hybrid Ring Control (HRC), FDDI-II.

30.17 Practical FDDI network – EaStMAN

The EaStMAN (Edinburgh and Stirling Metropolitan Area Network) consortium comprises seven institutions of Higher Education in the Edinburgh and Stirling area of Scotland. The main institutions are: the University of Edinburgh, the University of Stirling, Napier University, Heriot-Watt University, Edinburgh College of Art, Moray House Institute of Education and Queen Margaret College. FDDI and ATM networks have been installed around Edinburgh with an optical link to Stirling. It was funded jointly by the Scottish Higher Education Funding Council (SHEFC), the Joint Information Systems Committee (JISC) and the individual institutions.

Figure 30.13 shows the connections of Phase 1 of the project and Figure 30.14 shows the rings. The total circumference of the rings is 58 km (which is less than the maximum limit of 100 km).

The FDDI ring provides intercampus communications and also a link to the SuperJANET (Joint Academic NETwork) and the ATM ring is for future development. The ATM ring will be discussed in more detail in Chapter 31.

Future plans for the network are to link it to the FDDI networks of FaTMAN (Fife and Tayside), ClydeNet (Glasgow network), and AbMAN (Aberdeen network).

30.17.1 Fiber optic cables

The new optical fiber network is leased from Scottish Telecom (a subsidiary of Scottish Power). The FDDI ring is 58 km and, for the purpose of FDDI standardization across the MAN, the outer FDDI ring is driven anticlockwise and the inner ring clockwise. Figure 30.15 shows the attenuation rates and distance between each site (in dBs). Note that an extra 0.4 dB should be added onto each fiber connection to take into account the attenuation at the fiber termination. Thus the total attenuation between the New College and Moray House will be approximately 2.4 dB.

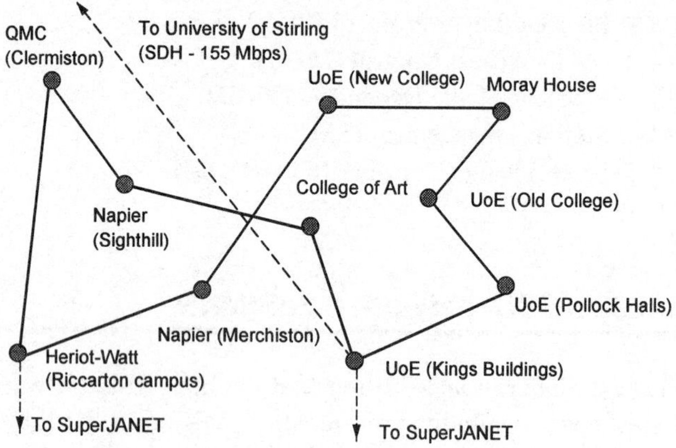

Figure 30.13 EaStMAN Phase 1 connections

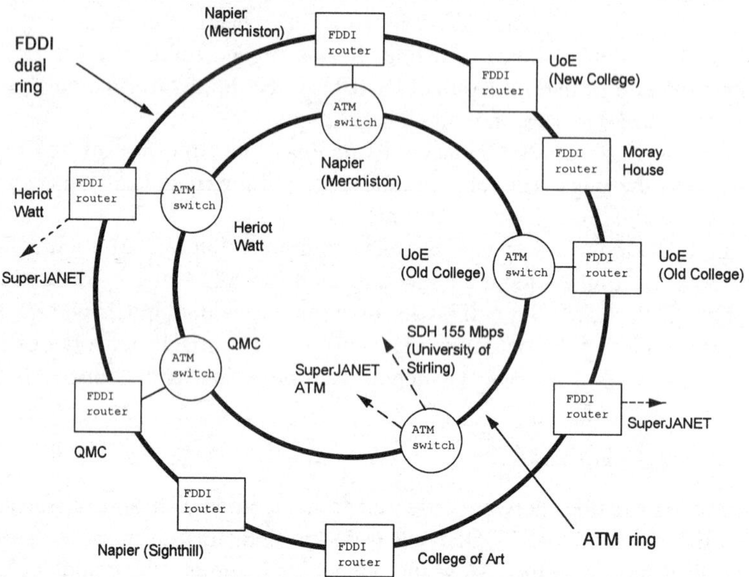

Figure 30.14 EaStMAN ring connections

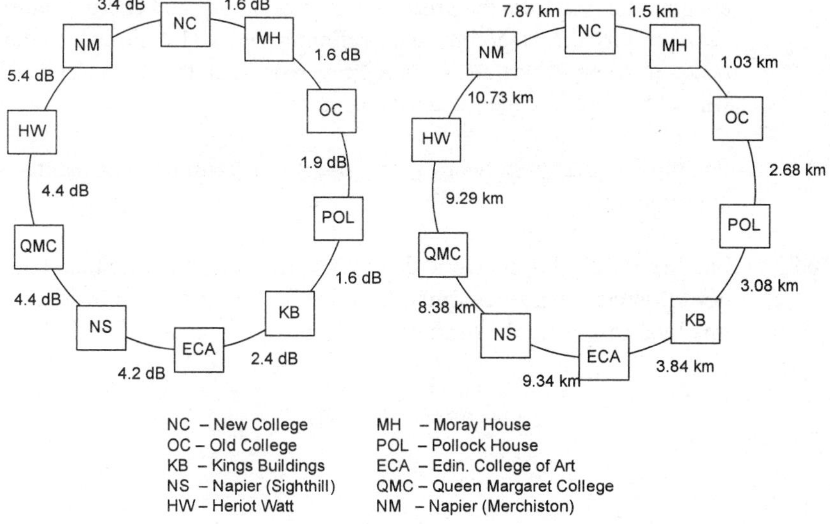

Figure 30.15 Attenuation and distances of EaStMAN Phase I

30.18 Exercises

30.1 Discuss the token-passing technique used in FDDI.

30.2 Explain how FDDI allows robust networking after a cable break.

30.3 For the input binary data given below determine the line code (after 4B/5B coding) on the line.

0000101011110011000100110001111111111111

30.4 For the line coding (before 5B/4B decoding) given below determine the output data.

1010101110110111110011101111 0110010

30.5 Identify the control characters and data from the following line code.

0000011111101000101110001011010111101110 11000

30.6 Using the FDDI token format given in Figure 30.5, show the line code on the line from the preamble, start delimiter and end delimiter. Explain why the preamble, start delimiter and end delimiter cannot occur anywhere else in a token frame or data frame (apart from where they are intended to appear).

30.7 For the EaStMAN network prove that the maximum ring length is not violated.

30.8 For the EaStMAN network determine the approximate attenuation per kilometer. Also determine the input/output power ratio of each link by transposing the formula:

$$\text{Attenuation (dB)} = 10\log_{10}\left(\frac{P_i}{P_o}\right)$$

31 Asynchronous Transfer Mode (ATM)

31.1 Introduction

Most of the networking technologies discussed so far are good at carrying computer-type data and they provide a reliable connection between two nodes. Unfortunately they are not as good at carrying real-time sampled data, such as digitized video or speech. Real-time data from speech and video requires constant sampling and these digitized samples must propagate through the network with the minimum of delay. Any significant delay in transmission can cause the recovered signal to be severely distorted or for the connection to be lost. Ethernet, Token Ring and FDDI simply send the data into the network without first determining whether there is a communication channel for the data to be transported.

Figure 31.1 shows some traffic profiles for sampled speech and computer-type data (a loading of 1 is the maximum loading). It can be seen that computer-type data tends to burst in periods of time. These bursts have a relatively heavy loading on the network. On the other hand, sampled speech has a relatively low loading on the network but requires a constant traffic throughput. It can be seen that if these traffic profiles were to be mixed onto the same network then the computer-type data would swamp the sampled speech data at various times.

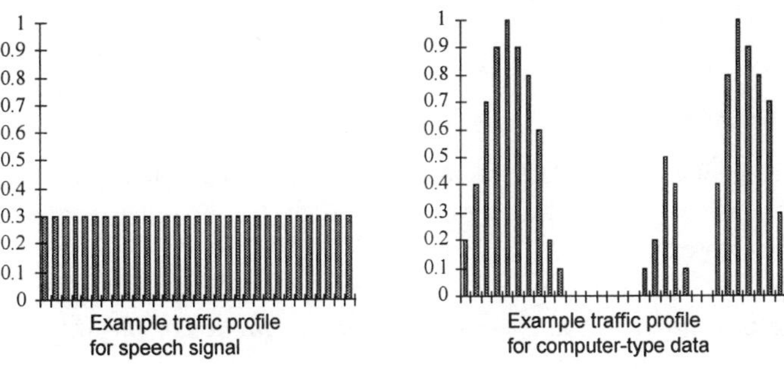

Figure 31.1 Traffic profiles for sampled speech and computer-type data

Asynchronous transfer mode (ATM) overcomes the problems of transporting computer-type data and sampled real-time data by:

- Analyzing the type of connection to be made. The type of data dictates the type of connection; for example, computer data requires a reliable connection, whereas real-time sampled data requires a connection with a low propagation time.
- Analyzing the type of data to be transmitted and knowing its traffic profile. Computer data tends to create bursts of traffic whereas real-time data will be constant traffic.
- Reserving a virtual path for the data to allow the data profile to be transmitted within the required quality of service.
- Splitting the data into small packets which have the minimum overhead in the number of extra bits. These 'fast-packets' traverse the network using channels which have been reserved for them.

ATM has been developed mainly by the telecommunications companies. Unfortunately two standards currently exist. In the USA the ANSI T1S1 subcommittee have supported and investigated ATM and in Europe it has been investigated by ETSI. There are small differences between the two proposed standards, but they may converge into one common standard. The CCITT has also dedicated study group XVIII to ATM-type systems with the objective of merging differences and creating one global standard for high-speed networks throughout the world.

31.2 Real-time sampling

Before introducing the theory of ATM, first consider the concept of how analogue signals (such as speech, audio or video) are converted into a digital form. The basic principle involves sampling theory and pulse code modulation.

31.2.1 Sampling theory

Recall from Section 1.6 that for a signal to be reconstructed as the original signal it must be sampled at a rate defined by the Nyquist criterion (Figure 31.2). This states:

the sampling rate must be twice the highest frequency of the signal

For telephone speech channels the maximum signal frequency is limited to 4 kHz and must therefore be sampled at least 8000 times per second (8 kHz). This gives one sample every 125 μs. Hi-fi quality audio has a maximum signal frequency of 20 kHz and must be sampled at least 40 000 times per second (many professional hi-fi sampling systems sample at 44.1 kHz). Video signals have a maximum frequency of 6 MHz, so a video signal must be sampled at 12 MHz (or once every 83.3 ns).

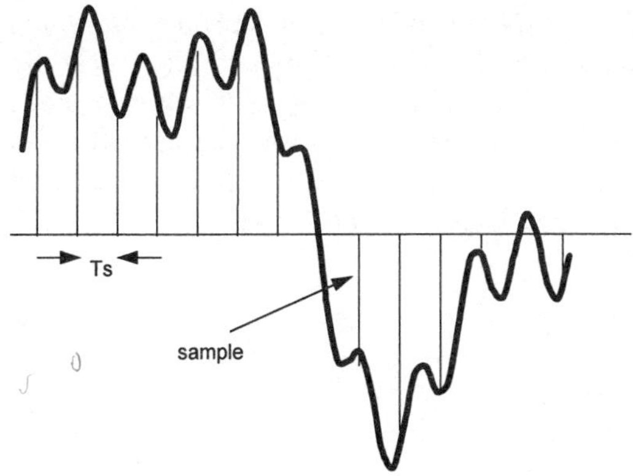

Figure 31.2 The sampling process

31.2.2 Pulse code modulation

Once analogue signals have been sampled for their amplitude they can be converted into a digital format using pulse code modulation (PCM). The digital form is then transmitted over the transmission media. At the receiver the digital code is converted back into an analogue form.

The accuracy of the PCM depends on the number of bits used for each analogue sample. This gives a PCM-based system a dependable response over an equivalent analogue system because an analogue system's accuracy depends on component tolerance, producing a differing response for different systems.

31.3 PCM-TDM systems and ISDN

ATM tries to integrate real-time data and computer-type data. The main technology currently used in transmitting digitized speech over the public switched telephone network (PSTN) is PCM-TDM (PCM time division multiplexing). PCM-TDM involves multiplexing the digitized speech samples in time. Each sample is assigned a time slot which is reserved for the total time of the connection, as illustrated in Figure 31.3. This example shows the connection of Telephone 1–4 to Telephone A–D, respectively. The digitizer, in this case, consists of a sampler (at 8 kHz) and an analogue-to-digital converter

(ADC). Four input channels are time-division multiplexed onto a signal line, the time between one sample from a certain channel and the next must be 125 μs.

The integrated services digital network (ISDN) is to be covered in Chapter 32. It uses a base bit rate of 64 kbps and can be used to transmit 8-bit samples at a rate of 8 kHz. ISDN is similar to a telephone connection but allows the direct connection of digital equipment. As with the PSTN the connection between the transmitter and the receiver is set up by means of a switched connection. On an ISDN network the type of data carried is transparent to the network. Higher bit rates are achieved by splitting the data into several channels and transmitting each channel at 64 kbps.

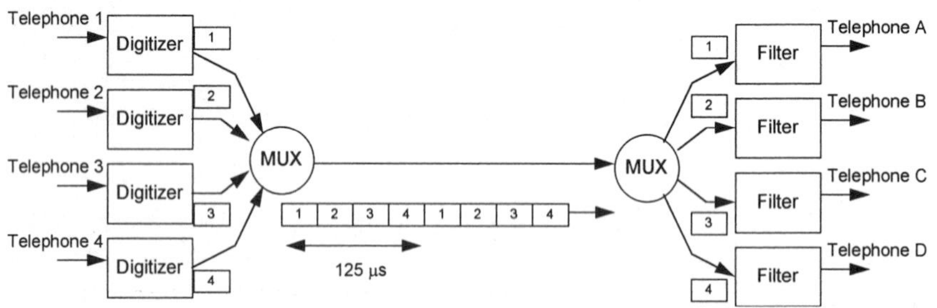

Figure 31.3 PCM-TDM with four connections

31.4 Objectives of ATM

The major objective of ATM is to integrate real-time data (such as voice and video signals) and non-real-time data (such as computer data and file transfer). Computer-type data can typically be transferred in non-real-time but it is important that the connection is free of errors. In many application programs a single bit error can cause serious damage. On the other hand, voice and video data require a constant sampling rate and low propagation delays, but are more tolerant to errors and any losses of small parts of the data.

An ATM network relies on user-supplied information to profile traffic flows so that a connection has the desired service quality. Table 31.1 gives four basic data types. These are further complicated by differing data types either sending data in a continually repeating fashion (such as telephone data) or with a variable frequency (such as interactive video). For example a high-resolution video image may need to be sent as several megabytes of data in a short time burst, but then nothing for a few seconds. For speech the signal must be sampled 8000 times per second.

Computer data will typically be sent in bursts. Sometimes a high transfer rate is required (perhaps when running a computer package remotely over a network) or a relatively slow transfer (such as when reading text information). Conventional circuit-switched technology (such as ISDN and PCM-TDM) are thus wasteful in their connection because they either allocate a switched circuit (ISDN) or reserve a fixed time slot (PCM-TDM), no matter whether there data is being transmitted at that time. And it may not be possible to service high burst rates by allocating either time slots or switched circuits when all of the other time slots are full, or because switched circuits are being used.

Table 31.1 Four basic categories of data

Data type	*Error or loss sensitive*	*Delay sensitive*
Real-time control system	yes	yes
Telephone/hi-fi music	no	yes
File transfer, application programs	yes	no
Teletex information	no	no

31.5 ATM versus ISDN and PCM-TDM

ISDN and PCM-TDM use a synchronous transfer mode (STM) technique where a connection is made between two devices by circuit switching. The transmitting device is assigned a given time slot to transmit the data. This time slot is fixed for the period of the transmission. The main problems with this type of transmission are:

- Not all the time slots are filled by data when there is light data traffic; this is wasteful in data transfer.
- When a connection is made between two endpoints a fixed time slot is assigned and data from that connection is always carried in that time slot. This is also wasteful because there may be no data being transmitted in certain time periods.

ATM overcomes these problems by splitting the data up into small fixed-length packets, known as cells. Each data cell is sent with its connection address and follows a fixed route through the network. The packets are small enough that, if they are lost, possibly due to congestion, they can either be requested (for high reliability) or cause little degradation of the signal (typically in voice and video traffic).

The address of devices on an ATM network are identifier by a virtual circuit identifier (VCI), instead of by a time slot as in an STM network. The VCI is carried in the header portion of the fast packet.

31.6 Statistical multiplexing

Fast packet switching attempts to solve the problem of unused time slots of STM. This is achieved by statistically multiplexing several connections on the same link based on their traffic characteristics. Applications, such as voice traffic, which require a constant data transfer are allowed safe routes through the network. Whereas several applications, which have bursts of traffic, may be assigned to the same link in the hope that statistically they will not all generate bursts of data at the same time. Even if some of them were to burst simultaneously, then their data could be buffered and sent at a later time. This technique is called statistical multiplexing and allows the average traffic on the network to be evened out over a relatively short time period. This is impossible on an STM network.

31.7 ATM user network interfaces (UNIs)

A user network interface (UNI) allows users to gain access to an ATM network. The UNI transmits data into the network with a set of agreed specifications and the network must then try to ensure the connection stays within those requirements. These requirements define the required quality of service for the entire duration of the connection.

It is likely that there will be several different types of ATM service provision. One type will provide an interface to one or more of the LAN standards (such as Ethernet or Token Ring) or FDDI. The conversion of the LAN frames to ATM cells will be done inside the UNI at the source and destination endpoints respectively. Typically it will be used as a bridge for two widely separated LANs. This provides a short-term solution to justifying the current investment in LAN technology and allows a gradual transition to a complete ISDN/ATM network.

The best long-term solution is to connect data communication equipment directly onto an ATM network. This allows computer equipment, telephones, video, and so on, to connect directly to a global network. The output from an ATM multiplexer interfaces with the UNI of a larger ATM backbone network.

A third type of ATM interface connects existing STM networks to ATM networks. This allows a slow migration of existing STM technology to ATM.

31.8 ATM cells

The ATM cell, as specified by the ANSI T1S1 subcommittee, has 53 bytes, as shown in Figure 31.4. The first 5 bytes are the header and the remaining bytes are the information field which can hold 48 bytes of data. Optionally the data can contain a 4-byte ATM adaptation layer and 44 bytes of actual data. A bit in the control field of the header sets the data to either 44 or 48 bytes. The ATM adaptation layer field allows for fragmentation and reassembly of cells into larger packets at the source and destination respectively. The control field also contains bits which specify whether this is a flow control cell or an ordinary data cell, a bit to indicate whether this packet can be deleted in a congested network, and so on.

The ETSI definition of an ATM cell also contains 53 bytes with a 5-byte header and 48 bytes of data. The main differences are the number of bits in the VCI field, the number of bits in the header checksum, and the definitions and position of the control bits.

The IEEE 802.6 standard for the MAC layer of the metropolitan area network (MAN) DQDB (distributed queue dual bus) protocol is similar to the ATM cell.

31.9 Routing cell within an ATM network

In STM networks, data can change its position in each time slot in the interchanges over the global network. This can occur in ATM where the VCI label changes between intermediate nodes in the route.

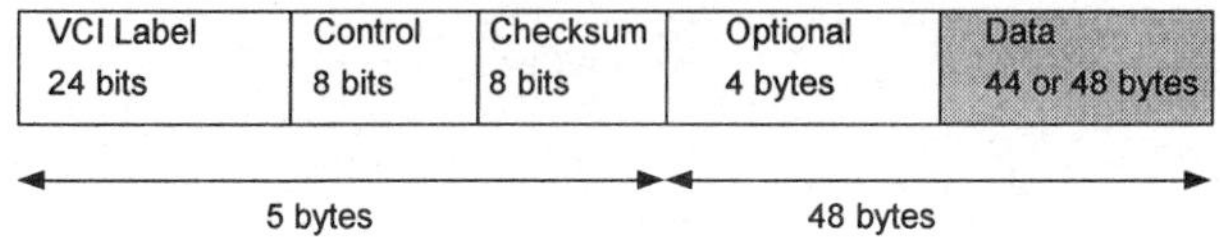

Figure 31.4 ATM cell

When a transmitting node wishes to communicate through the network it makes contact with the UNI and negotiates parameters such as destination, traffic type, peak and traffic requirements, delay and cell loss requirement, and so on. The UNI forwards this request to the network. From this data the network computes a route based on the specified parameters and determines which links on each leg of the route can best support the requested quality of service and data traffic. It sends a connection setup request to all the nodes in the path en route to the destination node.

Figure 31.5 shows an example of ATM routing. In this case User 1 connects to Users 2 and 3. The virtual path set up between User 1 and User 2 is through the ATM switches 2, 3 and 4, whereas User 1 and User 3 connect through ATM switches 1, 5 and 6. A VCI number of 12 is assigned to the path between ATM switches 1 to 2, in the connection between User 1 and User 2.

When ATM switch 2 receives a cell with a VCI number of 12 then it sends the cell to ATM switch 3 and gives it a new VCI number of 6. When it gets to ATM switch 3 it is routed to ATM switch 4 and given the VCI number of 22. The virtual circuit for User 1 to User 2 is through ATM switches 1, 5 and 6, and the VCI numbers used are 10 and 15. Once a connection is terminated the VCI labels assigned to the communications are used for other connections.

Certain users, or applications, can be assigned reserved VCI labels for special services that may be provided by the network. However, as the address field only has 24 bits it is unlikely that many of these requests would be granted. ATM does not provide for acknowledgments when the cells arrive at the destination.

Note that as there is a virtual circuit setup between the transmitting and receiving node then cells are always delivered in the same order as they are transmitted. This is because cells cannot take alternative routes to the destination. Even if the cells are buffered at a node, they will still be transmitted in the correct sequence.

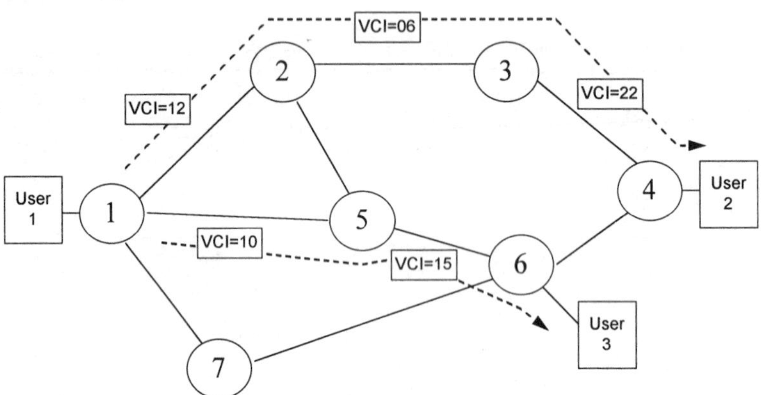

Figure 31.5 A virtual ATM virtual connection

31.9.1 VCI header

The 5-byte ATM user network cell header is split into six main fields:

- GFC (4 bits) which is the generic control bit. It is used only for local significance and is not transmitted from the sender to the receiver.
- VPI (8 bits) which is the path connection identifier (VPI). See next section for an explanation of virtual path.
- VCI (16 bits) which is the virtual path/channel identifier (VCI). Its usage was described in the previous section. Each part of the route is allocated VCI number.

- PT (3 bits) which is the payload type field. This is used to identify the higher-layer application or data type.
- CLP (1 bit) which is the cell loss priority bit and indicates if a cell is expendable. When the network is busy an expendable cell may be deleted.
- HEC (8 bits) which is the header error control field. This is an 8-bit checksum for the header.

Note that the user to network cell differs from the network to network cell. A network to network cell uses a 12-bit VPI field and does not have a GFC field. Otherwise it is identical.

31.10 Virtual channels and virtual paths

Virtual circuits are set up between two users when a connection is made. Cells then travel over this fixed path through a reserved path. Often several virtual circuits take the same path. These circuits can be grouped together to form a virtual path.

A virtual path is defined as a collection of virtual channels which have the same start and end points. These channels will take the same route. This makes the network administration easier and allows new virtual circuit, with the same route, to be set up easily.

Some of the advantages of virtual paths are:

- Network user groups or interconnected networks can be mapped to virtual paths and are thus easily administered.
- Simpler network architecture which consists of groups (virtual paths) with individual connections (virtual circuits).
- Less network administration and shorter connection times arise from fewer setup connections.

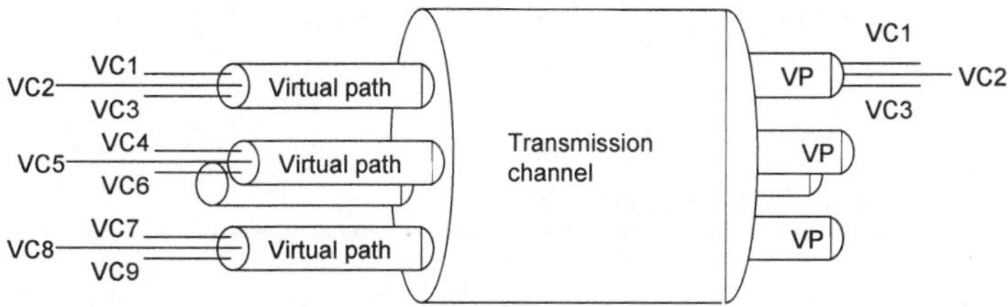

Figure 31.6 Virtual circuits and virtual paths

Virtual circuits and virtual paths allows two levels of cell routing through the network. A VC switch routes virtual circuits and a VP switch routes virtual paths. Figure 31.7 shows a VP switch and a VC switch. In this case the VP switch contains the routing table which maps VP1 to VP2, VP2 to VP3 and VP3 to VP1. This switch does not change the VCI number of the incoming virtual circuits (for example VC1 goes in as VC1 and exits as VC1).

The diagram shows the concepts between both types of switches. The VP switch of the left will redirect the contents of a virtual path to a different virtual path. The virtual connections it contains are unchanged. This is similar to switching an input cable to a different physical cable. In a VC switch the virtual circuits are switched. In the case of Figure 31.7 the routing table will contain VC1 mapped to VC5, VC2 to VC6, and so on. The VC switch thus ignores the VP number and only routes the VC number. Thus the input and output VP number can change.

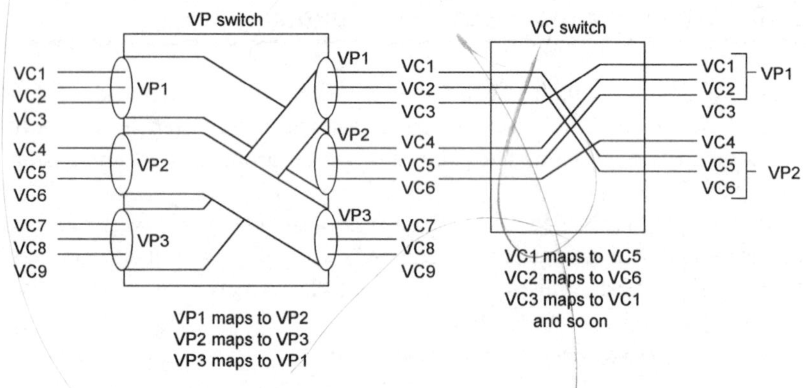

Figure 31.7 Virtual circuits and virtual paths

A connection is made by initially sending routing information cells through the network. When the connection is made, each switch in the route adds a link address for either a virtual path and or a virtual connection.

The combination of VP and VC addressing allows for the support of any addressing scheme, including subscriber telephone numbering or IP addresses. Each of these address can be broken down in a chain of VPI/VCI addresses.

31.11 ATM and the OSI model

The basic ATM cell fits roughly into the data link layer of the OSI model, but contains some network functions, such as end-to-end connection, flow control, and routing. It thus fits into layers 2 and 3 of the model, as shown in Figure 31.8. The layer 4 software layer, such as TCP/IP (as covered in Chapter 16), can communicate directly with ATM.

The ATM network provides a virtual connection between two gateways and the IP protocol fragments IP packets into ATM cells at the transmitting UNI which are then reassembled into the IP packet at the destination UNI.

With TCP/IP each host is assigned an IP address as is the ATM gateway. Once the connection has been made then the cells are fragmented into the ATM network and follow a predetermined route through the network. At the receiver the cells are reassembled using the ATM adaptation layer. This re-forms the original IP packet which is then passed to the next layer.

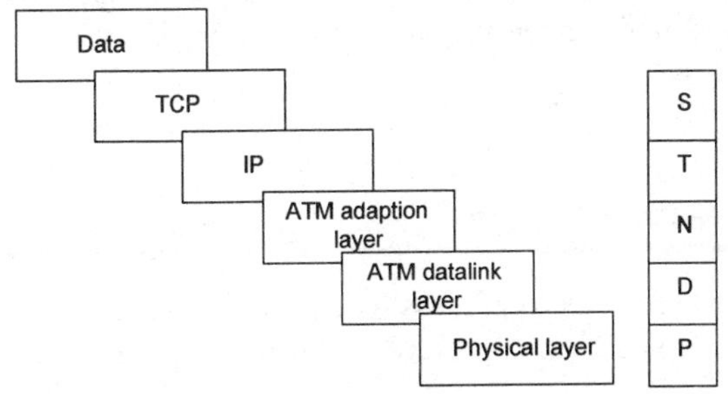

Figure 31.8 ATM and the OSI model

The functions of the three ATM layers are:

- ATM adaptation layer (AAL) – segmentation and reassembly of data into cells and vice-versa, such as convergence (CS) and segmentation (SAR). It is also involved with quality of service (QOS).
- ATM data link (ADL) – maintenance of cells and their routing through the network, such as generic flow control, cell VPI/VCI translation, cell multiplex and demultiplex.
- ATM physical layer (PHY) – transmission and physical characteristics, such as cell rate decoupling, HEC header sequence generation/verification, cell delineation, transmission frame adaptation, transmission frame generation/recovery.

31.12 ATM physical layer

The physical layer is not an explicit part of the ATM definition, but is currently being considered by the standards organizations. T1S1 has standardized on SONET (Synchronous Optical NETwork) as the preferred physical layer, with STS-3c at 155.5 Mbps, STS-12 at 622 Mbps and STS-48 at 2.4 Gbps.

The SONET physical layer specification provides a standard worldwide digital telecommunications network hierarchy, known internationally as the Synchronous Digital Hierarchy (SDH). The base transmission rate, STS-1, is 51.84 Mbps. This is then multiplexed to make up higher bit rate streams, such as STS-3 which is 3 times STS-1, STS-12 which is 12 times STS-1, and so on. The 155 Mbps stream is the lowest bit rate for ATM traffic and is also called STM-1 (synchronous transport module - level 1).

The SDH specifies a standard method on how data is framed and transported synchronously across fiber-optic transmission links without requiring that all links and nodes have the same synchronized clock for data transmission and recovery.

31.13 AAL service levels

The AAL layer uses cells to process data into a cell-based format and to provide information to configure the level of service required.

31.13.1 Processing data

The AAL performs two essential functions for processing data as shown: the higher-level protocols present a data unit with a specific format. This data frame is then converted using a convergence sublayer with the addition of a header and trailer that give information on how the data unit should be segmented into cells and reassembled at the destination. The data is then segmented into cells, together with the convergence subsystem information and other management data, and sent through the network.

31.13.2 AAL functionality

The AAL layer, as part of the process, also defines the level of service that the user wants from the connection. The following shows the four classes supported. For each class there is an associated AAL.

Timing information between source and destination
CLASS A: Required CLASS B: Required
CLASS C: Not required CLASS D: Not required

Bit rate characteristics
CLASS A: Constant CLASS B: Variable
CLASS C: Variable CLASS D: Variable

Connection mode
CLASS A: Connection-oriented CLASS B: Connection-oriented
CLASS C: Connection-oriented CLASS D: Connectionless

31.13.3 AAL services

Thus class A supports a constant bit rate with a connection and preserves timing information. This is typically used for voice transmission. Class B is similar to A but has a variable bit rate. Typically it is used for video/audio data. Class C also has a variable bit rate and is connection-oriented, although there is no timing information. This is typically used for non-real-time data, such as computer data. Class D is the same as class C but is connectionless. This means there is no connection between the sender and the receiver before the data is transmitted.

There are four AAL services: AAL1 (for class A), AAL2 (for class B), AAL3/4 or AAL5 (for class C) and AAL3/4 (for class D).

AAL1

The AAL1 supports class A and is intended for real-time voice traffic; it provides a constant bit rate and preserves timing information over a connection. The format of the 48 bytes of the data cell consists of 47 bytes of data, such as PCM or ADPCM code and a 1-byte header. Figure 31.9 shows the format of the cell, including the cell header. The 47-byte data field is described as the SAR-PDU (segmentation and reassembly protocol data units). The header consists of:

- SN (4 bits) which is a sequence number.
- SNP (4 bits) which is the sequence number protection.

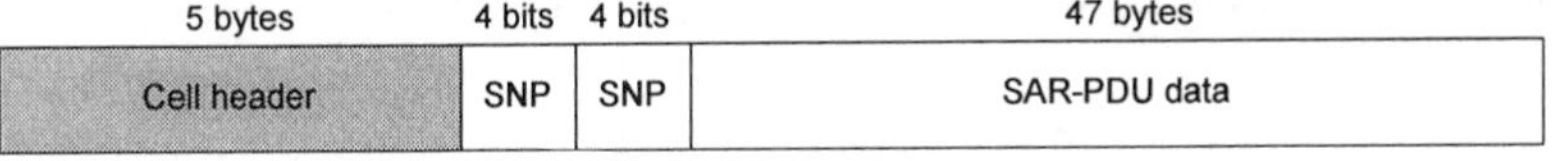

Figure 31.9 ATM and the OSI model

AAL type 2

AAL type 2 is under further study.

AAL type 3/4

Type 3/4 is connection-oriented where the bit rate is variable and there is no need for timing information (Figure 31.10). It uses two main formats:

- SAR (segment and reassemble) which is segments of CPCS PDU with a SAR header and trailer. The extra SAR fields allow the data to be reassembled at the receiver. When the CPCS PDU data has been reassembled the header and trailer are discarded. The fields in the SAR are:

- Segment type (ST) identifies has the SAR has been segmented. Figure 31.10 show how the CPCS PDU data has been segmented into five segments: one beginning segment, three continuation segments and one end segment. The ST field has 2 bits and can therefore contain one of four possible types:

    - SSM (single sequence message) identifies that the SAR contains the complete data.
    - BOM (beginning of message) identifies that it is the first SAR PDU in a sequence.
    - COM (continuation of message) identifies that it is an intermediate SAR PDU.
    - EOM (end of message) identifies that it is the last SAR PDU.

- Sequence number (SN) which is used to reassemble a SAR SDU and thus verify that all of the SAR PDUs have been received.
- Message identifier (MI) which is a unique identifier associated with the set of SAR PDUs that carry a single SAR SDU.
- Length indication (LI) which defines the number of bytes in the SAR PDU. It can have a value between 4 and 44. The COM and BOM types will always have a value of 44. If the EOM field contains fewer than 44 bytes, it is padded to fill the remaining bytes. The LI then indicates the number of value bytes. For example if the LI is 20; there are only 20 value bytes in the SAR PDU the other 24 are padding bytes.
- CRC which is a 10-bit CRC for the entire SAR PDU.

- CPCS (convergence protocol sublayer) takes data from the PDU. As this can be any length, the data is padded so it can be divided by 4. A header and trailer are then added and the completed data stream is converted into one or more SAR PDU format cells. Figure 31.11 shows the format of the CPCS-PDU for AAL type 3/4.

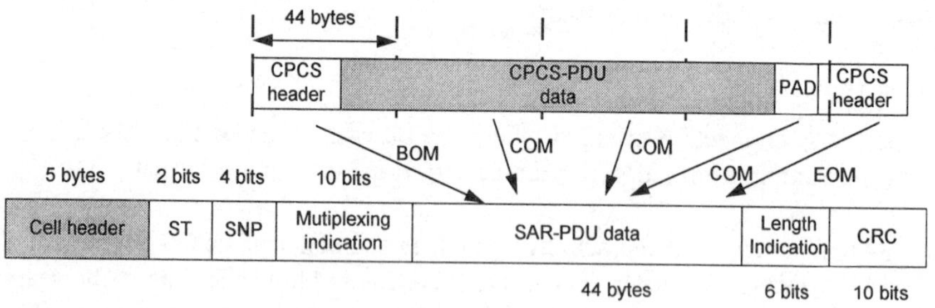

Figure 31.10 AAL type 3/4 cell format

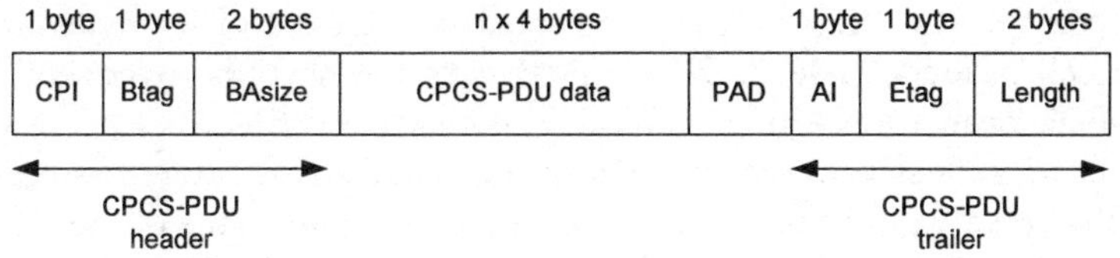

Figure 31.11 CPCS-PDU type 3/4 frame format

The fields in the CPCS-PDU are:

- CPI (common part indicator) which indicates how the remaining fields are interpreted (currently one version exists).
- Btag (beginning tag) which is a value associated with the CPCS-PDU data. The Etag has the same value as the Btag.
- BASize (buffer allocation size) which indicates the size of the buffer that must be reserved so that the completed message can be stored.
- AI (alignment) a single byte which is added to make the trailer equal to 32 bits.
- Etag (end tag) which is the same as the Btag value.
- Length which gives the length of the CPCS PDU data field.

AAL type 5

AAL type 5 is a connectionless service; it has no timing information and can have a variable bit rate. It assumes that one of the levels above the AAL can establish and maintain a connection. Type 5 provides stronger error checking with a 32-bit CRC for the entire CPCS PDU, whereas type 3/4 only allows a 10-bit CRC which is error checking for each SAR PDU. The type 5 format is given in Figure 31.12. The fields are:

- CPCS-UU (CPCS user-to-user) indication.
- CPI (common part indicator) which indicates how the remaining fields are interpreted (currently one version exists).
- Length which gives the length of the CPCS-PDU data.
- CRC which is a 32-bit CRC field.

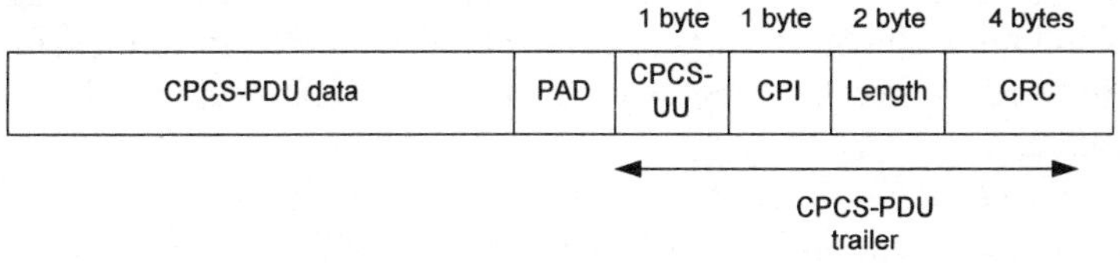

Figure 31.12 CPCS-PDU type 5 frame format

The type 5 CPCS-PDU is then segmented into groups of 44 bytes and the ATM cell header is added. Thus type 5 does not have the overhead of the SAR-PDU header and trailer (i.e. it does not have ST, SN, MID, LI or CRC). This means it does not contain any sequence numbers. It is thus assumed that the cells will always be received in the correct order and none of the cells will be lost. Types 3/4 and 5 can be summarized as follows:

Type 3/4: SAR-PDU overhead is 4 bytes, CPCS-PDU overhead is 8 bytes.
Type 5: SAR-PDU overhead is 0 bytes, CPCS-PDU overhead is 8 bytes.

Type 5 can be characterized as:

- Strong error checking.
- Lack of sequence numbers.
- Reduced overhead of the SAR-PDU header and trailer.

31.14 ATM flow control

ATM cannot provide for a reactive end-to-end flow control because by the time a message is returned from the destination to the source, large amounts of data could have been sent along the ATM pipe, possibly making the congestion worse. The opposite can occur when the congestion on the network has cleared by the time the flow control message reaches the transmitter. The transmitter will thus reduce the data flow when there is little need. ATM tries to react to network congestion quickly, and it slowly reduces the input data flow to reduce congestion.

This rate-based scheme of flow control involves controlling the amount of data to a specified rate, agreed when the connection is made. It then automatically changes the rate based on the past history of the connection as well as the present congestion state of the network.

Data input is thus controlled by early detection of traffic congestion through closely monitoring the internal queues inside the ATM switches, as shown in Figure 31.13. The network then reacts gradually as the queues lengthen and reduces the traffic into the network from the transmitting UNI. This is an improvement over imposing a complete restriction on the data input when the route is totally congested. In summary, anticipation is better than desperation.

A major objective of the flow control scheme is to try to affect only the streams which are causing the congestion, not the well-behaved streams.

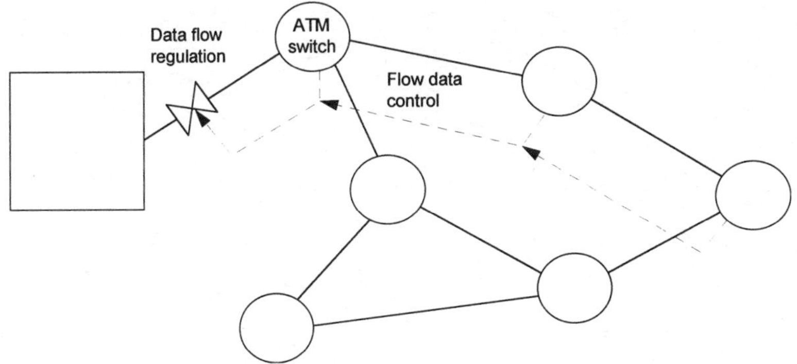

Figure 31.13 Flow control feedback from ATM switches

31.15 Practical ATM networks

As mentioned in Chapter 30 a metropolitan area network has been set-up around Edinburgh, UK. It consists of two rings on ATM and FDDI. The two rings of FDDI and ATM have been run around the Edinburgh sites. This also connects to the University of Stirling through a 155 Mbps SDH connection. Two connections on the ring are made to the SuperJANET network, connections at Heriot-Watt University and the University of Edinburgh.

The 100 Mbps FDDI dual rings link 10 Edinburgh city sites. This ring provides for IP traffic on SuperJANET and also for high-speed metropolitan connections. Initially a 155 Mbps SDH/STM-1 ATM network connects five Edinburgh sites and the University of Stirling. This also connects to the SuperJANET ATM pilot network. Figure 31.14 shows the FDDI and ATM connections.

The two different network technologies allow the universities to operate a two-speed network. For computer-type data the well-established FDDI technology provides good reliable communications and the ATM network allows for future exploitation of mixed voice, data and video transmissions.

The JANET and SuperJANET networks provide connections to all UK universities. A gateway out of the network to the rest of the world is located at University College London (UCL).

31.16 Tutorial

31.1 Discuss how ATM connections are more efficient in their transmission than ISDN and PCM-TDM.

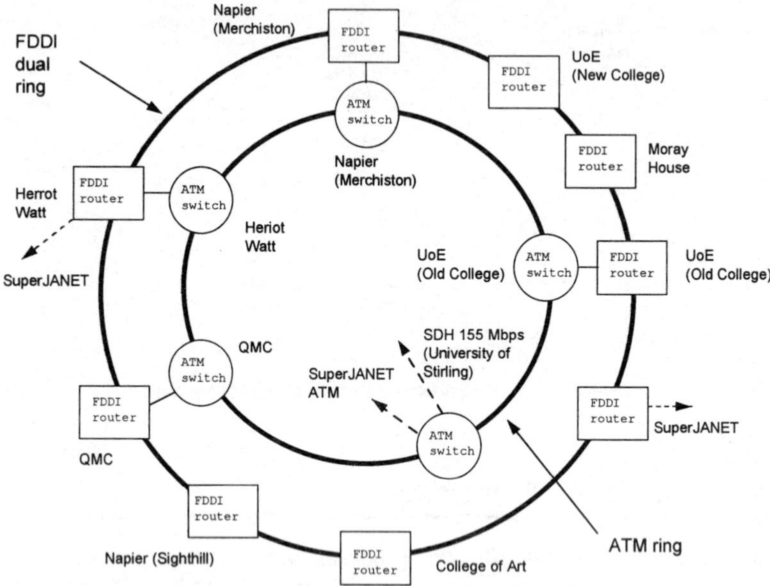

Figure 31.14 EaStMAN ring connections

31.2 Determine the time between samples for speech (maximum frequency 4 kHz), audio (maximum frequency 20 kHz) and for video (maximum frequency 6 MHz).

31.3 Explain the fields in an ATM cell.

31.4 Explain how ATM cells are routed through a network, highlighting why the cells always take a fixed route.

31.5 Explain virtual channels, virtual paths and their uses.

31.6 Explain the AAL service levels and give examples of data types which might use the service levels.

31.7 Explain why computer-type data normally differs in its traffic profile from speech and video data.

31.8 Investigate a local WAN; determine its network topology and its network technology.

31.9 If there is access to an Internet connection, use the WWW to investigate the current status of the EaStMAN network.

32 Integrated Services Digital Network (ISDN)

32.1 Introduction

A major problem in data communications and networks is the integration of real-time sampled data with non-real-time (normal) computer. Sampled data tends to create a constant traffic flow whereas computer-type data has bursts of traffic. And sampled data normally needs to be delivered at a given time but computer-type data needs reliable path where delays are relatively unimportant.

The basic rate for real-time data is speech. It is normally sampled at a rate of 8 kHz and each sample is coded with 8 bits. This leads to a transmission bit rate of 64 kbps. ISDN uses this transmission rate for its base transmission rate. Computer-type data can then be transmitted using this rate or can be split to transmit over several 64 kbps channels. The basic rate ISDN service uses two 64 kbps data lines and a 16 kbps control line, as illustrated in Figure 32.1. Table 32.1 summarizes the I series CCITT standards.

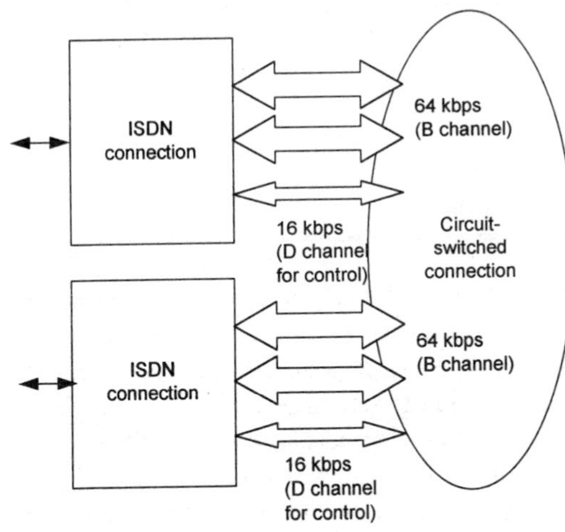

Figure 32.1 Basic rate ISDN services

Table 32.1 CCITT standards on ISDN

CCITT standard number	*Description*
I.1*XX*	ISDN terms and technology
I.2*XX*	ISDN services
I.3*XX*	ISDN addressing
I.430 and I.431	ISDN physical layer interface
I.440 and I.441	ISDN data layer interface
I.450 and I.451	ISDN network layer interface
I.5*XX*	ISDN internetworking
I.6*XX*	ISDN maintenance

Typically modems are used in the home for the transmission of computer-type data. Unfortunately modems have a maximum bit rate of 28.8 kbps. This is automatically increased, on a single channel, to 64 kbps. The connections made by a modem and by ISDN are circuit-switched.

The great advantage of an ISDN connection is that the type of data transmitted is irrelevant to the transmission and switching circuitry. Thus it can carry other types of digital data, such as facsimile, teletex, videotex and computer data. This reduces the need for modems, which convert digital data into an analogue form, only for the public telephone network to convert the analogue signal back into a digital form for transmission over a digital link. It is also possible to multiplex the basic rate of 64 kbps to give even higher data rates. This multiplexing is known as N × 64 kbps or Broadband ISDN (B-ISDN).

Another advantage of ISDN is that it is a circuit-switched connection where a permanent connection is established between two nodes. This connection is guaranteed for the length of the connection. It also has a dependable delay time and is thus suited to real-time data.

32.2 ISDN channels

ISDN uses channels to identify the data rate, each based on the 64 kbps provision. Typical channels are B, D, H0, H11 and H12. The B-channel has a data rate of 64 kbps and provides a circuit-switched connection between end-points. A D-channel operates at 16 kbps and it controls the data transfers over the B channels. The other channels provide B-ISDN for much higher data rates. Table 32.2 outlines the basic data rates for these channels.

The two main types of interface are the basic rate access and the primary rate access. Both are based around groupings of B- and D-channels. The basic rate access allows two B-channels and one 16 kbps D-channel.

Table 32.2 ISDN channels

Channel	*Description*
B	64 kbps
D	16 kbps signalling for channel B (ISDN)
	64 kbps signalling for channel B (B-ISDN)
H0	384 kbps (6 × 64 kbps) for B-ISDN
H11	1.536 Mbps (24 × 64 kbps) for B-ISDN
H12	1.920 Mbps (30 × 64 kbps) for B-ISDN

Primary rate provides B-ISDN, such as H12 which gives 30 B-channels and a 64 kbps D-channel. For basic and primary rates, all channels multiplex onto a single line by combining channels into frames and adding extra synchronization bits. Figure 32.3 gives examples of the basic rate and primary rate.

The basic rate ISDN gives two B-channels at 64 kbps and a signalling channel at 16 kbps. These multiplex into a frame and, after adding extra framing bits, the total output data rate is 192 kbps. The total data rate for the basic rate service is thus 128 kbps. One or many devices may multiplex their data, such as two devices transmitting at 64 kbps, a single device multiplexing its 128 kbps data over two channels (giving 128 kbps), or by several devices transmitting a sub-64 kbps data rate over the two channels. For example, four 32 kbps devices could simultaneously transmit their data, eight 16 kbps devices, and so on.

For H12, 30 × 64 kbps channels multiplex with a 64 kbps signalling channel, and with extra framing bits, the resulting data rate is 2.048 Mbps (compatible with European PCM-TDM systems). This means the actual data rate is 1.920 Mbps. As with the basic service this could contain a number of devices with a data rate of less than or greater than a multiple of 64 kbps.

For H11, 24 × 64 kbps channels multiplex with a 64 kbps signalling channel, and with extra framing bits, it produces a data rate of 1.544 Mbps (compatible with USA PCM-TDM systems). The actual data rate is 1.536 Mbps.

32.3 ISDN physical layer interfacing

The physical layer corresponds to layer 1 of the OSI 7-layer model and is defined in CCITT specifications I.430 and I.431. Pulses on the line are not coded as pure binary, they use a technique called alternate mark inversion (AMI).

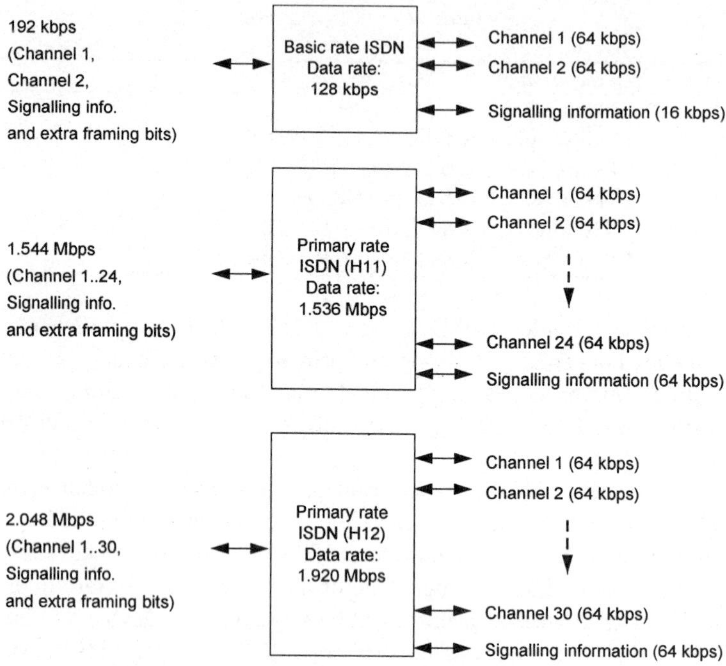

Figure 32.2 Basic rate, H11 and H12 ISDN services

32.3.1 Alternative mark inversion (AMI) line code

AMI line codes use three voltage levels. In pure AMI, 0 V represents a '0', and the voltage amplitude for each '1' is the inverse of the previous '1' bit. ISDN uses the inverse of this, i.e. 0 V for a '1' and an inverse in voltage for a '0', as shown in Figure 32.3. Normally the pulse amplitude is 0.75 V.

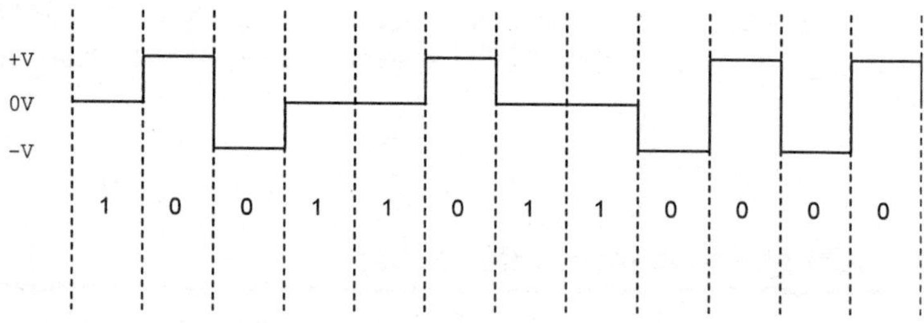

Figure 32.3 AMI used in ISDN

Inversion of the AMI signal (i.e. inverting a '0' rather than a '1') allows for timing information to be recovered when there are long runs of zeros, which is typical in the idle state. AMI line code also automatically balances the signal voltage, and the average voltage will be approximately zero even when there are long runs of 0's.

32.3.2 System connections

In basic rate connections up to eight devices, or items of termination equipment (TE), can connect to the network termination (NT). They connect over a common four-wire bus using two sets of twisted-pair cables. The transmit output (T_X) on each TE connects to the transmit output on the other TEs, and the receive input (R_X) on each TE connects to all other TEs. On the NT the receive input connects to the transmit of the TEs, and the transmit output of the NT connects to the receive input of the TEs. A contention protocol allows only one TE to communicate at a time.

An 8-pin ISO 8877 connector connects a TE to the NT; this is similar to the RJ-45 connector but has two extra pin connections. Figure 32.4 shows the pin connections. Pins 3 and 6 carry the T_X signal from the TE, pins 4 and 5 provide the R_X to the TEs. Pins 7 and 8 are the secondary power supply from the NT and pins 1 and 2 the power supply from the TE (if used). The T_X/R_X lines connect via transformers thus only the AC part of the bitstream transfers into the PCM circuitry of the TE and the NT. This produces a need for a balanced DC line code such as AMI, as the DC component in the bitstream will not pass through the transformers.

32.3.3 Frame format

Figures 33.5 and 33.6 show the ISDN frame formats. Each frame is 250 μs long and contains 48 bits; this give a total bit rate of 192 kbps ($48/250 \times 10^{-6}$) made up of two 64 kbps B channels, one 16 kbps D-channel and extra framing, DC balancing and synchronization bits.

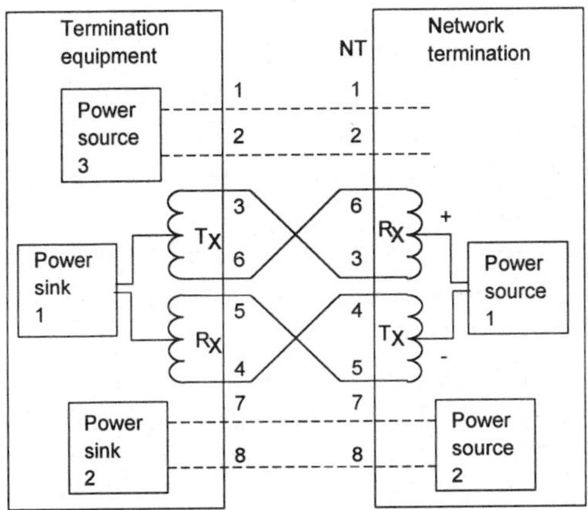

Figure 32.4 Power supplies between NT and TE

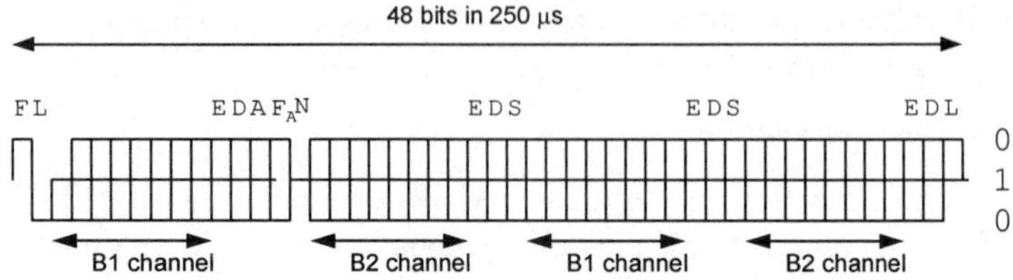

Figure 32.5 ISDN frame format for NT to TE

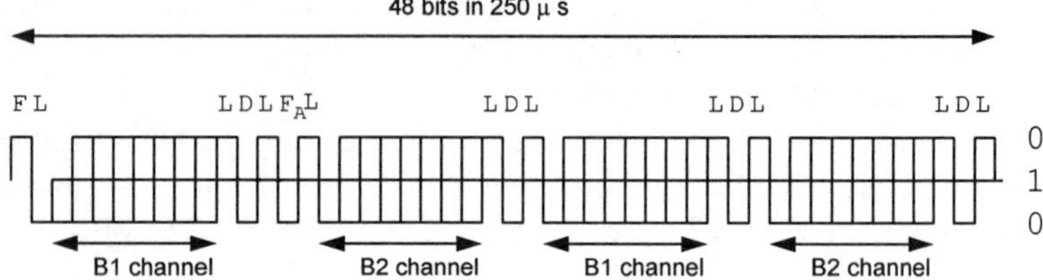

Figure 32.6 ISDN frame format for TE to NT

where

F –	framing bit	N –	set to a 1
L –	DC balancing bit	D –	D-channel bit
E –	D-echo channel bit	F_A –	auxiliary framing bit (=0)
S –	reserved for future use	A –	activation bit
M –	multiframing bit	B1–	bits for channel 1
B2–	bits for channel 2		

The F/L pair of bits identify the start of each transmitted frame. When transmitting from a TE to an NT there is a 10-bit offset in the return of the frame back to the TE. The E bits echo the D-channel bits back to the TE.

When transmitting from the NT to the TE, the bits after the F/L bits, in the B-channel, have a volition in the first 0. If any of these bits is a 0 then a volition will occur, but if they are 1's then no volition can occur. To overcome this the F_A bit forces a volition. Since it is followed by 0 (the N bit) it will not be confused with the F/L pair. The start of the frame can thus be traced backwards to find the F/L pair.

There are 16 bits for each B-channel, giving a basic data rate of 64 kbps $(16/250 \times 10^{-6})$ and there are 4 bits in the frame for the D-channel, giving a bit rate of 16 kbps $(4/250 \times 10^{-6})$.

The L bit balances the DC level on the line. If the number of 0's following the last balancing bit is odd then the balancing bit is a 0, else it is a 1. When synchronized the NT informs the TEs by setting the A bit.

32.4 ISDN data link layer

The data link layer uses a form known as the Link Access Procedure for the D-channel (LAPD). Figure 32.7 shows the frame format. The unique bit sequence `01111110` identifies the start and end of the frame. This bit pattern cannot occur in the rest of the frame due to zero bit-stuffing.

The address field contains information on the type of data contained in the frame (the service access point identifier) and the physical address of the ISDN device (the terminal endpoint identifier). The control field contains a supervisory, an unnumbered or an information frame. The frame check sequence provides error detection information.

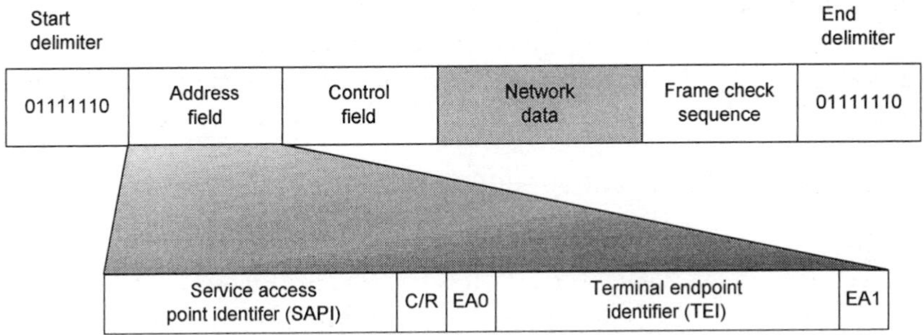

Figure 32.7 D-channel frame structure

32.4.1 Address field

The data link address only contains addressing information to connect the TE to the NT and does not have network addresses. Figure 32.7 shows the address field format. The SAPI identifies the type of ISDN services. For example a frame from a telephone would be identified as such, and only telephones would read the frame.

All TEs connect to a single multiplexed bus, thus each has a unique data link address, known as a terminal endpoint identifier (TEI). The user or the network sets this; the ranges of available addresses are:

0–63	non-automatic assignment TEIs
64–126	automatic assignment TEIs
127	global TEI

The non-automatic assignment involves the user setting the address of each of the devices connected to the network. When a device transmits data it inserts its own TEI address and only receives data which has its TEI address. In most cases devices should not have the same TEI address, as this would cause all devices with the same TEI address, and the SAPI, to receive the same data (although, in some cases, this may be a requirement).

The network allocates addresses to devices requiring automatic assignment before they can communicate with any other devices. The global TEI address is used to broadcast messages to all connected devices. A typical example is when a telephone call is incoming to a group on a shared line where all the telephones would ring until one was answered.

The `C/R` bit is the command/response bit and `EA0/EA1` are extended address field bits.

32.4.2 Bit stuffing

With zero bit-stuffing the transmitter inserts a zero into the bitstream when transmitting five consecutive 1's. When the receiver receives five consecutive 1's it deletes the next bit if it is a zero. This stops the unique `01111110` sequence occurring within the frame. For example if the bits to be transmitted are:

`101000101011111100001010001010000111110101010`

then the with the start and end delimiter this would be:

01111110101000101*01111110*0001010001010000111110101010**01111110**

It can be seen from this bitstream that the stream to be transmitted contains the delimiter within the frame. This zero bit-insertion is applied to give:

01111110101000101*01111*0*10*00010100010100001111100101010**0111111**

Notice that the transmitter has inserted a zero when five consecutive 1's occur. Thus the bit pattern 01111110 cannot occur anywhere in the bitstream. When the receiver receives five consecutive 1's it deletes the next bit if it is a zero. If it is a 1 then it is a valid delimiter. In the example the received stream will be:

011111101010001010111111000010100010100001111101010 10**01111110**

32.4.3 Control field

ISDN uses a 16-bit control field for information and supervisory frames and an 8-bit field for unnumbered frames, as illustrated in Figure 32.8. Information frames contain sequenced data. The format is `0SSSSSSSXRRRRRRR`, where `SSSSSSS` is the send sequence number and `RRRRRRR` is the frame sequence number that the sender expects to receive next (`X` is the poll/final bit). Since the extended mode uses a 7-bit sequence field then information frames are numbered from 0 to 127.

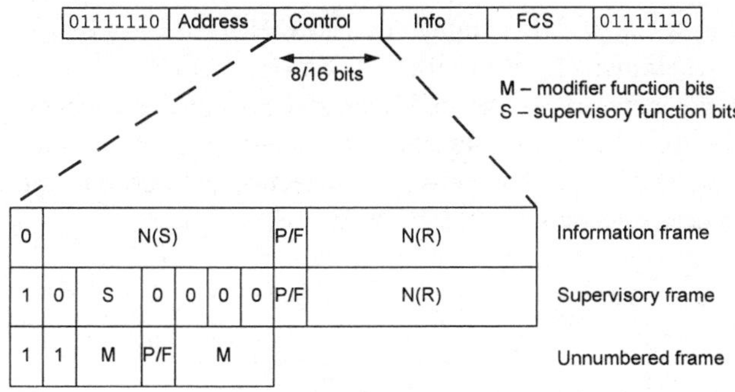

Figure 32.8 ISDN control field

Supervisory frames contain flow control data. Table 32.3 lists the supervisory frame types and the control field bit settings. The RRRRRRR value represent the 7-bit receive sequence number.

Table 32.3 Supervisory frame types and control field settings

Type	*Control field setting*
Receiver ready (RR)	10000000PRRRRRRR
Receiver not ready (RNR)	10100000PRRRRRRR
Reject (REJ)	10010000PRRRRRRR

Unnumbered frames set up and clear connections between a node and the network.. Table 32.4 lists the unnumbered frame commands and Table 32.5 lists the unnumbered frame responses.

Table 32.4 Unnumbered frame commands and control field settings

Type	*Control field setting*
Set asynchronous balance mode extended (SABME)	1111P110
Unnumbered information (UI)	1100F000
Disconnect mode (DISC)	1100P010

Table 32.5 Unnumbered frame responses and control field settings

Type	*Control field setting*
Disconnect mode (DM)	1111P110
Unnumbered acknowledgment (UA)	1100F000
Frame reject (FRMR)	1110P001

In ISDN all connected nodes and the network connection can send commands and receive responses. Figure 32.9 shows a sample connection of an incoming call to an ISDN node (address TEI_1). The SABME mode is set up initially using the SABME command (U[SABME,TEI_1,P=1]). followed by an acknowledgment from the ISDN node (U[UA,TEI_1,F=1]). At any time, either the network or the node can disconnect the connection. In this case the ISDN node disconnects the connection with the command U[DISC,TEI_1,P=1]. The network connection acknowledges this with an unnumbered acknowledgment (U[UA,TEI_1,F=1]).

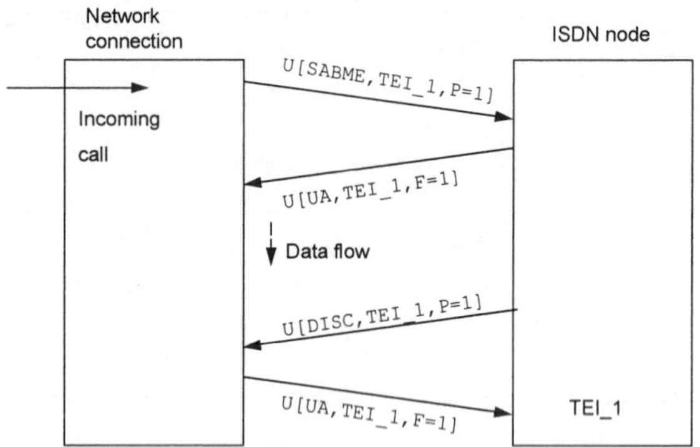

Figure 32.9 Example connection between a primary/secondary

32.4.4 D-channel contention

The D-channel contention protocol ensures that only one terminal can transmit its data at a time. This happens because the start and the end of the D-channel bits have the bitstream 01111110, as shown below:

11111**01111110**XXXXXXXXX...XXXXXXXX**01111110**1111

When idle, each TE floats to a high-impedance state, which is taken as a binary 1. To transmit, a TE counts the number of 1's in the D-channel. A 0 resets this count. After a predetermined number, greater than a predetermined number of consecutive 1's, the TE transmits its data and monitors the return from the NT. If it does not receive the correct D-channel bitstream returned through the E bits then a collision has occurred. When a TE detects a collision it immediately stops transmitting and monitors the line.

When a TE has finished transmitting data it increases its count value for the number of consecutive 1's by 1. This gives other TEs an opportunity to transmit their data.

32.4.5 Frame check sequence

The frame check sequence (FCS) field contains an error detection code based on cyclic redundancy check (CRC) polynomials. It uses the CCITT V.41 polynomial, which is $G(x) = x^{16} + x^{12} + x^{5} + x^{1}$.

32.5 ISDN network layer

The D-channel carriers network layer information within the LAPD frame. This information establishes and controls a connection. The LAPD frames contain no true data as this is carried in the B-channel. Its function is to set up and manage calls and to provide flow control between connections over the network.

Figure 32.10 shows the format of the layer 3 signalling message frame. The first byte is the protocol discriminator. In the future this byte will define different communications protocols. At present it is normally set to `0001000`. After the second byte the call reference value is defined. This is used to identify particular calls with a reference number. The length of the call reference value is defined within the second byte. As it contains a 4-bit value, up to 16 bytes can be contained in the call reference value field. The next byte gives the message type and this type defines the information contained in the proceeding field.

There are four main types of message: call establish, call information, call clearing and miscellaneous messages. Table 32.6 outlines the main messages. Figure 32.11 shows an example connection procedure. The initial message sent is SETUP. This may contain some of the following:

- Channel identification – identifies a channel with an ISDN interface.
- Calling party number.
- Calling party subaddress.
- Called party number.
- Called party subnumber.
- Extra data (2–131 bytes).

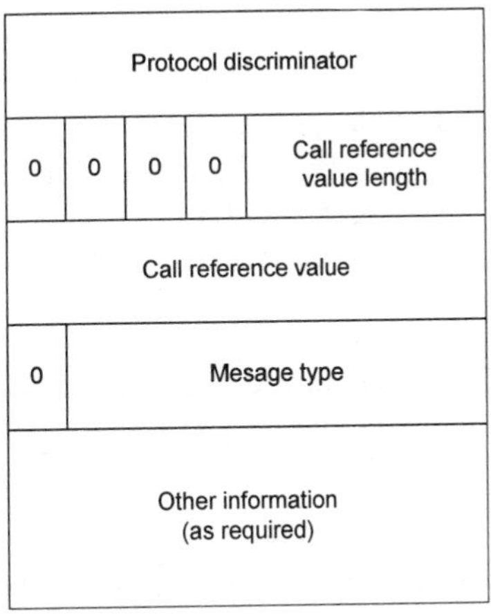

Figure 32.10 Signalling message structure

Table 32.6 ISDN network messages

Call establish	*Information messages*	*Call clearing*
ALERTING	RESUME	DISCONNECT
CALL PROCEEDING	RESUME ACKNOWLEDGE	RELEASE
CONNECT	RESUME REJECT	RELEASE COMPLETE
CONNECT ACKNOWLEDGE	SUSPEND	RESTART
PROGRESS	SUSPEND ACKNOWLEDGE	RESTART ACKNOWLEDGE
SETUP	SUSPEND REJECT	
SETUP ACKNOWLEDGE	USER INFORMATION	

After the calling TE has sent the SETUP message, the network then returns the SETUP ACK message. If there is insufficient information in the SETUP message then other information needs to flow between the called TE and the network. After this the network sends back a CALL PROCEEDING message and it also sends a SETUP message to the called TE. When the called TE detects its TEI address and SAPI, it sends back an ALERTING message. This informs the network that the node is alerting the user to answer the call. When it is answered, the called TE sends a CONNECT to the network. The network then acknowledges this with a CONNECT ACK message, at the same time it sends a CONNECT message to the calling TE. The calling TE then acknowledges this with a CONNECT ACK. The connection is then established between the two nodes and data can be transferred.

To disconnect the connection the DISCONNECT, RELEASE and RELEASE COMPLETE messages are used.

32.6 Exercises

32.1 Show why speech requires to be transmitted at 64 kbps.

32.2 If the bandwidth of hi-fi audio is 20 kHz and 16 bits are used to code each sample, determine the required bit rate for single-channel transmission.

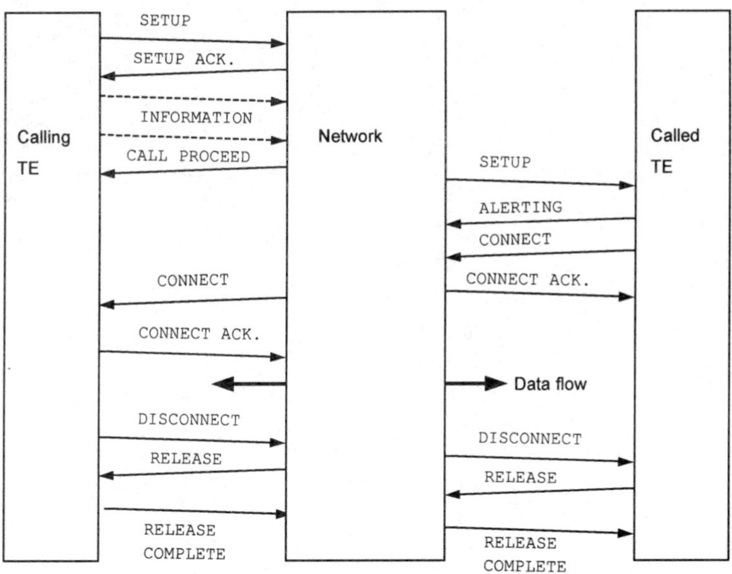

Figure 32.11 Call establishment and clearing

32.3 Explain the format of the ISDN frame.

32.4 Suppose that an ISDN frame has 48 bits and takes 250 μs to transmit. Show that the bit rate on each D-channel is 16 kbps and that the bit rate of the B-channel is 64 kbps.

32.5 Explain the different types of frames and show how a connection is made between ISDN nodes.

32.6 Show how supervisory frames are used to control the flow of data.

32.7 Discuss the format of the ISDN network layer packet.

32.8 How does an ISDN node set up and disconnect a network connection.

33 Modems

33.1 Introduction

Modems (MOdulator/DEModulator) connects digital equipment to a telephone line. It connects digital equipment to a speech bandwidth-limited communications channel. Typically, modems are used on telephone lines which have a bandwidth of between 400 Hz and 3.4 kHz. If digital pulses were applied directly these lines they would end up severely distorted.

Modem speeds range from 300 bps to 28.8 kbps. A modem normally transmits about 10 bits per character (each character has 8 bits). Thus the maximum rate of characters for a high-speed modem is 2880 characters per second. This chapter contains approximately 15 000 characters, thus to transmit the text in this chapter would take approximately 5 seconds. Text, itself, is relatively fast-transfer, unfortunately even compressed graphics can take some time to be transmitted. A compressed image of 20 kB (equivalent to 20 000 characters) will take nearly 6 seconds to load on the fastest modem.

The document that was used to store this chapter occupies, in an uncompressed form, 360 kB. Thus to download this document over a modem, on the fastest modem, would take:

$$\text{Time taken} = \frac{\text{Total file size}}{\text{Characters per second}} = \frac{360\,000}{2\,800} = 125 \text{ second}$$

A 14.4 kbps modem would take 250 seconds. Most home users connect to the Internet and WWW through a 14.4 kbps or 28.8 kbps modem (although increasingly ISDN is being used). The example above shows the need to compress files when transferring them over a modem. On the WWW, documents and large files are normally compressed into a ZIP file and images and video compressed in GIF and JPG.

Most modems are able to do the following:

- Automatically dial (known as Auto-dial) another modem using either touch-tone or pulse dialling.
- Automatically answer (known as Auto-answer) calls and make a connection with another modem.

- Disconnect a phone connection when data transfer has completed or if an error occurs.
- Automatically speed negotiation between the two modems.
- Convert bits into a form suitable for the line (modulator).
- Convert received signals back into bits (demodulator).
- Transfer data reliably with the correct type of handshaking.

Figure 33.1 shows how two computers connect to each other using RS-232 converters and modems. The RS-232 converter is normally an integral part of the computer, while the modem can either be external or internal to the computer. If it is externally connected then it is normally connected by a cable with a 25-pin male D-type connector on either end.

Modems are either synchronous or asynchronous. A synchronous modem recovers the clock at the receiver. There is no need for start and stop bits in a synchronous modem. Asynchronous modems are by far the most popular types. Synchronous modems have a maximum speed of 28.8 kbps whereas the maximum rate of asynchronous modems is 19.6 kbps. A measure of the speed of the modem is the baud rate or bps (bits per second).

There are two types of circuits available from the public telephone network: either direct dial or a permanent connection. The direct dial type is a dial-up network where the link is established in the same manner as normal voice calls with a standard telephone or some kind of an automatic dial/answer machine. They can use either touch-tones or pulses to make the connection. With private line circuits, the subscriber has a permanent dedicated communication link.

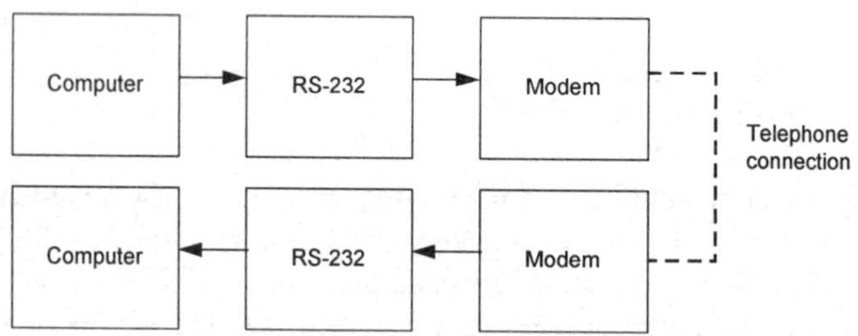

Figure 33.1 Data transfer using modems

33.2 RS-232 communications

The communication between the modem and the computer is via RS-232. RS-232 uses asynchronous communication which has a start-stop data format.

Each character is transmitted one at a time with a delay between characters. This delay is called the inactive time and is set at a logic level high as shown in Figure 33.2. The transmitter sends a start bit to inform the receiver that a character is to be sent in the following bit transmission. This start bit is always a '0'. Next 5, 6 or 7 data bits are sent as a 7-bit ASCII character, followed by a parity bit and finally either 1, 1.5 or 2 stop bits. The rate of transmission is set by the timing of a single bit. Both the transmitter and receiver need to be set to the same bit-time interval. An internal clock on both of them sets this interval. They only have to be roughly synchronized and approximately at the same rate as data is transmitted in relatively short bursts.

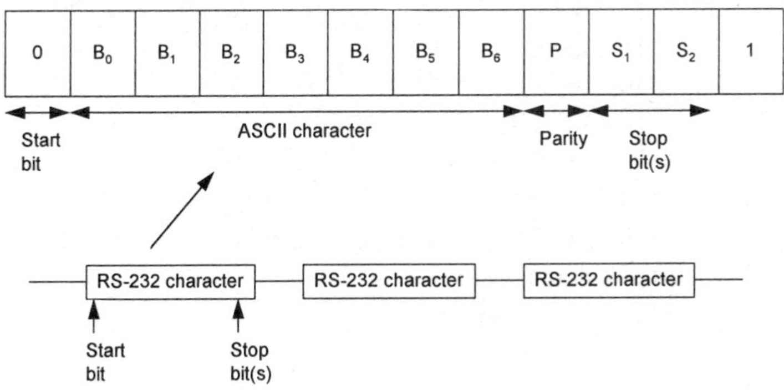

Figure 33.2 RS-232 frame format

33.2.1 Bit rate and the baud rate

One of the main parameters for specifying RS-232 communications is the rate at which data is transmitted and received. It is important that the transmitter and receiver operate at roughly the same speed.

For asynchronous transmission the start and stop bits are added in addition to the seven ASCII character bits and the parity. Thus a total of 10 bits are required to transmit a single character. With 2 stop bits, a total of 11 bits are required. If 10 characters are sent every second and if 11 bits are used for each character, then the transmission rate is 110 bits per second (bps). The fastest modem thus has a character transmission rate of 2880 characters per second.

In addition to the bit rate, another term used to describe the transmission speed is the baud rate. The bit rate refers to the actual rate at which bits are transmitted, whereas the baud rate is to the rate at which signaling elements, used to represent bits, are transmitted. Since one signalling element encodes one bit, the two rates are then identical. Only in modems does the bit rate differ from the baud rate.

33.3 Modem standards

The CCITT (now known as the ITU) has defined standards which relate to RS-232 and modem communications. Each uses a V. number to define their type. Modems tend to state all the standards they comply with. An example FAX/modem has the following compatibly:

- V.32bis (14.4 kbps).
- V.32 (9.6 kbps).
- V.22bis (2.4 kbps).
- V.22 (1.2 kbps).
- Bell 212A (1.2 kbps).
- Bell 103 (300 bps).
- V.17 (14.4 bps FAX).
- V.29 (9.6 kbps FAX).
- V.27ter (4.8 kbps FAX).
- V.21 (300 bps FAX - secondary channel).
- V.42bis (data compression).
- V.42 (error correction).
- MNP5 (data compression).
- MNP2-4 (error correction).

A 28.8 kbps modem also supports the V.34 standard.

33.4 Modem commands

Most modems are Hayes compatible. Hayes was the company that pioneered modems and defined a standard method of programming the mode of the modem with the AT command language. A computer gets the attention of the modem by sending an '`AT`' command. For example, '`ATDT`' is the dial command. Initially, a modem is in the command mode and accepts commands from the computer. These commands are sent at either 300 bps or 1200 bps (the modem automatically detects which of the speeds is being used).

Most commands are sent with the AT prefix. Each command is followed by a carriage return character (ASCII character 13 decimal); a command without a carriage return character is ignored (after a given time period). More than one command can be placed on a single line and, if necessary, spaces can be entered to improve readability. Commands can be sent either in upper or lower case. Some AT commands are listed in Table 33.1.

Table 33.1 Example AT modem commands

Command	*Description*
ATDT54321	Automatically phone number 54321 using touch-tone dialling. Within the number definition, a comma (,) represents a pause and a W waits for a second dial tone and an @ waits for a 5 second silence.
ATPT12345	Automatically phone number 12345 using pulse dialling.
AT S0=2	Automatically answer a call. The S0 register contains the number of rings the modem uses before it answers the call. In this case there will be two rings before it is answered. If S0 is zero then the modem will not answer a call.
ATH	Hang up telephone line connection.
+++	Disconnect line and return to on-line command mode.
AT A	Manually answer call.
AT E0	Commands are not echoed (ATE1 causes commands to be echoed). See Table 33.2.
AT L0	Low speaker volume (ATL1 gives medium volume and ATL2 gives high speaker volume)
AT MO	Internal speaker off (ATM1 gives internal speaker on until carrier detected, ATM2 gives the speaker always on, ATM3 gives speaker on until carrier detect and while dialling)
AT QO	Modem sends responses (ATQ1 does not send responses). See Table 33.2.
AT V0	Modem sends numeric responses (ATV1 sends word responses). See Table 33.2.

The modem can enter into one of two states: the normal state and the command state. In the normal state the modem transmits and/or receives characters from the computer. In the command state, characters sent to the modem are interpreted as commands. Once a command is interpreted the modem goes into the normal mode. Any characters sent to the modem are then sent along the line. To interrupt the modem so that it goes back into command mode, three consecutive '+' characters are sent, i.e. '+++'.

After the modem has received an AT command it responds with a return code. Some return codes are given in Table 33.2. For example if a modem calls another which is busy then the return code is 7. A modem dialling another modem returns the codes for OK (when the ATDT command is received), CONNECT (when it connects to the remote modem) and CONNECT 1200 (when it detects the speed of the remote modem). Note that the return code from the modem can be suppressed by sending the AT command '`ATQ1`'. The AT code for it to return the code is '`ATQ0`', normally this is the default condition

Table 33.2 Example return codes

Message	*Digit*	*Description*
OK	0	Command executed without errors
CONNECT	1	A connection has been made
RING	2	An incoming call has been detected
NO CARRIER	3	No carrier detected
ERROR	4	Invalid command
CONNECT 1200	5	Connected to a 1200 bps modem
NO DIALTONE	6	Dial-tone not detected
BUSY	7	Remote line is busy
NO ANSWER	8	No answer from remote line
CONNECT 600	9	Connected to a 600 bps modem
CONNECT 2400	10	Connected to a 2400 bps modem
CONNECT 4800	11	Connected to a 4800 bps modem
CONNECT 9600	13	Connected to a 9600 bps modem
CONNECT 14400	15	Connected to a 14 400 bps modem
CONNECT 19200	61	Connected to a 19 200 bps modem
CONNECT 28800	65	Connected to a 28 800 bps modem
CONNECT 1200/75	48	Connected to a 1200/75 bps modem

Figure 33.5 shows an example session when connecting one modem to another. Initially the modem is set up to receive commands from the computer. When the computer is ready to make a connection it sends the command '`ATDH 54321`' which makes a connection with telephone number 54321 using tone dialling. The modem then replies with an OK response (a 0 value) and the modem tries to make a connection with the remote modem. If it cannot make the connection it returns back a response of NO CARRIER (3), BUSY (7), NO DIALTONE (6) or NO ANSWER (8). If it does connect to the remote modem then it returns a connect response, such as CONNECT 9600 (13). The data can then be transmitted between the modem at the assigned rate (in this case 9600 bps). When the modem wants to end the connection it gets the modem's attention by sending it three '+' characters ('+++'). The modem will then wait for a command from the host computer. In this case the command is the hang-up the connection (ATH). The modem will then return an OK response when it has successfully cleared the connection.

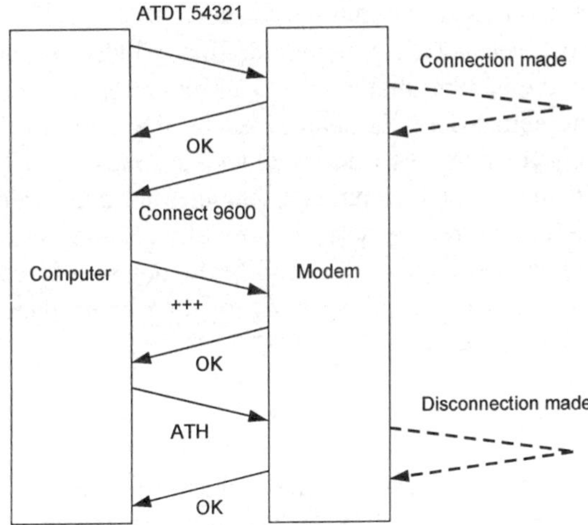

Figure 33.3 Commands and responses when making a connection

The modem contains various status registers called the S-registers which store modem settings. Table 33.3 lists some of these registers. The S0 register sets the number of rings that must occur before the modem answers an incoming call. If it is set to zero (0) then the modem will not answer incoming calls. The S1 register stores the number of incoming rings when the modem is rung. S2 stores the escape character, normally this is set to the '+' character and the S3 register stores the character which defines the end of a command, normally the CR character (13 decimal).

Table 33.3 Modem registers

Register	*Function*	*Range [typical default]*
S0	Rings to Auto-answer	0–255 rings [0 rings]
S1	Ring counter	0–255 rings [0 rings]
S2	Escape character	[43]
S3	Carriage return character	[13]
S6	Wait time for dial-tone	2–255 s [2 s]
S7	Wait time for carrier	1–255 s [50 s]
S8	Pause time for automatic dialling	0–255 [2 s]

33.5 Modem setups

Figure 33.4 shows a sample window from the Microsoft Windows Terminal program (in both Microsoft Windows 3.x and Windows 95). It shows the Modem commands window. In this case, it can be seen that when the modem dials a number the prefix to the number dialed is 'ATDT'. The hang-up command sequence is '+++ ATH'. A sample dialling window is shown in Figure 33.6. In this case the number dialed is 9,123456789. A ',' character represent a delay. The actual delay is determined by the value in the S8 register (see Table 33.2). Typically, this value is about 2 seconds.

On many private switched telephone exchanges a 9 must prefix the number if an outside line is required. A delay is normally required after the 9 prefix before dialling the actual number. To modify the delay to 5 seconds, dial the number 9 0112432 and wait 30 seconds for the carrier, then the following command line can be used:

```
ATDT 9,0112432 S8=5 S7=30
```

It can be seen in Figure 33.4 that a prefix and a suffix are sent to the modem. This is to ensure there is a time delay between the transmission prefix and the suffix string. For example, when the modem is to hang up the connection, the '+++' is sent followed by a delay then the 'ATH'.

In Figure 33.6 there is an option called Originater. This is the string that is sent initially to the modem to set it up. In this case the string is 'ATQ0V1E1S0=0'. The Q0 part informs the modem to return a send status code. The V1 part informs the modem that the return code message is to be displayed rather than just the value of the return code; for example, it displays CONNECT 1200 rather than the code 5 (V0 displays the status code). The E1 part enables the command message echo (E0 disables it).

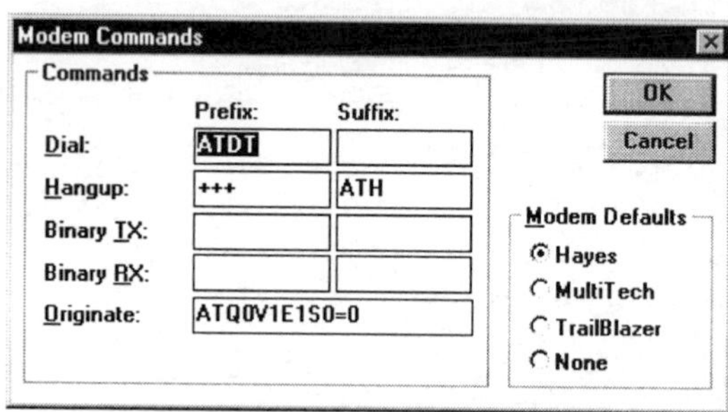

Figure 33.4 Modem commands

Figure 33.5 Dialling a remote modem

Figure 33.5 shows the modem setup windows for CompuServe access. The string in this case is:

```
ATSO=0 Q0 V1 &C1&D2^M
```

as previously seen, `S0` stops the modem from Auto-answering. `V1` causes the modem to respond with word responses. `&C1` and `&D2` set up the hardware signals for the modem. Finally `^M` represent Cntrl-M which defines the carriage return character.

The modem reset command in this case is `AT &F`. This resets the modem and restores the factor default settings.

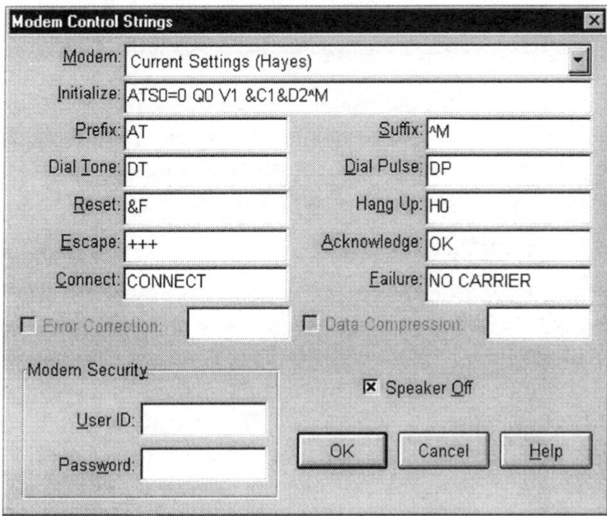

Figure 33.6 Example modem settings

33.6 Modem indicators

Most external modems have status indicators to inform the user of the current status of a connection. Typically, the indicator lights are:

- AA – is ON when the modem is ready to receive calls automatically. It flashes when a call is incoming. If it is OFF then it will not receive incoming calls. Note that if the S0 register is loaded with any other value than 0 then the modem goes into Auto-answer mode. The value stored in the S0 register determines the number of rings before the modem answers.

- CD – is ON when the modem detects the remote modem's carrier, else it is OFF.
- OH – is ON when the modem is on-hook, else it is OFF.
- RD – flashes when the modem is receiving data or is getting a command from the computer.
- SD – flashes when the modem is sending data.
- TR – shows that the DTR line is active (i.e. the computer is ready to transmit or receive data).
- MR – shows that the modem is powered up.

33.7 Digital Modulation

Digital modulation changes the characteristic of a carrier according to binary information. With a sine wave carrier the amplitude, frequency or phase can be varied. Figure 33.7 illustrates the three basic types: amplitude-shift keying (ASK), frequency-shift keying (FSK) and phase-shift keying (PSK).

33.7.1 Frequency shift keying (FSK)

FSK, in the most basic case, represents a 1 (a mark) by one frequency and a 0 (a space) by another. These frequencies lie within the bandwidth of the transmission channel.

On a V.21, 300 bps, full-duplex modem the originator modem uses the frequency 980 Hz to represent a mark and 1180 Hz a space. The answering modem transmits with 1650 Hz for a mark and 1850 Hz for a space. The four frequencies allow the caller originator and the answering modem to communicate at the same time; that is full-duplex communication.

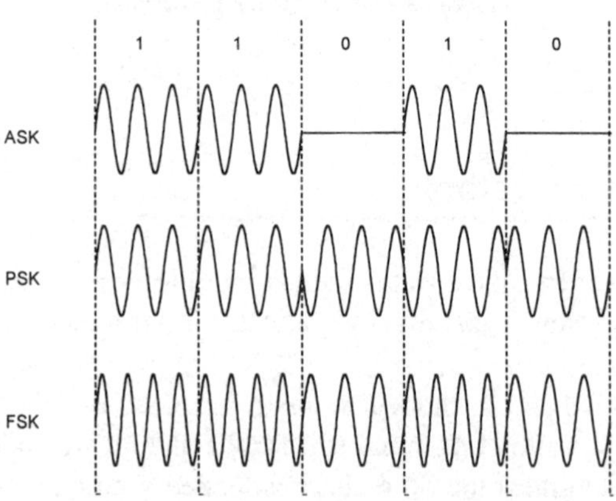

Figure 33.7 Waveforms for ASK, PSK and FSK

FSK modems are inefficient in their use of bandwidth, with the result that the maximum data rate over normal telephone lines is 1800 bps. Typically, for rates over 1200 bps, other modulation schemes are used.

33.7.2 Phase shift keying (PSK)

In coherent PSK a carrier gets no phase shift for a 0 and a 180° phase shift for a 1, as given next:

$$0 \Rightarrow 0°$$
$$1 \Rightarrow 180°$$

Its main advantage over FSK is that since it uses a single frequency it uses much less bandwidth. It is thus less affected by noise. It has an advantage over ASK because its information is not contained in the amplitude of the carrier, thus again it is less affected by noise.

33.7.3 M-ary modulation

With *M*-ary modulation a change in amplitude, phase of frequency represents one of *M* possible signals. It is possible to have *M*-ary FSK, *M*-ary PSK and *M*-ary ASK modulation schemes. This is where the baud rate differs from the bit rate. The bit rate is the true measure of the rate of the line, whereas the baud rate only indicates the signalling element rate, which might be a half or a quarter of the bit rate.

For four-phase differential phase-shift keying (DPSK) the bits are grouped into two and each group is assigned a certain phase shift. For 2 bits there are four combinations: a 00 is coded as 0°, 01 coded as 90°, and so on:

$$00 \Rightarrow 0° \qquad 01 \Rightarrow 90°$$
$$11 \Rightarrow 180° \qquad 10 \Rightarrow 270°$$

It is also possible to change a mixture of amplitude, phase or frequency. *M*-ary amplitude-phase keying (APK) varies both the amplitude and phase of a carrier to represent *M* possible bit patterns.

M-ary quadrature amplitude modulation (QAM) changes the amplitude and phase of the carrier. 16-QAM uses four amplitudes and four phase shifts, allowing it to code 4 bits at a time. In this case the baud rate will be a quarter of the bit rate.

Typical technologies for modems are:

FSK	– used up to 1200 bps
Four-phase DPSK	– used at 2400 bps
Eight-phase DPSK	– used at 4800 bps
16-QAM	– used at 9600 bps

33.8 Typical modems

Most modern modems operate with V.22bis (2400 bps), V.32 (9600 bps), V.32bis (14 400 bps), some standards are outlined in Table 33.4. The V.32 and V.32bis modems can be enhanced with echo cancellation. They also typically have built-in compression using either the V.42bis standard or MNP level 5.

Table 33.4 Example AT modem commands

ITU recommendation	*Bit rate (bps)*	*Modulation*
V.21	300	FSK
V.22	1 200	PSK
V.22bis	2 400	ASK/PSK
V.27ter	4 800	PSK
V.29	9 600	PSK
V.32	9 600	ASK/PSK
V.32bis	14 400	ASK/PSK
V.34	28 800	ASK/PSK

33.8.1 V.42bis and MNP compression

There are two main standards used in modems for compression. The V.42bis standard is defined by the ITU and the MNP (Microcom Networking Protocol) has been developed by a company named Microcom. Most modems will try to compress using V.42bis but if this fails they will try MNP level 5. V.42bis uses the Lempel-Ziv algorithm which builds dictionaries of code words for recurring characters in the data stream. These codes words normally take up fewer bits than the uncoded bits. V.42bis is associated with the V.42 standard which covers error correction.

33.8.2 V.22bis modems

V.22bis modems allow transmission at up to 2400 bps. It uses four amplitudes and four phases. Figure 33.8 shows the 16 combinations of phase and amplitude for a V.22bis modem. It can be seen that there are 12 different phase shifts and 4 different amplitudes. Each transmission is known as a symbol, thus each transmitted symbol contains four bits. The transmission rate for a symbol is 600 symbols per second (or 600 Baud) thus the bit rate will be 2 400 bps.

Trellis coding tries to ensure that consecutive symbols differ as much as possible.

33.8.3 V.32 modems

V.32 modems include echo cancellation which allows signals to be transmitted in both directions at the same time. Previous modems used different frequencies to transmit on different channels. Basically echo cancellation uses DSP (digital signal processing) to subtract the sending signal from the received signal.

V.32 modems use trellis encoding to enhance error detection and correction. They encode 32 signalling combinations of amplitude and phase. Each of the symbols contains four data bits and a single trellis bit. The basic symbol rate is 2400 bps thus the actual data rate will be 9600 bps. A V.32bis modem uses 7 bits per symbol thus the data rate will be 14 400 bps (2400 × 6).

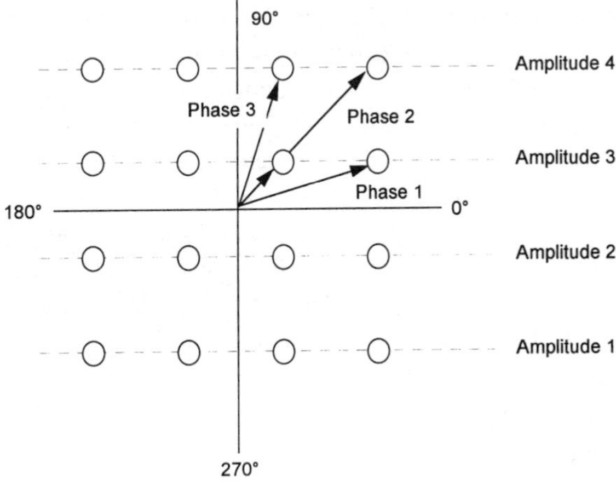

Figure 33.8 Phase and amplitude coding for V.32

33.9 Fax transmission

Facsimile (fax) transmission involves the transmission of images over a telephone line using a modem. A standalone fax basically consists of:

- An image scanner.
- A graphics printer (normally a thermal printer).
- A transmission/reception modem.

The fax scans an A4 image with 1142 scan lines (3.85 lines per millimeter) and 1728 pixels per line. The EIA and ITU originally produced the RS-328 standard for the transmission of analogue voltage levels to represent different brightnesses. The ITU recommendations are known as Group I and Group II standards. The Group III standard defines the transmission of faxes using digital transmission with 1142 × 1728 pixels of black or white. Group IV is an extension to Group III but allows different gray scales and also colour (unfortunately it requires a high bit rate, such as ISDN).

An A4 scan would consist of 1 976 832 (1142 × 1728) scanned elements. If each element is scanned for black and white, then, at 9600 bps, it would take over 205 seconds (2 minutes and 25 seconds) to transmit. This transmission time can be drastically reduced by using RLE coding.

33.9.1 Modified Huffman coding

Group III compression uses modified Huffman code to compress the transmitted bitstream. It uses a table of codes in which the most frequent run lengths are coded with a short code. Typically documents contain long runs of white or black. A compression ratio of over 10:1 is easily achievable (thus a single-page document can be sent in under 20 seconds, for a 9600 bps transmission rate). Table 33.5 shows some code runs of white and Table 33.6 shows some codes for runs of black. The transmitted code always starts on white code. The codes range from 0 to 63. Values from 64 to 2560 use two codes. The first gives the multiple of 64 followed by the normally coded remainder.

For example, if the data to be encoded is:

16 white, 4 black, 16 white, 2 black, 63 white, 10 black, 63 white

it would be coded as:

```
101010  011 101010  11 00110100  0000100  00110100
```

This would take 40 bits to transmit the coding, whereas it would take 304 bits. (i.e. 16+4+16+2+128+10+128). This results in a compression ratio of 7.6:1.

Table 33.5 White run-length coding

Run length	*Coding*	*Run length*	*Coding*	*Run length*	*Coding*
0	00110101	1	000111	2	0111
3	1000	4	1011	5	1100
6	1110	7	1111	8	10011
9	10100	10	00111	11	01000
12	001000	13	000011	14	110100
15	110101	16	101010	17	101011
18	0100111	19	0001100	61	00110010
62	00110011	63	00110100	EOL	00000000001

Table 33.6 Black run-length coding

Run length	*Coding*	*Run length*	*Coding*	*Run length*	*Coding*
0	0000110111	1	010	2	11
3	10	4	011	5	0011
6	0010	7	00011	8	000101
9	000100	10	0000100	11	0000101
12	0000111	13	00000100	14	00000111
15	000011000	16	0000010111	17	0000011000
18	0000001000	19	00001100111	61	000001011010
62	0000001100110	63	000001100111	EOL	00000000001

33.10 Exercises

33.1 Which modem indicators would be ON when a modem has made a connection and is receiving data? Which indicators would be flashing?

33.2 Which modem indicators would be ON when a modem has made a connection and is sending data? Which indicators will be flashing?

33.3 Investigate the complete set of AT commands by referring to a modem manual or reference book.

33.4 Investigate the complete set of S-registers by referring to a modem manual or reference book.

33.5 Determine the location of modems on a network or in a works building. If possible, determine the type of data being transferred and its speed.

33.6 Connect a modem to a computer and dial a remote modem.

33.7 Find a PC with Microsoft Windows and run the `Terminal` program (normally found in the Accessories group). Determine the following:

(a) the default RS-232 settings, such as baud rate, the parity, flow control, and so on (select Communications... from the Settings menu).

(b) the hang-up command for the modem (select Modem Command... from the Settings menu).

(c) the dial-up command for the modem (select Modem Com mand... from the Settings menu).

33.8 If possible connect two modems together and, using a program such as `Terminal`, transfer text from one computer to the another.

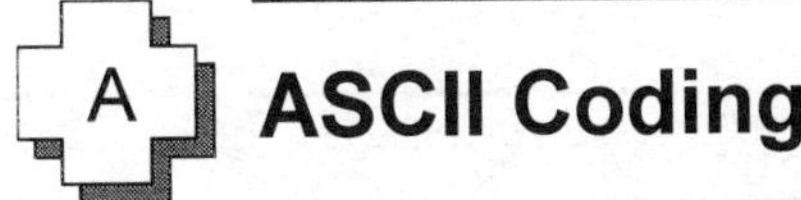

ASCII Coding

A.1 International alphabet No. 5

ANSI defined a standard alphabet known as ASCII. This has since been adopted by the CCITT as a standard, known as IA5 (International Alphabet No. 5). The following tables define this alphabet in binary, as a decimal value, as a hexadecimal value and as a character.

Binary	*Decimal*	*Hex*	*Character*	*Binary*	*Decimal*	*Hex*	*Character*
00000000	0	00	NUL	00010000	16	10	DLE
00000001	1	01	SOH	00010001	17	11	DC1
00000010	2	02	STX	00010010	18	12	DC2
00000011	3	03	ETX	00010011	19	13	DC3
00000100	4	04	EOT	00010100	20	14	DC4
00000101	5	05	ENQ	00010101	21	15	NAK
00000110	6	06	ACK	00010110	22	16	SYN
00000111	7	07	BEL	00010111	23	17	ETB
00001000	8	08	BS	00011000	24	18	CAN
00001001	9	09	HT	00011001	25	19	EM
00001010	10	0A	LF	00011010	26	1A	SUB
00001011	11	0B	VT	00011011	27	1B	ESC
00001100	12	0C	FF	00011100	28	1C	FS
00001101	13	0D	CR	00011101	29	1D	GS
00001110	14	0E	SO	00011110	30	1E	RS
00001111	15	0F	SI	00011111	31	1F	US

Binary	*Decimal*	*Hex*	*Character*	*Binary*	*Decimal*	*Hex*	*Character*
00100000	32	20	SPACE	00110000	48	30	0
00100001	33	21	!	00110001	49	31	1
00100010	34	22	"	00110010	50	32	2
00100011	35	23	#	00110011	51	33	3
00100100	36	24	$	00110100	52	34	4
00100101	37	25	%	00110101	53	35	5
00100110	38	26	&	00110110	54	36	6
00100111	39	27	/	00110111	55	37	7
00101000	40	28	(	00111000	56	38	8
00101001	41	29	)	00111001	57	39	9
00101010	42	2A	*	00111010	58	3A	:
00101011	43	2B	+	00111011	59	3B	;
00101100	44	2C	,	00111100	60	3C	<
00101101	45	2D	-	00111101	61	3D	=
00101110	46	2E	.	00111110	62	3E	>
00101111	47	2F	/	00111111	63	3F	?

Binary	*Decimal*	*Hex*	*Character*	*Binary*	*Decimal*	*Hex*	*Character*
01000000	64	40	@	01010000	80	50	P
01000001	65	41	A	01010001	81	51	Q
01000010	66	42	B	01010010	82	52	R
01000011	67	43	C	01010011	83	53	S
01000100	68	44	D	01010100	84	54	T
01000101	69	45	E	01010101	85	55	U
01000110	70	46	F	01010110	86	56	V
01000111	71	47	G	01010111	87	57	W
01001000	72	48	H	01011000	88	58	X
01001001	73	49	I	01011001	89	59	Y
01001010	74	4A	J	01011010	90	5A	Z
01001011	75	4B	K	01011011	91	5B	[
01001100	76	4C	L	01011100	92	5C	\
01001101	77	4D	M	01011101	93	5D	]
01001110	78	4E	N	01011110	94	5E	`
01001111	79	4F	O	01011111	95	5F	_

Binary	*Decimal*	*Hex*	*Character*	*Binary*	*Decimal*	*Hex*	*Character*
01100000	96	60		01110000	112	70	p
01100001	97	61	a	01110001	113	71	q
01100010	98	62	b	01110010	114	72	r
01100011	99	63	c	01110011	115	73	s
01100100	100	64	d	01110100	116	74	t
01100101	101	65	e	01110101	117	75	u
01100110	102	66	f	01110110	118	76	v
01100111	103	67	g	01110111	119	77	w
01101000	104	68	h	01111000	120	78	x
01101001	105	69	i	01111001	121	79	y
01101010	106	6A	j	01111010	122	7A	z
01101011	107	6B	k	01111011	123	7B	{
01101100	108	6C	l	01111100	124	7C	:
01101101	109	6D	m	01111101	125	7D	}
01101110	110	6E	n	01111110	126	7E	~
01101111	111	6F	o	01111111	127	7F	DEL

A.2 Extended ASCII code

The standard ASCII character has 7 bits and the basic set ranges from 0 to 127. This code is rather limited as it does not contains symbols such as Greek letters, lines, and so on. For this purpose the extended ASCII code has been defined. This fits into character numbers 128 to 255. The following four tables define a typical extended ASCII character set.

Binary	*Decimal*	*Hex*	*Character*	*Binary*	*Decimal*	*Hex*	*Character*
10000000	128	80	Ç	10010000	144	90	É
10000001	129	81	ü	10010001	145	91	æ
10000010	130	82	é	10010010	146	92	Æ
10000011	131	83	â	10010011	147	93	ô
10000100	132	84	ä	10010100	148	94	ö
10000101	133	85	à	10010101	149	95	ò
10000110	134	86	å	10010110	150	96	û
10000111	135	87	ç	10010111	151	97	ù
10001000	136	88	ê	10011000	152	98	ÿ
10001001	137	89	ë	10011001	153	99	Ö
10001010	138	8A	è	10011010	154	9A	Ü
10001011	139	8B	ï	10011011	155	9B	¢
10001100	140	8C	î	10011100	156	9C	£
10001101	141	8D	ì	10011101	157	9D	¥
10001110	142	8E	Ä	10011110	158	9E	₧
10001111	143	8F	Å	10011111	159	9F	ƒ

Binary	*Decimal*	*Hex*	*Character*	*Binary*	*Decimal*	*Hex*	*Character*
10100000	160	A0	á	10110000	176	B0	░
10100001	161	A1	í	10110001	177	B1	▒
10100010	162	A2	ó	10110010	178	B2	▓
10100011	163	A3	ú	10110011	179	B3	│
10100100	164	A4	ñ	10110100	180	B4	┤
10100101	165	A5	Ñ	10110101	181	B5	╡
10100110	166	A6	ª	10110110	182	B6	╢
10100111	167	A7	º	10110111	183	B7	╖
10101000	168	A8	¿	10111000	184	B8	╕
10101001	169	A9	⌐	10111001	185	B9	╣
10101010	170	AA	¬	10111010	186	BA	║
10101011	171	AB	½	10111011	187	BB	╗
10101100	172	AC	¼	10111100	188	BC	╝
10101101	173	AD	¡	10111101	189	BD	╜
10101110	174	AE	«	10111110	190	BE	╛
10101111	175	AF	»	10111111	191	BF	┐

Binary	*Decimal*	*Hex*	*Character*	*Binary*	*Decimal*	*Hex*	*Character*
11000000	192	C0	└	11010000	208	D0	╨
11000001	193	C1	┴	11010001	209	D1	╤
11000010	194	C2	┬	11010010	210	D2	╥
11000011	195	C3	├	11010011	211	D3	╙
11000100	196	C4	─	11010100	212	D4	╘
11000101	197	C5	┼	11010101	213	D5	╒
11000110	198	C6	╞	11010110	214	D6	╓
11000111	199	C7	╟	11010111	215	D7	╫
11001000	200	C8	╚	11011000	216	D8	╪
11001001	201	C9	╔	11011001	217	D9	┘
11001010	202	CA	╩	11011010	218	DA	┌
11001011	203	CB	╦	11011011	219	DB	█
11001100	204	CC	╠	11011100	220	DC	▄
11001101	205	CD	═	11011101	221	DD	▌
11001110	206	CE	╬	11011110	222	DE	▐
11001111	207	CF	╧	11011111	223	DF	▀

Binary	*Decimal*	*Hex*	*Character*	*Binary*	*Decimal*	*Hex*	*Character*
11100000	224	E0	α	11110000	240	F0	Ξ
11100001	225	E1	ß	11110001	241	F1	±
11100010	226	E2	Γ	11110010	242	F2	≥
11100011	227	E3	π	11110011	243	F3	≤
11100100	228	E4	Σ	11110100	244	F4	⌠
11100101	229	E5	σ	11110101	245	F5	⌡
11100110	230	E6	µ	11110110	246	F6	÷
11100111	231	E7	τ	11110111	247	F7	≈
11101000	232	E8	Φ	11111000	248	F8	°
11101001	233	E9	Θ	11111001	249	F9	•
11101010	234	EA	Ω	11111010	250	FA	·
11101011	235	EB	δ	11111011	251	FB	√
11101100	236	EC	φ	11111100	252	FC	^{n}
11101101	237	ED	ϕ	11111101	253	FD	^{2}
11101110	238	EE	E	11111110	254	FE	■
11101111	239	EF	Λ	11111111	255	FF	□

A.3 RS-232C interface

Table A.1 RS-232C connections

9-pin D-type	*25-pin D-type*	*Name*	*RS-232 name*	*Description*	*Signal direction on DCE*
	1		AA	Protective GND	
3	2	TXD	BA	Transmit Data	IN
2	3	RXD	BB	Receive Data	OUT
7	4	RTS	CA	Request to Send	IN
8	5	CTS	CB	Clear to Send	OUT
6	6	DSR	CC	Data Set Ready	OUT
5	7	GND	AB	Signal GND	
1	8	DCD	CF	Received Line Signal detect	OUT
	9		-	RESERVED	-
	10		-	RESERVED	-
	11			UNASSIGNED	-
	12		SCF	Secondary Received Line Signal Detector	OUT
	13		SCB	Secondary Clear to Send	OUT
	14		SBA	Secondary Transmitted Data	IN
	15		DB	Transmission Signal Element Detector	OUT
	16		SBB	Secondary Received Data	OUT
	17		DD	Receiver Signal Element Time	OUT
	18			UNASSIGNED	-
	19		SCA	Secondary Request to Send	IN
4	20	DTR	CD	Data Terminal Ready	IN
	21		CG	Signal Quality Detector	OUT
9	22	RI	CE	Ring Indicator	OUT
	23		CH/CI	Data Signal Rate Selector	IN/OUT
	24		DA	Transmit Signal Element Timing	IN
	25			UNASSIGNED	-

A.4 RS-449 interface

RS-449 defines a standard for the functional/mechanical interface of DTEs/DCEs for serial communications and is usually used with synchronous transmissions. Table A.2 lists the main connections.

Table A.2 RS-449 connections.

Pin number	*Mnemonic*	*Description*
1		Shield
2	SI	Signalling Rate Indicator
3,21		Spare
4,22	SD	Sending Time
5,23	ST	Receive Data
6,24	RD	Receive Data
7,25	RS	Request to Send
8,26	RT	Receive Timing
9,27	CS	Clear to Send
10	LL	Local Loopback
11,29	DM	Data Mode
12,30	TR	Terminal Ready
13,31	RR	Receiver Ready
14	RL	Remote Loopback
15	IC	Incoming Call
16	SF/SR	Select Frequency/ Signalling Rate Select
17,37	TT	Terminal Timing
18	TM	Test Mode
19	SG	Signal Ground
20	RC	Receive Common
28	IS	Terminal in Service
32	SS	Select Standby
33	SQ	Signal Quality
34	NS	New Signal
36	SB	Standby Indicator
37	SC	Send Common

Cable Specifications

B.1 Introduction

The cable type used to transmit a signal depends on several parameters, including:

- The signal bandwidth.
- The reliability of the cable.
- The maximum length between nodes.
- The possibility of electrical hazards.
- Power loss in the cables.
- Tolerance to harsh conditions.
- Expense and general availability of the cable.
- Ease of connection and maintenance.
- Ease of running cables, and so on.

The main types of networking cables are twisted-pair, coaxial and fiber-optic. Twisted-pair and coaxial cables transmit electric signals, whereas fiber-optic cables transmit light pulses. Twisted-pair cables are not shielded and thus interfere with nearby cables. Public telephone lines generally use twisted-pair cables. In LANs they are generally used up to bit rates of 10 Mbps and with maximum lengths of 100 m.

Coaxial cable has a grounded metal sheath around the signal conductor. This limits the amount of interference between cables and thus allows higher data rates. Typically they are used at bit rates of 100 Mbps for maximum lengths of 1 km.

The highest specification of the three cables is fiber-optic. This type of cable allows extremely high bit rates over long distances. Fiber-optic cables do not interfere with nearby cables and give greater security, give more protection from electrical damage by external equipment and greater resistance to harsh environments; they are also safer in hazardous environments.

B.1.1 Cable characteristics

The main characteristics of cables used in video communication are attenuation, crosstalk and characteristic impedance. Attenuation defines the reduction in the signal strength at a given frequency for a defined distance. It is normally defined in dB/100 m, which is the attenuation (in dB) for 100 m. An attenuation of 3 dB/100 m gives a signal voltage reduction of 0.5 for every 100 m. Table B.1 lists some attenuation rates and equivalent voltage ratios; they are illustrated in Figure B.1.

Table B.1 Attenuation rates as a ratio

dB	*Ratio*	*dB*	*Ratio*	*dB*	*Ratio*
0	1.000	10	0.316	60	0.001
1	0.891	15	0.178	65	0.000 6
2	0.794	20	0.100	70	0.000 3
3	0.708	25	0.056	75	0.000 2
4	0.631	30	0.032	80	0.000 1
5	0.562	35	0.018	85	0.000 06
6	0.501	40	0.010	90	0.000 03
7	0.447	45	0.005 6	95	0.000 02
8	0.398	50	0.003 2	100	0.000 01
9	0.355	55	0.001 8		

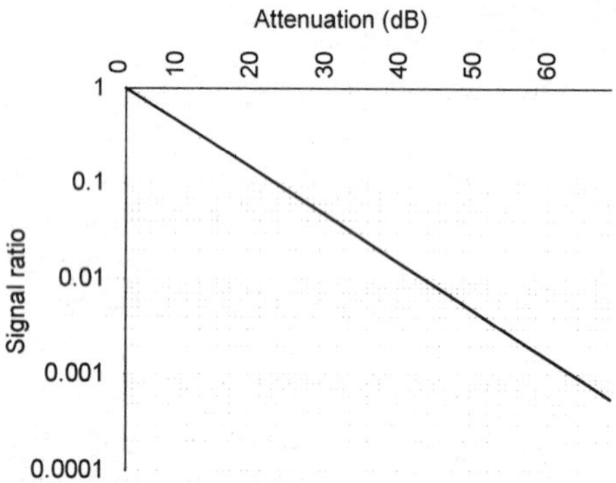

Figure B.1 Signal ratio related to attenuation

The characteristic impedance of a cable and its connectors are important as all parts of the transmission system need to be matched to the same impedance. This impedance is normally classified as the characteristic impedance of the cable. Any differences in the matching result in a reduction of signal power and also produce signal reflections (or ghosting).

Crosstalk is important as it defines the amount of signal that crosses from one signal path to another. This cause some of the transmitted signal to be received back where it was transmitted.

Capacitance (pF/100 m) defines the amount of distortion in the signal caused by each signal pair. The lower the capacitance value, the lower the distortion.

The main types of cable used in networking and data communications are:

- Coaxial cable – cables with an inner core and a conducting shield having a characteristic impedance of either 75 Ω for TV signal or 50 Ω for other types.
- Cat-3 UTP cable – level 3 cables have non-twisted-pair cores with a characteristic impedance of 100 Ω (±15 Ω)and a capacitance of 59 pF/m. Conductor resistance is around 9.2 Ω/100 m.
- Cat-4 UTP cable – level 4 cables have twisted-pair cores with a characteristic impedance of 100 Ω (±15 Ω) and a capacitance of 49.2 pF/m. Conductor resistance is around 9 Ω/100 m.
- Cat-5 UTP cable – level 5 cables have twisted-pair cores with a characteristic impedance of 100 Ω (±15 Ω) and a capacitance of 45.9 pF/m. Conductor resistance is around 9 Ω/100 m.

The Electrical Industries Association (EIA) has defined five main types of cables. Levels 1 and 2 are used for voice and low-speed communications (up to 4 Mbps). Level 3 is designed for LAN data transmission up to 16 Mbps and level 4 is designed for speeds up to 20 Mbps. Level 5 cables, have the highest specification of the UTP cables and allow data speeds of up to 100 Mbps. The main EIA specification on these types of cables is EIA/TIA568 and the ISO standard is ISO/IEC11801.

Coaxial cables have an inner core separated from an outer shield by a dielectric. They have an accurate characteristic impedance (which reduces reflections), and because they are shielded they have very low crosstalk levels.

UTPs (unshielded twisted-pair cables) have either solid cores (for long cable runs) or are stranded patch cables (for shorts run, such as connecting to workstations, patch panels, and so on). Solid cables should not be flexed, bent or twisted repeatedly, whereas stranded cable can be flexed without damaging the cable. Coaxial cables use BNC connectors while UTP cables use either the RJ-11 (small connector which is used to connect the handset to the telephone) or the RJ-45 (larger connector which is used to connect LAN networks to a hub).

Table B.2 and Figure B.2 show typical attenuation rates (dB/100 m) for the Cat-3, Cat-4 and Cat-5 cables. Notice that the attenuation rates for Cat-4 and Cat-5 are approximately the same. These two types of cable have lower attenuation rates than equivalent Cat-3 cables. Notice that the attenuation of the cable increases as the frequency increases. This is due to several factors, such as the skin effect, where the electrical current in the conductors becomes concentrated around the outside of the conductor, and the fact that the insulation (or dielectric) between the conductors actual starts to conduct as the frequency increases.

The Cat-3 cable produces considerable attenuation over a distance of 100 m. The table shows that the signal ratio of the output to the input at 1 MHz, will be 0.76 (2.39 dB), then, at 4 MHz it is 0.55 (5.24 dB), until at 16 MHz it is 0.26. This differing attenuation at different frequencies produces not just a reduction in the signal strength but also distorts the signal (because each frequency is affected differently by the cable. Cat-4 and Cat-5 cables also produce distortion but their effects will be lessened because attenuation characteristics have flatter shapes.

Coaxial cables tend to have very low attenuation, such as 1.2 dB at 4 MHz. They also have a relatively flat response and virtually no crosstalk (due to the physical structure of the cables and the presence of a grounded outer sheath).

Table B.3 and Figure B.3 show typical near end crosstalk rates (dB/100 m) for Cat-3, Cat-4 and Cat-5 cables. The higher the figure, the smaller the crosstalk. Notice that Cat-3 cables have the most crosstalk and Cat-5 have the least, for any given frequency. Notice also that the crosstalk increases as the frequency of the signal increases. Thus high-frequency signals have more crosstalk than lower-frequency signals.

Table B.2 Attenuation rates (dB/100 m) for Cat-3, Cat-4 and Cat-5 cable

	Attenuation rate (dB/100m)		
Frequency (MHz)	*Cat-3*	*Cat-4*	*Cat-5*
1	2.39	1.96	2.63
4	5.24	3.93	4.26
10	8.85	6.56	6.56
16	11.8	8.2	8.2

Table B.3 Near-end crosstalk (dB/100 m) for Cat-3, Cat-4 and Cat-5 cable

	Near end crosstalk (dB/100m)		
Frequency (MHz)	*Cat-3*	*Cat-4*	*Cat-5*
1	13.45	18.36	21.65
4	10.49	15.41	18.04
10	8.52	13.45	15.41
16	7.54	12.46	14.17

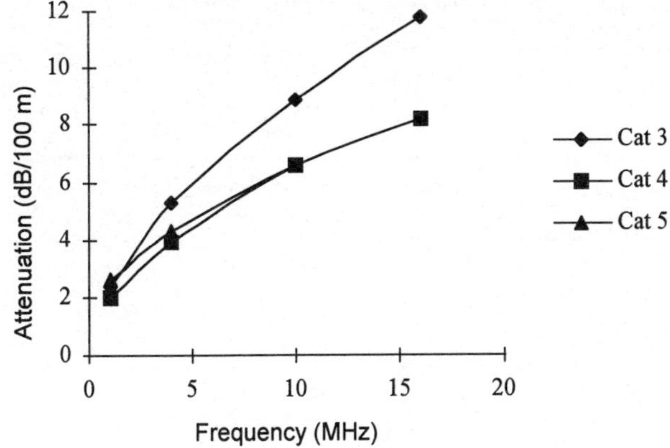

Figure B.2 Attenuation characteristics for Cat-3, Cat-4 and Cat-5 cables

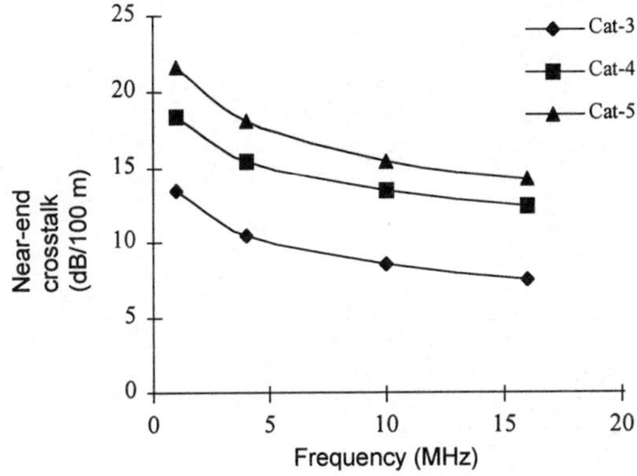

Figure B.3 Near-end crosstalk characteristics for Cat-3, Cat-4 and Cat-5 cables

C RLE Program

C.1 RLE program

Program C.1 is a very simple program which scans a file `IN.DAT` and, using RLE, stores to a file `OUT.DAT`. The special character sequence is:

ZZ*cxx*

where ZZ is the flag sequence, *c* is the repeditive character and *xx* the number of times the character occurs. The ZZ flag sequence is choosen because, in a text file, it is unlikely to occur within the file. File listing C.1 shows a sample `IN.DAT` and File listing C.2 shows the RLE encoded file (`OUT.DAT`).

Program C.1

```
/* ENCODE.C        */
#include <stdio.h>

int   main(void)
{
FILE *in,*out;
char previous,current;
int      count;

   if ((in=fopen("in.dat","r"))==NULL)
   {
      printf("Cannot open <in.dat>");
      return(1);
   }
   if ((out=fopen("out.dat","w"))==NULL)
   {
      printf("Cannot open <out.dat>");
      return(1);
   }

   do
   {
      count=1;
      previous=current;
      current=fgetc(in);
      do
      {
```

```
            previous=current;
            current=fgetc(in);
            if (previous!=current) ungetc(current,in);
            else count++;
         } while (previous==current);

         if (count>1)
              printf(out,"ZZ%c%02d",previous,count);
         else fprintf(out,"%c",previous);
      } while (!feof(in));

      fclose(in);
      fclose(out);
      return(0);
}
```

File list C.1

```
The     bbbbbbboy stood onnnnn the burning
deck         and still did.
1.000000000
3.000000010
5.000000000
```

File list C.2

```
TheZZ 05ZZb07oy stZZo02d oZZn05 the burning
deckZZ 09and stiZZl02 did.
1.ZZ009
3.ZZ00710
5.ZZ009
```

Program C.2 gives a simple C program which unencodes the RLE file produced by the previous program.

Program C.2

```
/* UNENCODE.C    */
#include <stdio.h>

int  main(void)
{
FILE *in,*out;
char ch;
int  count,i;

   if ((in=fopen("out.dat","r"))==NULL)
   {
      printf("Cannot open <out.dat>");
      return(1);
   }
   if ((out=fopen("in1.dat","w"))==NULL)
   {
      printf("Cannot open <in1.dat>");
      return(1);
   }
```

```
   do
   {
      ch=fgetc(in);

      if (ch=='Z')
      {
         ch=fgetc(in);
         if (ch=='Z')
         {
            fscanf(in,"%c%02d",&ch,&count);
            for (i=0;i<count;i++)
               fprintf(out,"%c",ch);
         }
         else ungetc(ch,in);
      }
      else fprintf(out,"%c",ch);

   } while (!feof(in));
   fclose(in);
   fclose(out);
   return(0);
}
```

The `ZZ` flag sequence is inefficient as it uses two characters to store the flag; a better flag could be an 8-bit character that cannot occur, such as `11111111b` or `ffh`. Program C.3 shows an example of this and Program C.4 shows the decoder.

Program C.3

```
#include <stdio.h>

#define FLAG 0xff /* 1111 1111b*/

int  main(void)
{
FILE *in,*out;
char previous,current;
int  count;


 ;;; ;;;;;

      if (count>1)
         fprintf(out,"%c%c%02d",FLAG,previous,count);
      else fprintf(out,"%c",previous);
   } while (!feof(in));

   fclose(in);
   fclose(out);
   return(0);
}
```

Program C.4

```
/* UNENCODE.C    */
#include <stdio.h>

#define FLAG 0xff /* 1111 1111b*/

int  main(void)
{
FILE *in,*out;
char ch;
int  count,i;

   ;;; ;;;;
   ;;; ;;;;
   do
   {
      ch=fgetc(in);

      if (ch==FLAG)
      {
         ch=fgetc(in);
         fscanf(in,"%c%02d",&ch,&count);
         for (i=0;i<count;i++)
                fprintf(out,"%c",ch);
      }
      else fprintf(out,"%c",ch);

   } while (!feof(in));
   fclose(in);
   fclose(out);
   return(0);
}
```

In a binary file any bit sequence can occur. To overcome this, a flag sequence, such as 10101010 can be used to identify the flag. If this sequence occurs within the data, it will be coded with two flags two consecutive flags in the data are coded with three flags and so on. For example:

```
00000000 10101010 10101010 00011100 01001100
```

would be encoded as:

```
00000000 10101010 10101010 10101010 00011100 01001100
```

thus when the three flags are detected, one of them is deleted.

SNR for PCM

D.1 SNR

If a waveform has a maximum signal amplitude of V, then the relative signal power will be:

$$\text{Signal power} = v_{rms}^2 = \left(\frac{V}{\sqrt{2}}\right)^2 = \frac{V^2}{2}$$

If n-bit PCM coding is used then there will be 2^n different levels, as illustrated in Figure D.1.

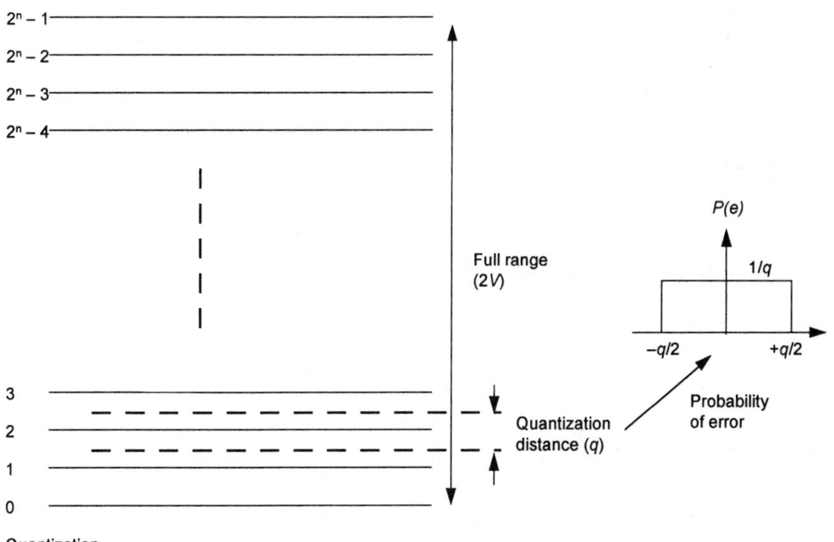

Figure D.1 Quantization

Thus, if the input signal ranges between $+V$ and $-V$, the error in the quantization signal will range from:

$$+\frac{q}{2} \quad \text{to} \quad -\frac{q}{2}$$

where q is the quantization distance and is given by:

$$q = \frac{2V}{2^n}$$

Figure D.1 shows that the probability of error will be constant from between $-q/2$ to $+q/2$. The area over the interval the $P(e)$ should equal unity, hence the y-axis value for $P(e)$ will be $1/q$. Thus the noise power will be:

$$\text{Noise power} = \int_{-\infty}^{+\infty} v^2 P(v)dv = 2\int_{0}^{+q/2} \frac{v^2}{q} dv$$

$$= \frac{2}{q}\int_{0}^{+q/2} v^2 dv = \frac{2}{q}\left[\frac{v^3}{3}\right]_0^{+q/2} = \frac{2}{3q}\left(\frac{q^3}{2^3}\right)$$

$$= \frac{q^2}{12}$$

The signal-to-noise ratio will thus be:

$$\text{SNR(dB)} = 10\log_{10}\left(\frac{\text{Signal power}}{\text{Noise power}}\right) = 10\log_{10}\left(\frac{V^2/2}{q^2/12}\right) = 10\log_{10}\left(\frac{6V^2}{q^2}\right)$$

$$= 10\log_{10}\left(\frac{6V^2}{(2V)^2/(2^n)^2}\right) = 10\log_{10}\left(\frac{3}{2}(2^n)^2\right)$$

$$= 10\log_{10}(1.5) + 20\log_{10}(2^n) = 10\log_{10}(1.5) + 20\log_{10}(2^n)$$

$$= 1.76 + 20\,n\log_{10}(2)$$

$$= 6.02\,n + 1.76$$

E RFC Standards

The IAB (Internet Advisor Board) has published many documents on the TCP/IP protocol family. They are known as RFC (request for comment) and can be obtained using FTP from the Internet Network Information Center (NIC) at `nic.ddn.mil`, or one of several other FTP sites. The main RFC documents are:

RFC768	User Datagram Protocol
RFC775	Directory-Oriented FTP Commands
RFC781	Specification of the Internet Protocol Timestamp Option
RFC783	TFTP Protocol
RFC786	User Datagram Protocol (UDP)
RFC791	Internet Protocol(IP)
RFC792	Internet Control Message Protocol(ICMP)
RFC793	Transmission Control Protocol(TCP)
RFC799	Internet Name Domains
RFC813	Window and Acknowledgment in TCP
RFC815	IP Datagram Reassembly Algorithms
RFC821	Simple Mail-Transfer Protocol(SMTP)
RFC822	Standard for the Format of ARPA Internet Text Messages
RFC823	DARPA Internet Gateway
RFC827	Exterior Gateway Protocol (EGP)
RFC877	Standard for the Transmission of IP Datagrams over Public Data Networks
RFC879	TCP Maximum Segment Size and Related Topics
RFC886	Proposed Standard for Message Header Munging
RFC893	Trailer Encapsulations
RFC894	Standard for the Transmission of IP Datagrams over Ethernet Networks
RFC895	Standard for the Transmission of IP Datagrams over Experimental Ethernet Networks
RFC896	Congestion Control in TCP/IP Internetworks
RFC903	Reverse Address Resolution Protocol
RFC904	Exterior Gateway Protocol Formal Specifications
RFC906	Bootstrap Loading Using TFTP
RFC919	Broadcast Internet Datagram
RFC920	Domain Requirements
RFC932	Subnetwork Addressing Schema
RFC949	FTP Unique-Named Store Command
RFC950	Internet Standard Subnetting Procedure
RFC951	Bootstrap Protocol
RFC959	File Transfer Protocol
RFC974	Mail Routing and the Domain System

RFC980	Protocol Document Order Information
RFC1009	Requirements for Internet Gateways
RFC1011	Official Internet Protocol
RFC1013	X Windows System Protocol
RFC1014	XDR: External Data Representation Standard
RFC1027	Using ARP to Implement Transparent Subnet Gateways
RFC1032	Domain Administrators Guide
RFC1033	Domain Administrators Operation Guide
RFC1034	Domain Names - Concepts and Facilities
RFC1035	Domain Names - Implementation and Specifications
RFC1041	Telnet 3270 Regime Option
RFC1042	Standard for the Transmission of IP Datagrams over IEEE 802 Networks
RFC1043	Telnet Data Entry Terminal Option
RFC1044	Internet Protocol on Network System's HYPERchannel
RFC1053	Telnet X.3 PAD Option
RFC1055	Nonstandard for Transmission of IP Datagrams over Serial Lines
RFC1056	PCMAIL: A Distributed Mail System for Personal Computers
RFC1058	Routing Information Protocol
RFC1068	Background File Transfer Program (BFTP)
RFC1072	TCP Extensions of Long-Delay Paths
RFC1073	Telnet Window Size Option
RFC1074	NSFNET Backbone SPF-based Interior Gateway Protocol
RFC1079	Telnet Terminal Speed Option
RFC1080	Telnet Remote Flow Control Option
RFC1084	BOOTP Vendor Information Extensions
RFC1088	Standard for the Transmission of IP Datagrams over NetBIOS Network
RFC1089	SNMP over Ethernet
RFC1091	Telnet Terminal-Type Option
RFC1094	NFS: Network File System Protocol Specification
RFC1101	DNS Encoding of Network Names and Other Types
RFC1102	Policy Routing in Internet Protocols
RFC1104	Models of Policy-Based Routing
RFC1112	Host Extension for IP Multicasting
RFC1122	Requirement for Internet Hosts - Communication Layers
RFC1123	Requirement for Internet Hosts - Application and Support
RFC1124	Policy Issues in Interconnecting Networks
RFC1125	Policy Requirements for Inter-Administrative Domain Routing
RFC1127	Perspective on the Host Requirements RFC
RFC1129	Internet Time Protocol
RFC1143	Q Method of Implementing Telnet Option Negotiation
RFC1147	FYI on a Network Management Tool Catalog
RFC1149	Standard for the Transmission of IP Datagrams over Avian Carriers
RFC1155	Structure and Identification of Management Information for TCP/IP-Based Internets
RFC1156	Management Information Base for Network Management of TCP/IP-Based Internets
RFC1157	Simple Network Management Protocol (SNMP)
RFC1163	Border Gateway Protocol (BGP)
RFC1164	Application of the Border Gateway Protocol in the Internet
RFC1166	Internet Numbers
RFC1171	Point-to-Point Protocol for the Transmission of Multi-Protocol Datagrams
RFC1172	Point-to-Point Protocol Initial Configuration Options
RFC1173	Responsibilities of Host and Network Managers

RFC1175 FYI on Where to Start: A Bibliography of Internetworking Information
RFC1178 Choosing a Name For Your Computer
RFC1179 Line Printer Daemon Protocol
RFC1184 Telnet Linemode Option
RFC1187 Bulk Table Retrieval with the SNMP
RFC1188 Proposed Standard for the Transmission of TP Datagrams over FDDI Networks
RFC1195 Use of OSI IS-IS for Routing in TCP/IP and Dual Environments
RFC1196 Finger User Information Protocol
RFC1198 FYI on the X Windows System
RFC1201 Transmitting IP Traffic over ARCNET Networks
RFC1205 520 Telnet Interface
RFC1208 Glossary of Networking Terms
RFC1209 Transmission of IP Datagrams over the SMDS Service
RFC1212 Concise MIB Definitions
RFC1213 MIB for Network Management of TCP/IP-Based Internets
RFC1214 OSI Internet Management: Management Information Base
RFC1215 Convention for Defining Traps for Use with the SNMP
RFC1219 On the Assignment of Subnet Numbers
RFC1220 Point-to-Point Protocol Extensions for Bridges
RFC1224 Techniques for Managing Asynchronous Generated Alerts
RFC1227 SNMP MUX Protocol and MIB
RFC1228 SNMP-DPI: Simple Network Management Protocol Distributed Program Interface
RFC1229 Extensions to the Generic-interface MIB
RFC1230 IEEE 802.4 Token Bus MIB
RFC1231 IEEE 802.5 Token Ring MIB
RFC1232 Definitions of Managed Objects for the DS1 Interface Type
RFC1233 Definitions of Managed Objects for the DS3 Interface Type
RFC1236 IP to X.121 Address Mapping for DDN IP
RFC1238 CLNS MIB for Use with Connectionless Network Protocol
RFC1239 Reassignment of Experiment MIBs to Standard MIBs
RFC1243 Appletalk Management Information Base
RFC1245 OSPF Protocol Analysis
RFC1246 Experience with the OSPF Protocol
RFC1247 OSPF Version 2
RFC1253 OSPF Version 2: Management Information Base
RFC1254 Gateway Congestion Control Survey
RFC1267 A Border Gateway Protocol (BGP-3)
RFC1271 Remote Network Monitoring Management Information Base
RFC1321 The MD5 Message-Digest Algorithm
RFC1340 Assigned Numbers
RFC1341 MIME Mechanism for Specifying and Describing the Format of Internet Message Bodies
RFC1360 IAB Official Protocol Standards
RFC1522 MIME (Multipurpose Internet Mail Extensions)Part Two : Message Header Extensions for Non-ASCII Text
RFC1521 MIME (Multipurpose Internet Mail Extensions) Part One : Mechanisms for Specifying and Describing the Format of Internet Mail Message Bodies)
RFC1583 OSPF Version 2
RFC1630 Universal Resource Identifiers in WWW
RFC1738 Uniform Resource Identifiers (URL)
RFC1752 The Recommendation for the IP Next-Generation Protocol
RFC1771 A Border Gateway Protocol 4 (BGP-4)
RFC1808 Relative Uniform Resource Identifiers

RFC1809	Using the Flow Label in IPv6
RFC1825	Security Architecture for the Internet Protocol
RFC1826	IP Authentication Header
RFC1827	IP Encapsulating Security Payload (ESP)
RFC1828	IP Authentication Using Keyed MD5
RFC1829	The ESP DES-CBC Transform
RFC1883	Internet Protocol, Version 6 Specification
RFC1884	IP Version 6 Addressing Architecture
RFC1885	Internet Control Message Protocol (ICMPv6) for the Internet Protocol Version-6 (IPv6) Specification
RFC1886	DNS Extensions to Support IP Version 6
RFC1887	An Architecture for IPv6 Unicast Address Allocation
RFC1901	Introduction to Community-Based SNMPv2
RFC1902	Structure of Management Information for SNMPv2
RFC1903	Textual Conventions for SNMPv2
RFC1904	Conformance Statements for SNMPv2
RFC1905	Protocol Operations for SNMPv2
RFC1906	Transport Mappings for SNMPv2
RFC1907	Management Information Base for SNMPv2
RFC1908	Coexistence Between Version 1 and Version 2 of the Internet-Standard Network Management Framework

UNIX Network Startup Files

F.1 netnfsrc file

This appendix documents a typical netnfsrc (NFS startup file) file. In the script portion given below the NFS_CLIENT is set to a 1 if the host is set to a client (else it will be 0) and the NFS_SERVER parameter is set to a 1 if the host is set to a server (else it will be 0). Initially the NFS clients and servers are started. Note that a host can be a client, a server, both or neither.

Next the mountd daemon is started, after which the NFS daemons (nfsd) are started (only on servers). After this the biod daemon is run.

```
NFS_CLIENT=1
NFS_SERVER=1
START_MOUNTD=0
##
#       Read in /etc/exports
##
if [ $LFS -eq 0 -a $NFS_SERVER -ne 0 -a -f /etc/exports ] ; then
    > /etc/xtab
    /usr/etc/exportfs -a  && echo "    Reading in /etc/exports"
    set_return
fi

if [ $NFS_SERVER -ne 0 -a $START_MOUNTD -ne 0 -a -f /usr/etc/rpc.mountd ] ;
then
    /usr/etc/rpc.mountd && echo "starting up the mountd" && echo
"\t/usr/etc/rpc.mountd"
    set_return
fi
##
if [ $LFS -eq 0 -a $NFS_SERVER -ne 0 -a -f /etc/nfsd ] ; then
    /etc/nfsd 4 && echo "starting up the NFS daemons" && echo "\t/etc/nfsd 4"
    set_return
fi
##
if [ $NFS_CLIENT -ne 0 ] ; then
   if [ -f /etc/biod ] ; then
        /etc/biod 4 && echo
             "starting up the BIO daemons" && echo "\t/etc/biod 4"
          set_return
fi
   /bin/cat /dev/null > /etc/nfs.up
fi
```

The next part of the `netnfsrc` file deals with the NIS services. There are three states: `NIS_MASTER_SERVER`, `NIS_SLAVE_SERVER` and `NIS_CLIENT`. A host can either be a master server or a slave server, but cannot be both. All NIS servers must also be NIS clients, so the `NIS_MASTER_SERVER` or `NIS_SLAVE_SERVER` parameters shoud be set to 1. Initally the domain name is set using the command `domainname` (in this case it is `eece`).

```
NIS_MASTER_SERVER=1
NIS_SLAVE_SERVER=0
NIS_CLIENT=1
NISDOMAIN=eece
NISDOMAIN_ERR=""

if [ "$NISDOMAIN" -a -f /bin/domainname ] ; then
    echo "\t/bin/domainname $NISDOMAIN"
    /bin/domainname $NISDOMAIN
    if [ $? -ne 0 ] ; then
   echo "Error:  NIS domain name not set" >&2
   NISDOMAIN_ERR=TRUE
    fi
else
    echo "\tNIS domain name not set"
    NISDOMAIN_ERR=TRUE
fi
```

Next portmap is started for ARPA clients.

```
if [ -f /etc/portmap ] ; then
    echo "\t/etc/portmap"
    /etc/portmap
    if [ $? -ne 0 ] ; then
   echo  "Error:  NFS portmapper NOT powered up"  >&2
   exit 1
    fi
fi
```

Next the NIS is started.

```
if [ "$NISDOMAIN_ERR" -o \( $NIS_MASTER_SERVER -eq 0 -a $NIS_SLAVE_SERVER -eq
0\
   -a $NIS_CLIENT -eq 0 \) ] ; then
    echo "    Network Information Service not started."
else
    echo "    starting up the Network Information Service"

    HOSTNAME=`hostname`

    if [ $NIS_MASTER_SERVER -ne 0 -o $NIS_SLAVE_SERVER -ne 0 ]; then
   NIS_SERVER=TRUE
    fi

    if [ $NIS_MASTER_SERVER -ne 0 -a $NIS_SLAVE_SERVER -ne 0 ]; then
   echo "NOTICE:both NIS_MASTER_SERVER and NIS_SLAVE_SERVER variables set;"
   echo "\t$HOSTNAME will be only a NIS slave server."
```

```
    NIS_MASTER_SERVER=0
     fi

     if [ $NIS_CLIENT -eq 0 ]; then
    echo "NOTICE:$HOSTNAME will be a NIS server, but the NIS_CLIENT variable is"
    echo "\tnot set; $HOSTNAME will also be a NIS client."
    NIS_CLIENT=1
     fi
```

Next the yp services are started.

```
##
#  The verify_ypserv function determines if it is OK to start ypserv(1M)
#  (and yppasswdd(1M) for the master NIS server).  It returns its result
#  in the variable NISSERV_OK - if non-null, it is OK to start ypserv(1M);
#  if it is null, ypserv(1M) will not be started.
#
#  First, the filesystem containing /usr/etc/yp is examined to see if it
#  supports long or short filenames.  Once this is known, the proper list
#  of standard NIS map filenames is examined to verify that each map exists
#  in the NIS domain subdirectory.  If any map is missing, verify_ypserv
#  sets NISSERV_OK to null and returns.
##

verify_ypserv() {
   ##
   #  LONGNAMES are the names of the NIS maps on a filesystem that
   #  supports long filenames.
   ##

   LONGNAMES="group.bygid.dir group.bygid.pag group.byname.dir \
         group.byname.pag hosts.byaddr.dir hosts.byaddr.pag \
         hosts.byname.dir hosts.byname.pag networks.byaddr.dir \
         networks.byaddr.pag networks.byname.dir networks.byname.pag \
         passwd.byname.dir passwd.byname.pag passwd.byuid.dir \
         passwd.byuid.pag protocols.byname.dir protocols.byname.pag \
         protocols.bynumber.dir protocols.bynumber.pag \
         rpc.bynumber.dir rpc.bynumber.pag services.byname.dir \
         services.byname.pag ypservers.dir ypservers.pag"

   ##
   #  SHORTNAMES are the names of the NIS maps on a filesystem that
   #  supports only short filenames (14 characters or less).
   ##

   SHORTNAMES="group.bygi.dir group.bygi.pag group.byna.dir \
          group.byna.pag hosts.byad.dir hosts.byad.pag \
          hosts.byna.dir hosts.byna.pag netwk.byad.dir \
          netwk.byad.pag netwk.byna.dir netwk.byna.pag \
          passw.byna.dir passw.byna.pag passw.byui.dir \
          passw.byui.pag proto.byna.dir proto.byna.pag \
          proto.bynu.dir proto.bynu.pag rpc.bynu.dir \
          rpc.bynu.pag servi.byna.dir servi.byna.pag \
          ypservers.dir ypservers.pag"

   NISSERV_OK=TRUE

   if `/usr/etc/yp/longfiles`; then
      NAMES=$LONGNAMES
   else
      NAMES=$SHORTNAMES
```

```
   fi

   for NAME in $NAMES ; do
      if [ ! -f /usr/etc/yp/$NISDOMAIN/$NAME ] ; then
         NISSERV_OK=
         return
      fi
   done
}
```

Next ypserv and ypbind are started.

```
    if [ "$NIS_SERVER" -a -f /usr/etc/ypserv ] ; then
   verify_ypserv
   if [ "$NISSERV_OK" ] ; then
      /usr/etc/ypserv && echo "\t/usr/etc/ypserv"
                set_return
   else
      echo "\tWARNING:  /usr/etc/ypserv not started:  either"
      echo "\t          -  the directory /usr/etc/yp/$NISDOMAIN does not exis
or"
      echo "\t          -  some or all of the $NISDOMAIN NIS domain's"
      echo "\t             maps are missing."
      echo "\tTo initialize $HOSTNAME as a NIS server, see ypinit(1M)."
                returnstatus=1
   fi
    fi
    if [ $NIS_CLIENT -ne 0 -a -f /etc/ypbind ] ; then
   /etc/ypbind  && echo "\t/etc/ypbind "
        set_return

        ##
        #   check if the NIS domain is bound. If not disable NIS
        ##
        CNT=0;
        MAX_NISCHECKS=2
        NIS_CHECK=YES
   echo "   Checking NIS binding."
        while [ ${CNT} -le ${MAX_NISCHECKS} -a "${NIS_CHECK}" = "YES" ]; do
       /usr/bin/ypwhich 2>&1 | /bin/fgrep 'not bound ypwhich' > /dev/null

       if [ $? -eq 0 ]; then
           CNT=`expr $CNT + 1`
           if [ ${CNT} -le 2 ]; then
          sleep 5
           else
          echo "  Unable to bind to NIS server using domain ${NISDOMAIN}."
          echo "  Disabling NIS"
          /bin/domainname ""
          /bin/ps -e | /bin/grep ypbind | \
            kill -15 `/usr/bin/awk '{ print $1 }'`
          NIS_CHECK=NO
          returnstatus=1
          break;
           fi
       else
           echo " Bound to NIS server using domain ${NISDOMAIN}."
           NIS_CHECK=NO
       fi
        done
    fi
```

```
    ##
    if [ $NIS_MASTER_SERVER -ne 0 -a -f /usr/etc/rpc.yppasswdd ] ; then
   if [ "$NISSERV_OK" ] ; then
         echo "\t/usr/etc/rpc.yppasswdd"
         /usr/etc/rpc.yppasswdd /etc/passwd -m passwd PWFILE=/etc/passwd
      set_return
   else
      echo "\tWARNING:  /usr/etc/rpc.yppasswdd not started:  refer to the"
      echo "\t          reasons listed in the WARNING above."
           returnstatus=1
   fi
    fi
fi
```

Finally the PC-NFS daemons (`pcnfsd`) and the lock manager daemon (`rpc.lockd`) status monitor daemon (`rpc.statd`) are started.

```
PCNFS_SERVER=1
if [ $LFS -eq 0 -a $PCNFS_SERVER -ne 0 -a -f /etc/pcnfsd ] ; then
    /etc/pcnfsd && echo "starting up the PC-NFS daemon" && echo "\t/etc/pcnfsd"
    set_return
fi

if [ $NFS_CLIENT -ne 0 -o $NFS_SERVER -ne 0 ] ; then
    if [ -f /usr/etc/rpc.statd ] ; then
   /usr/etc/rpc.statd && echo "starting up the Status Monitor daemon" && echo
"\t/usr/etc/rpc.statd"
        set_return
    fi
    if [ -f /usr/etc/rpc.lockd ] ; then
   /usr/etc/rpc.lockd && echo "starting up the Lock Manager daemon" && echo
"\t/usr/etc/rpc.lockd"
        set_return
    fi
fi
exit $returnstatus
```

F.2 rc file

The `rc` file is executed when the UNIX node starts. It contains a number of functions (such as `localrc()`, `hfsmount()`, and so on) which are called from a main section. The example script given next contains some of the functions defined in Table F.1.

Table F.1 Sample rc functions

Function	*Description*
`localrc()`	Add local configuration to the node. In the example script the Bones-Licensing 2.4 is started locally on the node. This part of the script will probably be the only function which is different on different nodes.

`hfsmount()`	Mounts local disk drives
`map_keyboard()`	Loads appropriate keymap
`syncer_start()`	The syncer helps to minimize file damage when this is a power failure or a system crash
`lp_start()`	Starts the lp (line printer) scheduler
`net_start()`	Starts networking through `netlinkrc`
`swap_start()`	Starts swapping on alternate swap devices

```
initialize()
{
   if [ "$SYSTEM_NAME" = "" ]
   then
   SYSTEM_NAME=pollux
      export SYSTEM_NAME
   fi
}

localrc()
{

#%%CSIBeginFeature: Bones-Licensing 2.4
   DESIGNERHOME=/win/designer-2.0
   export DESIGNERHOME

   echo -n "Starting Bones-Licensing 2.4 ..."
   if [ -f ${DESIGNERHOME}/bin/start-lmgrd ]; then
      ${DESIGNERHOME}/bin/start-lmgrd
      echo " lmgrd."
   else
      echo " failed."
   fi
}

set_date()
{
  if [ $SET_PARMS_RUN -eq 0 ] ; then
   if [ $TIMEOUT -ne 0 ] ; then
      echo "\007Is the date `date` correct? (y or n, default: y) \c"
      reply=`line -t $TIMEOUT`
      echo ""

      if [ "$reply" = y -o "$reply" = "" -o "$reply" = Y ]
      then
         return
      else
         if [ -x /etc/set_parms ]; then
            /etc/set_parms time_only
         fi
      fi
   fi


   fi # if SET_PARMS_RUN
}
```

```
hfsmount()
{
   # create /etc/mnttab with valid root entry
   /etc/mount -u >/dev/null

   # enable quotas on the root file system
   # (others are enabled by mount)
   [ -f /quotas -a -x /etc/quotaon ] && /etc/quotaon -v /

   # Mount the HFS volumes listed in /etc/checklist:
   /etc/mount -a -t hfs -v
   # (NFS volumes are mounted via net_start() function)

   # Uncomment the following mount command to mount CDFS's
   /etc/mount -a -t cdfs -v

   # Preen quota statistics
   [ -x /etc/quotacheck ] && echo checking quotas && /etc/quotacheck -aP
}

map_keyboard()
{
#
itemap_option=""
if [ -f /etc/kbdlang ]
then
   read MAP_NAME filler < /etc/kbdlang
   if [ $MAP_NAME ]
   then
      itemap_option="-l $MAP_NAME"
   fi
fi

if [ -x /etc/itemap ]
then
   itemap -i -L $itemap_option -w /etc/kbdlang
fi
}

syncer_start()
{
   if /usr/bin/rtprio 127 /etc/syncer
   then
      echo syncer started
   fi
}

lp_start()
{
   if [ -s /usr/spool/lp/pstatus ]
   then
      lpshut > /dev/null 2>&1
      rm -f /usr/spool/lp/SCHEDLOCK
      lpsched
      echo line printer scheduler started
   fi
}

clean_ex()
{
   if [ -x /usr/bin/ex ]
   then
      echo "preserving editor files (if any)"
      ( cd /tmp; expreserve -a )
   fi
```

```
}

clean_uucp()
{
   if [ -x /usr/lib/uucp/uuclean ]
   then
      echo "cleaning up uucp"
      /usr/lib/uucp/uuclean -pSTST -pLCK -n0
   fi
}

net_start()
{
   if [ -x /etc/netlinkrc ] && /etc/netlinkrc
   then
      echo NETWORKING started.
   fi
}

swap_start()
{
   if /etc/swapon -a
   then
      echo 'swap device(s) active'
   fi
}

cron_start()
{
   if [ -x /etc/cron ]
   then
      if [ -f /usr/lib/cron/log ]
      then
         mv /usr/lib/cron/log /usr/lib/cron/OLDlog
      fi
      /etc/cron && echo cron started
   fi
}

audio_start ()
{
   # Start up the audio server
   if [ -x /etc/audiorc ] && /etc/audiorc
   then
      echo "Audio server started"
   fi
}

#
# The main section of the rc script
#

# Where to find commands:
PATH=/bin:/usr/bin:/usr/lib:/etc

# Set termio configuration for output device.
stty clocal icanon echo opost onlcr ixon icrnl ignpar

if [ ! -f /etc/rcflag ]    # Boot time invocation only
then
   # /etc/rcflag is removed by /etc/brc at boot and by shutdown
   touch /etc/rcflag

   hfsmount
   map_keyboard
```

```
    setparms
    initialize
    switch_over
    uname -S $SYSTEM_NAME
    hostname $SYSTEM_NAME

    swap_start
    syncer_start
    lp_start
    clean_ex
    clean_uucp
    net_start
    audio_start
    localrc
fi
```

Ethernet Monitoring System

G.1 Ethernet receiver

The next page gives a schematic for an Ethernet monitoring system. Its component values are:

R1 = 1 kΩ
R2 = 500 Ω
R3 = 10 MΩ
R4 = 39 Ω
R5 = 1.5 kΩ
R6 = 10 Ω

L1 = 1:1, 200 μH

C1 = 100 mF
C2 = 0.1 μF
C3 = 1.5 pF

XTAL = 20 MHz

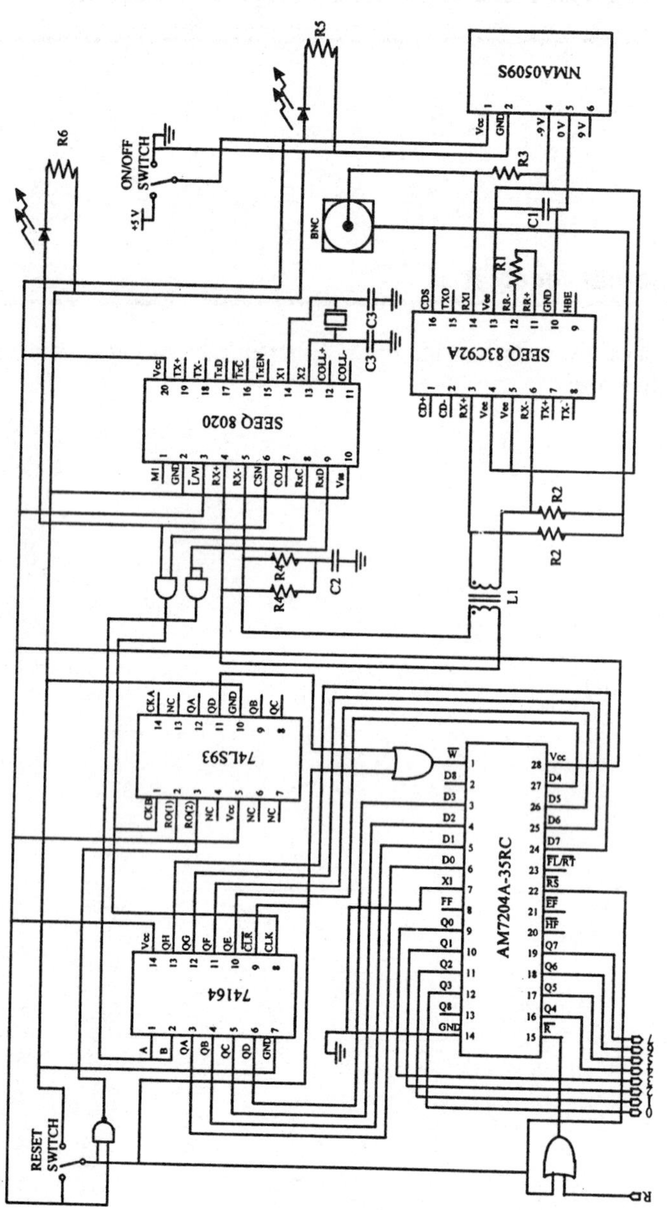

Ethernet Monitoring System

Encryption

H.1 Cracking the code

A cryptosystem converts plaintext into cipertext using a key. There are several methods that a hacker can use to crack a code, including:

- Known plaintext attack. Where the hacker knows part of the ciphertext and the corresponding plaintext. The known cipertext and plaintext can then be used to decrypt the rest of the cipertext.
- Chosen-cipertext. Where the hacker sends a message to the target, this is then encrypted by the target's private-key and the hacker then analyses the encrypted message. For example, a hacker may send an email to the encryption file server and the hacker spies on the delivered message.
- Exhaustive search. Where the hacker uses brute force to decrypt the cipertext and tries every possible key.
- Active attack. Where the hacker inserts or modifies messages.
- Man in the middle. Where the hacker is hidden between two parties and impersonates each of them to the other.
- The replay system. Where the hacker takes a legitimate message and sends it into the network at some future time.
- Cut and paste. Where the hacker mixes parts of two different encrypted messages and, sometimes, is able to create a new message. This message is likely to make no sense, but may trick the receiver into doing something that helps the hacker.
- Time resetting. Some encryption schemes use the time of the computer to create the key. Resetting this time or determining the time that the message was created can give some useful information to the hacker.

H.2 Random number generators

One way to crack a code is to exploit a weakness in the generation of the encryption key. The hacker can then guess which keys are more likely to occur.

This is known as a statistical attack.

Many programming languages use a random number generator which is based on the current system time (such as `rand()`). This method is no good in data encryption as the hacker can simply determine the time that the message was encrypted and the algorithm used.

An improved source of randomness is the time between two keystrokes (as used in PGP – pretty good privacy). However this system has been criticized as a hacker can spy on a user over a network and determine the time between keystrokes. Other sources of true randomness have also been investigated, including noise from an electronic device and noise from an audio source.

H.3 Survey of private-key cryptosystems

The most commonly used private-key cryptosystems are covered in Chapter 15.

H.3.1 DES

DES is a block cipher scheme which operates on 64-bit block sizes. The private key has only 56 useful bit as eight of its bits are used for parity. This gives 2^{56} or 10^{17} possible keys. DES uses a complex series of permutations and substitutions, the result of these operations is XOR'ed with the input. This is then repeated 16 times using a different order of the key bits each time. DES is a very strong code and has never been broken, although several high-powered computers are now available which, using brute force, can crack the code. A possible solution is 3DES (or triple DES) which uses DES three times in a row. First to encrypt, next to decrypt and finally to encrypt. This system allows a key-length of more than 128 bits.

H.3.2 MOSS

MOSS (MIME object security service) is an Internet RFC and is typically used for sound encryption. It uses symmetric encryption and the size of the key is not specified. The only public implementation is TIS/MOSS 7.1 which is basically an implementation of 56-bit DES code with a violation.

H.3.3 IDEA

IDEA (International Data Encryption Algorithm) is discussed in Section 15.2.2 and is similar to DES. It operates on 64-bit blocks of plaintext and uses a 128-bit key.

IDEA operates over 17 rounds with a complicated mangler function. During decryption this function does not have to be reversed and can simply be applied in the same way as during encryption (this also occurs with DES).

IDEA uses a different key expansion for encryption and decryption, but every other part of the process is identical. The same keys are used in DES decryption but in the reverse order.

The key is devised in eight 16-bit blocks; the first six are used in the first round of encryption the last two are used in the second run. It is free for use in non-commercial version and appears to be a strong cipher.

H.3.4 RC4/RC5

RC4 is a cipher designed by RSA Data Security, Inc and was a secret until information on it appeared on the Internet. The Netscape secure socket layer (SSL) uses RC4. RC4 uses a pseudo random number generator where the output of the generator is XOR'ed with the plaintext. It is has a fast algorithm and can use any key-length. Unfortunately the same key cannot be used twice. Recently a 40-bit key version was broken in 8 days without special computer power.

RC5 is a fast block cipher designed by Rivest for RSA Data Security. It has a parameterized algorithm with a variable block size (either 32, 64 or 128 bits), a variable key size (0 to 2048 bits) and a variable number of rounds (0 to 255).

It has a heavy use of data dependent rotations and the mixture of different operations. This assures that RC5 is secure. Kaliski and Yin found that RC5 with a 64-bit block size and 12 or more rounds gives good security.

H.3.5 SAFER

SAFER (Secure and Fast Encryption Routine) is a non-proprietary block-cipher developed by Massey in 1993. It operates on a 64-bit block size and has a 64-bit or 128-bit key size. SAFER has up to 10 rounds (although a minimum of 6 is recommended). Unlike most recent block ciphers, SAFER has a slightly different encryption and decryption procedure. The algorithm operates on single bytes at a time and it thus can be implemented on systems with limited processing power, such as on smart-cards applications.

A typical implementation is SAFER K-64 which uses a 40-bit key and has been shown that it is immune from most attacks when the the number of rounds is greater than 6.

H.3.6 SKIPJACK

Skipjack is new block cipher which operates on 64-bit blocks. It uses an 80-bit key and has 32 rounds. The NSA have classified details of Skipjack and its algorithm is only available in hardware implementation called Clipper Chips. The name Clipper derives from an earlier implementation of the algorithm. Each transmission contains the session key encrypted in the header. The licensing of clipper chips allows US government to decrypt all SKIPJACK messages.

H.4 Public-key cryptosystems

The most commonly used public-key cryptosystems are covered in Chapter 15.

H.4.1 RSA

RSA stands for Rivest, Shamir and Adelman, and is the most commonly used public-key cryptosystem. It is patented only in the USA and is secure for key-length of over 728 bits. The algorithm relies of the fact that it is difficult to factorize large numbers. Unfortunately, it is particularly vulnerable to chosen plaintext attacks and a new timing attack (spying on keystroke time) was announced on the 7 December 1995. This attack would be able to break many existing implementations of RSA.

H.4.2 Elliptic curve

Elliptic curve is a new kind of public-key cryptosystem. It suffers from speed problems, but this has been overcome with modern high-speed computers.

H.4.3 DSS

DSS (digital signature standard) is related to the DSA (digital signature algorithm). This standard has been selected by the NIST and the NSA, and is part of the Capstone project. It uses 512-bit or 1024-bit key size. The design presents some lack in key-exchange capability and is slow for signature-verification.

H.4.4 Diffie-Hellman

Diffie-Hellman is commonly used for key-exchange. The security of this cipher relies on both the key-length and the discrete algorithm problem. This problem is similar to the factorizing of large numbers. Unfortunately, the code can be cracked and the prime number generator must be carefully chosen.

H.4.5 LUC

LUC was developed by Peter Smith and is a public-key cipher that uses Lucas functions instead of exponentiation. Four other algorithms have also been developed:

- LUCDIF (a key-negotiation method).
- LUCELG PK (equivalent to EL Gamel encryption).
- LUCELG DS (equivalent to EL GAMEL data signature system).
- LUCDSA (equivalent to the DSS).

Digital Line Codes

I.1 Line codes

Data bits often have to be encoded before they can be stored or transmitted. This is typically done when:

- The bitstream needs to contain timing information, as a long run of the same digital value does not give much timing information.
- The Hamming distance between encoded values needs to be increased. For example, using a code which uses all the encoded bit values will mean that a single bit in error will lead to another valid code.
- To reduce DC wander, where the average value of the encoded bitstream is almost zero.
- To reduce power dissipation. Most of the information in a digital waveform is contained in the transitions, thus the DC component of the waveform does not carry much information. So the DC component of the signal leads to a waste of power. Typically, also, the power to remote equipment can be sent along the transmission line as the DC component. The signal line is then coupled to the transmission line through an isolated transformer.

I.2 NRZI

Non-return to zero (NRZI) is typically used in the transmission and storage of bits as it helps to reduces the DC content of the bitstream and also aids clock recovery. A 1 is coded with an alternative level, i.e. either a high (H) or a low (L). For example, the bitstream `110100` would be coded as:

Bitstream:	1	1	0	1	0	0
Waveform:	H	L	L	H	H	H

which has 2 lows and 4 highs, and the bitstream `0100010010` would be encoded as:

Bitstream:	0	1	0	0	0	1	0	0	1	0
Waveform:	L	H	H	H	H	L	L	L	H	H

which has 4 lows and 6 highs. This is illustrated in Figure I.1.

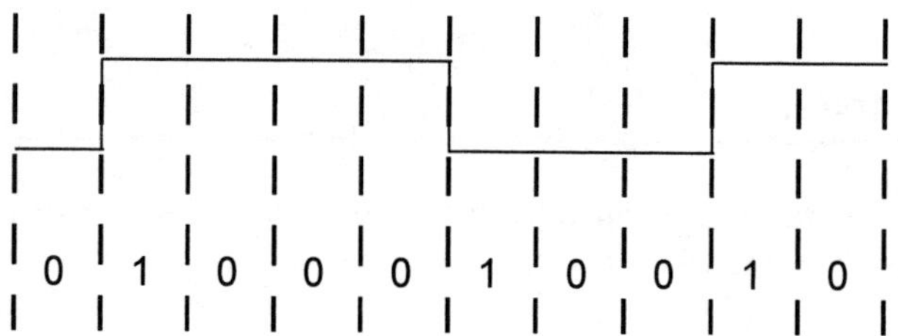

Figure I.1 NRZI waveform for 0100010010

Another way of viewing the NRZI encoding is that 1's are coded with alternative transitions, from low-to-high or a high-to-low. In this case the viewpoint of the waveform is at transitional points. For example, the bitstream 110100 would be viewed as:

Data:	1	1	0	1	0	0
Waveform:	LH	HL	LL	LH	HH	HH

which has 5 lows and 7 highs, and the bitstream `0100010010` would be encoded as:

Data:	0	1	0	0	0	1	0	0	1	0
Waveform:	LL	LH	HH	HH	HH	HL	LL	LL	LH	HH

9 lows and 11 highs. This is illustrated in Figure I.2.

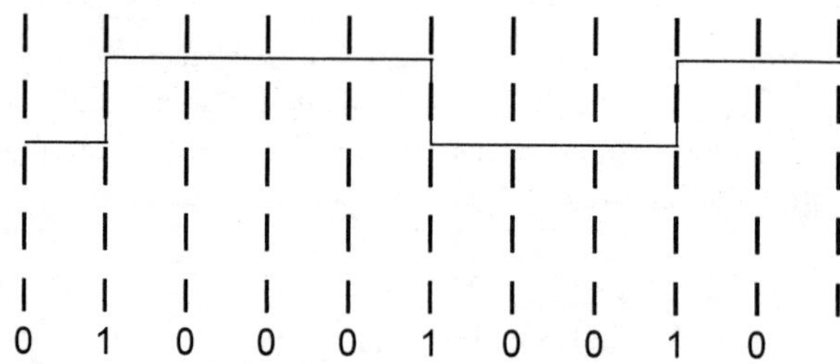

Figure I.2 Viewpoint of NRZI waveform for 0100010010 around transitions

I.3 DSV

An important measure of the DC content of a digital waveform is the DSV (digital sum value). This parameter keeps a running total of the deviation of the encoded waveform away from the zero level. Every high level adds one onto the value and a low takes one away from it. Thus taking the example of:

Bitstream:	1	1	0	1	0	0
Waveform:	L	H	L	H	H	H
	–	+	–	+	+	+

gives a DSV of +2. The bitstream `0100010010` would be encoded as follows:

Bitstream:	0	1	0	0	0	1	0	0	1	0
Waveform:	L	H	H	H	H	L	L	L	H	H
	–	+	+	+	+	–	–	–	+	+

which gives a DSV of +2. Note that if the first bit had a low level then the DSV would be −2, as shown next:

Bitstream:	0	1	0	0	0	1	0	0	1	0
Waveform:	H	L	L	L	L	H	H	H	L	L
	+	–	–	–	–	+	+	+	–	–

Thus it can be seen that this code will always have a DSV magnitude of 2. The technique of alternating inverting the levels can be used to either produce a positive or a negative DSV, which can be used to help the current DSV value.

I.4 4B/5B code

The 4B/5B is used in many applications, including FDDI and MADI (multichannel audio digital interface). It encodes 4 bits into 5 bit and the encoding table has been chosen so that they have the least DC component combined with a high clock content. The maximum run of zero's is limited to 3.

It can be seen that the DSV for the encoded bitstream for `11110` is:

Bitstream:	1	1	1	1	0
Waveform:	L	H	L	H	H
	–	+	–	+	+

is +2 (or −2 if the previous 1 bit is a Low).

Table I.1 4B/5B encoding

4-bit data	5-bit encoded data	4-bit data	5-bit encoded data
0000	11110	0001	01001
0001	10100	0011	10101
0100	01010	0101	01011
0110	01110	0111	01111
1000	10010	1001	10011
1010	10110	1011	10111
1100	11010	1101	11011
1110	11100	1111	11101

I.5 EFM (eight-to-fourteen modulation)

The EFM code is used in Compact Discs where 8-bit symbols are represented by a 14-bit code. There are 256 combinations of 8-bit symbols out of a possible 16 384 (2^{14}) different encoded values. Table I.2 shows a portion of the codetable. The code has been chosen so that the maximum run-length so that the The maximum run-length of 0's will never be greater that 11.

Table I.2 EFM encoding

Hex	*8-bit data*	*14-bit encoded data*	*Hex*	*8-bit data*	*14-bit encoded data*
64	01100100	01000100100010	72	01110010	11000100100010
65	01100101	00000000100010	73	01110011	00100001000010
66	01100110	01000000100100	74	01110100	00000010000010
67	01100111	00100100100010	75	01110101	00000010000010
68	01101000	01001001000010	76	01110110	00010010000010
69	01101001	10000001000010	77	01110111	00100010000010
6A	01101010	10010001000010	78	01111000	01001000000010
6B	01101011	10001001000010	79	01111001	00001001001000
6C	01101100	01000001000010	7A	01111010	10010000000010
6D	01101101	00000001000010	7B	01111011	10001000000010
6E	01101110	00010001000010	7C	01111100	01000000000010
6F	01101111	00100001000010	7F	01111101	00001000000010
70	01110000	10000000100010	7E	01111110	00010000000010
71	01110001	10000010000010	7F	01111111	00100000000010

For example, the DSV for the first three codes is:

14-bit encoded data	*Waveform*	*DSV*
01000100100010	LHHHHLLLHHHHLL	+2
00000000100010	LLLLLLLLLHHHHLL	−6
01000000100100	LHHHHHHHLLLHHH	+6

It can be seen that these codes give a DSV between +6 and −6. For this purpose CD discs have 3 extra packing bits added, these bits try to bring the DSV back to 0. It also inverts the code to give the inverse (for example the 01000100100010 can give either a DSV of +2 (LH HHHL LLHH HHLL) or −2 (HL LLLH HHLL LLHH).

I.6 5B5B

The 100VG-AnyLAN standard uses 5B6B to transmit an Ethernet frame between the hub and the node. This code is used so that there is an increase the number of transitions in the transmitted waveform.

In 100VG-AnyLAN, a 100 Mbps bitstream is multiplexed onto four 25 Mbps streams and transmitted over the 4 twisted-pair cables. The encoding process thus increases the bit rate on each twisted pair cable to 30Mbps (as 6 encoded bits are sent for every 5 bitstream bits). This is illustrated in Figure I.3 and Table I.3 gives the 5B/6B encoding.

Unfortunately, it is not possible to code each one of the 6-bit encoded values with an equal number of 0's and 1's as there are only 20 encoded values which have an equal number of 0's and 1's, these are highlighted in Table I.3. Thus two modes are used with the other 12 values having either two 0's and four 1's or four 0's and two 1's. The data is then transmitted in two modes:

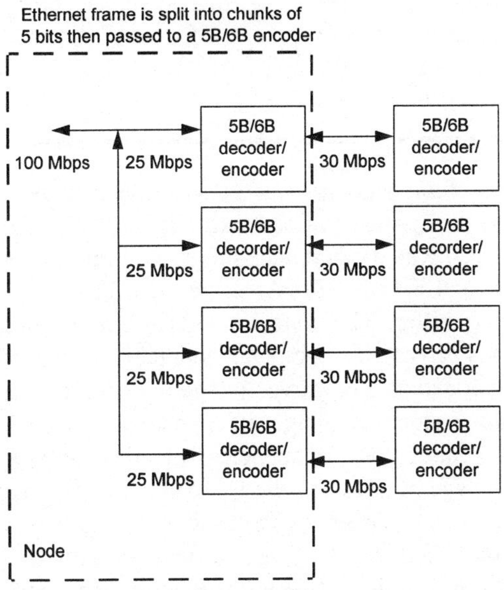

Figure I.3 Encoding of the bitstream in 100VG-AnyLAN

Table I.3 5B/6B encoding

4-bit data	*Mode 2 encoding*	*Mode 4 encoding*	*4-bit data*	*Mode 2 encoding*	*Mode 4 encoding*
00000	001100	110011	10000	000101	111010
00001	**101100**	101100	10001	**100101**	100101
00010	100010	101110	10010	001001	110110
00011	**001101**	001101	10011	**010110**	010110
00100	001010	110101	10100	**111000**	111000
00101	**010101**	010101	10101	011000	100111
00110	**001110**	001110	10110	**011001**	011001
00111	**001011**	001011	10111	100001	011110
01000	**000111**	000111	11000	**110001**	110001
01001	**100011**	100011	11001	**101010**	101010
01010	**100110**	100110	11010	010100	101011
01011	000110	111001	11011	**110100**	110100
01100	101000	010111	11100	**011100**	011100
01101	**011010**	011010	11101	**010011**	010011
01110	100100	100100	11110	010010	101101
01111	**101001**	101001	11111	**110010**	110010

- Mode 2. Where the encoded data has either an equal number of 0's and 1's or has four 1's and two 0's.
- Mode 4. Where the encoded data has either an equal number of 1's and 0's or has four 0's and two 1's.

These modes alternate and this gives, on average, DSV of zero.

I.7 8B6T

100BASE-4T uses four separate Cat-3 twisted-pair wires. The maximum clock rate that can be applied to Cat-3 cable is 30 Mbps. Thus some mechanism must be devised which can reduce the line bit rate to under 30 Mbps but give a symbol rate of 100 Mbps. This is achieved with a 3-level code (+, – and 0) and is known as 8B6T. The code converts 8 binary digits into 6 ternary symbols. Table I.4 gives the complete codetable. Thus the bit sequence 00000000 will be coded as a negative voltage, a positive voltage, a zero voltage, a zero voltage, a negative voltage and a positive voltage.

Apart from reducing the frequencies with the digital signal, the 8B6T code has the advantage of reducing the DC content of the signal. Most of the codes contain the same number of positive and negative voltages. This is because only 256 of the possible 729 (3^6) codes are actually used. The codes are also chosen to have at least two transitions in every code word, thus the clock information is embedded into signal.

Table I.4 8B6T code

8-bit data	*Encoded data*	*8-bit data*	*Encoded data*
00000000	+-00+-	00010000	+0+--0
00000001	0+-+-+	00010001	++0-0
00000010	+-0+-0	00010010	+0+-0-
00000011	-0++-0	00010011	0++-0-
00000100	-0+0+-	00010100	0++--0
00000101	0+--0+	00010101	++00--
00000110	+-0-0+	00010110	+0+0--
00000111	-0+-0+	00010111	0++0--
00001000	-+00+-	00011000	0+-0+-
00001001	0-++-0	00011001	0+-0-+
00001010	-+0+-0	00011010	0+-++-
00001011	+0-+-0	00011011	0+-00+
00001100	+0-0+-	00011100	0-+00+
00001101	0-+-0+	00011101	0-+++-
00001110	-+0-0+	00011110	0-+0-+
00001111	+0--0+	00011111	0-+0+-

Unfortunately it is not possible to have all codes with the same number of negative voltages as positive voltages. Thus there are some codes which have a different number of negatives and positives these include:

```
0100 0001   +0–00++
0111 1001   +++–0–
```

The technique used to overcome this is to invert consecutive codes which have a weighing of +1. For example suppose the line code were:

```
+0++––     ++0+––     +++––0   +++––0
```

it would actually be coded as:

```
+0++––     ––0–++     +++––0   –––++0
```

The receiver detects the –1 weighted codes as an inverted pattern.

I.8 8/10 code

The 8/10 code is used in DAT (digital audio tape) recordings where 8-bit symbols are represented by a 10-bit code. There are 256 combinations of 8-bit symbols out of a possible 1024 (2^{10}) different encoded values. Table I.4 shows a portion of the codetable. It is used to zero DC and suppress low frequencies.

Table I.5 Portion of the 8/10 encoding table

8-bit data	*10-bit encoded*	*Alternative 10-bit encoded*
00010000	1101010010	
00010001	0100010010	01110011
00010010	0101010010	
00010011	0101110010	
00010100	1101110001	01110111
00010101	1101110011	01111000
00010110	1101110110	01111001

Some of the codes have a maximum DSV of +2. For example, `0100010010` and a previous 1 of a Low will give:

```
Bitstream: 1  1  0  1  0  1  0  0  1  0
Waveform:  H  L  L  H  H  L  L  L  H  H
           +  –  –  +  +  –  –  –  +  +
```

which gives a DSV of 0. If the previous 1 was a High then:

```
Bitstream: 1  1  0  1  0  1  0  0  1  0
Waveform:  L  H  H  L  L  H  H  H  L  L
           –  +  +  –  –  +  +  +  –  –
```

which also gives a DSV of 0, thus there is no DC offset. Taking the `0100010010` as an example and if the previous 1 was a Low then:

```
Bitstream: 0  1  0  0  0  1  0  0  1  0
Waveform:  L  H  H  H  H  L  L  L  H  H
           –  +  +  +  +  –  –  –  +  +
```

which gives a DSV is +2. It can be shown that if the previous 1 was a High then the DSV will be −2. The alternative code to this is `1100010010` which is encoded as:

```
Bitstream: 1  1  0  0  0  1  0  0  1  0
Waveform:  H  L  L  L  L  H  H  H  L  L
           +  –  –  –  –  +  +  +  –  –
```

which has a DSV of −2 and can be used to offset a +2 DSV. It can be seen that the only difference between the two codes is the first bit which is either a 0 or a 1. Thus:

- If the current DSV is +2 and the last 1 bit was a Low then the

`1100010010` can be used to bring the DSV back to 0.

- If the current DSV is −2 and the last 1 bit was a Low then the `0100010010` can be used to bring the DSV back to 0.
- If the current DSV is +2 and the last 1 bit was a High then the `0100010010` can be used to bring the DSV back to 0.
- If the current DSV is −2 and the last 1 bit was a High then the `1100010010` can be used to bring the DSV back to 0.

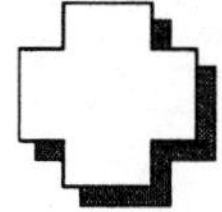

Common Abbreviations

AA	auto-answer
AAN	autonomously attached network
ABM	asynchronous balanced mode
AC	access control
ACK	acknowledge
ACL	access control list
ADC	analogue-to-digital converter
ADPCM	adaptive delta pulse code modulation
AES	audio engineering society
AFI	authority and format identifier
AM	amplitude modulation
AMI	alternative mark inversion
ANSI	American National Standard Institute
APCM	adaptive pulse code modulation
API	application program interface
ARM	asynchronous response mode
ARP	address resolution protocol
ASCII	American standard code for information exchange
ASK	amplitude-shift keying
AT	attention
ATM	asynchronous transfer mode
AUI	attachment unit interface
BCC	blind carbon copy
BCD	binary coded decimal
BIOS	basic input/output system
B-ISDN	broadband ISDN
BMP	bitmapped
BNC	British Naval Connector
BOOTP	boot protocol
BPDU	bridge protocol data units
bps	bits per second
CAD	computer-aided design
CAN	concentrated area network
CASE	common applications service elements
CATNIP	common architecture for the Internet
CCITT	International Telegraph and Telephone Consultative
CD	carrier detect
CDE	common desktop environment
CD-R	CD-recordable

CD-ROM	compact disk - read-only memory
CGI	common gateway interface
CGM	computer graphics metafile
CIF	common interface format
CMC	common mail call
CMOS	complementary MOS
CPU	central processing unit
CRC	cyclic redundancy
CRLF	carriage return, line feed
CRT	cathode ray tube
CSDN	circuit-switched data network
CSMA	carrier sense multiple access
CSMA/CA	CSMA with collision avoidance
CSMA/CD	CSMA with collision detection
CS-MUX	circuit-switched multiplexer
CSPDN	circuit-switched public data network
CTS	clear to send
DA	destination address
DAC	digital-to-analogue converter
DAC	dual attachment concentrator
DARPA	Defense Advanced Research Projects Agency
DAS	dual attachment station
DAT	digital audio tape
dB	decibel
DBF	NetBEUI frame
DC	direct current
DCC	digital compact cassette
DCD	data carrier detect
DCE	data circuit-terminating equipment
DCT	discrete cosine transform
DD	double density
DDE	dynamic data exchange
DES	data encryption standard
DHCP	dynamic host configuration program
DIB	directory information base
DISC	disconnect
DLC	data link control
DLL	dynamic link library
DM	disconnect mode
DNS	domain name server
DOS	disk operating system
DPCM	differential PCM
DPSK	differential phase-shift keying
DQDB	distributed queue dual bus
DR	dynamic range
DRAM	dynamic RAM
DSP	domain specific part

DSS	digital signature standard
DTE	data terminal equipment
DTR	data terminal ready
EaStMAN	Edinburgh/Stirling MAN
EBCDIC	extended binary coded decimal interchange code
EBU	European broadcast union
EEPROM	electrically erasable PROM
EF	empty flag
EFM	eight-to-fourteen modulation
EGP	exterior gateway protocol
EIA	Electrical Industries Association
EISA	extended international standard interface
ENQ	enquiry
EOT	end of transmission
EPROM	erasable PROM
EPS	encapsulated postscript
ETB	end of transmitted block
ETX	end of text
FAT	file allocation table
FAX	facsimile
FC	frame control
FCS	frame check sequence
FDDI	fiber distributed data interface
FDM	frequency division multiplexing
FDX	full duplex
FEC	forward error correction
FM	frequency modulation
FRMR	frame reject
FSK	frequency-shift keying
FTP	file transfer protocol
GFI	group format identifier
GGP	gateway-gateway protocol
GIF	graphics interface format
GUI	graphical user interface
HD	high density
HDB3	high-density bipolar code no. 3
HDLC	high-level data link control
HDTV	high-definition television
HDX	half duplex
HF	high frequency
HMUX	hybrid multiplexer
HPFS	high performance file system
HTML	hypertext mark-up language
HTTP	hypertext transfer protocol
Hz	Hertz
I/O	input/output
IA5	international alphabet no. 5

IAB	internet advisory board
IAP	internet access provider
ICMP	internet control message protocol
ICP	internet connectivity provider
IDEA	international data encryption algorithm
IDI	initial domain identifier
IDP	initial domain part
IEEE	Institute of Electrical and Electronic Engineers
IEFF	internet engineering task force
IGP	interior gateway protocol
ILD	injector laser diode
IMAC	isochronous MAC
IP	internet protocol
IPP	internet presence provider
IPX	internet packet exchange
ISA	international standard interface
ISDN	integrated services digital network
IS-IS	immediate system to intermediate system
ISO	International Standards Organisation
ISP	internet service provider
ITU	International Telecommunications Union
JANET	joint academic network
JFIF	jpeg file interchange format
JISC	Joint Information Systems Committee
JPEG	Joint Photographic Expert Group
LAN	local area network
LAPB	link access procedure balanced
LAPD	link access procedure
LCN	logical channel number
LD-CELP	low-delay code excited linear prediction
LED	light emitting diode
LGN	logical group number
LIP	large IPX packets
LLC	logical link control
LRC	longitudinal redundancy check
LSL	link support level
LSP	link state protocol
LZ	Lempel-Ziv
LZW	LZ-Welsh
MAC	media access control
MAN	metropolitan area network
MAU	multi-station access unit
MDCT	modified discrete cosine transform
MDI	media dependent interface
MHS	message handling service
MIC	media interface connector
MIME	multi-purpose internet mail extension

MODEM	modulation/demodulator
MOS	metal oxide semiconductor
MPEG	motion picture experts group
NAK	negative acknowledge
NCP	netware control protocols
NCSA	National Center for Supercomputer Applications
NDIS	network device interface standard
NETBEUI	NetBIOS extended user interface
NIC	network interface card
NIS	network information system
NLSP	netware link-state routing protocol
NRZI	non-return to zero with inversion
NSAP	network service access point
NSCA	National Center for Supercomputer Applications
NTE	network terminal equipment
NTFS	NT file system
NTP	network time protocol
NTSC	National Television Standards Committee
ODI	open data-link interface
OH	off-hook
OSI	open systems interconnection
OSPF	open shortest path first
OUI	originators unique identifier
PA	point of attachment
PAL	phase alternation line
PC	personal computer
PCM	pulse code modulation
PDN	public data network
PHY	physical layer protocol
PING	packet internet gopher
PISO	parallel-in-serial-out
PKP	public key partners
PLL	phase-locked loop
PLS	physical signaling
PMA	physical medium attachment
PMD	physical medium dependent
PPP	point-to-point protocol
PPSDN	public packet-switched data network
PS	postscript
PSDN	packet-switched data network
PSE	packet switched exchange
PSK	phase-shift keying
PSTN	public-switched telephone network
QAM	quadrature amplitude modulation
QCIF	quarter common interface format
QIC	quarter inch cartridge
QT	quicktime

RAID	redundant array of inexpensive disks
RAM	random-access memory
RD	receive data
REJ	reject
RFC	request for comment
RGB	red, green and blue
RI	ring in
RIP	routing information protocol
RLE	run-length encoding
RNR	receiver not ready
RO	ring out
ROM	read-only memory
RPC	remote procedure call
RR	receiver ready
RSA	Rivest, Shamir and Adleman
RTF	rich text format
RTMP	routing table maintenance protocol
S/PDIF	Sony/Philips digital interface format
SABME	set asynchronous balanced mode extended
SAC	single attachment concentrator
SAP	service advertising protocol
SAPI	service access point identifier
SAS	single attachment station
SB-ADCMP	sub-band ADPCM
SCMS	serial copy management system
SCSI	small computer systems interface
SD	sending data
SDH	synchronous digital hierarchy
SDIF	Sony digital interface
SDLC	synchronous data link control
SECAM	séquential couleur à mémoire
SEL	selector/extension local address
SHEFC	Scottish Higher Education Funding Council
SIPO	serial-in parallel-out
SIPP	simple internet protocol plus
SMDS	switched multi-bit data stream
SMP	symmetrical multiprocessing
SMT	station management
SMTP	simple message transport protocol
SNA	systems network architecture (IBM)
SNMP	simple network management protocol
SNR	signal-to-noise ratio
SONET	synchronous optical network
SPX	sequenced packet exchange
QTV	studio-quality television
SRAM	static RAM
STA	spanning-tree architecture

STM	synchronous transfer mode
STP	shielded twisted-pair
SVGA	super VGA
TCP	transmission control protocol
TDAC	time-division aliasing cancellation
TDM	time-division multiplexing
TEI	terminal equipment identifier
TIFF	tagged input file format
TR	transmit data
TUBA	TCP and UDP with bigger addresses
UDP	user datagram protocol
UI	unnumbered information
UNI	universal network interface
UPS	uninterruptable power supplies
URI	universal resource identifier
URL	uniform resource locator
UTP	unshielded twisted pair
UV	ultra violet
VCI	virtual circuit identifier
VCR	video cassette recorder
VGA	variable graphics adapter
VIM	vendor-independent messaging
VLC-LZW	variable-length-code LZW
VLM	virtual loadable modules
VRC	vertical redundancy check
WAIS	wide area information servers
WAN	wide area network
WIMPs	windows, icons, menus and pointers
WINS	windows internet name service
WINSOCK	windows sockets
WORM	write-once read many
WWW	World Wide Web
XDR	external data representation
XOR	exclusive-OR

Index

—0—

—1—

—2—

—3—

—4—

—5—

—6—

—7—

—8—

—9—

—A—

—D—

—I—

—J—

—K—

—L—

—N—

—T—

—U—

—Y—

—Z—